Advanced Research in Plant Genomics

Advanced Research in Plant Genomics

Editor: Isabelle Nickel

Callisto Reference
www.callistoreference.com

Callisto Reference,
118-35 Queens Blvd., Suite 400,
Forest Hills, NY 11375, USA

Visit us on the World Wide Web at:
www.callistoreference.com

ISBN: 978-1-63239-800-0 (Hardback)

Printed in China.

Cataloging-in-publication Data

Advanced research in plant genomics / edited by Isabelle Nickel.
p. cm.
Includes bibliographical references and index.
ISBN 978-1-63239-800-0
1. Plant genomes. 2. Genomics. 3. Plant genetics. I. Nickel, Isabelle.
QK981 .A38 2017
581.35--dc23

Table of Contents

Preface

Plant genomics is the study of plant genes and their characteristics. It encompasses the study of plant genomes and examines the genes necessary for mutations. It also comprises the study of plant breeding, genetic modification, DNA sequencing, etc. This book discusses the fundamentals as well as modern approaches of plant genomics. It provides interesting topics for research which readers can take up. Different evaluations, methodologies and advanced studies on plant genomics have been included in the text. Students, researchers, experts and all associated with this area will benefit alike from this book. It will serve as a valuable resources guide for the readers.

This book is the end result of constructive efforts and intensive research done by experts in this field. The aim of this book is to enlighten the readers with recent information in this area of research. The information provided in this profound book would serve as a valuable reference to students and researchers in this field.

At the end, I would like to thank all the authors for devoting their precious time and providing their valuable contributions to this book. I would also like to express my gratitude to my fellow colleagues who encouraged me throughout the process.

Editor

1

Genome-Wide Analysis of the Expansin Gene Superfamily Reveals Grapevine-Specific Structural and Functional Characteristics

Silvia Dal Santo[1◑], Alessandro Vannozzi[2◑], Giovanni Battista Tornielli[1], Marianna Fasoli[1], Luca Venturini[1], Mario Pezzotti[1], Sara Zenoni[1*]

1 Department of Biotechnology, University of Verona, Verona, Italy, **2** Department of Agronomy, Food, Natural Resources, Animals and Environment, University of Padova, Legnaro, Padova, Italy

Abstract

Background: Expansins are proteins that loosen plant cell walls in a pH-dependent manner, probably by increasing the relative movement among polymers thus causing irreversible expansion. The expansin superfamily (EXP) comprises four distinct families: expansin A (EXPA), expansin B (EXPB), expansin-like A (EXLA) and expansin-like B (EXLB). There is experimental evidence that EXPA and EXPB proteins are required for cell expansion and developmental processes involving cell wall modification, whereas the exact functions of EXLA and EXLB remain unclear. The complete grapevine (*Vitis vinifera*) genome sequence has allowed the characterization of many gene families, but an exhaustive genome-wide analysis of expansin gene expression has not been attempted thus far.

Methodology/Principal Findings: We identified 29 EXP superfamily genes in the grapevine genome, representing all four EXP families. Members of the same EXP family shared the same exon–intron structure, and phylogenetic analysis confirmed a closer relationship between EXP genes from woody species, i.e. grapevine and poplar (*Populus trichocarpa*), compared to those from *Arabidopsis thaliana* and rice (*Oryza sativa*). We also identified grapevine-specific duplication events involving the EXLB family. Global gene expression analysis confirmed a strong correlation among EXP genes expressed in mature and green/vegetative samples, respectively, as reported for other gene families in the recently-published grapevine gene expression atlas. We also observed the specific co-expression of EXLB genes in woody organs, and the involvement of certain grapevine EXP genes in berry development and post-harvest withering.

Conclusion: Our comprehensive analysis of the grapevine EXP superfamily confirmed and extended current knowledge about the structural and functional characteristics of this gene family, and also identified properties that are currently unique to grapevine expansin genes. Our data provide a model for the functional characterization of grapevine gene families by combining phylogenetic analysis with global gene expression profiling.

Editor: Cynthia Gibas, University of North Carolina at Charlotte, United States of America

Funding: This work was supported by the "Completamento del Centro di Genomica Funzionale e Vegetale" project founded by the Cariverona bank foundation. The funders had no role in study design, data collection and analysis, decision to publish, or preparation of the manuscript.

Competing Interests: The authors have declared that no competing interests exist.

* E-mail: sara.zenoni@univr.it

◑ These authors contributed equally to this work.

Introduction

Expansins are cell wall-loosening proteins that regulate cell wall expansion and cell enlargement in a pH-dependent manner [1]. The first expansins were found in cucumber (*Cucumis sativus*) [2] but many others have subsequently been identified throughout the plant kingdom [3] and related proteins exist in bacteria, nematodes and mollusks, probably helping to digest plant biomass [4–7]. Genome-wide analysis has revealed a large expansin superfamily (EXP) in plants comprising four families [8] defined as expansin A (EXPA), expansin B (EXPB), expansin-like A (EXLA) and expansin-like B (EXLB) according to a standardized nomenclature [9]. Only the largest families (EXPA and EXPB, the latter originally identified as group I pollen allergens) have been shown empirically to loosen cell walls, whereas the expansin-like proteins (EXLA and EXLB) are recognized only by their conserved sequences and their precise functions remain unknown [10,11].

Expansins are typically 250–275 amino acids in length and comprise a 20-residue N-terminal signal peptide that targets the protein to the cell wall and two major domains, one homologous to glycoside hydrolase family 45 although its activity has yet to be confirmed (domain 1) and another thought to bind cell-wall polysaccharides (domain 2) [8]. The mechanism of action is still largely uncharacterized but may involve the acid-induced disruption of hydrogen bonds linking cellulose and hemicellulose fibers, promoting slippage between the polymers and allowing the cell to absorb water, ultimately leading to cell wall expansion [1,12]. The shortage of functional information reflects the scarcity of informative structural data, given that the only solved crystal structures represent one bacterial expansin [4] and the pollen-specific EXLB in maize [13]. Although no structural data exist for plant EXPA and EXPB proteins, the developmental and physiological functions of expansins have been studied in many plant species, organs and developmental stages, showing that these proteins are involved in processes such as root and shoot apical meristem development [14,15], the definition of plant architecture [16], xylem formation [17], abscission [18,19], seed germination [20], several types of plant–microbe interactions [21–23] and fruit softening [24–27].

Grapevine (*Vitis* spp., family Vitaceae) is an important perennial woody species because of its economically-valuable role in fruit and wine production, thus it was the first fruit crop selected for genome sequencing [28,29]. The relatively small grapevine genome (475 Mb) and the absence of recent whole genome duplication (WGD) events means it is also ideal for the genome-wide characterization of gene families. Recent studies have identified many gene families and superfamilies in grapevine, including stilbene synthases [30], dehydrins [31], pathogen-related proteins (PR-10) [32], aldehyde dehydrogenases [33] and transcription factors such as TIFY [34] and NAC [35]. There has been no systematic analysis of grapevine expansins thus far, although the expression profiles of several expansin genes have been studied during berry development [36,37] and one expansin was identified as minor grape allergen [38]. Recently, certain expansin transcripts have been identified by microarray analysis as markers of berry development [39–41].

We screened the V1 annotation of the 12X grapevine genome (PN40024 genotype; http://srs.ebi.ac.uk/) and identified 29 expansin genes representing all four families. This is the first genome-wide study of the grapevine expansin superfamily and we propose a systematic nomenclature based on chromosome position and existing guidelines [9]. We carried out a global analysis of differential expansin gene expression by mining publicly-available microarray data from the recently-published grapevine transcriptome atlas [42] and combining this with data from quantitative real-time RT-PCR experiments. Finally, we focused on the role of expansins in post-harvest withering, a process used in the manufacture of straw wines. Our data provide a basis for the further evolutionary and functional characterization of expansin genes and our exhaustive genome-wide expression analysis has helped to determine the precise role of expansins in grapevine, and thus their potential functions in other fruit crops.

Results and Discussion

Identification of expansin genes in grapevine by genome-wide analysis

The genome of the near-homozygous PN40024 genotype of *V. vinifera* cv Pinot Noir was screened for EXP gene sequences. Previously-identified expansin sequences from *Arabidopsis thaliana*, rice (*Oryza sativa*) and poplar (*Populus trichocarpa*) were used as BLASTP queries against the 12X V1 assembly using the JIGSAW software platform [43] at the CRIBI in Padova, Italy (http://genomes.cribi.unipd.it/grape/). This led to the identification of 29 expansin or expansin-like sequences, representing all four expansin families [8]. We identified 24 genes containing the HFD motif which is found only in the EXPA and EXPB families, including 20 in which domain 1 included the characteristic large insertion (α-insertion) and deletion (α-deletion) that is unique to the EXPA family, suggesting that the grapevine genome contains 20 EXPA genes and four EXPB genes (Figure S1). Among the remaining sequences, one was classified as an EXLA gene, based on the absence of the HDF domain and the presence of a characteristic EXLA extension at the C-terminus, and the other four were classified as EXLB genes (Figure S1). All 29 genes were named based on family identity and, within each subgroup, their chromosomal position. The proposed nomenclature for grapevine expansin genes, together with identifiers in the 12X V1 annotation and the length of each open reading frame (ORF), are listed in Table 1. We compared the number of grapevine genes in each expansin family to the other available angiosperm genomes, i.e. Arabidopsis, rice and poplar (Table 2). The EXPA family is the largest, with 20 members in grapevine (*VvEXPA1–VvEXPA20*), 26 in Arabidopsis [6], 34 in rice and 27 in poplar [44,45]. The EXPB family is similar in grapevine, Arabidopsis and poplar, but has undergone a grass-specific expansion in rice [8]. EXLA was the smallest family with only one member in each species, whereas the EXLB family appeared to have expanded specifically in woody species (grapevine and poplar) but not in Arabidopsis and rice.

Experimental evidence has shown that EXPA and EXPB proteins can loosen cell walls [11,46] whereas EXLA and EXLB have been identified by homology and their activity has not been demonstrated. The similar phylogenetic profile of EXLA and EXLB genes in grapevine and poplar may suggest they perform similar roles in these two perennial woody species.

The chromosomal distribution of grapevine expansin genes is presented in Figure 1, which shows that paralogs are present on all chromosomes except 5, 10, 11 and 16. The last released grapevine genome assembly did not allow us to assign the positions of three expansin gene models (*VvEXLB2–VvEXLB4*) so these could potentially be located on the chromosomes listed above. The distribution of expansin genes is similar in grapevine and poplar, with no more than three expansin genes present on a single chromosome (Figure S2). In contrast, expansin genes tend to be clustered in Arabidopsis and rice, including the presence of 18 expansin genes on rice chromosome 3, reflecting the large number of tandem duplications present in the rice genome (Figure S2).

The expansins do not appear to have undergone such extensive phylogenetic expansion and diversification as other superfamilies, e.g. stilbene synthases, terpene synthases and other genes related to wine characteristics [28]. As proposed for Arabidopsis, the grapevine expansin superfamily may result from massive gene loss following genome duplication.

Phylogenetic and structural analysis of the grapevine expansin genes

We constructed a phylogenetic tree including the 29 grapevine expansin proteins based on their deduced amino acid sequences, clustering those in the same families (Figure 2A). To gain more insight into the structural evolution of the expansin gene superfamily, the previously-defined exon–intron boundaries [8] were investigated, revealing that genes from the same family were generally characterized by similar exon–intron structures (Figure 2B). The grapevine EXPA genes generally contained two introns at positions A and B, except *VvEXPA9*, *VvEXPA15* and *VvEXPA10* (which lacked the B intron), *VvEXP4* (which featured an unusually long B intron), and *VvEXPA6*, which contained a third intron at position H, downstream from the BOX 2 site (Figure 2B). The grapevine EXPB genes generally contained three introns at positions A, C and B, except *VvEXPB2* which featured an intron at position F instead of C. Finally, it is noteworthy that the most phylogenetically-divergent EXLB gene (*VvEXLB1*) displayed an exon–intron structure dissimilar to the other *EXLB* genes, but identical to that of the only grapevine EXLA gene (*VvEXLA1*) with four introns at positions A, C, B and F. The highly-conserved

Table 1. Grapevine expansin genes identified in the PN40024 12X V1 prediction.

Proposed nomenclature	PN40024 12Xv1 ID	RefSeq	Coordinates	Strand	ORF (aa)	Gene (bp)
EXPA						
VvEXPA1	VIT_01s0026g02620	XM_002272843.2	chr1: 12260499..12261641	minus	246	1143
VvEXPA2	VIT_02s0025g04700	XM_002275577.1	chr2: 4249304...4250807	minus	264	1504
VvEXPA3	VIT_03s0038g04060	XM_002283109.1	chr3: 2934516..2935485	plus	250	970
VvEXPA4	VIT_04s0079g00420		chr4: 10905332..10909691	plus	271	4360
VvEXPA5	VIT_06s0004g00070		chr6: 173176..174251	plus	247	1076
VvEXPA6	VIT_06s0004g04860		chr6: 5816980..5819510	minus	277	2531
VvEXPA7	VIT_06s0004g07970		chr6: 8696274..8697795	minus	253	1522
VvEXPA8	VIT_07s0005g02310	XM_002273247.2	chr7: 4682986..4683933	plus	256	948
VvEXPA9	VIT_07s0005g04850		chr7: 8110218..8111014	minus	240	797
VvEXPA10	VIT_08s0058g01200		chr8: 10683123..10683967	minus	249	845
VvEXPA11	VIT_08s0007g00440		chr8: 14733277..14734895	minus	258	1619
VvEXPA12	VIT_08s0007g04680	XM_002266207.1	chr8: 18624417..18625435	plus	272	1019
VvEXPA13	VIT_13s0019g01650	XM_002283322.1	chr13: 3120586..3122103	minus	258	1518
VvEXPA14	VIT_13s0067g02930	XM_002280264.2	chr13: 1578700..1580132	plus	252	1433
VvEXPA15	VIT_14s0068g00600	XM_002276604.1	chr14: 24335233..24336088	minus	251	856
VvEXPA16	VIT_14s0108g01020	XM_002283705.2	chr14: 29649689..29650636	minus	240	948
VvEXPA17	VIT_17s0000g06360	XM_002284822.1	chr17: 6888734..6889740	minus	248	983
VvEXPA18	VIT_17s0053g00990	XM_002269481.1	chr17: 16772521..16773735	minus	246	1215
VvEXPA19	VIT_18s0001g01130	XM_002285855.2	chr18: 1779003..1779933	plus	256	931
VvEXPA20	VIT_19s0014g02910		chr19: 3020095..3021255	plus	257	1161
EXPB						
VvEXPB1	VIT_09s0002g08510	XM_002269154.2	chr9: 9571601..9574374	minus	279	2774
VvEXPB2	VIT_12s0059g00190	XM_002267091.2	chr12: 5089208..5091217	plus	259	2010
VvEXPB3	VIT_15s0021g02670	XM_002275521.2	chr15: 13673386..13674732	minus	273	1356
VvEXPB4	VIT_15s0021g02700	XM_002278523.2	chr15: 13735960..13737451	plus	273	1492
EXLA						
VvEXLA1	VIT_03s0038g03430	XM_002281815.2	chr3:2487139..2488660	minus	259	1522
EXLB						
VvEXLB1	VIT_02s0012g00830	XM_002278881.1	chr2:6705411..6707155	plus	255	1745
VvEXLB2	VIT_00s0309g00050	XM_002270139.2	chrUn: 22613478..22614823	minus	251	1345
VvEXLB3	VIT_00s0309g00090	XM_002269059.2	chrUn: 22639761..22641104	minus	251	1344
VvEXLB4	VIT_00s1455g00010	XM_002273860.1	chrUn:39210490..39211840	plus	251	1571

Genes were named based on the expansin family designation and their chromosomal locus. The corresponding identifier in the 12X V1 prediction and open reading frame length are also provided.

Table 2. Sizes of the four expansin families in different plants.

	EXPA	*EXPB*	*EXLA*	*EXLB*
Arabidopsis	26	6	3	1
Rice	33	18	4	1
Poplar	27	3	2	4
Grapevine	20	4	1	4

The number of expansin genes in each family is listed in grapevine, Arabidopsis, poplar and rice.

sequences and similar exon–intron structures within each family suggest that the families may have undergone independent gene duplications to generate paralogs with partially or completely overlapping functions.

The four expansin gene families in angiosperms were previously subdivided into 17 clades based on a reconstruction of the last common ancestor of monocots and eudicots, which was predicted to contain 10–12 EXPA genes, two EXPB genes, one EXLA gene and two EXLB genes, making 15–17 in total [8]. Each clade would therefore include all of the genes descending from a single gene present in the last common ancestor, constituting a monocot–eudicot orthologous group. We therefore constructed a phylogenetic tree containing all the grapevine expansins and additional Arabidopsis, rice and poplar expansins that represent all the proposed monocot-eudicot clades [8] (Figure S3). We found that

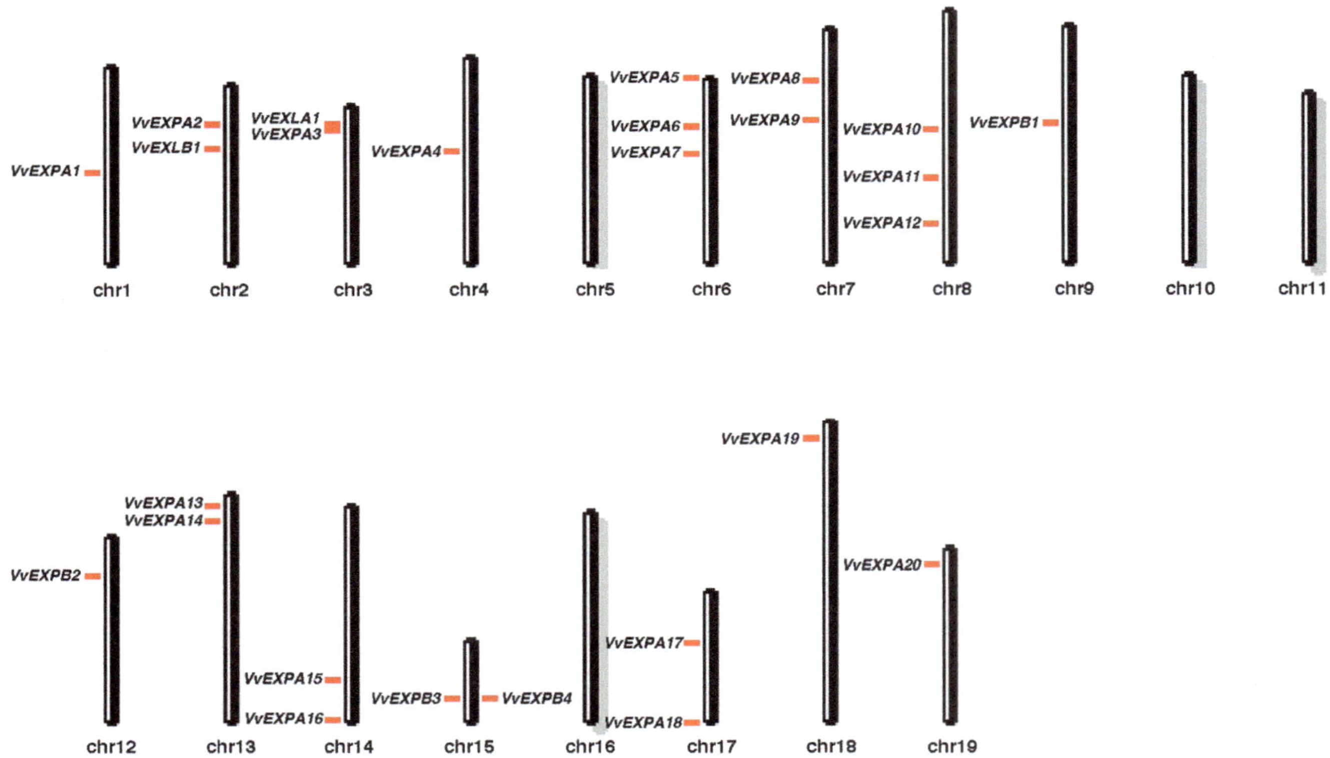

Figure 1. Chromosomal distribution of grapevine expansin genes. Chromosomes bearing no expansin genes are highlighted with gray shading.

all 17 clades were present in the grapevine genome, including EXLB-II and EXPA-IX, which were lost in Arabidopsis but retained in rice and poplar. Based on the minimal ancestral exon–intron structures for each clade [8], we found that most of the grapevine expansins had maintained the ancestral pattern (Figure S3). The exceptions were *VvEXPA12* (clade EXPA-XI), which featured two introns (A and B) instead of a single intron at position B as reported for poplar [45], suggesting the ancestral exon–intron structure of clade EXPA-XI may need to be revised. The EXPA-IX clade includes *VvEXPA4*, which has lost intron A, potentially representing a grapevine-specific intron loss as also reported for some members of clade EXPA-VIII in poplar [45]. Furthermore, an extra intron was found in *VvEXPA6* (clade EXPA-IV) whereas *VvEXPA10* (also clade EXPA-IV) had lost the intron at position B (Figure S3), suggesting either a unique exon–intron structure for certain grapevine expansins or the presence of an independent monocot-eudicot clade which has been lost in Arabidopsis, rice and poplar.

We identified five pairs of grapevine expansin genes apparently reflecting a recent lineage-specific gene duplication event, given their short branches (*VvEXPA1*:*VvEXPA18*, *VvEXPA13*:*VvEXPA6*, *VvEXPA9*:*VvEXPA15*, *VVEXB4*:*VvEXPB3* and *VvEXLB2*:*V-vEXLB3*). The duplication of EXLB genes from clade EXLB-I (*VvEXLB2*, *VvEXLB3* and *VvEXLB4*) is particularly remarkable because this appears to be a grapevine-specific event. The majority of grapevine expansin genes appear more closely related to poplar expansins than those from Arabidopsis and rice, which is consistent with the more recent divergence of the grapevine and poplar genomes from a common ancestor than from the lineages leading to monocots and Arabidopsis [28].

Tandem and segmental duplication across different plant lineages has led to the species-dependent expansion of expansin gene families in rice, Arabidopsis and poplar [45]. Grapevine has not undergone a recent whole genome duplication event, but it is unclear whether the haploid grapevine genome (which originated from the contribution of three ancestral genomes) is the product of a true hexaploidization event or successive genome duplications [28]. Recently, it was proposed that the extant triplicate state of the grapevine genome could have arisen from the fusion of two chromosome groups approximate 60 million years ago or by allopolyploidy following eudicot radiation [47]. Therefore, it will be necessary to investigate microsyntenic relationships in order to better clarify the evolution of the grapevine expansins.

Global analysis of grapevine expansin gene expression

The recently-published *V. vinifera* cv Corvina gene expression atlas [42] allowed us to characterize the expression of the expansin superfamily in different grapevine organs and developmental stages. We analyzed the fluorescence intensity values of the 29 expansin transcripts, generating a biclustered heat map in which the data were normalized based on the mean center genes/rows adjustment (Figure 3A). The different expansin families shared a common expression profile, with the EXPA, EXPB and EXLA transcripts detected in a wide range of tissues, organs and developmental stages without any particular correlation with the phylogenetic tree. The exception was the EXLB family, whose expression was restricted to organs such as the winter dormant bud (E-L 1), swelling bud (E-L 2), woody stem, root and mature stages of the rachis (V, MR and R).

Pearson's distance correlation analysis was used in a previous study to show that 54 diverse grapevine samples clustered

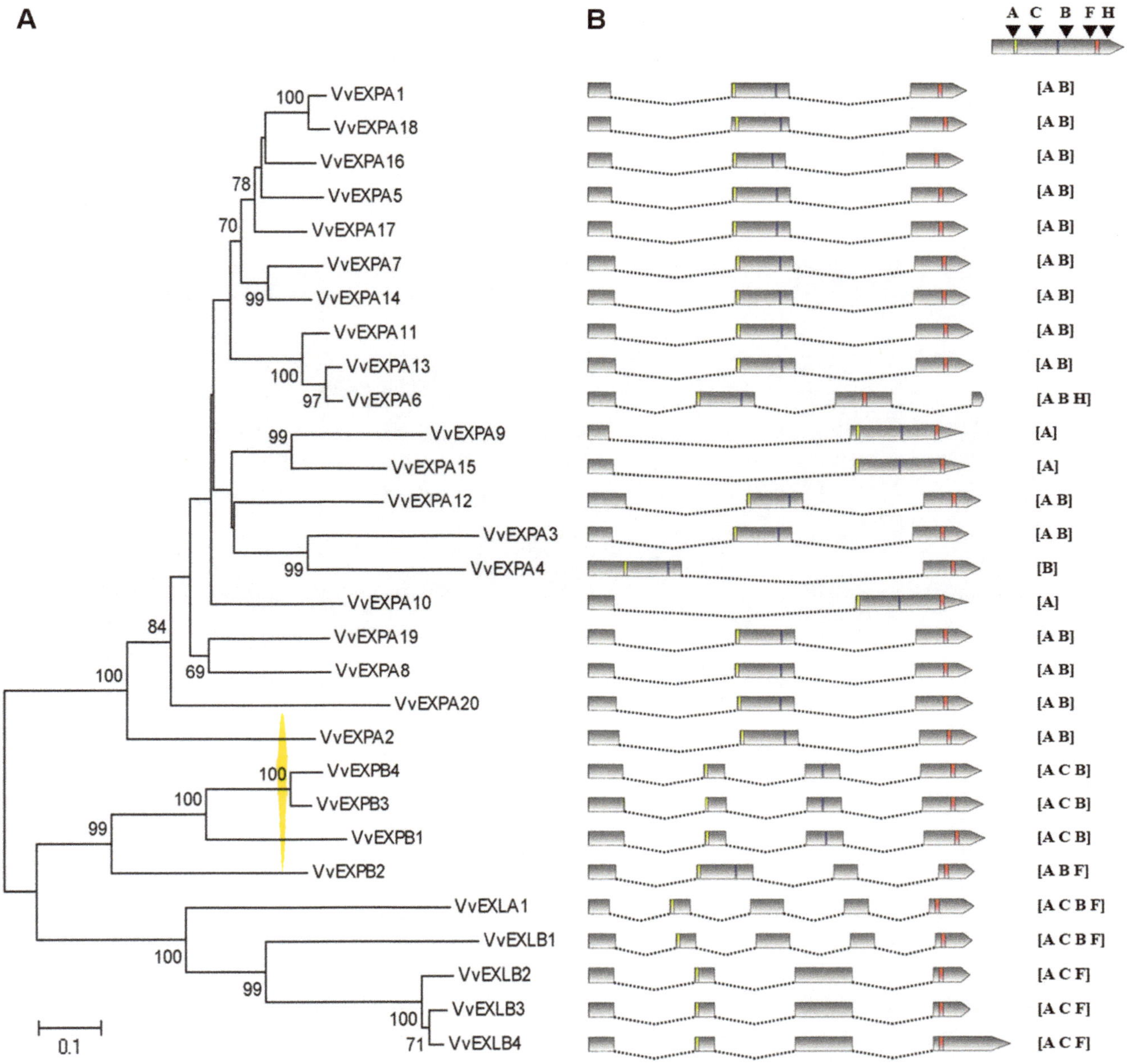

Figure 2. Phylogenetic analysis and exon–intron structures of grapevine expansin genes. (**A**) The deduced amino acid sequences of the grapevine expansins were aligned with MAFFT v6.0 software and the phylogenetic tree was constructed using the neighbor joining method in MEGA v4.0. The reliability of the predicted tree was tested by bootstrapping with 1000 replicates. The percentage of neighbor-joining bootstrap replications (>60%) is shown above each node. (**B**) Exon–intron structures of the grapevine expansin genes. On the left, exons are represented as boxes (drawn to scale) whereas introns are represented as lines (not drawn to scale). Blue, yellow and red boxes in exons indicate the HFD, Box 1 and Box2 motifs, respectively. On the right, the intron structure for each expansin gene is indicated, with positions as previously defined [8]. The intron in position H was defined in the present work.

predominantly in relation to temporal dynamics, distinguishing particularly between mature/woody and vegetative/green samples, whereas the organ identity appeared to be less important [42]. We therefore used the same method to cluster the 29 sets of fluorescence data representing the grapevine expansin genes, and the resulting dendrogram was strikingly similar to the one produced in the earlier study (Figure 3B). Temporal-dynamic clustering was preserved in the expansin-specific correlation analysis, highlighting the distinction between mature and vegetative/green samples. These observations indicate that different expansins are predominantly expressed in one or the other grapevine temporal-dynamic cluster and that different organs are characterized by the expression of more than one expansin during its development. Indeed, when grapevine expansin gene expression was analyzed in individual organs, it was still possible to detect the separation between vegetative/green and mature phases (Figure S4). However, the expansin-specific dendrogram also contained a third branch representing woody organs (plus pollen and senescing leaf), which was well-separated from the mature samples. This discrepancy between the expansin dendrogram and

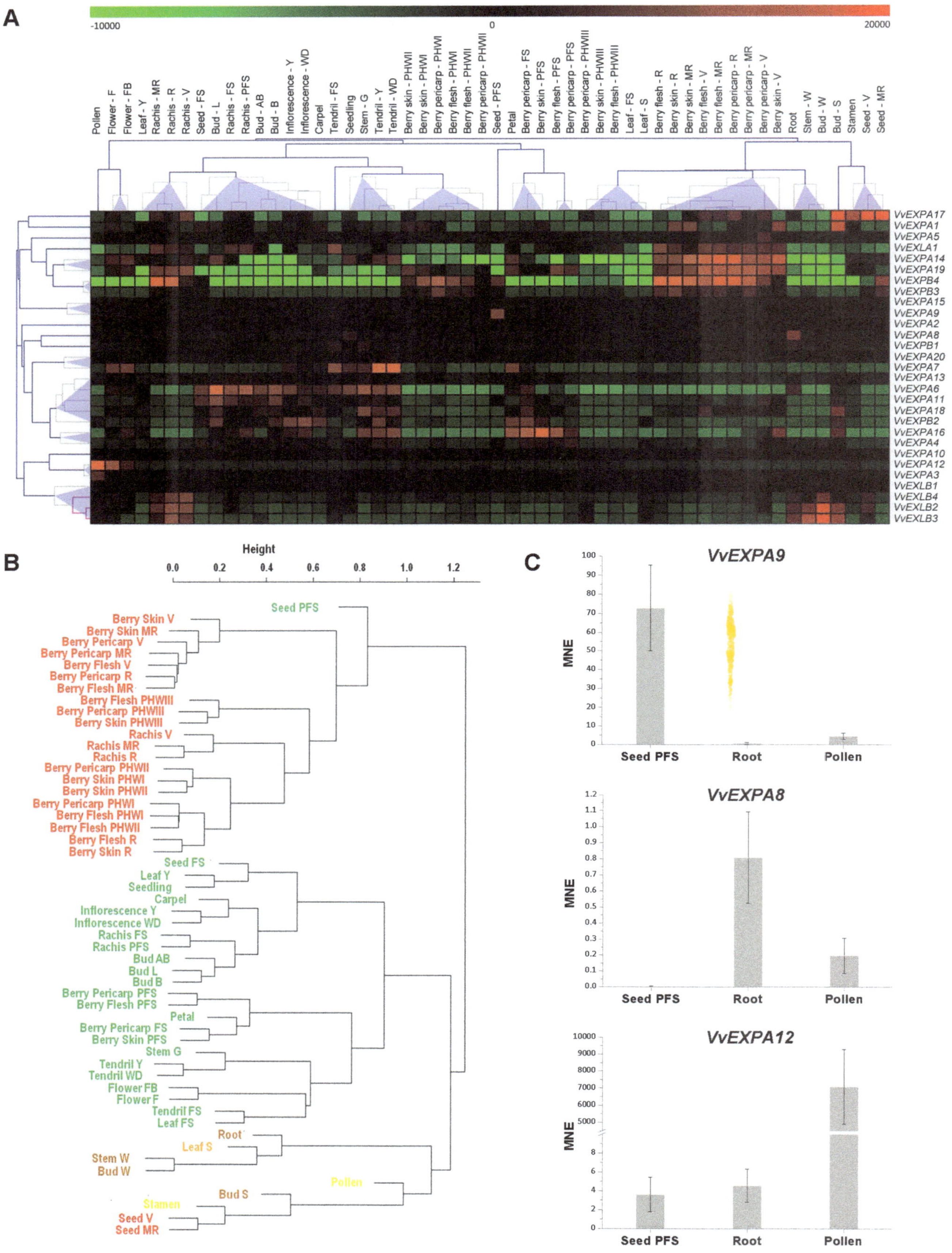
A
-10000
0
20000
Pollen
Flower - F
Flower - FB
Leaf - Y
Rachis - MR
Rachis - R
Rachis - V
Seed - FS
Bud - L
Rachis - FS
Rachis - PFS
Bud - AB
Bud - B
Inflorescence - Y
Inflorescence - WD
Carpel
Tendril - FS
Seedling
Stem - G
Tendril - Y
Tendril - WD
Berry skin - PHWII
Berry skin - PHWI
Berry pericarp - PHWI
Berry flesh - PHWI
Berry flesh - PHWII
Berry pericarp - PHWII
Seed - PFS
Petal
Berry pericarp - FS
Berry skin - PFS
Berry flesh - PFS
Berry pericarp - PFS
Berry pericarp - PHWIII
Berry skin - PHWIII
Berry flesh - PHWIII
Leaf - FS
Leaf - S
Berry flesh - R
Berry skin - R
Berry skin - MR
Berry flesh - V
Berry flesh - MR
Berry pericarp - R
Berry pericarp - MR
Berry pericarp - V
Berry skin - V
Root
Stem - W
Bud - W
Bud - S
Stamen
Seed - V
Seed - MR
VvEXPA17
VvEXPA1
VvEXPA5
VvEXLA1
VvEXPA14
VvEXPA19
VvEXPB4
VvEXPB3
VvEXPA15
VvEXPA9
VvEXPA2
VvEXPA8
VvEXPB1
VvEXPA20
VvEXPA7
VvEXPA13
VvEXPA6
VvEXPA11
VvEXPA18
VvEXPB2
VvEXPA16
VvEXPA4
VvEXPA10
VvEXPA12
VvEXPA3
VvEXLB1
VvEXLB4
VvEXLB2
VvEXLB3
B
Height
0.0
0.2
0.4
0.6
0.8
1.0
1.2
Seed PFS
Berry Skin V
Berry Skin MR
Berry Pericarp V
Berry Pericarp MR
Berry Flesh V
Berry Pericarp R
Berry Flesh MR
Berry Flesh PHWIII
Berry Pericarp PHWIII
Berry Skin PHWIII
Rachis V
Rachis MR
Rachis R
Berry Pericarp PHWII
Berry Skin PHWI
Berry Skin PHWII
Berry Pericarp PHWI
Berry Flesh PHWI
Berry Flesh PHWII
Berry Flesh R
Berry Skin R
Seed FS
Leaf Y
Seedling
Carpel
Inflorescence Y
Inflorescence WD
Rachis FS
Rachis PFS
Bud AB
Bud L
Bud B
Berry Pericarp PFS
Berry Flesh PFS
Petal
Berry Pericarp FS
Berry Skin PFS
Stem G
Tendril Y
Tendril WD
Flower FB
Flower F
Tendril FS
Leaf FS
Root
Leaf S
Stem W
Bud W
Pollen
Bud S
Stamen
Seed V
Seed MR
C
VvEXPA9
MNE
100
90
80
70
60
50
40
30
20
10
0
Seed PFS
Root
Pollen
VvEXPA8
MNE
1.2
1.1
1.0
0.9
0.8
0.7
0.6
0.5
0.4
0.3
0.2
0.1
0.0
Seed PFS
Root
Pollen
VvEXPA12
MNE
10000
9000
8000
7000
6000
5000
8
6
4
2
0
Seed PFS
Root
Pollen

Figure 3. Expression of grapevine expansins in the V. vinifera cv Corvina atlas. (**A**) Expression data were normalized based on the mean expression value of each gene in all tissues/organs. Different organs/tissues are displayed vertically above each column. Gene names are displayed to the right of each row. The color scheme used to represent expression level is red/green: black boxes indicate a low variation in expression, green boxes indicate a fold decrease and red boxes indicate a fold increase with respect to the mean value. Samples and genes were hierarchically clustered based on the average Pearson's distance. Abbreviations indicating developmental stages are listed under Materials and Methods. (**B**) Cluster dendrogram of the 54 grapevine atlas samples, based on the expression profiles within the grapevine expansin superfamily. Pearson's correlation values were converted into distance coefficients to define the height of the dendrogram. (**C**) Organ-specific expansins *VvEXPA8*, *VvEXPA9* and *VvEXPA12* were analyzed by real time RT-PCR in a subset of organs (seed PFS, roots and pollen). Transcripts were normalized to the expression of ubiquitin (*UBQ*). Bars indicate standard error (SE) in three technical replicates.

the whole-transcriptome atlas reflects the strong co-expression of the EXLB family in woody tissues.

Certain expansin genes were characterized by unique expression profiles in particular tissues, as confirmed by real time RT-PCR (Figure 3C). *VvEXPA9* was expressed uniquely in the seed after fruit setting (Figure 3C, upper panel) and this sample also represents an outlier in the co-expression analysis (Figure 3B), suggesting a highly-specific expansin gene expression profile in seeds at this stage. *VvEXPA8* was expressed predominantly in the root (Figure 3C, middle panel), a phenomenon also observed in other species because expansins are specifically involved in processes such as root hair initiation in Arabidopsis [14] and root elongation in soybean (*Glycine max*) and rice [48,49]. Finally, *VvEXPA12* was expressed strongly in pollen (Figure 3C, lower panel) and to a lesser extent in the flower at the onset of flowering (10% caps off; Flower - FB) and during flowering (50% caps off; Flower - F), indicating a role in pollen development. No EXPB genes were expressed in grapevine pollen, whereas this family plays a prominent role in monocot pollen development [10]. Our data suggest that only the EXLB family shows tissue-dependent functional specialization in grapevine.

EXLB genes are expressed in woody tissues

EXLA and EXLB genes contain both domains that characterize the expansin superfamily but the sequence has diverged substantially from the EXPA and EXPB families and no function has yet been established for EXLAs and EXLBs [9].

The EXLB genes showed a unique expression profile in the grapevine gene expression atlas, characterized by little or no expression in the tissues that express EXPA and EXPB genes, such as berries, tendrils, flowers and leaves (Figure 3). Although *VvEXLB1* was not modulated in any of the tissues we analyzed, the remaining EXLB genes (*VvEXLB2–4*) were expressed strongly during the mature stages of rachis development (V, MR and R), in winter and swelling buds (bud–W and bud–S), in woody stems (stem–W) and, to a lesser extent, in roots and seeds (Figure 4A). *VvEXLB2* was strongly expressed in all the woody tissues listed above, *VvEXLB3* was induced in the seed and rachis–MR samples, and *VvEXLB4* was induced in woody stems and swelling buds. All these tissues are characterized by various degrees of lignification: by the time berries reach full maturity, the rachis is already brown, dehydrated and lignified, the seed contains 44% lignin (as well as 7% cellulose and 31% hemicellulose) [50], the winter and swelling buds are still enclosed within protective lignified tissues known as perulae, and the woody stem and roots have secondary cell walls.

The microarray expression profiles were validated by real time RT-PCR analysis of individual EXLB genes (*VvEXLB3* and *VvEXLB4*) in rachis, bud and stem tissues (Figure 4B). To emphasize the relationship between EXLB gene expression and secondary cell wall lignification, *VvEXLB3* and *VvEXLB4* transcript accumulation was compared in vegetative/green tissues (green stem, latent bud and rachis at fruit setting) and the same tissues during the mature/woody phase (winter stem, winter bud and rachis at the ripening stage), confirming that *VvEXLB3* and *VvEXLB4* are strongly induced in all woody tissues (Figure 4B).

Expansins are thought to be required for cell wall loosening during primary wall growth [46], which occurs during many different developmental processes in growing and non-growing organs. However, certain expansin genes are expressed specifically in the xylem secondary wall, suggesting a role in secondary wall formation. In Chinese fir (*Cunninghamia lanceolata*), two recently-identified EXPA genes (*ClEXPA1* and *ClEXPA2*) are strongly expressed in wood-forming tissues and preferentially in the cambium [51]. Expansins are also expressed during primary xylem formation in zinnia (*Zinnea elegans*) [52]. In hybrid aspen (*Populus* spp.), three EXPA genes (*PttEXPA1*, *PttEXPA2* and *PttEXPA8*) and one EXPB gene (*PttEXPB1*) are expressed in the cambium and adjacent radial zone of mature stem tissues, supporting a role for expansins in secondary cell wall formation [17]. A recent transcriptome-wide analysis of the transition from primary to secondary stem development in *Populus trichocarpa* identified four EXPA genes and one EXLA gene whose expression profiles were linked to secondary growth [53]. The lack of functional data for EXLB genes means their precise role is difficult to establish. However, an intriguing observation is that the expansin superfamily in the moss *Physcomitrella patens* is similar in number to that in Arabidopsis, but lacks EXLA and EXLB genes [54]. Generally, bryophytes are land plants that do not have vascular tissues and do not contain lignin. This observation supports a relationship between the presence/action of expansin-like sequences such as EXLA and EXLB and the range of processes linked to lignification. Furthermore, a recent investigation of vascular overgrowth in Arabidopsis following treatment with an auxin transport inhibitor identified two EXLB sequences induced in vascular tissues and hydathodes [55]. Taken together, these data clearly indicate that EXLA and EXLB are specifically involved in secondary wall formation and support our proposal that at least *VvEXLB2*, *VvEXLB3* and *VvEXLB4* play specific functional roles in grapevine woody tissues and organs.

Berry growth and ripening require the activities of different expansin genes

The economical importance of grapevine depends largely on the quality of its berries, which undergo a long process of development and ripening involving a series of dramatic physical and biochemical changes. These include several cell wall structural modifications which are required for cell expansion and softening [56]. During berry development, there are two distinct periods of rapid growth [57]. The first growth phase involves an increase in cell number and volume, whereas the second involves an increase in volume only and is also characterized by softening.

We investigated the role of expansins in berry development by focusing on the transcriptional changes during fruit-set (FS) and post-fruit set (PFS) in the first growth phase, and during veraison (V), mid-ripening (MR) and ripening (R) in the second growth phase. The expression profiles of 14 grapevine expansin genes that show significant modulation of expression in at least one of the five

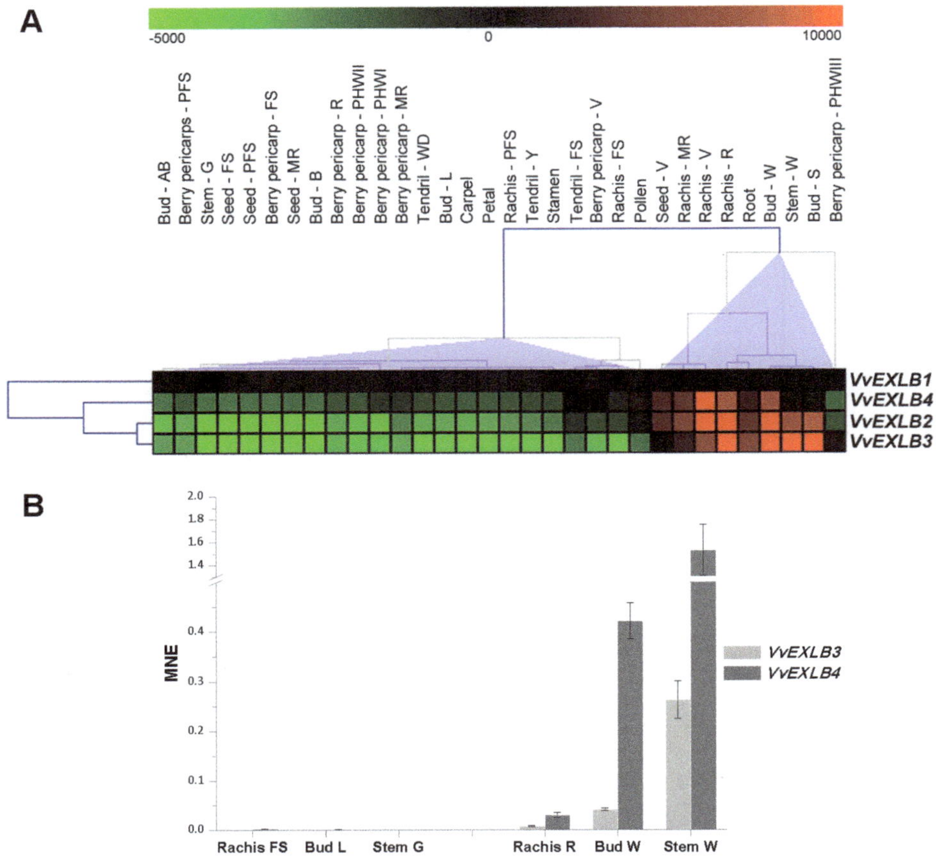

Figure 4. Expression of the grapevine EXLB gene family in woody organs. (**A**) Expression data were normalized based on the mean expression value of each gene in a selected subset of tissues/organs. The color scheme used to represent expression level is red/green: black boxes indicate a low variation in expression, green boxes indicate a fold decrease and red boxes indicate a fold increase with respect to the mean value. Samples were hierarchically clustered based on the average Pearson's distance. (**B**) *VvEXLB3* and *VvEXLB4* were analyzed by real time RT-PCR in rachis, bud and stem both at the green and woody stages to validate data obtained from the expression atlas. Transcripts were normalized to the expression of ubiquitin (*UBQ*). Bars indicate standard error (SE) in three technical replicates.

samples (encompassing all expansin families except EXLB) are shown in Figure 5A. We identified 10 EXPA genes in clades EXPA-I and EXPA-II, suggesting these are particularly relevant during fruit development, as well as three EXPB genes with berry-specific expression (*VvEXPB2–4*) and also *VvEXLA1*, whose global expression profile suggests a role in berry ripening and rachis development (Figure 3A). These expression profiles divide the samples in two clear groups: berries before veraison and berries during and after veraison (Figure 5A). This is consistent with the contrast between vegetative/green and mature samples on the basis of expansin expression profiles (Figure 3B).

The expansins expressed before veraison were *VvEXPA4* and *VvEXPA16* (validated by real-time RT-PCR as shown in Figure 5B), as well as *VvEXPA6*, *VvEXPB2* and *VvEXPA18*. Interestingly, *VvEXPA11* was expressed in the FS and V samples and only at low levels, suggesting a marginal involvement in berry development (Figure 5A). We identified eight expansin genes whose expression was upregulated during the second growth phase. The majority of these (including *VvEXPA19*, *VvEXLA1*, *VvEXPB4*, *VvEXPA14* and *VvEXPA17*) had similar expression profiles albeit with different degrees of induction. *VvEXPB3* and *VvEXPA1* showed single expression peaks in the MR and R samples, respectively, whereas *VvEXPA5* was expressed only at veraison, suggesting a specific role during the second phase of berry growth. The expression profiles of *VvEXPA19* and *VvEXPA5* were also validated by RT-PCR (Figure 5B).

We noted that some members of the expansin superfamily were expressed specifically either in the berry skin or pulp. For example, *VvEXPA1* was expressed only in the skin whereas *VvEXPA17* was found predominantly in the flesh (Figure 3A). This suggests that common expansin substrates are modified specifically within each tissue and that different expansins contribute to the fine tuning of cell wall metabolism in the berry flesh and skin.

Previous studies have shown that some cell wall-modifying enzymes, including expansins, are expressed in a bimodal fashion in which high expression levels coincide with the two periods of rapid berry growth and cell expansion [37]. Transcriptome analysis has also demonstrated that the abundance of most expansin transcripts increases or decreases steadily during berry development [39,58] although some show characteristic expression profiles matching the second phase of fruit enlargement and softening [39,59,60]. One expansin transcript was also shown to be a potential negative marker (i.e. conspicuous by its absence) of the pre-ripening/ripening phases [41]. Expansin genes expressed during the first rapid period of berry growth period may be required for primary cell wall loosening and the incorporation of new cell wall material, whereas those expressed after veraison may regulate contact between cell wall polymers and the enzymes that degrade them, facilitating cell wall disassembly and

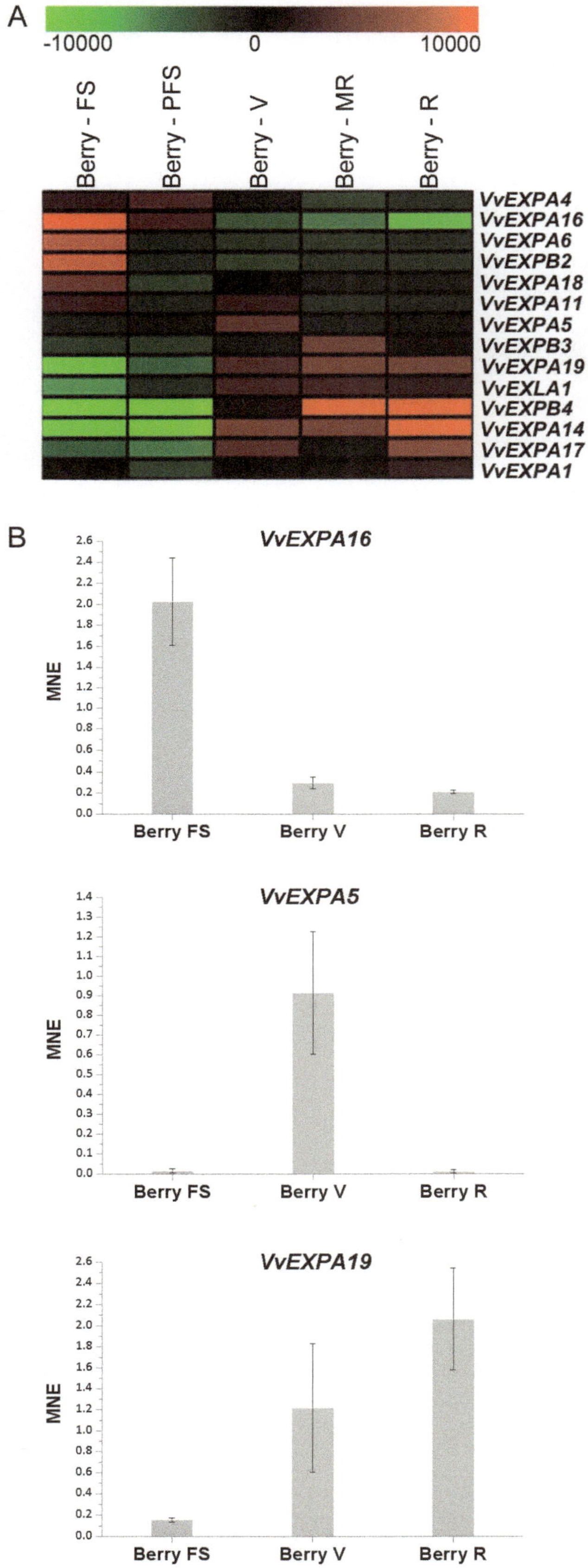

Figure 5. Grapevine expansin gene expression during berry pericarp growth and ripening. Expression data were restricted to berry development from fruit set to ripening. Genes were variance-filtered eliminating 50% of the expansin genes showing low variation in expression among the different stages. (**B**) The expression of selected expansin genes, specifically expressed in different berry developmental stages, was validated by real time RT-PCR. Transcripts were normalized to the expression of ubiquitin (*UBQ*). Bars indicate standard error (SE) in three technical replicates.

fruit softening [36,37]. The classification of grapevine expansin genes involved in the modulation of berry cell wall metabolism, which is a key molecular event in berry development and ripening, could be used to improve berry quality traits and thus the economic value of grapevine.

Grapevine expansins are involved in water-deficit stress responses

A link has been established between expansins and water-deficit stress, as demonstrated by the induction of several expansin genes by dehydration stress [61,62] (as well as the induction of others by heat, shading, anoxia and pathogens [63–67]) and the ability of expansin transgenes to confer drought tolerance, e.g. in transgenic tobacco (*Nicotiana tabacum*) expressing *TaEXPB23* [68,69] and in transgenic rose (*Rosa* spp.) expressing *RhEXPA4* [70]. In this context, expansins may regulate adaptive changes in the cell wall structure to increase flexibility and thus prevent damage caused by the folding of rigid cell walls as the cell volume is reduced.

Samples representing controlled biotic and/or abiotic stress were not included in the grapevine gene expression atlas, but three stages of post-harvest berry withering were included [42]. This is a controlled form of post-harvest dehydration which is used to improve certain berry quality traits for the production of premium wines. During withering, the berries undergo several chemical and physical changes representing metabolic processes such as the modification of cell wall polymers in the skin [71,72]. A recent transcriptomic survey of withering in Raboso Piave berry skins [73] showed that *VvEXPA1* transcript levels declined during this period. We therefore used the grapevine expression atlas data to investigate the role of grapevine expansins during post-harvest withering (Figure 6A). In Corvina berries, 12 expansin genes were differentially modulated during withering and we analyzed their expression profiles separately in the flesh and skin. *VvEXPA14*, *VvEXPB3* and *VvEXPB4* were expressed in both tissues and were downregulated during withering. We validated the *VvEXPA14* expression profile by real-time RT PCR (Figure 6B). The other expansins showed a more complex expression profile, e.g. *VvEXPA17* was upregulated at the second withering time point (PHWII) specifically in the berry skin, as confirmed by real-time RT PCR (Figure 6B). *VvEXPA19* and *VvEXPA1* were modulated from the onset of the stress treatment (PHWI) and their expression was also skin-specific (Figure 6A, 6B). Interestingly, *VvEXLA1* expression was induced only in the berry flesh at the last stage of the dehydration process, when weight loss approached 30% (Figure 6A, 6B). The expansins that were induced during withering represent candidate dehydration stress response markers. The coordinated activity of pectin methylesterase (PME) and polygalacturonase (PG) has recently been shown to degrade pectins in Corvina berry skin [74]. Therefore, the skin-specific expansins induced during dehydration (e.g. *VvEXPA19* and *VvEXPA1*) may be involved in pectin metabolism, probably by regulating contact between pectins and pectin-degrading enzymes. Our data suggest that expansins contribute to post-harvest dehydration stress responses by regulating cell wall metabolism in berries.

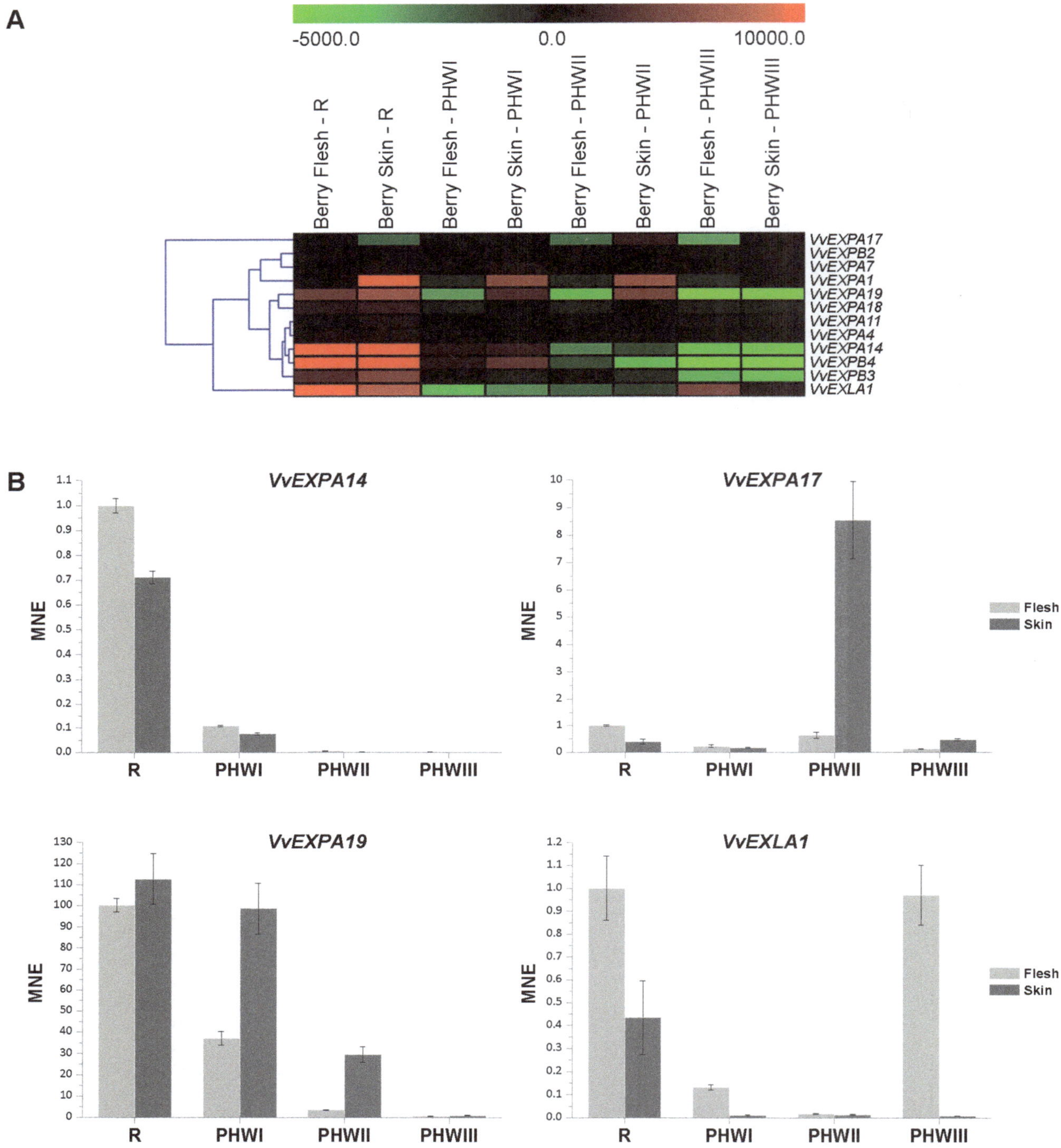

Figure 6. Grapevine expansin gene expression in berry tissues during post-harvest withering process. Expression data from the ripening stage were included for comparison. Genes were variance-filtered eliminating 50% of the expansin genes showing low variation between samples. Selected expansin genes (*VvEXLA1*, *VvEXPA19*, *VvEXPA17* and *VvEXPA14*) were validated by real time RT-PCR. Transcripts were normalized to the expression of ubiquitin (*UBQ*). Bars indicate standard error (SE) in three technical replicates.

Conclusion

We carried out a genome-wide survey of the expansin superfamily in grapevine to gain insight into the chromosomal organization, phylogenetic relationships and expression profiles of these genes. The grapevine genome contains 20 EXPA genes, four EXPB genes, one EXLA gene and four EXLB genes, representing all 17 monocot-eudicot clades proposed for Arabidopsis and rice as well as a potential novel clade not identified thus far in other species, represented by *VvEXPA10*. Most of the grapevine expansin genes retained the ancestral exon–intron structure, although some showed unique structures that may be grapevine specific. A comparison of the grapevine, Arabidopsis, rice and poplar expansin genes revealed that grapevine and poplar expansins were the most closely related and that the grapevine expansins had not undergone significant phylogenetic expansion. However, integrated phylogenetic characterization and gene colinearity analysis within and between plant genomes will be necessary to understand the evolution of the grapevine expansin superfamily in more detail.

Genome-wide gene expression analysis indicated that members of the same expansin family did not necessarily share common expression profiles although most grapevine EXLB genes were expressed strongly in woody tissues and organs, suggesting a specific role in the formation of secondary cell walls. Correlative analysis of expansin gene expression profiles suggested temporal-dynamic clustering in which mature samples and vegetative/green samples were clearly separated.

We also identified the expansins whose expression was significantly modulated during berry development, particularly in the context of the two distinctive phases of berry growth. All expansin families except EXLB were found to be involved in berry development, supporting previous reports that cell wall modification is a critical component of fruit growth and maturation.

Finally, we investigated expansin expression during post-harvest withering, which showed that some expansins are induced by dehydration, indicating a potential role in the control of adaptive cell wall modification under stress. Our data lay the groundwork for the further structural, functional and phylogenetic characterization of expansin genes and their developmental role in commercially valuable fruit crops.

Materials and Methods

Database searches, analysis of gene structure and chromosome mapping

Previously-identified Arabidopsis and rice expansin protein sequences (AtEXPA1, AtEXPB1, ATEXLA1, ATEXLB1, OsEXPA1, OsEXPA24, OsEXPB2, OsEXPB15, OsEXLA1 and OsEXLB1) were used as BLAST queries against the Genoscope Grape Genome Database (http://www.genoscope.cns.fr) representing the 8.4X and 12X V0 assemblies of the PN40024 genotype. The search was extended by matching expansin genes and their chromosomal coordinates against an uploaded version of the PN40024 12X assembly V1 (http://genomes.cribi.unipd.it/), which merges the 12 V0 prediction based on GAZE software [75] and a gene prediction performed with JIGSAW software [43] at the CRIBI Genomics in Padova, Italy. Sequences were processed using Vector NTI v9 (Invitrogen) and the gene structure was deduced from the Genoscope and CRIBI annotations or by manual editing based on comparisons with corresponding expressed sequence tags and deduced protein sequences from paralogous expansin genes. The exon–intron structures were presented using the online tool FancyGene [76]. The chromosomal loci were inferred from the genome browsers listed above and rendered using NCBI MapViewer (http://www.ncbi.nlm.nih.gov/).

Phylogenetic analysis

Multiple alignments of the 29 deduced expansin protein sequences from grapevine, together with 16 from rice, 21 from poplar and 21 from Arabidopsis, were created using the E-INSI tool in the MAFFT v6.0 software suite, which takes into account the possibility of large gaps in the alignments [77]. An unrooted phylogenetic tree was generated using the neighbor joining method [78] in MEGA v5.0 [79]. The reliability of the tree was tested by bootstrapping with 1000 replicates, and the resulting tree was edited and modified with MEGA v5.0 and Treedyn software (http://treedyn.org). An additional tree was produced with the same tools considering only the grapevine expansin gene sequences predicted from the 12X V1 assembly of the PN40024 genotype.

Analysis of the *V. vinifera* cv Corvina expression atlas

The expression profiles of grapevine expansin genes were analyzed in the global *V. vinifera* cv Corvina (clone 48) gene expression atlas of different organs at various developmental stages [42]. The gene expression microarray data were obtained by hybridization to a NimbleGen microarray 090818 Vitis exp HX12 (Roche, NimbleGen), which contains probes targeting 29,549 predicted grapevine genes (http://ddlab.sci.univr.it/FunctionalGenomics/), representing ~98.6% of the genes predicted from the V1 annotation of the 12× grapevine genome sequence (http://srs.ebi.ac.uk/) and 19,091 random probes as negative controls. The Corvina expression atlas comprises the following tissues: *in vitro* roots, green stem, buds after bud burst (rosette of leaf tips visible), young leaves (leaves collected from shoots with only five leaves), senescing leaves (leaves at the beginning of leaf-fall), berry rachis (from fruit-set to ripening), flowers (50% cap-fall). In particular, berry pericarp was collected at five developmental stages: the first stage (15 days after flowering (daf); E–L 29) corresponds to the fruit set (FS), when young berries are enlarging (>3 mm diameter); the second stage (35 daf; E–L 32) is the post-fruit set stage (PFS), when berries (>7 mm diameter) start touching; the third stage (70 daf; E–L 35) is the veraison (V), when berries begin to change colour and enlarge (10.4° Brix); the fourth stage (84 daf; E–L 36) corresponds to the mid-ripening (MR) stage (15.5° Brix); and the final stage (115 daf; E–L 38) represents complete ripening (R) (20.0° Brix). The atlas also contains a set of expression data obtained from berries that have undergone post-harvesting withering for 1–3 months after harvest: At the first withering stage (WI), berry weight was 76.4% the ripe value and the sugar content was 24.5° Brix. The second stage (WII) was characterized by 69.7% berry weight and 25.9° Brix, and the last stage (WIII) was characterized by 67.3% berry weight and 26.7° Brix. The expression data were analyzed using T-MeV v4.81 [80], normalized based on the mean center genes/rows adjustment and (in the analysis of individual organs) were also variance filtered with a 50% cut-off. Abbreviations after organ names indicate the developmental stage. FS, fruit set; PFS, post fruit set; V, veraison; MR, mid-ripening; R, ripening; PHWI, post-harvest withering (1st month); PHWII, post-harvest withering (2nd month); PHWIII, post-harvest withering (3rd month), Bud - L, latent bud; Bud - W, winter bud; Bud - S, bud swell; Bud - B, bud burst; Bud - AB, bud after burst; Inflorescence - Y, young inflorescence with single flowers separated; Inflorescence - WD, well developed inflorescence; Flower - FB, flowering begins; Flower - F, flowering; Tendril - Y, young tendril; Tendril - WD, well developed tendril; Tendril - FS, mature tendril; Leaf - Y, young leaf; Leaf - FS, mature leaf; Leaf - S, senescing leaf; Stem - G, green stem; Stem - W, woody stem.

Correlation analysis

A correlation matrix was prepared using R software and Pearson's correlation coefficient as the statistical metric to compare the values of the 29 members of grapevine expansin genes in all 54 samples [42], using the average expression value of the three biological replicates. Correlation values were converted into distance coefficients to define the height scale of the dendrogram.

RNA extraction, cDNA synthesis and real-time RT-PCR

Total RNA was isolated using the Spectrum™ Total RNA kit (SIGMA-Aldrich, St. Louis, MO), with sample-specific modifications of the extraction protocol, as previously described [42]. First strand cDNA was synthesized using 1 μg of total RNA as the template and the Improm-II™ Reverse Transcription system (Promega), according to the manufacturer's instructions. The total RNA was derived from *Vitis vinifera* cv. Corvina organs harvested at the same phenological phase as described in the grapevine gene expression atlas [42], and all RNA samples were first treated with DNase I (Promega). Gene-specific primers were targeted, when possible, to the 3′-UTR, using ubiquitin (VIT_16s0098g01190) as the reference gene (Table S1). Primers and cDNA were mixed with the Power SYBR® Green PCR Master Mix (Applied Biosystems, Foster City, CA, USA) and the reaction was carried out on a Stratagene MX 3000 P™ QPCR System (Agilent, Technologies, CA) using the following cycling conditions: 95°C hold for 10 min followed by 45 cycles at 95°C for 30 s, 55°C for 30 s and 72°C for 20 s. Nonspecific PCR products were identified by inspecting dissociation curves. In addition, the specificity of EXLB PCR products was verified by cloning the relative amplicons in the pGem®-T Easy Vector (Promega), sequencing and aligning them onto the reference genome. Amplification efficiency was calculated from raw data using LingRegPCR software [81]. Standard error (SE) values were calculated as described [82].

Supporting Information

Figure S1 Alignment of grapevine expansin protein sequences. Alignment of all grapevine expansin deduced protein sequences representative of the VvEXPA, VvEXPB, VvEXLA and VvEXLB families. The alignment was created using MAFFT v6.0 and edited with GeneDoc. The family-specific conserved motifs and active sites are indicated.

Figure S2 Gene frequency per chromosome of expansins in Arabidopsis, poplar, rice and grapevine. Arabidopsis data were obtained from TAIR, whereas poplar and rice data were deduced by consulting the expansin center site (https://homes.bio.psu.edu/expansins/index.htm).

Figure S3 Phylogenetic tree of grapevine expansins together with a set of sequences obtained from Arabidopsis, rice and poplar that represent all of the monocot-eudicot clades present in all the three species proposed [45]. The sequences were aligned with CLUSTALW and a neighbor-joining tree was constructed with MEGA v4.0. Clades are indicated on the right with black bars, numbered using the conventional nomenclature [45].

Figure S4 Expression of expansin genes in individual organs. The expression of expansin genes was considered separately in (A) bud, (B) leaf and (C)rachis. Genes were variance-filtered eliminating 50% of the expansin genes showing low variation between samples. (TIF) Table S1 List of real time RT PCR primers.

Acknowledgments

The authors thank Dr. Sara Sordo for technical support with real time RT-PCRs and Prof. Margherita Lucchin for helpful suggestions.

Author Contributions

Conceived and designed the experiments: SZ SDS AV. Performed the experiments: SDS AV MF. Analyzed the data: SDS AV. Contributed reagents/materials/analysis tools: AV MF LV. Wrote the paper: SDS AV GBT SZ. Provided guidance of the whole study: MP.

References

1. Cosgrove DJ (2000) Loosening of plant cell walls by expansins. Nature 407: 321–326.
2. Shcherban TY, Shi J, Durachko DM, Guiltinan MJ, McQueen-Mason SJ, et al. (1995) Molecular cloning and sequence analysis of expansins-a highly conserved, multigene family of proteins that mediate cell wall extension in plants. Proc Natl Acad Sci U S A 92: 9245–9249.
3. Cosgrove DJ (2000) New genes and new biological roles for expansins. Curr Opin Plant Biol 3: 73–78.
4. Kerff F, Amoroso A, Herman R, Sauvage E, Petrella S, et al. (2008) Crystal structure and activity of Bacillus subtilis YoaJ (EXLX1), a bacterial expansin that promotes root colonization. Proc Natl Acad Sci U S A 105: 16876–16881.
5. Kudla U, Qin L, Milac A, Kielak A, Maissen C, et al. (2005) Origin, distribution and 3D-modeling of Gr-EXPB1, an expansin from the potato cyst nematode Globodera rostochiensis. FEBS Lett 579: 2451–2457.
6. Li Y, Darley CP, Ongaro V, Fleming A, Schipper O, et al. (2002) Plant expansins are a complex multigene family with an ancient evolutionary origin. Plant Physiol 128: 854–864.
7. Xu B, Janson JC, Sellos D (2001) Cloning and sequencing of a molluscan endo-beta-1,4-glucanase gene from the blue mussel, Mytilus edulis. Eur J Biochem 268: 3718–3727.
8. Sampedro J, Cosgrove DJ (2005) The expansin superfamily. Genome Biol 6: 242.
9. Kende H, Bradford K, Brummell D, Cho HT, Cosgrove D, et al. (2004) Nomenclature for members of the expansin superfamily of genes and proteins. Plant Mol Biol 55: 311–314.
10. Cosgrove DJ, Bedinger P, Durachko DM (1997) Group I allergens of grass pollen as cell wall-loosening agents. Proc Natl Acad Sci U S A 94: 6559–6564.
11. McQueen-Mason S, Durachko DM, Cosgrove DJ (1992) Two endogenous proteins that induce cell wall extension in plants. Plant Cell 4: 1425–1433.
12. Whitney SE, Gidley MJ, McQueen-Mason SJ (2000) Probing expansin action using cellulose/hemicellulose composites. Plant J 22: 327–334.
13. Li LC, Bedinger PA, Volk C, Jones AD, Cosgrove DJ (2003) Purification and characterization of four beta-expansins (Zea m 1 isoforms) from maize pollen. Plant Physiol 132: 2073–2085.
14. Cho HT, Cosgrove DJ (2002) Regulation of root hair initiation and expansin gene expression in Arabidopsis. Plant Cell 14: 3237–3253.
15. Zenoni S, Fasoli M, Tornielli GB, Dal Santo S, Sanson A, et al. (2011) Overexpression of PhEXPA1 increases cell size, modifies cell wall polymer composition and affects the timing of axillary meristem development in Petunia hybrida. New Phytol 191: 662–677.
16. Dal Santo S, Fasoli M, Cavallini E, Tornielli GB, Pezzotti M, et al. (2011) PhEXPA1, a Petunia hybrida expansin, is involved in cell wall metabolism and in plant architecture specification. Plant Signal Behav 6: 2031–2034.
17. Gray-Mitsumune M, Mellerowicz EJ, Abe H, Schrader J, Winzell A, et al. (2004) Expansins abundant in secondary xylem belong to subgroup A of the alpha-expansin gene family. Plant Physiol 135: 1552–1564.
18. Belfield EJ, Ruperti B, Roberts JA, McQueen-Mason S (2005) Changes in expansin activity and gene expression during ethylene-promoted leaflet abscission in Sambucus nigra. J Exp Bot 56: 817–823.
19. Cho HT, Cosgrove DJ (2000) Altered expression of expansin modulates leaf growth and pedicel abscission in Arabidopsis thaliana. Proc Natl Acad Sci U S A 97: 9783–9788.
20. Chen F, Dahal P, Bradford KJ (2001) Two tomato expansin genes show divergent expression and localization in embryos during seed development and germination. Plant Physiol 127: 928–936.
21. Dermatsev V, Weingarten-Baror C, Resnick N, Gadkar V, Wininger S, et al. (2010) Microarray analysis and functional tests suggest the involvement of expansins in the early stages of symbiosis of the arbuscular mycorrhizal fungus Glomus intraradices on tomato (Solanum lycopersicum). Mol Plant Pathol 11: 121–135.
22. Giordano W, Hirsch AM (2004) The expression of MaEXP1, a Melilotus alba expansin gene, is upregulated during the sweetclover-Sinorhizobium meliloti interaction. Mol Plant Microbe Interact 17: 613–622.
23. Wieczorek K, Golecki B, Gerdes L, Heinen P, Szakasits D, et al. (2006) Expansins are involved in the formation of nematode-induced syncytia in roots of Arabidopsis thaliana. Plant J 48: 98–112.
24. Brummell DA, Harpster MH, Civello PM, Palys JM, Bennett AB, et al. (1999) Modification of expansin protein abundance in tomato fruit alters softening and cell wall polymer metabolism during ripening. Plant Cell 11: 2203–2216.

25. Civello PM, Powell AL, Sabehat A, Bennett AB (1999) An expansin gene expressed in ripening strawberry fruit. Plant Physiol 121: 1273–1280.
26. Hayama H, Ito A, Moriguchi T, Kashimura Y (2003) Identification of a new expansin gene closely associated with peach fruit softening. Postharvest Biology and Technology 29: 1–10.
27. Powell ALT, Kalamaki MS, Kurien PA, Gurrieri S, Bennett AB (2003) Simultaneous Transgenic Suppression of LePG and LeExp1 Influences Fruit Texture and Juice Viscosity in a Fresh Market Tomato Variety. J Agric Food Chem 51: 7450–7455.
28. Jaillon O, Aury JM, Noel B, Policriti A, Clepet C, et al. (2007) The grapevine genome sequence suggests ancestral hexaploidization in major angiosperm phyla. Nature 449: 463–467.
29. Velasco R, Zharkikh A, Troggio M, Cartwright DA, Cestaro A, et al. (2007) A high quality draft consensus sequence of the genome of a heterozygous grapevine variety. PLoS One 2: e1326.
30. Vannozzi A, Dry IB, Fasoli M, Zenoni S, Lucchin M (2012) Genome-wide analysis of the grapevine stilbene synthase multigenic family: genomic organization and expression profiles upon biotic and abiotic stresses. BMC Plant Biol 12: 130.
31. Yang Y, He M, Zhu Z, Li S, Xu Y, et al. (2012) Identification of the dehydrin gene family from grapevine species and analysis of their responsiveness to various forms of abiotic and biotic stress. BMC Plant Biol 12: 140.
32. Lebel S, Schellenbaum P, Walter B, Maillot P (2010) Characterisation of the Vitis vinifera PR10 multigene family. BMC Plant Biol 10: 184.
33. Zhang Y, Mao L, Wang H, Brocker C, Yin X, et al. (2012) Genome-wide identification and analysis of grape aldehyde dehydrogenase (ALDH) gene superfamily. PLoS One 7: e32153.
34. Zhang Y, Gao M, Singer SD, Fei Z, Wang H, et al. (2012) Genome-Wide Identification and Analysis of the TIFY Gene Family in Grape. PLoS One 7: e44465.
35. Wang N, Zheng Y, Xin H, Fang L, Li S (2013) Comprehensive analysis of NAC domain transcription factor gene family in Vitis vinifera. Plant Cell Rep 32: 61–75.
36. Ishimaru M, Smith DL, Gross KC, Kobayashi S (2007) Expression of three expansin genes during development and maturation of Kyoho grape berries. J Plant Physiol 164: 1675–1682.
37. Schlosser J, Olsson N, Weis M, Reid K, Peng F, et al. (2008) Cellular expansion and gene expression in the developing grape (Vitis vinifera L.). Protoplasma 232: 255–265.
38. Vassilopoulou E, Zuidmeer L, Akkerdaas J, Tassios I, Rigby NR, et al. (2007) Severe immediate allergic reactions to grapes: part of a lipid transfer protein-associated clinical syndrome. Int Arch Allergy Immunol 143: 92–102.
39. Deluc LG, Grimplet J, Wheatley MD, Tillett RL, Quilici DR, et al. (2007) Transcriptomic and metabolite analyses of Cabernet Sauvignon grape berry development. BMC Genomics 8: 429.
40. Guillaumie S, Fouquet R, Kappel C, Camps C, Terrier N, et al. (2011) Transcriptional analysis of late ripening stages of grapevine berry. BMC Plant Biol 11: 165.
41. Zamboni A, Di Carli M, Guzzo F, Stocchero M, Zenoni S, et al. (2010) Identification of putative stage-specific grapevine berry biomarkers and omics data integration into networks. Plant Physiol 154: 1439–1459.
42. Fasoli M, Dal Santo S, Zenoni S, Tornielli GB, Farina L, et al. (2012) The grapevine expression atlas reveals a deep transcriptome shift driving the entire plant into a maturation program. Plant Cell 24: 3489–3505.
43. Allen JE, Salzberg SL (2005) JIGSAW: integration of multiple sources of evidence for gene prediction. Bioinformatics 21: 3596–3603.
44. Geisler-Lee J, Geisler M, Coutinho PM, Segerman B, Nishikubo N, et al. (2006) Poplar carbohydrate-active enzymes. Gene identification and expression analyses. Plant Physiol 140: 946–962.
45. Sampedro J, Carey RE, Cosgrove DJ (2006) Genome histories clarify evolution of the expansin superfamily: new insights from the poplar genome and pine ESTs. J Plant Res 119: 11–21.
46. Cosgrove DJ (2005) Growth of the plant cell wall. Nat Rev Mol Cell Biol 6: 850–861.
47. Malacarne G, Perazzolli M, Cestaro A, Sterck L, Fontana P, et al. (2012) Deconstruction of the (paleo)polyploid grapevine genome based on the analysis of transposition events involving NBS resistance genes. PLoS One 7: e29762.
48. Lee DK, Ahn JH, Song SK, Choi YD, Lee JS (2003) Expression of an expansin gene is correlated with root elongation in soybean. Plant Physiol 131: 985–997.
49. ZhiMing Y, Bo K, XiaoWei H, ShaoLei L, YouHuang B, et al. (2011) Root hair-specific expansins modulate root hair elongation in rice. Plant J 66: 725–734.
50. Moldes D, Gallego PP, Rodriguez Couto S, Sanroman A (2003) Grape seeds: the best lignocellulosic waste to produce laccase by solid state cultures of Trametes hirsuta. Biotechnol Lett 25: 491–495.
51. Wang G, Gao Y, Wang J, Yang L, Song R, et al. (2011) Overexpression of two cambium-abundant Chinese fir (Cunninghamia lanceolata) alpha-expansin genes ClEXPA1 and ClEXPA2 affect growth and development in transgenic tobacco and increase the amount of cellulose in stem cell walls. Plant Biotechnol J 9: 486–502.
52. Im KH, Cosgrove DJ, Jones AM (2000) Subcellular localization of expansin mRNA in xylem cells. Plant Physiol 123: 463–470.
53. Dharmawardhana P, Brunner AM, Strauss SH (2010) Genome-wide transcriptome analysis of the transition from primary to secondary stem development in Populus trichocarpa. BMC Genomics 11: 150.
54. Carey RE, Cosgrove DJ (2007) Portrait of the expansin superfamily in Physcomitrella patens: comparisons with angiosperm expansins. Ann Bot 99: 1131–1141.
55. Wenzel CL, Hester Q, Mattsson J (2008) Identification of Genes Expressed in Vascular Tissues Using NPA-Induced Vascular Overgrowth in Arabidopsis. Plant and Cell Physiology 49: 457–468.
56. Nunan KJ, Sims IM, Bacic A, Robinson SP, Fincher GB (1998) Changes in cell wall composition during ripening of grape berries. Plant Physiol 118: 783–792.
57. Coombe BG (1995) Growth Stages of the Grapevine: Adoption of a system for identifying grapevine growth stages. Australian Journal of Grape and Wine Research 1: 104–110.
58. Pilati S, Perazzolli M, Malossini A, Cestaro A, Dematte L, et al. (2007) Genome-wide transcriptional analysis of grapevine berry ripening reveals a set of genes similarly modulated during three seasons and the occurrence of an oxidative burst at veraison. BMC Genomics 8: 428.
59. da Silva FG, Iandolino A, Al-Kayal F, Bohlmann MC, Cushman MA, et al. (2005) Characterizing the grape transcriptome. Analysis of expressed sequence tags from multiple Vitis species and development of a compendium of gene expression during berry development. Plant Physiol 139: 574–597.
60. Glissant D, Dedaldechamp F, Delrot S (2008) Transcriptomic analysis of grape berry softening during ripening. J Int Sci Vigne vin 42: 1–13.
61. Wu Y, Thorne ET, Sharp RE, Cosgrove DJ (2001) Modification of expansin transcript levels in the maize primary root at low water potentials. Plant Physiol 126: 1471–1479.
62. Zhu J, Alvarez S, Marsh EL, LeNoble ME, Cho I-J, et al. (2007) Cell Wall Proteome in the Maize Primary Root Elongation Zone. II. Region-Specific Changes in Water Soluble and Lightly Ionically Bound Proteins under Water Deficit. Plant Physiol 145: 1533–1548.
63. Colmer TD, Peeters AJ, Wagemaker CA, Vriezen WH, Ammerlaan A, et al. (2004) Expression of alpha-expansin genes during root acclimations to O2 deficiency in Rumex palustris. Plant Mol Biol 56: 423–437.
64. Ding X, Cao Y, Huang L, Zhao J, Xu C, et al. (2008) Activation of the indole-3-acetic acid-amido synthetase GH3-8 suppresses expansin expression and promotes salicylate- and jasmonate-independent basal immunity in rice. Plant Cell 20: 228–240.
65. Fudali S, Janakowski S, Sobczak M, Griesser M, Grundler FM, et al. (2008) Two tomato alpha-expansins show distinct spatial and temporal expression patterns during development of nematode-induced syncytia. Physiol Plant 132: 370–383.
66. Sasidharan R, Chinnappa CC, Voesenek LA, Pierik R (2008) The regulation of cell wall extensibility during shade avoidance: a study using two contrasting ecotypes of Stellaria longipes. Plant Physiol 148: 1557–1569.
67. Xu J, Tian J, Belanger FC, Huang B (2007) Identification and characterization of an expansin gene AsEXP1 associated with heat tolerance in C3 Agrostis grass species. J Exp Bot 58: 3789–3796.
68. Li F, Han Y, Feng Y, Xing S, Zhao M, et al. (2012) Expression of wheat expansin driven by the RD29 promoter in tobacco confers water-stress tolerance without impacting growth and development. J Biotechnol.
69. Li F, Xing S, Guo Q, Zhao M, Zhang J, et al. (2011) Drought tolerance through over-expression of the expansin gene TaEXPB23 in transgenic tobacco. J Plant Physiol 168: 960–966.
70. Dai F, Zhang C, Jiang X, Kang M, Yin X, et al. (2012) RhNAC2 and RhEXPA4 Are Involved in the Regulation of Dehydration Tolerance during the Expansion of Rose Petals. Plant Physiol 160: 2064–2082.
71. Rolle L, Torchio F, Giacosa S, Gerbi V (2009) Modifications of mechanical characteristics and phenolic composition in berry skins and seeds of Mondeuse winegrapes throughout the on-vine drying process. Journal of the Science of Food and Agriculture 89: 1973–1980.
72. Vicens A, Fournand D, Williams P, Sidhoum L, Moutounet M, et al. (2009) Changes in Polysaccharide and Protein Composition of Cell Walls in Grape Berry Skin (Cv. Shiraz) during Ripening and Over-Ripening. J Agric Food Chem 57: 2955–2960.
73. Rizzini FM, Bonghi C, Tonutti P (2009) Postharvest water loss induces marked changes in transcript profiling in skins of wine grape berries. Postharvest Biology and Technology 52: 247–253.
74. Zoccatelli G, Zenoni S, Savoi S, Dal Santo S, Tononi P, et al. (2013) Skin Pectin Metabolism During The Postharvest Dehydration Of Berries From Three Distinct Grapevine Cultivars. Australian Journal of Grape and Wine Research. pp. In press.
75. Howe KL, Chothia T, Durbin R (2002) GAZE: a generic framework for the integration of gene-prediction data by dynamic programming. Genome Res 12: 1418–1427.
76. Rambaldi D, Ciccarelli FD (2009) FancyGene: dynamic visualization of gene structures and protein domain architectures on genomic loci. Bioinformatics 25: 2281–2282.
77. Katoh K, Toh H (2010) Parallelization of the MAFFT multiple sequence alignment program. Bioinformatics 26: 1899–1900.
78. Saitou N, Nei M (1987) The neighbor-joining method: a new method for reconstructing phylogenetic trees. Mol Biol Evol 4: 406–425.

79. Tamura K, Peterson D, Peterson N, Stecher G, Nei M, et al. (2011) MEGA5: molecular evolutionary genetics analysis using maximum likelihood, evolutionary distance, and maximum parsimony methods. Mol Biol Evol 28: 2731–2739.
80. Saeed AI, Bhagabati NK, Braisted JC, Liang W, Sharov V, et al. (2006) TM4 microarray software suite. Methods Enzymol 411: 134–193.
81. Ramakers C, Ruijter JM, Deprez RH, Moorman AF (2003) Assumption-free analysis of quantitative real-time polymerase chain reaction (PCR) data. Neurosci Lett 339: 62–66.
82. Pfaffl MW (2001) A new mathematical model for relative quantification in real-time RT-PCR. Nucleic Acids Res 29: e45.

2

SNP-Associations and Phenotype Predictions from Hundreds of Microbial Genomes without Genome Alignments

Barry G. Hall*

Bellingham Research Institute, Bellingham, Washington, United States of America

Abstract

SNP-association studies are a starting point for identifying genes that may be responsible for specific phenotypes, such as disease traits. The vast bulk of tools for SNP-association studies are directed toward SNPs in the human genome, and I am unaware of any tools designed specifically for such studies in bacterial or viral genomes. The PPFS (Predict Phenotypes From SNPs) package described here is an add-on to ***kSNP***, a program that can identify SNPs in a data set of hundreds of microbial genomes. PPFS identifies those SNPs that are non-randomly associated with a phenotype based on the χ^2 probability, then uses those diagnostic SNPs for two distinct, but related, purposes: (1) to predict the phenotypes of strains whose phenotypes are unknown, and (2) to identify those diagnostic SNPs that are most likely to be causally related to the phenotype. In the example illustrated here, from a set of 68 *E. coli* genomes, for 67 of which the pathogenicity phenotype was known, there were 418,500 SNPs. Using the phenotypes of 36 of those strains, PPFS identified 207 diagnostic SNPs. The diagnostic SNPs predicted the phenotypes of all of the genomes with 97% accuracy. It then identified 97 SNPs whose probability of being causally related to the pathogenic phenotype was >0.999. In a second example, from a set of 116 *E. coli* genome sequences, using the phenotypes of 65 strains PPFS identified 101 SNPs that predicted the source host (human or non-human) with 90% accuracy.

Editor: Michael Edward Zwick, Emory University School Of Medicine, United States of America

Funding: This author has no support or funding to report.

Competing Interests: The author has declared that no competing interests exist.

* E-mail: barryghall@gmail.com

Introduction

kSNP [1,2], is a program that identifies SNPs in bacterial and viral genomes (finished genomes, genome assemblies, or raw reads) without the necessity for genome alignment or reference genomes. *kSNP* v2 can analyze over 100 bacterial genomes in under 16 hours on a typical Linux cluster and in under 22 hours on a desktop Macintosh computer. *kSNP* has the further advantage of automatically generating Parsimony, Neighbor Joining, and Maximum Likelihood phylogenetic trees of the genomes. Several versions of the phylogenetic trees are provided, including those in which nodes are labeled with the number of SNPs that are found in all of the descendants of that node. kSNP automatically annotates the identified SNPs based upon the annotations in GenBank files. The number of SNPs that are identified can be enormous: over 100,000 SNPs from 288 genomes of the Rhabdoviridiae family, over 430,000 SNPS for 119 *Escherichia coli* genomes, over 1,388,000 SNPS in 207 Acinetobacter genomes [2].

SNP-association studies are the starting point for (1) identifying genes that may be responsible for specific phenotypes, such as disease traits, and (2) identifying potential gene-gene interactions. In bacteria and viruses they have the additional potential of identifying SNPs that might be incorporated into PCR probes for the purpose of quickly and inexpensively assessing phenotypes that are difficult, expensive or time consuming to determine by classical means, such as virulence or drug resistance in slow-growing organisms such as *Mycobacterium tuberculosis*.

Here I describe the PPFS package that identifies SNPs that are non-randomly associated with a phenotype. It is an "add on" to ***kSNP*** and has the following characteristics: (1) it runs seamlessly with ***kSNP***, (2) not all SNPs have to be present in all genomes, (3) not all genomes need to have a known phenotype, (4) it takes advantage of the fact that *kSNP* can provide annotations of most SNPS and incorporates those annotations into its analysis, (5) it is fast, requiring less than 10 minutes to analyze over 418,500 SNPs. The PPFS package of executables for Mac OS X and Linux operating systems, source code, and a detailed User Guide are freely available at https://sourceforge.net/projects/ppfs.

As a proof-of-principle I illustrate the uses of the PPFS package by identifying 207 SNPs associated with the phenotype "pathogenic" for a set of 68 *Escherichia coli* genomes. Pathogenic *E. coli* include EHEC (enterohemorrhagic), ETEC (enterotoxigenic) EPEC (enteropathogenic), ExPec (extraintestinal pathogenic), EAEC (entero-adherent), AIEC (adherent-invasive), UPEC (uropathogenic), and *Shigella*. Accession numbers, phenotypes and references for those phenotypes are included in Table S1A. I also show that when the phenotypes of only 36 of those genomes are known to the PPFS package, it predicts the phenotypes of the remaining strains with >94% accuracy. Finally, it identifies the 97 SNPs that are likely to be causally related to the pathogenic phenotype. In a second example, from a set of 116 *E. coli* genome sequences, using the phenotypes of 65 strains PPFS identified 101 SNPs that predicted the source host (human or non-human) with 90% accuracy. Accession numbers are included in Table S1B.

Methods

The PPFS Package consists of five programs: ***PPFS, PickPhenotypeSubset, GetSNPprobs, DiagnosticSNPs*** and ***CausalSNPs***. It uses the output from *kSNP*, and must be run from with the *kSNP* output-files folder (directory). Using the full PPFS package requires that MEGA5 or MEGA 6 (available free at http://www.megasoftware.net/) be used to estimate ancestral states at internal nodes of an ML tree.

The user must provide a manually constructed file (.pheno file) in which the phenotype of each genome is listed either as positive (1), negative (0), or unknown (?).

PPFS serves as a "front end" that takes the command line arguments, then runs ***PickPhenotypeSubset, GetSNPprobs,*** and ***DiagnosticSNPs*** without direct interaction with the user. After running those programs ***PPBS*** collects the output files of those programs into a folder (directory) named PPFS.

PickPhenotypeSubset is used to select, from those genomes whose phenotypes are known, a random subset of genomes that will be used by *GetSNPprobs* to calculate, for each SNP, the probability that the alleles of that SNP are randomly associated with the phenotype. Through ***PPFS*** the user chooses the approximate fraction of strains to include in that subset. If the user sets the fraction to be included at 0.5, then for each strain whose phenotype is known in the.pheno file there is a 50% probability that its phenotype will be changed to unknown in the output subset.pheno file. That file thus provides a set of "unknown phenotype" genomes whose phenotype can be predicted by the *DiagnosticSNPs* program, allowing those predicted phenotype to be compared with the phenotypes in the.pheno file in order to assess the accuracy with which phenotypes are predicted.

GetSNPprobs uses the output file from *PredictPhenotypeSubset* to calculate, for each SNP, the probability that the alleles of that SNP are randomly associated with the phenotype. Not all SNP sites are equally informative. For some sites the SNP alleles are distributed randomly with respect to the phenotype, with others the SNP alleles are biased. An extreme bias would be one in which all of the positive strains had an A and all of the negative strains had a G. A χ^2 test is used to determine the probability that a SNP is distributed randomly with respect to phenotype.

Sites with low probabilities of the SNPs being randomly distributed with respect to phenotype are much more strongly correlated with the phenotype and have much more resolving power in predicting phenotypes than do sites with high probabilities. The χ^2 probabilities of each SNP are written to an output file.

GetSNPprobs writes two output files: (1) a.SNPs file that is the same as the input.fasta file except in a different format, and (2) a.ppcs (posterior probabilities chi-square) file. The.ppcs file lists, for each SNP, the probability that the alleles is distributed randomly with respect to the phenotype, and for each possible character state (ACGT-) the probability that a genome is positive and the probability that it is negative given that character state for that SNP.

DiagnosticSNPs uses the original.pheno file, the subset.pheno file written by *PredictPhenotypeSubset*, and the output files written by *GetSNPprobs*, together with several files written by ***kSNP*** to identify the set of SNPs that most accurately predicts the phenotypes. Those SNPs are the "diagnostic SNPs".

The SNPs are sorted according to the χ^2 probability (p), then starting with the SNP with the lowest p, ***DiagnosticSNPs*** calculates the accuracy of the phenotype predictions. It then adds the next SNP in the sorted list and again calculates accuracy. A minimum of 50 SNPs are added, then SNP addition continues until one of several termination conditions is met: The accuracy with which positive strains are predicted (PPV) reaches a user-defined maximum (defaults to 0.98), or a user-defined number of SNPs is added (defaults to 1000), or the PPV (positive predictive value) declines below a user-defined fraction of the maximum PPV value so far (defaults to 0.97).

To understand the termination conditions it helps to understand the terms "accuracy", "positive predictive value" (PPV) and "negative predictive value" (NPV). By comparing the known with predicted phenotypes the number of true positives (TP), True Negatives (TN), False Positive (FP) and False Negatives (FN) is determined. Accuracy is (TP+TN)/(TP+FP+TN+FN). Often we are actually more interested in knowing the probability that a strains that is predicted to be positive is actually positive. That probability is PPV, TP/(TP+FP). Similarly NPV, TN/(TN+FN), is the probability that a strain predicted to be negative is actually negative.

As SNPs with gradually increasing p values are added the PPV at first increases, but as p increases the quality of the SNPs for predictions declines, so that eventually a point is reached where addition of more lower-quality SNPs reduces the PPV. PPV does fluctuate somewhat, so addition is not terminated until PPV declines enough below its maximum value to be confident that the decline is real.

DiagnosticSNPs writes several output files: (1) a DiagnosticSNPs.summary file (for an example see Table S2) that shows accuracy, PPV and NPV at each step as new SNPs are added; the final accuracy, PPV and NPV for those genomes whose phenotypes are known but were shown as unknown in the subset.pheno file; the predicted phenotype for each genome; and the accuracy, PPV and NPV averaged over all of the genomes. (2) a DiagnosticSNPs.call file (for an example see Table S3) that gives the predicted phenotype for each genome, but groups those predicted to be positive and those predicted to be negative. (3) a DiagnosticSNPs.list file that provides some annotation information for each diagnostic SNP (SNP ID, in a protein or not, synonymous or non-synonymous, protein accession number, and description of protein function). and (4) a DiagnosticSNPs.info file that provides the same information as the.list file but is restricted to non-synonymous substitution SNPs. (5 & 6) a SNPset.fasta and a SNPtranlations.txt file that are used by *CausalSNPs* and are discussed below.

CausalSNPs identifies those diagnostic SNPs that are most likely to be causal with respect to the phenotype. Although predictive, the diagnostic SNPs may not be causally related to phenotype, and may only reflect evolutionary history. For instance, a SNP that arose along the same branch that a change in phenotype occurred may be perfectly associated with that phenotype but be completely unrelated to the reason that phenotype occurred. ***CausalSNPs*** is run from within the PPFS folder that was created by ***PPFS***.

It is often important to use SNPs to identify genes that are causally related to a phenotype. *CausalSNPs* is based on the assumption that diagnostic SNPs that are causally related to the phenotype are more likely than the average diagnostic SNP to change allele state along the same branches along which the phenotype changes. In order to determine the branches along which SNPs and the phenotype change it is necessary to infer the ancestral state of each SNP and of the phenotype at each internal node.

MEGA 5 [3] is used to estimate an ML tree from the SNP data, and then to calculate the ancestral state of each SNP at each internal node in order to identify the change in SNP allele across a branch from an ancestral to a descendant node. MEGA 5 cannot handle datasets with hundreds of genomes and hundreds of thousands of SNPs, so a reduced data set is written to the SNPset.fasta file by ***DiagnosticSNPs***. In that fasta file the sequence for each genome consists of the diagnostic SNPS, followed by a character that indicates the *predicted* phenotype of that genome, followed by a random sample of the SNPs. The number of random SNPs is 25 times the number of diagnostic SNPs, so if there are 100 diagnostic SNPs each sequence in the SNPset.fasta file will be 2601 characters, with the final 2500 having been randomly drawn from the original full set of SNPs. MEGA is run from within the PPFS folder that was created by PPFS.

In order to calculate ancestral states the ML tree estimated by MEGA must be rooted so that it has direction from ancestral to descendant nodes [4]. ML trees are unrooted, so by default MEGA roots the tree at its midpoint. Midpoint rooting assumes a molecular clock, an assumption that is often invalid [4]. It is far preferable to root the tree manually in MEGA using an outgroup genome. That outgroup may either be included in the kSNP analysis, or a separate KSNP analysis of the same data plus an outgroup can be used to produce an ML tree and that tree can be used to identify the branch along which the root should be manually placed in MEGA (virtual rooting). A brief protocol for estimating phylogenetic trees with MEGA 5 has recently been published [5], and an extensive and detailed description of its use is included in [4].

MEGA writes an output file of Most Probable Sequences that is used by *CausalSNPs*. See the PPFS User Guide for detailed instructions for using MEGA to estimate ancestral states from the SNPset.fasta file.

CausalSNPs parses the MEGA output file to identify the changes in the state of the allele of each diagnostic SNP across each branch, and to identify the change in phenotype across each branch. For each SNP *CausalSNPs* determines the χ^2 probability p that the SNP changed randomly across the same branches across which the phenotype changed. It is necessary to consider the direction of the allele change in calculating those probabilities. For instance, if the phenotype changed from negative to positive across 10 branches, and the allele state of a particular diagnostic SNP changed along those same branches it could reasonably be inferred that the change in allele caused the change in phenotype. Howsoever, if five of those changes were from A to G, and the other 5 from G to A, then it would be unlikely that that SNP caused those changes. It is therefore necessary to classify each change as a "positive" or "negative" change. Changes from any base to the absence of the allele are considered negative, and base changes are considered positive if the value at the descendant node is greater than that at the ancestral node and negative if lower. The values of bases are A<C<G<T. The ordering of the values is arbitrary. Each SNP has only three possible states: one of two bases, or missing. For the purpose of calculating χ^2 p values it is irrelevant whether a change from A to G is considered positive or negative, it only matters that the determination is consistent.

The diagnostic SNPs are ordered from lowest to highest p, with lower p values indicating a higher likelihood that the SNP is causal.

CausalSNPs writes four output files that are collected into a folder (directory) named "CausalSNPs Output Files":

(1) CausalSNP.report (for an example see Table S4) gives the identities of the branches across which the phenotype changed and direction of those phenotype changes. It then gives, for each diagnostic SNP, the SNP number (the position of the SNP in the sequences in SNPset.fasta), the corresponding SNP ID in the kSNP files (the SNPtranslations.txt file written by *DiagnosticSNPs* keeps track of the SNP number and the corresponding SNP ID), the p that allele changes were random over the branches along which the phenotype changed, and whether the SNP is synonymous or non-synonymous, the protein accession number, and a brief description of the protein function.
(2) A "ChiSquare_details.txt" file (for an example see Table S5) that gives for each SNP the observed and expected number of positive changes, negative changes, no changes, deletions, and additions of the SNP across the branches where the phenotype changed, the χ^2 value, the degrees of freedom and p.
(3) A "Changes.txt" file (for an example see Table S6) that give the change of each diagnostic SNP at each branch.
(4) A "Branches.txt" file (for an example see Table S7) that gives the node number (as listed by MEGA) at each end of each branch.

Genomes used in this Study

The Genbank accession numbers, corresponding ID on the tree in Figure 1, phenotype and reference for that phenotype are in Table S1.

Results and Discussion

SNP Analysis using kSNP

Data set 1. A set of 68 finished *E. coli* and *Shigella* genomes was analyzed on a Macintosh iMac computer with a 3.4 GHz Intel core i7 processor and with 16 GB of RAM as described in the User Guide to *kSNP*. All of those genomes were listed in a file of finished genome names, enabling *kSNP* to automatically retrieve and utilize the annotations for those genomes. The *kSNP* analysis required 10.3 hours, identified 418,500 SNPs, and is described in detail in [2].

In order to root the tree in MEGA *kSNP* was used to analyze the same 68 *E. coli* and *Shigella* genomes together with the finished genome of *Escherichia fergusonii* strain ATCC35469. Accession numbers, phenotypes and references for the genomes are in Table S1A of Supplementary Materials.

Data set 2. A set of 116 *Escherichia coli* genomes, including 8 finished genomes and 108 genome assemblies, was analyzed on the same Mac computer. The analysis required 15.6 hours and identified 470,806 SNPs. Accession numbers are in Table S1B.

PPFS Analysis of Data Set 1 in which the Phenotypes are Pathogenic vs Commensal

In the file Eco68_patho.pheno human pathogenic strains were listed as the positive phenotype (1), others were listed as the negative phenotype (0), and one strain had an unknown phenotype. *PickPhenotypeSubsets* generated a subset of 36 strains whose phenotypes were given in the Eco68_patho_50%.pheno file (see Methods). That file also included 31 strains whose phenotype were changed to unknown in the Eco68_patho_50%.pheno file and that were used to calculate accuracies of phenotype predictions.

GetSNPprobs was used to calculate, for each of the 418,500 SNPS the χ^2 probability that the alleles were distributed randomly with respect to phenotype.

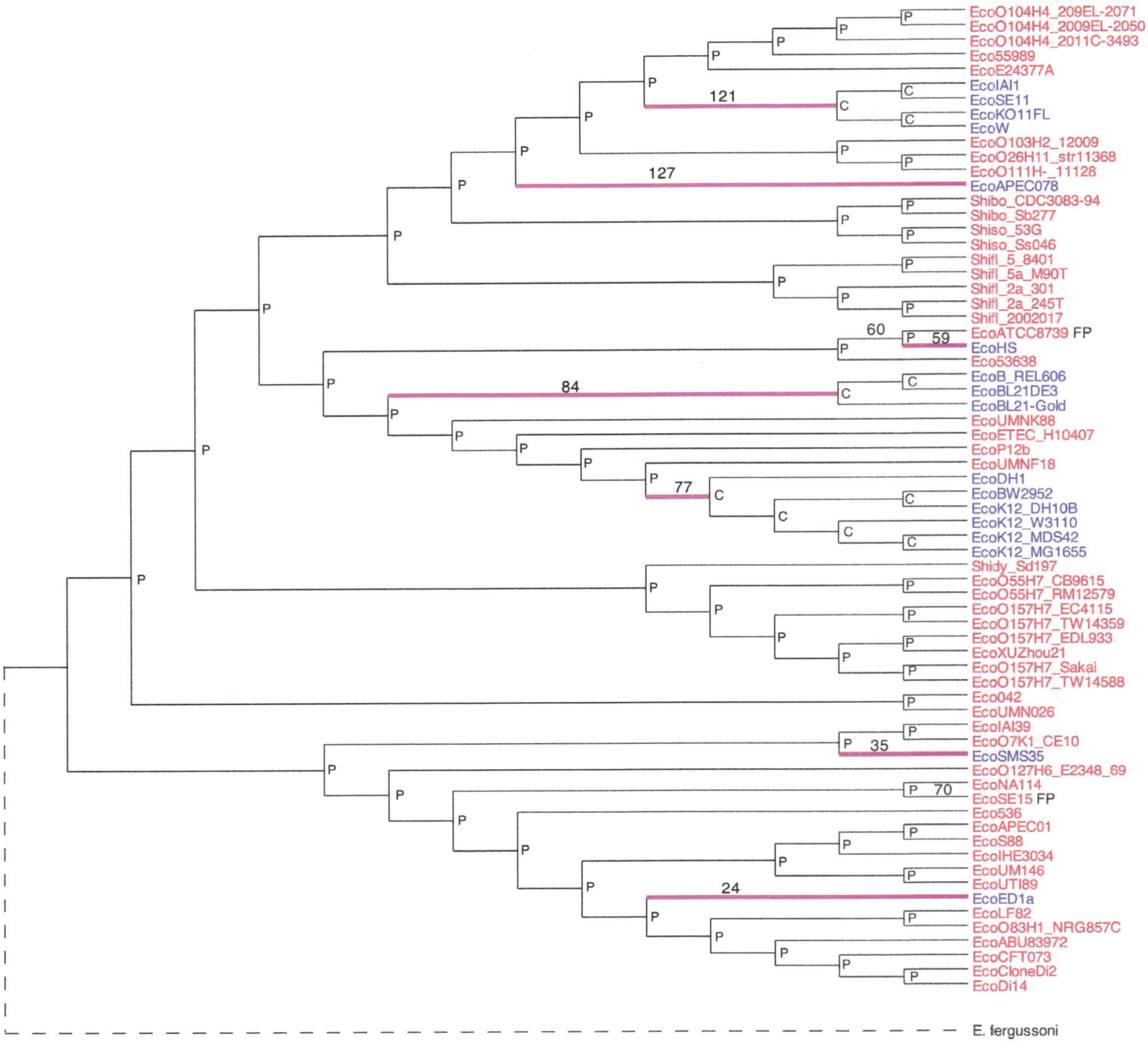

Figure 1. ML tree estimated by MEGA 5. The *predicted* phenotypes are indicated by the text color of the taxa: Red = pathogenic, Blue = commensal. The ancestral phenotypes are indicated by P (pathogenic) or C (commensal) at the internal nodes. Branches along which the phenotype changed are indicated in magenta. Numbers above some branches indicate the branch number given in Table S7. The dashed line indicates the virtual rooting with *Escherichia fergusonii*.

DiagnosticSNPs identified 207 diagnostic SNPS that predicted the pathogenic phenotype for the 31 genomes whose phenotypes were unknown in the Eco68_patho_50%.pheno file. The accuracy of those predictions was 0.944, PPV was 0.931 and NPV was 1.0. Over all, of the 67 genomes whose phenotype is known, accuracy was 0.970, PPV was 0.96 and NPV was 1.0. Out of those 67 genomes there were two false positives (Table S3).

MEGA 5 analysis. An ML tree was estimated by MEGA 5 from the SNPset.fasta file of 5,384 SNPs per genome using the GTR+G model with 5 discrete Gamma categories. It was 94.5% topologically congruent with the ML tree estimated by *kSNP* from 418,500 SNPs as determined by the Compare2Trees web application (http://www.mas.ncl.ac.uk/~ntmwn/compare2trees/index.html) which is an updated version of the application reported in [6]. The tree was virtually rooted using *Escherichia fergusonii* strain ATCC35459, and ancestral states of the diagnostic SNPS and the phenotype were estimated (Fig. 1).

The analysis of ancestral phenotypes makes it clear that pathogenicity was the *E. coli* ancestral state. All seven changes were from pathogenic to commensal. Figure 1 suggests that pathogenicity is the ancestral state of *E. coli*, but that conclusion may be subject to sampling bias. Most of the sequenced *E. coli* strains were chosen specifically because they are pathogenic or because they are well known laboratory strains. Sequencing of a large number of commensals, including samples from healthy non-humans, plus environmental samples, would cast light on the concern with sampling bias.

CausalSNPs identified 97 SNPs whose probability of changing randomly across the branches was <0.001 (Table S4). That table lists the diagnostic SNPs in increasing order of p, the probability that the SNP allele changed randomly over the eight branches along which the pathogenicity phenotype changed. $1-p$ can be taken as the probability that the SNP is causally related to the pathogenicity phenotype.

The list of likely causal SNPs in Table S4 should not be treated as definitive, but should instead be read with some judgment. The first SNPs in that list, #81, is from a bacteriophage lambda tail assembly protein, for which no obvious role in pathogenicity leaps to mind. The next SNP, #8, is in a nitrogen assimilation protein, accession number CBG35053.1. Base changes at that SNP are synonymous, but examination of the χ^2 details in Table S5 shows that on 6 of the 7 branches where phenotype changed the SNP change was deletion of that SNP. Furthermore, there were a total of nine diagnostic SNPs in that protein, all of which had p values < 0.001. Similarly, the next SNP in the list, #26, is in a putative invasin, CBG35046.1. There are 10 diagnostic SNPs in that protein (Table S4), seven of which have p values <0.001. It appears quite reasonable to conclude that loss of invasin and the nitrogen assimilation protein are directly related to loss of pathogenicity.

Effects of false positives. Table S3 shows that there were two false positives, strains EcoATCC8739 and EcoSE15, indicated on Fig. 1 by appending FP to the genome ID. It is clear that the change from pathogenic to commensal phenotype occurred along branch 60, not branch 59; and that phenotype changed along branch 70 as well. Both false positives were strains that had been designated as unknown phenotype in the Eco68_patho_50%.-pheno file. For such strains the accuracy was only 0.944. False positive and false negatives affect the correct assignment of the branches along which phenotypes change, and that in turn affects correct calculation of the χ^2 probabilities. The reliability with which causal SNPs are identified is directly related to the accuracy with which "unknown" strains are called by the *DiagnosticSNPs* program. Undoubtedly, there will be many data sets for which the phenotypes of unknown strains cannot be called reliably. When the accuracy of those calls is <0.9 I do not suggest that identification of causal SNPs should be taken seriously.

The PPFS package helps identify SNPs that are associated with phenotypes, but does not attempt to detect interactions among those SNPs (epistasis). Multifactor Dimensionality Reduction (MDR) is a statistical approach to detecting such interactions [7] and is implemented by the free open-source MDR software package http://www.multifactordimensionalityreduction.org/. However "When you have more than 100 SNPs an exhaustive search may not be practical unless you are willing to wait days or even weeks for a run to finish. When the number of SNPs exceeds 10,000 an exhaustive search of all 3-way and higher combinations may be infeasible, even with a parallel computer." [8] The PPFS package is a useful tool to reduce the number of SNPs to a level suitable for MDR analysis. MDR has a feature to downselect SNPs based on Chi-square probabilities, but it only considers core SNPs, those present in all strains, and only considers strains that have a phenotype, so does not make predictions about un-typed strains, limiting the SNPs and strains that can be included in an MDR analysis.

It is somewhat surprising that a phenotype as vague as "pathogenic" can be accurately predicted on the basis of a small set of diagnostic SNPs in a set of sequenced genomes. There may appear to be little point in predicting the phenotypes of completely sequenced genomes; after all, given the effort and expense of sequencing a microbial genome, the phenotype is likely to be known already. There are three reasons that phenotype prediction can be valuable today and is likely to become more valuable in the future. First, phenotypes are rarely reported in genome annotations and many genomes are submitted without reference to papers where those phenotypes might be available. The PPFS package permits prediction of the phenotypes of those genomes. Second, even when phenotypes are reported they are not necessarily the phenotype of interest to any particular investigator. Third, when phenotypes are difficult or time consuming to determine; e.g. for very slow growing organisms such as *M. tuberculosis* or for virulence phenotypes of viruses, PCR based assays can be designed to detect allele variants and thus to predict phenotypes. If a sufficient number of genome sequences are known to allow reliable genome predictions to be made, then those predictions can be applied to complete genomes, genome assemblies and raw-read genomes in the various databases. The cost of bacterial genome sequencing now approaches $100 US per genome and is likely to decrease significantly in the near future. It is quite feasible for a laboratory to characterize a phylogenetically diverse collection of a species with respect to a clinically important phenotype such as virulence, then to sequence 25–30 strains of each phenotype and to analyze those sequences with *kSNP* and the PPFS package. That analysis could then be used to predict the virulence of strains already in the database. If sequencing costs continue to decrease it may even become practical to routinely sequence genomes in epidemiological investigations. If the phenotype of interest is either expensive or time consuming to determine, the results of genome sequencing together with the information provided by a *kSNP*-PPFS analysis could provide a rapid assessment of the probability of exhibiting the clinically important phenotype.

PPFS Analysis of Data Set 2 in which the Phenotypes are Human vs Non-human Host Source

E. coli is often used as a surrogate to indicate the presence of human pathogens in water resources [9]. When analyzing water resources it can be important to distinguish human (sewage, septic tank effluent, etc) source of contamination from non-human sources such as local fauna and agricultural resources (R. Norris and T. N. Fields, personal communication). Availability of DNA signatures for *E. coli* from human vs non-human sources would be valuable for distinguishing *E. coli* that indicate human waste contamination from those that are normally resident in soils near streams [10]. I have analyzed a set of 116 *E. coli* genomes in which the source host is specified in the annotations to determine whether there are SNPs that can predict human vs non-human origin.

In the Human.pheno file 61 strains were listed as the positive phenotype (1), i.e. coming from a human host, and 55 were listed as the negative (0) phenotype, i.e. coming from a non-human host. PickPhenotypeSubsets generated a subset of 65 strains whose phenotypes were given in the Human_50%.pheno file (see Methods). That file also included 51 strains whose phenotype were changed to unknown in theH_50%.pheno file and that were used to calculate accuracies of phenotype predictions. GetSNP-probs calculated, for each of the 470,806 SNPS the χ^2 probability that the alleles were distributed randomly with respect to phenotype. DiagnosticSNPs identified 101 diagnostic SNPS that predicted the phenotype for the 51 genomes whose "source" phenotypes were unknown in the Human_50%.pheno file. The accuracy of those predictions was 0.845, PPV was 0.793 and NPV was 0.91. Over all, of the 116 genomes accuracy was 0.897, PPV was 0.877 and NPV was 0.923. Out of those 116 genomes there were seven false positives.

Although accuracy, PPV and NPV are not as high as in data set 1, for environmental sampling purposes where assessments would be based on determinations on many individual isolates, accuracies of about 90% are probably high enough to be useful.

Conclusions

SNP association studies are not an end in themselves, but are a means to identify candidate genes for experimental studies related to the phenotype of interest. With the release of *kSNP* for SNP identification and the PPFS package for SNP association studies, the major barrier to SNP association studies in microorganisms is no longer the availability of appropriate tools; it is the culture of Genomic science itself. There are very few microorganisms for which both complete genome sequences and biological information are available in the databases. There are a plethora of databases that include or point to microbial genome sequences; e.g. NCBI's Genome database, SRA database, Bioprojects database, bacterial genome assembly database, but it is rare for any phenotypic information to be included in any of those files and most do not even refer to published papers that might shed light on phenotypes. The Broad Institute *E. coli* antibiotic resistance database http://www.broadinstitute.org/annotation/genome/escherichia_antibiotic_resistance/MultiHome.html provides detailed information on genotypes of hundreds of strains, but does not release the antibiotic resistance phenotypes. On the other side of the coin, the Network on Antimicrobial Resistance in *Staphylococcus aureus* (NARSA) http://www.narsa.net/control/member/allapprovedisolates provides antibiotic resistance phenotypes for hundreds of *S. aureus* strains, but genome sequences are available for only about a dozen of those strains. Genome sequencing has become so fast and so inexpensive that we seem to have lost track of the reasons for which those individual strains were sequenced. Obviously a complete description of the biological properties of each sequenced genome is impossible, but at the very least there should be a paragraph explaining why the particular strain was chosen to be sequenced. Was it from a sick or healthy individual, from an environmental sample (what environment), etc?

Ideally, an international repository would be created for maintenance of all strains (not strain information, but the organisms themselves) whose genomes are added to the genome sequence databases. Availability of the organisms would permit investigators to determine the phenotypes of interest of strains that have already been sequenced, and thus to conduct valuable SNP association studies. Failing that, the annotations of all genomes in the database should include contact information for an individual who is committed to providing a sample of the strain upon request.

Acknowledgments

I am grateful to Shea Gardner for encouraging me to develop PPFS and for testing early versions of the PPFS package. I am grateful to Rebecca Norris and Tina N. Fields for bringing the issue of human vs non-human source *E. coli* to my attention.

Author Contributions

Conceived and designed the experiments: BGH. Performed the experiments: BGH. Analyzed the data: BGH. Contributed reagents/materials/analysis tools: BGH. Wrote the paper: BGH.

References

1. Gardner SN, Slezak T (2010) Scalable SNP analyses of 100+ bacterial or viral genomes. J Forensic Science 1: 107–111.
2. Gardner SN, Hall BG (2013) When whole-genome alignments just won't work: kSNP v2 software for alignment-free SNP discovery and phylogenetics of hundreds of microbial genomes. PLoS One 8.
3. Tamura K, Peterson D, Peterson N, Stecher G, Nei M, et al. (2011) MEGA5: molecular evolutionary genetics analysis using maximum likelihood, evolutionary distance, and maximum parsimony methods. Mol Biol Evol 28: 2731–2739.
4. Hall BG (2011) Phylogenetic Trees Made Easy: A How-To Manual, Fourth Edition. Sunderland, MA: Sinauer Assoc.
5. Hall BG (2013) Building Phylogenetic Trees from Molecular Data with MEGA. Mol Biol Evol 30: 1229–1235.
6. Nye TM, Lio P, Gilks WR (2006) A novel algorithm and web-based tool for comparing two alternative phylogenetic trees. Bioinformatics 22: 117–119.
7. Moore JH, Gilbert JC, Tsai CT, Chiang FT, Holden T, et al. (2006) A flexible computational framework for detecting, characterizing, and interpreting statistical patterns of epistasis in genetic studies of human disease susceptibility. J Theor Biol 241: 252–261.
8. Moore JH (2006) Epistasis Blog. http://compgen.blogspot.com/2006_11_01_archive.html
9. Sinclair RG, Rose JB, Hashsham SA, Gerba CP, Haas CN (2012) Criteria for selection of surrogates used to study the fate and control of pathogens in the environment. Appl Environ Microbiol 78: 1969–1977.
10. Luo C, Walk ST, Gordon DM, Feldgarden M, Tiedje JM, et al. (2011) Genome sequencing of environmental Escherichia coli expands understanding of the ecology and speciation of the model bacterial species. Proc Natl Acad Sci U S A 108: 7200–7205.

3

TASSEL-GBS: A High Capacity Genotyping by Sequencing Analysis Pipeline

Jeffrey C. Glaubitz[1]*, Terry M. Casstevens[1], Fei Lu[1], James Harriman[1], Robert J. Elshire[1¤], Qi Sun[2], Edward S. Buckler[1,3]

1 Institute for Genomic Diversity, Cornell University, Ithaca, New York, United States of America, **2** Biotechnology Resource Center Bioinformatics Facility, Cornell University, Ithaca, New York, United States of America, **3** USDA Agricultural Research Service, Ithaca, New York, United States of America

Abstract

Genotyping by sequencing (GBS) is a next generation sequencing based method that takes advantage of reduced representation to enable high throughput genotyping of large numbers of individuals at a large number of SNP markers. The relatively straightforward, robust, and cost-effective GBS protocol is currently being applied in numerous species by a large number of researchers. Herein we describe a bioinformatics pipeline, TASSEL-GBS, designed for the efficient processing of raw GBS sequence data into SNP genotypes. The TASSEL-GBS pipeline successfully fulfills the following key design criteria: (1) Ability to run on the modest computing resources that are typically available to small breeding or ecological research programs, including desktop or laptop machines with only 8–16 GB of RAM, (2) Scalability from small to extremely large studies, where hundreds of thousands or even millions of SNPs can be scored in up to 100,000 individuals (e.g., for large breeding programs or genetic surveys), and (3) Applicability in an accelerated breeding context, requiring rapid turnover from tissue collection to genotypes. Although a reference genome is required, the pipeline can also be run with an unfinished "pseudo-reference" consisting of numerous contigs. We describe the TASSEL-GBS pipeline in detail and benchmark it based upon a large scale, species wide analysis in maize (*Zea mays*), where the average error rate was reduced to 0.0042 through application of population genetic-based SNP filters. Overall, the GBS assay and the TASSEL-GBS pipeline provide robust tools for studying genomic diversity.

Editor: Nicholas A. Tinker, Agriculture and Agri-Food Canada, Canada

Funding: This work was supported by the National Science Foundation (www.nsf.gov) under the Plant Genome Research Program (PGRP) (grant numbers DBI-0820619 and IOS-1238014) and the Basic Research to Enable Agricultural Development (BREAD) project (ID:IOS-0965342), as well as by the USDA-ARS (www.usda.gov). The funders had no role in study design, data collection and analysis, decision to publish, or preparation of the manuscript.

Competing Interests: The authors have declared that no competing interests exist.

* E-mail: jcg233@cornell.edu

¤ Current address: AgResearch Limited, Grasslands Research Centre, Palmerston North, New Zealand

Introduction

The advent of next generation sequencing has elicited a revolution in biology buoyed by an advancing tidal wave of raw sequence data[1–4]. By combining the power of next generation sequencing with reduced representation[5], which focuses sequencing resources on the ends of restriction fragments, it is now possible to quickly genotype unprecedented numbers of samples even in large genome species[6–8]. Several low cost, high throughput methods that combine next generation sequencing with reduced-representation have been developed (e.g., [9–19]). Because of its relative simplicity and robustness, the genotyping by sequencing (GBS) method of Elshire et al. [12] or close derivatives thereof have already been applied in numerous species by many researchers (e.g., [7,17,20–31]). For the first time, generation of copious quantities of genotypic data for genetic experiments is no longer a bottleneck. Instead, the new bottleneck is the efficient bioinformatics analysis of the vast and ever-expanding sea of data. Opportunities to apply markers to breeding or conservation biology are now often limited only by the availability of appropriate bioinformatics tools.

To address this bioinformatics bottleneck, we implemented a GBS analysis pipeline in the Java program TASSEL[32] (version 4) that is specifically tailored to the GBS protocols of Elshire et al. [12] or Poland et al. [20]. However, the TASSEL-GBS pipeline is not limited to the specific restriction enzymes utilized in those protocols: it currently accepts 15 single restriction enzymes and 15 restriction enzyme pairs, and new enzymes are easily added. Furthermore, the TASSEL-GBS pipeline should work on nearly any restriction enzyme and barcoding approach (e.g., [33]), provided that sequence reads commence with the barcode immediately followed by the remnant of the restriction enzyme cut site (Figure 1A). Compared to other available pipelines for similar purposes [16,25,34–38] the TASSEL-GBS pipeline is specifically designed to efficiently handle large quantities of data from large numbers of samples: to date, we have analyzed more than 45,000 maize samples. The TASSEL-GBS pipeline was designed for species with a reference genome; however, it is possible to use incomplete genome assemblies consisting of numerous contigs as a pseudo-reference. For species without a reference genome, an alternative approach, appropriate for small to medium scale studies, has already been implemented in TASSEL [22].

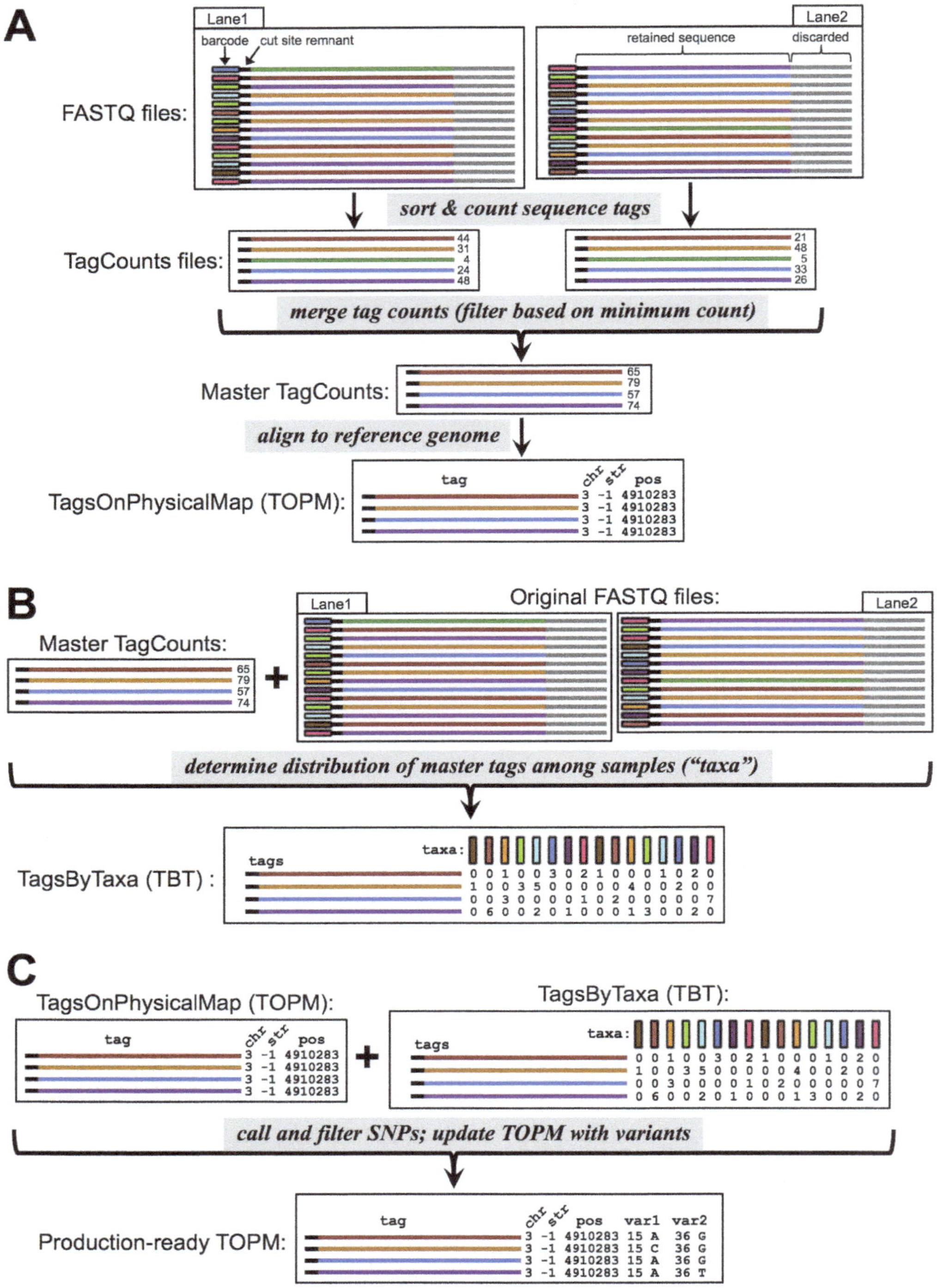

Figure 1. Schematic representation of the TASSEL-GBS Discovery Pipeline. (A) Barcoded sequence reads are processed and collapsed into a set of unique sequence tags, with one TagCounts file produced per input FASTQ file. The separate TagCounts files are then merged to form a "master" TagCounts file, which retains only those tags present at or above an experiment-wide minimum count. This master tag list is then aligned to the reference genome and a TagsOnPhysicalMap (TOPM) file is generated, containing the genomic position of each tag with a unique, best alignment. (B) The barcode information in the original FASTQ files is then used to tally the number of times each tag in the master tag list is observed in each sample ("taxon") and these counts are stored in a TagsByTaxa (TBT) file. (C) The information recorded in the TOPM and TBT is then used to discover SNPs at each "TagLocus" (set of tags with the same genomic position) and filter the SNPs based upon the proportion of taxa covered by the TagLocus, minor allele frequency, and inbreeding coefficient (F_{IT}). For each retained SNP, the allele represented by each tag in the corresponding TagLocus is recorded in the TOPM file, along with its relative position in the locus. The end product of the Discovery Pipeline is a "production-ready" TOPM that can then be used by the Production Pipeline to call SNPs.

Herein, we describe the TASSEL-GBS pipeline in detail, as implemented in TASSEL 4, and we benchmark the pipeline based upon a large-scale, species-wide analysis of maize (*Zea mays*). There were three primary motivations behind the development of this software: (1) Support the use of GBS in high-throughput, accelerated plant breeding, (2) Accommodate the high genomic diversity that is frequently encountered in species critical to agriculture and conservation, and (3) Provide an analysis platform that can be run in many contexts and with modest computational resources, such as those typically available in the developing world.

Terminology

A read is a single sequence in the FASTQ output file generated by the GBS assay

A ***good, barcoded read*** is a sequence read with a perfect match to one of the barcodes provided in a barcode key file and with no N's in the sequence following the barcode up to the trim length. Under the current implementation, reads are trimmed to 64 bp (not including the barcode).

A ***tag*** refers to a unique sequence (excluding the barcode) up to a specified length (currently 64 bp) from one or more "good, barcoded reads". A given tag is typically observed in numerous good, barcoded reads of identical sequence (up to the trim length).

For our purposes, a ***taxon*** refers to a nameable entity from which one or more DNA samples can be taken.

Design Considerations

Separation of SNP discovery and production SNP calling

Genomics-assisted, accelerated plant breeding usually consists of two separate phases of analysis: a survey of genetic diversity within a species or large breeding program to discover useful markers, followed by usage of these markers to rapidly advance generations. During the advancement cycle, time is of the essence, as thousands of samples need to be processed as quickly as possible so that decisions can be made for the next breeding cycle. The TASSEL-GBS software, by its division into Discovery (Figure 1) and Production pipelines (Figure 2), mirrors the two phases of accelerated plant breeding.

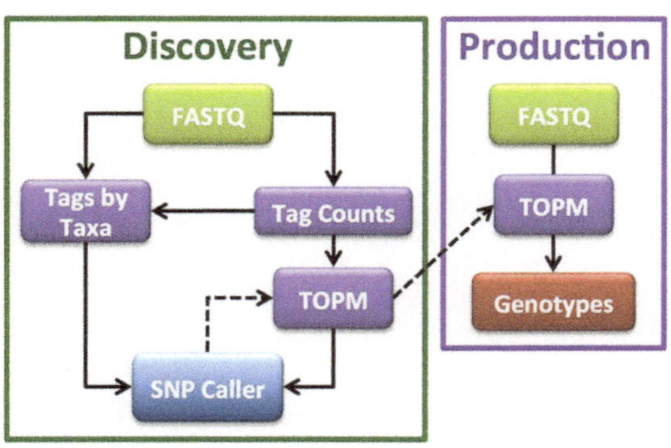

Figure 2. Relationship between the TASSEL-GBS Discovery and Production pipelines. The Discovery Pipeline is run periodically on all FASTQ files generated to date in a species, and the ascertained and filtered SNPs are stored in a "production-ready" TOPM. The Production pipeline utilizes this production-ready TOPM to quickly call SNPs either for the original samples in the Discovery Build, or for subsequent, post-Discovery samples.

The aim of the Discovery Pipeline (Figure 1) is to use the cumulative sequence data from all available samples run to date in a species (or breeding population) to discover SNPs. These SNPs are stored in a "TagsOnPhysicalMap" (TOPM) data structure, containing all of the potentially useful, unique, sequence tags, the genomic positions of the subset of the tags with unique best alignment positions, and the alleles that each useful tag represents for each discovered SNP. The "production-ready" TOPM (populated with variants for each useful tag) can then be in used in the single-step Production Pipeline to quickly produce genotypes, by determining which useful tags are present in each sample. A production-ready TOPM will be applicable to a breeding population as long as the genetic diversity present in the founders of the breeding population is well represented in the individuals that comprised the corresponding Discovery Build.

Our general approach is to periodically perform a comprehensive "Discovery Build" including all samples run to date in our study species. Performing a Discovery Build on a large number of samples is a multistep process that, depending on the number of samples, can require considerable computing resources (or time). For example, the most recent Discovery Build that we performed in maize (AllZeaGBSv2.6), comprising 31,978 samples (plus 758 blank negative controls, where TE buffer was substituted in place of a DNA sample), took 495 CPU-hours on 64 core Linux machine with 512GB of RAM (where each core was a 1.4 GHz AMD Opteron Processor 6272), plus additional time for staging of all of the input FASTQ files, etc. The Production Pipeline, in contrast, provides an avenue by which genotypes for the set of SNPs discovered in the most recent Discovery Build can be quickly generated for new samples. Running the Production Pipeline on a single FASTQ file containing sequence reads from 95 or 383 samples (plus a blank negative control) requires approximately 1 CPU-hour on a MacBook Pro with a 2.6 GHz Intel Core i7 processor and 16GB of RAM running OS X.

Favoring calling a large number of SNPs versus depth per SNP

In a GBS assay, the tradeoff, for a given genome size, between number of SNPs genotyped and the depth of coverage at each SNP is controlled by the level of multiplexing and the choice of restriction enzyme(s) [12,20]. One of our primary motivations for performing GBS (in maize and other organisms) is to enable GWAS, which requires a high density of markers, so that each causative polymorphism stands a reasonable chance of being in LD with one or more markers [39]. Hence, we favor increasing the number of markers at the expense of depth and thus designed the TASSEL-GBS pipeline with low coverage data in mind. The resultant missing data and under-calling of heterozygotes can be compensated for by redundant coverage of haplotypes at high marker density, facilitating imputation.

The likelihood of success of this imputation-based strategy depends on the number and length of homozygous, identical by descent (IBD) segments present in the study population. This, in turn, depends on the demographic history of the population. Imputation is least challenging in biparental populations consisting of RILs [40]. Imputation can also be relatively straightforward in a set of homozygous inbred individuals descending from a limited number of founders (e.g., modern maize lines; [24]). Even in outcrossing species, extensive homozygous IBD stretches can be present if a population bottleneck of sufficient severity occurred at some point in the demographic history of the study population

[41]. Such bottlenecks are common in outcrossing crop species, associated with domestication (e.g., [42]), modern improvement (e.g., [43]), or the development of a breeding population [44]. For unrelated (or distantly related) individuals (either inbred or outcrossed), large sample sizes improve the prospect of successful imputation: the more individuals genotyped, the more likely a given haplotype will be present in multiple individuals in homozygous form. Marker density and depth of coverage per marker need only be high enough to permit recognition of these homozygous IBD segments.

The TASSEL-GBS pipeline is thus optimized for low sequencing depth (0.5 to 3×) over a large number of markers in a large sample of individuals. However, it is flexible enough for the analysis of higher coverage data either from a small genome species, or from a large genome species where a lower level of multiplexing, replicate runs of the same library preps, and/or a less frequently cutting restriction enzyme (or enzyme combination) are used.

Capacity for large numbers of markers and samples

Our analysis strategy favoring large numbers of markers scored at low depth in a large sample of individuals requires the TASSEL-GBS pipeline to be able to handle very large data structures (Table 1). The largest data structure is the "TagsByTaxa" (TBT) object, which records the observed depth in each individual for each potentially useful sequence tag (where "taxa" in "TagsBy-Taxa" refers to an individual sample from a particular GBS library prep). Our most recent Discovery Build in maize (AllZeaGBSv2.6) comprised 97.5 million potentially useful tags from 31,978 samples (plus 758 blank negative controls), with the depth for each tag in each sample recorded as a single byte (prior to compression). In order to efficiently store and retrieve data from a matrix of this size (3.2 TB of uncompressed, raw data), we used the HDF5 storage format (http://www.hdfgroup.org). We also implemented a rapid and efficient run length compression algorithm to further decrease the storage size. At low depth and with high genetic diversity (numerous sequence tags per locus), the TBT is a sparse data matrix consisting mostly of zeros; our run length compression algorithm takes advantage of this. The sequence tags themselves are stored in a binary format requiring only 16 bytes to hold 64 bases (2 bits per base), prior to compression. Thus, 64 base tags can be conveniently held in two "longs" in Java. The 64 base upper limit on tag lengths in the current implementation of TASSEL-GBS will be lifted in the near future, which will be helpful for study organisms with limited diversity.

Capability to run on modest computing infrastructure

GBS provides an unprecedented opportunity for genomic markers to be used by researchers working in numerous species, including researchers in the developing world. Many of these potential users do not have access to big memory computers or large clusters. The TASSEL-GBS Discovery and Production pipelines can be run on a Linux, Mac or Windows computer with 8–16 GB of RAM. The main demand with respect to the amount of RAM required is the number good, barcoded reads in a typical FASTQ file produced from GBS, which currently ranges from 200–300 million. The small memory footprint required by the TASSEL-GBS pipeline, in relation to its high capacity in terms of number of markers and individual samples, renders it useful to a broad array of users without access to sophisticated computing infrastructure, a target user group which may include breeders carrying out genomic selection experiments.

Avoiding redundant alignment of identical reads

Genotyping approaches based on whole genome, rather than reduced-representation, sequencing typically align all or most of the reads produced to the reference genome prior to calling SNPs (e.g., [36]). In contrast, the TASSEL-GBS pipeline first collapses all of the reads into a master tag list containing all of the sequence tags present at or above a user-specified minimum count, tallied across all of the samples in the Discovery Build (Figure 1A). Each tag in this master tag list is then aligned to the reference genome. This strategy dramatically reduces the computation time devoted to alignment, and permits the use of more computationally expensive alignment algorithms. As the master tag list for our most recent maize Discovery Build (AllZeaGBSv2.6) consisted of 97.5 million tags, distilled from more than 46.8 billion sequence reads, a 392-fold reduction in computational time devoted to alignment was achieved in our case.

Favoring allelic redundancy over quality scores

Quality scores produced by the Illumina base caller are strongly negatively correlated with position in the read. In most whole genome sequencing approaches the position of a particular SNP in different reads is essentially random. In contrast, a GBS SNP has a consistent position in each read, as all of the GBS tags from a particular genomic location (a "TagLocus") that are used to discover and call SNPs originate from the same restriction enzyme cut site and have the same strand orientation. Consequently, the more distal SNPs in a GBS tag tend to have lower quality scores. If quality scores were used to filter GBS reads, these more distal SNPs might end up with lower depth of coverage. Furthermore, the quality scores frequently are not indicative of true quality [45,46]. Hence, rather than using quality scores to filter out bad reads, the TASSEL-GBS pipeline instead relies on the number of times a given tag has been observed as an indicator of sequence quality. GBS sequence tags that occur in a minimum, user-specified, number of reads across all of the samples in a Discovery Build are deemed as potentially useful and are kept for further processing (alignment to the genome and SNP calling). Illumina

Table 1. Size of the key data structures used by the TASSEL-GBS pipeline for a recent maize "Discovery Build" (AllZeaGBSv2.6).

Data Structure	Data Points	Compressed Size	Uncompressed size
Sequencing Files	4,679 Gnt[1]	3.9 TB	11.6 TB
Tags by Taxa (TBT)	3.2 trillion[2]	82.0 GB	3,198 GB
Tags on Physical Map (TOPM)	10.2 billion[3]	6.44 GB	14.0 GB

[1]Giganucleotides
[2]Read depths for 97,502,532 tags across 32,736 taxa (including 758 blank negative controls.
[3]105 data points per tag (with each base counted as one data point) times 97,502,532 tags.

quality scores are ignored, and therefore do not need to be tracked throughout the pipeline.

Population genetic-based filtering of putative SNPs

Putative SNPs from GBS may be of low quality for multiple reasons. The sequencing error rate for a SNP may be high because of its distance from the read start and/or its immediate sequence context [47,48]. Alternatively, paralogous sequence tags from different loci may be mistakenly aligned to a single TagLocus, resulting in spurious SNPs. To detect and filter out error-prone SNPs, the TASSEL-GBS pipeline relies on population-genetic parameters such as the minor allele frequency (MAF) and, in particular, the inbreeding coefficient (or "index of panmixia"), F_{IT}. Filtering based upon minimum MAF can remove spurious SNPs arising solely from sequencing error. Artifactual SNPs originating from paralogous tags will tend to be excessively heterozygous and can thus be distinguished on the basis of low F_{IT}.

These population-genetic filters are most powerful if a substantial proportion of the samples consist of inbred lines. Among inbred samples, both error-prone SNPs and spurious SNPs originating from paralogous tags will appear to be excessively heterozygous. The Discovery SNP caller in the TASSEL-GBS pipeline allows the user to specify which samples are highly inbred, and uses this subset of inbreds to calculate F_{IT} and apply the minimum F_{IT} filter. Additional, related, filters can also be applied enforcing a minimum "inbred coverage" (proportion of the inbred samples to be non-missing at the SNP) and maximum "inbred heterozygosity score" (= *nInbredHets*/[*nInbredsGT1ReadHomoMin* + *nInbredHets* + 0.5], where *nInbredHets* is the number of inbred taxa that are scored as heterozygous for the SNP, and *nInbredsGT1ReadHomoMin* is the number of inbred taxa with a read depth >1 for the SNP that are scored as minor allele homozygotes).

Implementation

Discovery Pipeline

The Discovery Pipeline consists of multiple steps, with each step (with the exception of alignment to the reference genome) being carried out by a TASSEL "plugin" that can be run from the TASSEL 4 Standalone command line interface. Detailed documentation on the function and usage of each individual plugin is available at www.maizegenetics.net/tassel/docs/TasselPipelineGBS.pdf. Rather than describe each individual plugin, here we describe the main functions of the pipeline (Figure 1) and their key features.

In order to have maximal power to discover and filter SNPs, we advocate running the Discovery Pipeline (i.e., performing a "Discovery Build") at the species-wide level, with all samples sequenced to date, across multiple FASTQ files. Each FASTQ file contains GBS data from multiple samples distinguished by DNA barcodes at the beginning of each read [12]. We currently perform GBS at 384 plex (3072 samples per flowcell), and run each GBS library prep on one or more flowcell lanes depending on the desired sequencing depth.

Collapsing reads into a master tag list. The TASSEL-GBS Discovery Pipeline first reads through each of the available FASTQ files and generates one output "TagCount" file per input FASTQ file (Figure 1A). Each output, binary TagCount file contains a sorted list of all the unique sequence tags observed in the corresponding FASTQ file, the length of each tag in bases, and the number of times each tag was observed. This list can be used as a key-value map where the tag sequence and length together serve as the key and the corresponding tag count serves as the value. Each nucleotide is encoded in two bits allowing tag sequences to be stored in Java longs (64 bits, or 32 bases per long). Degenerate bases or N's are not permitted and quality scores are not retained. Potentially chimeric sequences are eliminated by trimming the sequence at the corresponding restriction enzyme site, if present.

After this initial pass through all of the available FASTQ files, the individual TagCounts files are merged into a single, master TagCount file containing a list of all tags of interest for a species. Only tags occurring at or above a (user-specified) minimum number of reads across all of the FASTQ files in the experiment are retained in the output master tag list. The more times a particular tag has been observed, the less likely it contains a sequencing error. The minimum tag count controls the tradeoff between the amount of sequencing errors admitted into the analysis versus the minimum allele frequency of interest. We do not expect to eliminate all sequencing errors at this step. We are usually able to filter out or correct most of them in subsequent steps of the pipeline, or in further, downstream processing customized to the biology of the study population(s). Furthermore, tags containing one or more sequencing errors can still be useful to score SNPs at other, non-error positions.

Alignment of tags to the reference genome. Alignment of each tag of interest in the master tag list to the reference genome is carried out with third party software. To facilitate this, the master tag list file is converted from TagCount format into FASTQ format (with fake, uniformly high quality scores). Currently, SAM format [36] output alignment files produced by the free software programs Bowtie2 [49] or BWA [36] can be read by the TASSEL-GBS pipeline and converted into a "TagsOnPhysicalMap" (TOPM) file that can be used for SNP calling.

The TOPM contains all of the tags present in the master TagCount file and genomic positions for the subset of tags that align to a unique best position in the genome. Retention of all of the master tags in the TOPM, even those without unique best genomic positions, affords future incorporation of additional information regarding the positions of unplaced tags, either from alternative aligners or from genetic mapping evidence. The TOPM is sorted by tag sequence and functions as a key-value map where a tag sequence can be used as a key to retrieve its corresponding physical position. It is also possible to programmatically traverse the tags within a TOPM in their physical position order through the use of an in-memory, primitive treemap. At the SNP calling step of the Discovery Pipeline (see below), the alleles represented by each useful tag at each useful SNP ("variants") are added to the TOPM. Three file formats of the TOPM are supported: text, binary, and HDF5.

Determining the distribution of tags across individual samples. With a TOPM file available, the next ingredient needed to discover and call SNPs is a matrix that records the number of times each tag in the master tag list was observed in each DNA sample, which we refer to as a TagsByTaxa (TBT) file (Figure 1B). In order to maximize the capacity of our pipeline, we currently store the TBT in HDF5 format (http://www.hdfgroup.org). The HDF5 format facilitates extremely fast read and write access to large data sets. To construct a TBT, the DNA sample of origin of each good, barcoded read in the set of input FASTQ files is determined based upon its barcode. If a good, barcoded read matches one of the tags in the master tag list, its depth in the appropriate taxon is incremented in the output TBT, up to a maximum depth of 127. Since the matrix is often extremely sparse, a custom run length encoding compression algorithm was implemented that provides a high level of compression, with minimal reduction in access speed.

SNP discovery and initial filtering. SNP discovery (Figure 1C) is performed for each set of tags that align to the exact same starting genomic position and strand, where the starting genomic position of a tag is defined by the cut site remnant at the beginning of the tag. Such tags, originating from the same restriction enzyme cut site and with the same orientation (but not necessarily of the same length), collectively comprise a "TagLocus". To call SNPs and ensure that indels are handled consistently, a *de novo* multiple sequence alignment of all the tags in each TagLocus is performed using the BioJava 3.0 API [50], which implements the CLUSTAL W algorithm [51]. For each SNP in the resulting "TagLocusAlignment", the allele represented by each tag is determined and the TBT file is consulted to tally the observed depths of each allele in each taxon. The genotype of the SNP in each taxon is then determined either by a binomial likelihood ratio method of quantitative SNP calling (the default; for details see Supplementary Text S1) or, optionally, following the method of Hohenlohe et al. [52].

After genotypes are obtained for a potential SNP, initial filtering is then performed based upon user settings for minimum minor allele frequency, and for a minimum coefficient of panmixia, or inbreeding relative to the entire population, F_{IT}, (where $F_{IT} = 1 - Ho/He$, Ho = observed heterozygosity, He = expected heterozygosity = $2q(1\text{-}q)$, and q = minor allele frequency). Error-prone SNPs and spurious SNPs from paralogy often appear excessively heterozygous, with lower F_{IT} than expected. If the user supplies a "pedigree file" that indicates the expected inbreeding coefficient (F) of each taxon, then only inbred taxa, with an expected inbreeding coefficient greater than or equal to the user-specified minimum coefficient of panmixia (*minF* parameter), are used in the calculation of F_{IT}. Inbred samples, available in many crop species and model organisms, can greatly add to the power of this filter. If enough inbred samples are available, then, additional filtering of SNPs can then be optionally performed enforcing a minimum "inbred coverage" (proportion of the inbred samples to be non-missing at the SNP) and a maximum "inbred heterozygosity score" (defined above).

To illustrate the effectiveness of these population genetic-based SNP filters, as applied in our most recent maize Discovery Build (AllZeaGBSv2.6, with 31,978 samples), we focused on the subset of 5,254 samples from the maize Nested Association Mapping (NAM) population [53]. The NAM population is a series of 25 biparental, F_2-derived RIL families all with a common female parent, the inbred line B73. Error-prone SNPs can be identified in a biparental family through their tendency, when they are in fact not segregating, to *appear* to be weakly polymorphic, but with segregation ratios significantly deviating from the 1:1 expectation. In contrast, non-segregating SNPs in a given family that display no spurious polymorphism are free of genotypic error in that family. The availability of the 25 biparental NAM families provides tremendous power to detect error-prone SNPs, and thus to examine the effectiveness of our filters.

We used the 5,254 NAM RIL samples to estimate error rates for three alternative sets of chromosome 10 SNPs discovered in the full set of 31,978 maize samples comprising our AllZeaGBSv2.6 Discovery Build. The three sets of chromosome 10 SNPs were obtained after application of three different filtering regimes: (1) no filters other than MAF $>= 0.001$, (2) a minimal filter only for MAF $>= 0.01$, and (3) our "standard" maize Discovery Build filters of MAF $>= 0.001$, minimum F_{IT} in inbred samples of 0.8, inbred coverage >0.15, and inbred heterozygosity score <0.21. The SNPs were discovered and filtered based upon all 31,978 maize samples (including the NAM RILs) and their allele frequencies were then separately calculated in each of the 25 NAM families. To minimize sampling error, allele frequency was only estimated for a particular SNP-NAM family combination if at least 19 RILs in that family had non-missing genotypes for the SNP ($n >= 19$). Thus, by "SNP-NAM family combination" we mean a particular SNP (e.g., "S10_2918") in a particular NAM family (e.g., "B73 × B97") that has at least 19 non-missing genotypes, regardless of whether it is polymorphic or not within that family. Hence, for each SNP, allele frequencies were calculated in 25 or fewer families, depending on the amount of missing data in each family. NAM family-specific minor allele calls for a SNP were classified as errors if the family-specific MAF was greater than zero but less than 0.25, and the SNP significantly deviated from 1:1 segregation in that family at $p<0.001$ (binomial test). The overall error rate for a SNP was then estimated as the total number of these error calls divided by the total number of calls for that SNP in NAM families with $n >= 19$ where the SNP significantly deviated from 1:1 segregation at $p<0.001$ (including the monomorphic SNPs). A pictorial explanation of this method of estimating error rates, using a single NAM family for illustrative purposes, is provided in Supplementary Text S2.

Compared to non-filtered SNPs (Figures 3A and 3B), application of our "standard" maize Discovery Build filters (MAF $>= 0.001$, minimum F_{IT} in inbred samples of 0.8, inbred coverage >0.15, inbred heterozygosity score <0.21) (Figures 3E and 3F) greatly increased the proportion of SNP-NAM family combinations displaying either appropriate 1:1 segregation (Figure 3F), or no polymorphism at all (Figure 3E). In contrast, a minimal filter based only on MAF (MAF $>= 0.01$) (Figures 3C and 3D) was far less effective at removing error-prone SNPs than our standard filters. Furthermore, our standard filters clearly shifted the distribution of error rates (estimated from the NAM samples) toward zero (Figure 4) and reduced the mean error rate (Table 2) relative to either no filtering (other than MAF $>= 0.001$) or minimal filtering (only for MAF >0.01). Extremely low estimates of mean and median error rates after application of our standard filters (0.0042 and zero respectively; Table 2) indicate that, for the most part, highly reliable SNP genotypes are produced by the GBS assay and the TASSEL-GBS pipeline, at least for inbred samples (where under-calling of heterozygotes due to low coverage is not an issue).

Error prone SNPs that are not removed by our standard filters (e.g., polymorphic SNPs with family specific MAFs <0.25 in Figure 3F) can be easily removed from biparental RIL families by filtering for appropriate allele frequencies and/or based on their relatively low levels of linkage disequilibrium (LD) with neighboring SNPs. In addition, error-prone SNPs identified in biparental families can be excluded from analyses of the remaining, non-biparental samples in a Discovery Build. If biparental populations are not available in your study species, it should be possible to use half-sib (e.g., open pollinated) families, or any population having a prior expected allele frequency range for polymorphic markers, to filter out error-prone SNPs. Alternatively, recently bottlenecked populations with high levels of extended LD can be used to filter out error-prone SNPs, which should display relatively low levels of LD with their neighboring SNPs.

After SNP calling and filtering, the input TOPM is updated with variants, and this "production-ready" TOPM is then saved to a new file (Figure 1C). For each SNP that passed the filtering step, the allele that is represented by each tag in the corresponding TagLocus is recorded in the "production-ready" TOPM, as well as the relative position of the SNP with respect to the genomic position of the TagLocus. The production TOPM produced by our AllZeaGBSv2.6 Discovery Build contained 955,690 useful SNPs.

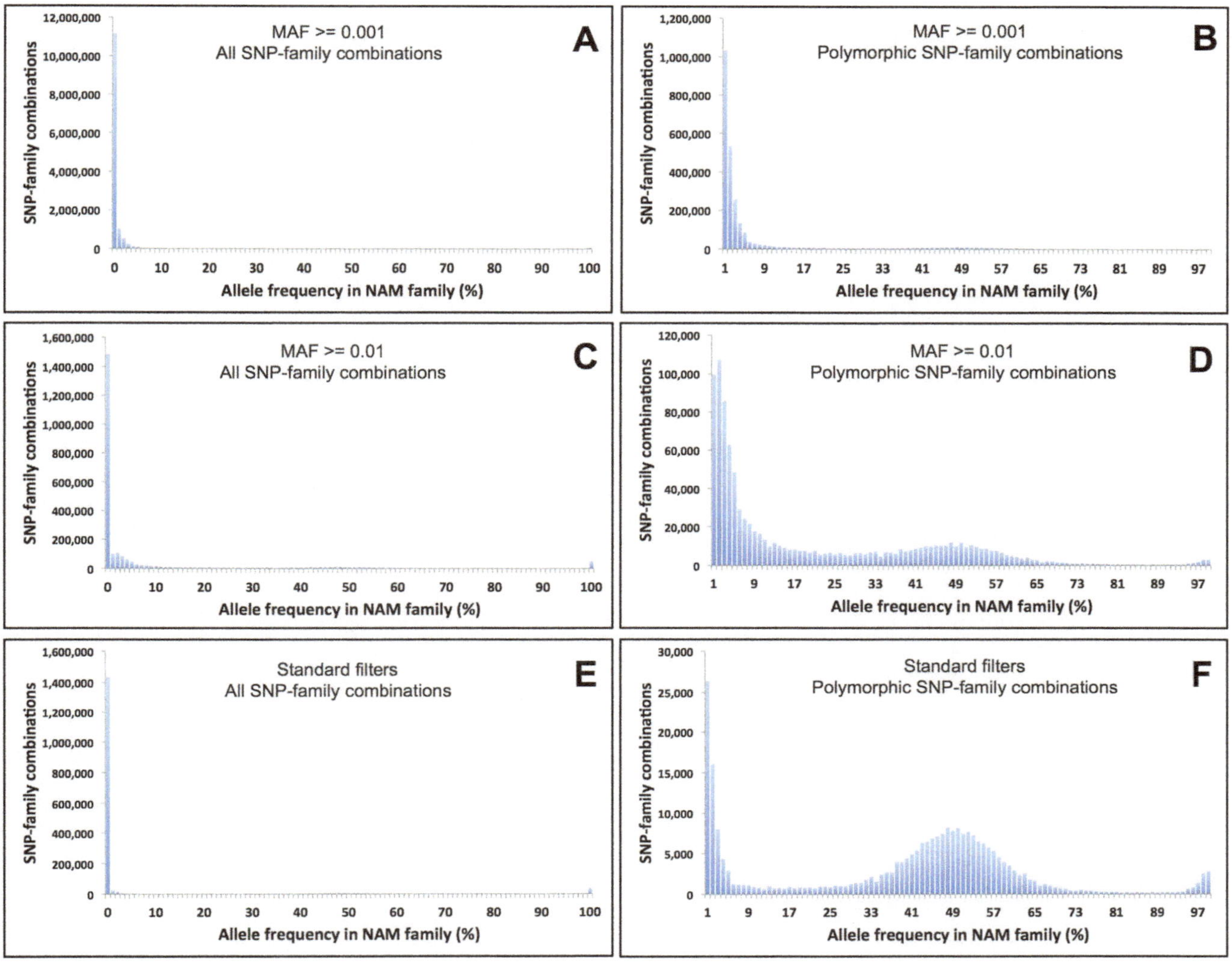

Figure 3. Within NAM family allele frequency distributions of chromosome 10 SNPs after different levels of filtering. Allele frequencies were calculated in each of the 25 Nested Association Mapping (NAM) families (collectively comprising 5,254 RILs) after application of the filters to the entire set of 31,978 maize samples in the AllZeaGBSv2.6 build. Allele frequencies were only estimated in a NAM family if at least 19 RILs had non-missing genotypes. Each histogram shows the allele frequency distribution for all the SNP-NAM family combinations with n >= 19. (A, B) No filter other than minimum MAF of 0.001. (C, D) A minimal filter only for MAF >= 0.01. (E, F) "Standard" maize build filters of MAF >= 0.001, minimum F_{IT} in inbred samples of 0.8, inbred coverage >0.15, and inbred heterozygosity score <0.21. (A, C, E) All SNP-family combinations: the error-free, monomorphic SNP-family combinations dwarf the segregating SNPs in all three cases. (B, D, F) Polymorphic SNP-family combinations only: omitting the monomorphic SNP-family combinations permits visualization of the remaining allele frequency distribution.

Production Pipeline

In contrast to the multiple-step Discovery Pipeline, the Production Pipeline consists of a single step, utilizing the production-ready TOPM generated by the Discovery Pipeline to produce genotypes directly from input FASTQ files (Figure 2). The Production Pipeline determines the taxon of origin of each good, barcoded sequence read in each input FASTQ file and then checks if the read matches one of the useful tags in the production-ready TOPM. In this manner, allelic depths for each useful SNP in the TOPM are recorded for each taxon, allowing quantitative SNP calling to be performed, again either by our own binomial likelihood ratio method (for details see Supplementary Text S1) or, optionally, according to the method of Hohenlohe et al. [52]. If the GBS library preps for some samples have been run in replicate on multiple flow cell lanes (to increase depth), the corresponding allelic depth information is tallied across the replicates prior to SNP calling. Genotype files are produced in HapMap format as well as in or our own custom HDF5 format (which also records allelic depth). The ability to convert from this custom HDF5 format into VCF format [54], which also retains allelic depth, will be added to the TASSEL GUI in the near future.

Downstream Processing

Further processing of the genotype files, such as sub-setting out specific taxa or genomic regions of interest, filtering SNPs or taxa based upon coverage, or filtering of SNPs based on minor allele frequency, can be performed either with the TASSEL 4 GUI or the TASSEL 4 Standalone command line interface (the ability to filter for SNPs that are in LD with their neighbors will be added soon). Depending on the genome size, the exact molecular protocol and restriction enzyme used, and the sequence depth obtained, missing data can be common and actual heterozygotes

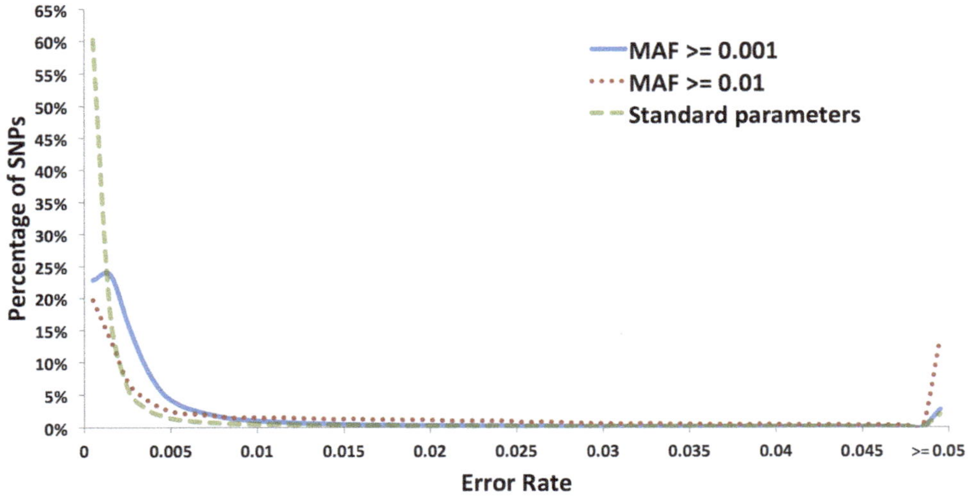

Figure 4. Error rate distribution of chromosome 10 SNPs for different levels of filtering. Error rates in the AllZeaGBSv2.6 Discovery build SNP calls were estimated using the NAM biparental families. NAM family-specific minor allele calls were defined as errors if the family-specific MAF was greater than zero but less than 0.25, and the SNP significantly deviated from 1:1 segregation in that family at $p<0.001$.

can be substantially under-called as homozygotes. Hence, imputation of missing data (and, possibly, phasing of the final genotypes) is usually desirable. Since the optimal imputation approach depends greatly upon the biology of the species and the experimental design, this topic is beyond the scope of this paper. Numerous, general purpose imputation tools are already available [55]. Custom imputation approaches that we are developing for our maize experimental populations will be the subject of future publications.

Hardware Needs and Installation

The TASSEL-GBS pipeline is part of the TASSEL package, which is written in Java, so it can be run on Linux, Mac, or Windows operating systems. A minimum of 8 GB of RAM is required (at least 16 GB is recommended). Detailed installation instructions are provided at www.maizegenetics.net/tassel as well as in the TASSEL-GBS pipeline documentation (www.maizegenetics.net/tassel/docs/TasselPipelineGBS.pdf). The source code is available from sourceforge.net/p/tassel/code/ci/master/tree, JUnit tests at sourceforge.net/p/tassel/maizegenetics4-test/ci/master/tree and a relatively small test data set at www.maizegenetics.net/tassel/GBSTestData.tar. The current version of TASSEL-GBS as described herein is implemented in TASSEL V4.3.5. Although we generally recommend using the latest version, users can revert to V4.3.5 or other versions by following the instructions posted in the following document www.maizegenetics.net/tassel/docs/UpdatingTasselStandaloneUsingGit.pdf, under the heading "To Update Packages to Older Releases…". The TASSEL-GBS pipeline will soon be available in TASSEL 5; major improvements to the pipeline (e.g., full VCF format support, allowing tag lengths greater than 64 bp, storage of tag depths per individual up to 10,000 rather than the current maximum of 127) will be implemented there.

Table 2. Comparison of error rates for chromosome 10 SNPs from the AllZeaGBSv2.6 build for different levels of filtering by the Discovery SNP caller.

Filter[1]	nSNPs	nSNPsTested[2]	avgErrorRate[3]	avg nSegregating[4]
MAF >= 0.001	694,517	680,623	0.00681	12,899
MAF >= 0.01	149,480	136,296	0.02218	12,818
Standard[5]	78,627	78,506	0.00420	7,192

[1]Filters applied to the entire build (31,978 non-blank samples)
[2]Minimum sample size of 19 in at least one maize Nested Association Mapping (NAM) family
[3]Average error rates estimated from 5,254 NAM RILs. Median error rates were zero for all three filters.
[4]Average number of chromosome 10 SNPs with n >= 19 and MAF between 0.25 and 0.75 across the 25 NAM families.
[5]MAF >= 0.001, minimum F_{IT} in inbred samples of 0.8, inbred coverage >0.15, inbred heterozygosity score <0.21.

Strengths and Weaknesses

Strengths

The strengths of GBS and the TASSEL-GBS pipeline are the large number of markers potentially produced (depending on the biology of the species and the choice of restriction enzymes), low cost and minimal startup cost, and integration of SNP discovery with SNP calling.

The potentially large number of markers available from GBS makes GWAS feasible in study populations where linkage disequilibrium (LD) extends far enough so that causative polymorphisms stand a reasonable chance of being in LD with one or more markers. Alternatively, the large number of markers facilitates accurate projection of haplotypes from a set of more densely genotyped reference haplotypes [56] derived from whole genome sequencing (WGS) (e.g., [57,58]). This projection strategy is especially effective if the WGS reference haplotypes are representative of the founders of the study population [59].

Compared to alternative high-density marker technologies such as SNP arrays, GBS is relatively inexpensive, particularly if low coverage data suffices for the purpose of your study. Startup costs for GBS are minimal, as startup involves only (1) testing that your one of your candidate restriction enzymes (or enzyme pairs) produces a suitable GBS library, and (2) optimization of the ratio of sample DNA to the PCR adapters [12]. In contrast, startup for a SNP array involves ascertainment of SNPs in a small discovery panel and assay design for each individual SNP.

The common practice of using of a small panel of individuals to discover SNPs for inclusion in a SNP array introduces an ascertainment bias that can severely distort estimates of key population genetic parameters gauging genetic diversity, sub-population differentiation and relatedness among individuals [60–62]. In contrast, the TASSEL-GBS Discovery pipeline integrates SNP discovery with SNP calling, using all available samples to date, and thus avoids the ascertainment bias that would arise from a small discovery panel. This type of ascertainment bias will also be minimal for new samples run through the Production Pipeline, provided that their genetic diversity is well-represented among the samples included in the Discovery Build. However, there might still be some subtle biases in either pipeline, caused by factors such as null alleles [63,64], alignment to the reference, and the use of inbreds only (rather than the full set of samples) to filter SNPs for F_{IT}.

Weaknesses

The main weakness of the GBS assay, when conducted at low coverage, is the amount of missing data. However, numerous imputation approaches are currently available [55] and yet more are currently in development, for a wide range of biological scenarios. As discussed above, the most appropriate imputation method and the probability of imputation success depends upon the biology of the study population. For some purposes, such as estimation of population allele frequencies [65], kinship, relatedness, and genetic distance, phylogenetic reconstruction [22], or germplasm quality control [24], imputation of missing data is usually not necessary.

Conclusions

The TASSEL-GBS pipeline for identifying and calling SNPs from next-generation, genotyping by sequencing data fulfills our design criteria better than any existing pipeline. It has a capacity for very large analyses involving tens of thousands of samples, yet can also be run at smaller scales. The pipeline permits rapid processing of the data, yet has a relatively modest memory footprint, allowing it to be run on desktop or laptop computers. This increases its usability by researchers in developing countries who may lack access to sophisticated computing resources. The separation of SNP discovery and genotyping into two phases reduces potential ascertainment biases and, more importantly, makes the TASSEL-GBS pipeline highly suitable for use in a genomics-assisted, accelerated breeding context, where rapid turnaround times from tissue collection to genotypes are essential. Furthermore, the high density of markers potentially available from the GBS assay should enable accurate genomic prediction over multiple generations.

Acknowledgments

We thank Yang (Jon) Zhang for contributions to the code for VCF output of genotypes and Sara J. Miller for assistance with manuscript formatting. We also thank the numerous users of the TASSEL-GBS pipeline (Katie Hyma in particular) who have made useful suggestions or have reported bugs.

Author Contributions

Conceived and designed the experiments: ESB JCG QS. Performed the experiments: ESB JCG TMC FL JH QS RJE. Analyzed the data: JCG ESB TMC QS RJE. Wrote the paper: JCG ESB.

References

1. Metzker ML (2010) Sequencing technologies - the next generation. Nat Rev Genet 11: 31–46. doi:10.1038/nrg2626.
2. Shendure J, Lieberman Aiden E (2012) The expanding scope of DNA sequencing. Nat Biotechnol 30: 1084–1094. Available: http://www.ncbi.nlm.nih.gov/pubmed/23138308.
3. Edwards D, Batley J, Snowdon RJ (2013) Accessing complex crop genomes with next-generation sequencing. Theor Appl Genet 126: 1–11. Available: http://www.ncbi.nlm.nih.gov/pubmed/22948437. Accessed 11 November 2013.
4. Kilpinen H, Barrett JC (2013) How next-generation sequencing is transforming complex disease genetics. Trends Genet 29: 23–30. Available: http://www.ncbi.nlm.nih.gov/pubmed/23103023. Accessed 11 November 2013.
5. Altshuler D, Pollara VJ, Cowles CR, Van Etten WJ, Baldwin J, et al. (2000) An SNP map of the human genome generated by reduced representation shotgun sequencing. Nature 407: 513–516. Available: http://www.ncbi.nlm.nih.gov/pubmed/11029002.
6. Davey JW, Hohenlohe PA, Etter PD, Boone JQ, Catchen JM, et al. (2011) Genome-wide genetic marker discovery and genotyping using next-generation sequencing. Nat Rev Genet 12: 499–510. Available: http://www.ncbi.nlm.nih.gov/pubmed/21681211. Accessed 6 November 2013.
7. Poland JA, Rife TW (2012) Genotyping-by-Sequencing for Plant Breeding and Genetics. Plant Genome J 5: 92. Available: https://www.crops.org/publications/tpg/abstracts/5/3/92. Accessed 11 November 2013.
8. Narum SR, Buerkle CA, Davey JW, Miller MR, Hohenlohe PA (2013) Genotyping-by-sequencing in ecological and conservation genomics. Mol Ecol 22: 2841–2847. Available: http://www.ncbi.nlm.nih.gov/pubmed/23711105. Accessed 7 November 2013.
9. Van Orsouw NJ, Hogers RCJ, Janssen A, Yalcin F, Snoeijers S, et al. (2007) Complexity reduction of polymorphic sequences (CRoPS): a novel approach for large-scale polymorphism discovery in complex genomes. PLoS One 2: e1172. Available: http://www.pubmedcentral.nih.gov/articlerender.fcgi?artid=2048665&tool=pmcentrez&rendertype=abstract. Accessed 11 November 2013.
10. Van Tassell C, Smith T, Matukumalli L (2008) SNP discovery and allele frequency estimation by deep sequencing of reduced representation libraries. Nat Methods 5: 247–252. Available: http://www.nature.com/nmeth/journal/v5/n3/abs/nmeth.1185.html. Accessed 2013 Nov 11.
11. Baird NA, Etter PD, Atwood TS, Currey MC, Shiver AL, et al. (2008) Rapid SNP discovery and genetic mapping using sequenced RAD markers. PLoS One 3: e3376. Available: http://www.pubmedcentral.nih.gov/articlerender.fcgi?artid=2557064&tool=pmcentrez&rendertype=abstract. Accessed 2013 Nov 7.
12. Elshire RJ, Glaubitz JC, Sun Q, Poland JA, Kawamoto K, et al. (2011) A Robust, Simple Genotyping-by-Sequencing (GBS) Approach for High Diversity Species. PLoS One 6: e19379. Available: http://www.pubmedcentral.nih.gov/articlerender.fcgi?artid=3087801&tool=pmcentrez&rendertype=abstract. Accessed 2011 Jul 18.
13. Andolfatto P, Davison D, Erezyilmaz D, Hu TT, Mast J, et al. (2011) Multiplexed shotgun genotyping for rapid and efficient genetic mapping. Genome Res 21: 610–617. Available: http://www.pubmedcentral.nih.gov/articlerender.fcgi?artid=3065708&tool=pmcentrez&rendertype=abstract. Accessed 2013 Nov 8.
14. Wang S, Meyer E, McKay JK, Matz M V (2012) 2b-RAD: a simple and flexible method for genome-wide genotyping. Nat Methods 9: 808–810. Available: http://www.ncbi.nlm.nih.gov/pubmed/22609625. Accessed 2013 Nov 8.
15. Truong HT, Ramos AM, Yalcin F, de Ruiter M, van der Poel HJA, et al. (2012) Sequence-based genotyping for marker discovery and co-dominant scoring in germplasm and populations. PLoS One 7: e37565, . Available: http://www.pubmedcentral.nih.gov/articlerender.fcgi?artid=3360789&tool=pmcentrez&rendertype=abstract. Accessed 2013 Nov 8.
16. Monson-Miller J, Sanchez-Mendez DC, Fass J, Henry IM, Tai TH, et al. (2012) Reference genome-independent assessment of mutation density using restriction enzyme-phased sequencing. BMC Genomics 13: 72, . Available: http://www.pubmedcentral.nih.gov/articlerender.fcgi?artid=3305632&tool=pmcentrez&rendertype=abstract. Accessed 2013 Nov 11.
17. Chen Q, Ma Y, Yang Y, Chen Z, Liao R, et al. (2013) Genotyping by genome reducing and sequencing for outbred animals. PLoS One 8: e67500. Available: http://www.pubmedcentral.nih.gov/articlerender.fcgi?artid=3715491&tool=pmcentrez&rendertype=abstract. Accessed 2013 Nov 11.

18. Morishige DT, Klein PE, Hilley JL, Sahraeian SM, Sharma A, et al. (2013) Digital genotyping of sorghum – a diverse plant species with a large repeat-rich genome. BMC Genomics 14: 448. Available: http://www.pubmedcentral.nih.gov/articlerender.fcgi?artid = 3716661&tool = pmcentrez&rendertype = abstract. Accessed 2013 Nov 10.
19. Stolle E, Moritz RFA (2013) RESTseq–efficient benchtop population genomics with RESTriction Fragment SEQuencing. PLoS One 8: e63960. Available: http://www.pubmedcentral.nih.gov/articlerender.fcgi?artid = 3656931&tool = pmcentrez&rendertype = abstract. Accessed 2013 Nov 11.
20. Poland JA, Brown PJ, Sorrells ME, Jannink J-L (2012) Development of high-density genetic maps for barley and wheat using a novel two-enzyme genotyping-by-sequencing approach. PLoS One 7: e32253. Available: http://www.pubmedcentral.nih.gov/articlerender.fcgi?artid = 3289635&tool = pmcentrez&rendertype = abstract. Accessed 2013 Nov 11.
21. Morris GP, Ramu P, Deshpande SP, Hash CT, Shah T, et al. (2013) Population genomic and genome-wide association studies of agroclimatic traits in sorghum. Proc Natl Acad Sci U S A 110: 453–458. Available: http://www.pubmedcentral.nih.gov/articlerender.fcgi?artid = 3545811&tool = pmcentrez&rendertype = abstract. Accessed 2013 Nov 8.
22. Lu F, Lipka AE, Elshire RJ, Glaubitz JC, Cherney J, et al. (2013) Switchgrass genomic diversity, ploidy and evolution: novel insights from a network-based SNP discovery protocol. PLoS Genet 9: e1003215.
23. Maron LG, Guimarães CT, Kirst M, Albert PS, Birchler JA, et al. (2013) Aluminum tolerance in maize is associated with higher MATE1 gene copy number. Proc Natl Acad Sci U S A 110: 5241–5246. Available: http://www.pubmedcentral.nih.gov/articlerender.fcgi?artid = 3612656&tool = pmcentrez&rendertype = abstract. Accessed 2013 Nov 11.
24. Romay MC, Millard MJ, Glaubitz JC, Peiffer JA, Swarts KL, et al. (2013) Comprehensive genotyping of the USA national maize inbred seed bank. Genome Biol 14: R55. Available: http://www.pubmedcentral.nih.gov/articlerender.fcgi?artid = 3707059&tool = pmcentrez&rendertype = abstract. Accessed 2013 Nov 7.
25. Sonah H, Bastien M, Iquira E, Tardivel A, Légaré G, et al. (2013) An improved genotyping by sequencing (GBS) approach offering increased versatility and efficiency of SNP discovery and genotyping. PLoS One 8: e54603. Available: http://www.pubmedcentral.nih.gov/articlerender.fcgi?artid = 3553054&tool = pmcentrez&rendertype = abstract. Accessed 2013 Nov 8.
26. De Donato M, Peters SO, Mitchell SE, Hussain T, Imumorin IG (2013) Genotyping-by-sequencing (GBS): a novel, efficient and cost-effective genotyping method for cattle using next-generation sequencing. PLoS One 8: e62137. Available: http://www.pubmedcentral.nih.gov/articlerender.fcgi?artid = 3656875&tool = pmcentrez&rendertype = abstract. Accessed 2013 Nov 11.
27. Rutkoski JE, Poland J, Jannink J-L, Sorrells ME (2013) Imputation of unordered markers and the impact on genomic selection accuracy. G3 (Bethesda) 3: 427–439. Available: http://www.pubmedcentral.nih.gov/articlerender.fcgi?artid = 3583451&tool = pmcentrez&rendertype = abstract. Accessed 2013 Nov 11.
28. Kim S-I, Tai TH (2013) Identification of SNPs in closely related Temperate Japonica rice cultivars using restriction enzyme-phased sequencing. PLoS One 8: e60176, . Available: http://www.pubmedcentral.nih.gov/articlerender.fcgi?artid = 3608622&tool = pmcentrez&rendertype = abstract. Accessed 2013 Nov 11.
29. Ly D, Hamblin M, Rabbi I, Melaku G, Bakare M, et al. (2013) Relatedness and Genotype × Environment Interaction Affect Prediction Accuracies in Genomic Selection: A Study in Cassava. Crop Sci 53: 1312. Available: https://www.crops.org/publications/cs/abstracts/53/4/1312. Accessed 2013 Nov 11.
30. Saintenac C, Jiang D, Wang S, Akhunov E (2013) Sequence-based mapping of the polyploid wheat genome. G3 (Bethesda) 3: 1105–1114. Available: http://www.ncbi.nlm.nih.gov/pubmed/23665877. Accessed 2013 Nov 7.
31. White TA, Perkins SE, Heckel G, Searle JB (2013) Adaptive evolution during an ongoing range expansion: the invasive bank vole (Myodes glareolus) in Ireland. Mol Ecol 22: 2971–2985. Available: http://www.ncbi.nlm.nih.gov/pubmed/23701376. Accessed 2013 Nov 7.
32. Bradbury PJ, Zhang Z, Kroon DE, Casstevens TM, Ramdoss Y, et al. (2007) TASSEL: software for association mapping of complex traits in diverse samples. Bioinformatics 23: 2633–2635. Available: http://www.ncbi.nlm.nih.gov/pubmed/17586829. Accessed 2011 Aug 9.
33. Mascher M, Wu S, Amand PS, Stein N, Poland J (2013) Application of Genotyping-by-Sequencing on Semiconductor Sequencing Platforms: A Comparison of Genetic and Reference-Based Marker Ordering in Barley. PLoS One 8: e76925. Available: http://www.pubmedcentral.nih.gov/articlerender.fcgi?artid = 3789676&tool = pmcentrez&rendertype = abstract. Accessed 2013 Nov 7.
34. Catchen JM, Amores A, Hohenlohe P, Cresko W, Postlethwait JH (2011) Stacks: building and genotyping Loci de novo from short-read sequences. G3 (Bethesda) 1: 171–182. Available: http://www.pubmedcentral.nih.gov/articlerender.fcgi?artid = 3276136&tool = pmcentrez&rendertype = abstract. Accessed 2013 Nov 7.
35. Catchen J, Bassham S, Wilson T, Currey M, O'Brien C, et al. (2013) The population structure and recent colonization history of Oregon threespine stickleback determined using restriction-site associated DNA-sequencing. Mol Ecol 22: 2864–2883. Available: http://www.ncbi.nlm.nih.gov/pubmed/23718143. Accessed 2013 Nov 11.
36. Li H, Handsaker B, Wysoker A, Fennell T, Ruan J, et al. (2009) The Sequence Alignment/Map format and SAMtools. Bioinformatics 25: 2078–2079. Available: http://www.pubmedcentral.nih.gov/articlerender.fcgi?artid = 2723002&tool = pmcentrez&rendertype = abstract. Accessed 2013 Nov 6.
37. Li R, Li Y, Fang X, Yang H, Wang J, et al. (2009) SNP detection for massively parallel whole-genome resequencing. Genome Res 19: 1124–1132. Available: http://www.pubmedcentral.nih.gov/articlerender.fcgi?artid = 2694485&tool = pmcentrez&rendertype = abstract. Accessed 2013 Nov 6.
38. DePristo MA, Banks E, Poplin R, Garimella KV, Maguire JR, et al. (2011) A framework for variation discovery and genotyping using next-generation DNA sequencing data. Nat Genet 43: 491–498. Available: http://www.pubmedcentral.nih.gov/articlerender.fcgi?artid = 3083463&tool = pmcentrez&rendertype = abstract. Accessed 2013 Nov 6.
39. Rafalski JA (2010) Association genetics in crop improvement. Curr Opin Plant Biol 13: 174–180. Available: http://www.ncbi.nlm.nih.gov/pubmed/20089441. Accessed 2013 Nov 8.
40. Xie W, Feng Q, Yu H, Huang X, Zhao Q, et al. (2010) Parent-independent genotyping for constructing an ultrahigh-density linkage map based on population sequencing. Proc Natl Acad Sci U S A 107: 10578–10583. Available: http://www.pubmedcentral.nih.gov/articlerender.fcgi?artid = 2890813&tool = pmcentrez&rendertype = abstract. Accessed 2013 Nov 11.
41. Li H, Durbin R (2011) Inference of human population history from individual whole-genome sequences. Nature 475: 493–496. Available: http://www.pubmedcentral.nih.gov/articlerender.fcgi?artid = 3154645&tool = pmcentrez&rendertype = abstract. Accessed 2013 Nov 6.
42. Harter AV, Gardner KA, Falush D, Lentz DL, Bye RA, et al. (2004) Origin of extant domesticated sunflowers in eastern North America. Nature 430: 201–205. Available: http://www.ncbi.nlm.nih.gov/pubmed/15241413.
43. Ofori A, Becker HC, Kopisch-Obuch FJ (2008) Effect of crop improvement on genetic diversity in oilseed Brassica rapa (turnip-rape) cultivars, detected by SSR markers. J Appl Genet 49: 207–212. Available: http://www.ncbi.nlm.nih.gov/pubmed/18670055.
44. Cowling WA (2013) Sustainable plant breeding. Plant Breed 132: 1–9. Available: http://doi.wiley.com/10.1111/pbr.12026. Accessed 2013 Nov 11.
45. Dohm JC, Lottaz C, Borodina T, Himmelbauer H (2008) Substantial biases in ultra-short read data sets from high-throughput DNA sequencing. Nucleic Acids Res 36: e105. Available: http://www.pubmedcentral.nih.gov/articlerender.fcgi?artid = 2532726&tool = pmcentrez&rendertype = abstract. Accessed 2013 Nov 6.
46. Eren AM, Vineis JH, Morrison HG, Sogin ML (2013) A filtering method to generate high quality short reads using illumina paired-end technology. PLoS One 8: e66643. Available: http://www.pubmedcentral.nih.gov/articlerender.fcgi?artid = 3684618&tool = pmcentrez&rendertype = abstract. Accessed 2013 Nov 6.
47. McElroy KE, Luciani F, Thomas T (2012) GemSIM: general, error-model based simulator of next-generation sequencing data. BMC Genomics 13: 74. Available: http://www.pubmedcentral.nih.gov/articlerender.fcgi?artid = 3305602&tool = pmcentrez&rendertype = abstract. Accessed 2013 Nov 11.
48. Allhoff M, Schönhuth A, Martin M, Costa IG, Rahmann S, et al. (2013) Discovering motifs that induce sequencing errors. BMC Bioinformatics 14 Suppl 5: S1. Available: http://www.pubmedcentral.nih.gov/articlerender.fcgi?artid = 3622629&tool = pmcentrez&rendertype = abstract. Accessed 2013 Nov 11.
49. Langmead B, Salzberg SL (2012) Fast gapped-read alignment with Bowtie 2. Nat Methods 9: 357–359. Available: http://www.pubmedcentral.nih.gov/articlerender.fcgi?artid = 3322381&tool = pmcentrez&rendertype = abstract. Accessed 2013 Nov 7.
50. Prlić A, Yates A, Bliven SE, Rose PW, Jacobsen J, et al. (2012) BioJava: an open-source framework for bioinformatics in 2012. Bioinformatics 28: 2693–2695. Available: http://www.pubmedcentral.nih.gov/articlerender.fcgi?artid = 3467744&tool = pmcentrez&rendertype = abstract. Accessed 2013 Nov 10.
51. Thompson JD, Higgins DG, Gibson TJ (1994) CLUSTAL W: improving the sensitivity of progressive multiple sequence alignment through sequence weighting, position-specific gap penalties and weight matrix choice. Nucleic Acids Res 22: 4673–4680. Available: http://www.pubmedcentral.nih.gov/articlerender.fcgi?artid = 308517&tool = pmcentrez&rendertype = abstract.
52. Hohenlohe PA, Bassham S, Etter PD, Stiffler N, Johnson EA, et al. (2010) Population genomics of parallel adaptation in threespine stickleback using sequenced RAD tags. PLoS Genet 6: e1000862. Available: http://www.pubmedcentral.nih.gov/articlerender.fcgi?artid = 2829049&tool = pmcentrez&rendertype = abstract. Accessed 2013 Nov 7.
53. McMullen MD, Kresovich S, Villeda HS, Bradbury P, Li H, et al. (2009) Genetic properties of the maize nested association mapping population. Science 325: 737–740. Available: http://www.ncbi.nlm.nih.gov/pubmed/19661427. Accessed 2011 Jun 14.
54. Danecek P, Auton A, Abecasis G, Albers CA, Banks E, et al. (2011) The variant call format and VCFtools. Bioinformatics 27: 2156–2158. Available: http://www.pubmedcentral.nih.gov/articlerender.fcgi?artid = 3137218&tool = pmcentrez&rendertype = abstract. Accessed 2013 Nov 6.
55. Marchini J, Howie B (2010) Genotype imputation for genome-wide association studies. Nat Rev Genet 11: 499–511. Available: http://www.ncbi.nlm.nih.gov/pubmed/20517342. Accessed 2013 Nov 6.

56. Howie B, Marchini J, Stephens M (2011) Genotype imputation with thousands of genomes. G3 (Bethesda) 1: 457–470. Available: http://www.pubmedcentral.nih.gov/articlerender.fcgi?artid = 3276165&tool = pmcentrez&rendertype = abstract. Accessed 2013 Nov 11.
57. Abecasis GR, Auton A, Brooks LD, DePristo MA, Durbin RM, et al. (2012) An integrated map of genetic variation from 1,092 human genomes. Nature 491: 56–65. Available: http://www.pubmedcentral.nih.gov/articlerender.fcgi?artid = 3498066&tool = pmcentrez&rendertype = abstract. Accessed 2013 Nov 6.
58. Chia J-M, Song C, Bradbury PJ, Costich D, de Leon N, et al. (2012) Maize HapMap2 identifies extant variation from a genome in flux. Nat Genet 44: 803–807. Available: http://dx.doi.org/10.1038/ng.2313. Accessed 2012 Nov 6.
59. Tian F, Bradbury PJ, Brown PJ, Hung H, Sun Q, et al. (2011) Genome-wide association study of leaf architecture in the maize nested association mapping population. Nat Genet 43: 159–162. Available: http://www.ncbi.nlm.nih.gov/pubmed/21217756. Accessed 2011 Jun 13.
60. Eller E (2001) Effects of ascertainment bias on recovering human demographic history. Hum Biol 73: 411–427. Available: http://muse.jhu.edu/journals/hub/summary/v081/81.5-6.eller.html. Accessed 2013 Nov 11.
61. Clark AG, Hubisz MJ, Bustamante CD, Williamson SH, Nielsen R (2005) Ascertainment bias in studies of human genome-wide polymorphism. Genome Res 15: 1496–1502. Available: http://www.pubmedcentral.nih.gov/articlerender.fcgi?artid = 1310637&tool = pmcentrez&rendertype = abstract. Accessed 2013 Nov 6.
62. Albrechtsen A, Nielsen FC, Nielsen R (2010) Ascertainment biases in SNP chips affect measures of population divergence. Mol Biol Evol 27: 2534–2547. Available: http://www.pubmedcentral.nih.gov/articlerender.fcgi?artid = 3107607&tool = pmcentrez&rendertype = abstract. Accessed 2013 Nov 11.
63. Gautier M, Gharbi K, Cezard T, Foucaud J, Kerdelhué C, et al. (2013) The effect of RAD allele dropout on the estimation of genetic variation within and between populations. Mol Ecol 22: 3165–3178. Available: http://www.ncbi.nlm.nih.gov/pubmed/23110526. Accessed 2013 Nov 11.
64. Arnold B, Corbett-Detig RB, Hartl D, Bomblies K (2013) RADseq underestimates diversity and introduces genealogical biases due to nonrandom haplotype sampling. Mol Ecol 22: 3179–3190. Available: http://www.ncbi.nlm.nih.gov/pubmed/23551379. Accessed 2013 Nov 7.
65. Alex Buerkle C, Gompert Z (2013) Population genomics based on low coverage sequencing: how low should we go? Mol Ecol 22: 3028–3035. Available: http://www.ncbi.nlm.nih.gov/pubmed/23174005. Accessed 2013 Nov 6.

Genome-Wide Identification, Evolution and Expression Analysis of *mTERF* Gene Family in Maize

Yanxin Zhao, Manjun Cai, Xiaobo Zhang, Yurong Li, Jianhua Zhang, Hailiang Zhao, Fei Kong, Yonglian Zheng, Fazhan Qiu*

National Key Laboratory of Crop Genetic Improvement, Huazhong Agricultural University, Wuhan, China

Abstract

Plant mitochondrial transcription termination factor (*mTERF*) genes comprise a large family with important roles in regulating organelle gene expression. In this study, a comprehensive database search yielded 31 potential *mTERF* genes in maize (*Zea mays* L.) and most of them were targeted to mitochondria or chloroplasts. Maize *mTERF* were divided into nine main groups based on phylogenetic analysis, and group IX represented the mitochondria and species-specific clade that diverged from other groups. Tandem and segmental duplication both contributed to the expansion of the *mTERF* gene family in the maize genome. Comprehensive expression analysis of these genes, using microarray data and RNA-seq data, revealed that these genes exhibit a variety of expression patterns. Environmental stimulus experiments revealed differential up or down-regulation expression of maize *mTERF* genes in seedlings exposed to light/dark, salts and plant hormones, respectively, suggesting various important roles of maize *mTERF* genes in light acclimation and stress-related responses. These results will be useful for elucidating the roles of *mTERF* genes in the growth, development and stress response of maize.

Editor: Rogerio Margis, Universidade Federal do Rio Grande do Sul, Brazil

Funding: This research was supported by Project No.91335205 supported by National Natural Science Foundation of China, the National High Technology Research and Development Program of China (863 Program) (2012AA101104) and the National Basic Research Program of China (973 Program) (2014CB138203). The funders had no role in study design, data collection and analysis, decision to publish, or preparation of the manuscript.

Competing Interests: The authors have declared that no competing interests exist.

* E-mail: qiufazhan@gmail.com

Introduction

Mitochondria and chloroplasts possess their own genetic materials containing a few dozen genes required for gene expression, photosynthesis and the electron transport chain, since most genes of these organelles have been either lost or transferred to the nucleus during evolution from their bacterial progenitors in different plant species [1]. Most of approximately 2000 and 2600 proteins located in plant mitochondria and chloroplasts, respectively, are encoded in the nuclear genome [2]. Despite their small genomes, the mitochondria and plastids in higher plant have complex transcription machineries. Several components of the transcriptional machinery have been reported, such as nuclear-encoded phage-type RNA polymerases for mitochondria and chloroplasts, and plastid-encoded cyanobacterial-type RNA polymerases and nuclear-encoded sigma-like factors for chloroplasts; however, additional unidentified auxiliary factors are required for organellar transcription [3,4].

Recently, *mTERF* homologs predicted to be imported in both chloroplasts and mitochondria were identified in *Arabidopsis thaliana* [5,6] as putative organellar transcription factors, since mTERF is a transcription terminator in animal mitochondria. The mitochondrial Transcription tERmination Factor (*mTERF*) family consists of a group of proteins in human. These proteins have a modular architecture based on repetitions of a 32-amino acid (aa) mTERF motif which contains three leucine zipper-like elements [7,8], and this has been suggested to be the basis of their ability to bind DNA [9]. The molecular function of *mTERFs* has so far only been described for metazoan members of the protein family. Vertebrates have four *mTERF* paralogs. In humans, mTERF1 is a sequence-specific DNA-binding protein responsible for mitochondrial transcription termination at the 3′-end of the 16S rRNA gene, promoting termination of transcripts from the first transcription initiation site (H1) [10,11]. mTERF2 can bind to mitochondrial DNA [12] and, at least in mouse, seems to influence transcription [13]. mTERF3 acts as a specific repressor of mammalian mtDNA transcription initiation *in vivo* [14]. mTERF3 binds to the mtDNA promoter region, and depletion of mTERF3 increases transcription initiation on both mtDNA strands. mTERF4 can form a stoichiometric complex with the ribosomal RNA methyltransferase NSUN4 to directly control mitochondrial ribosomal biogenesis and translation [15,16].

mTERF genes have been found in monocotyledonous and dicotyledonous nuclear genomes, in the moss *Physcomitrella patens*, but are apparently absent in fungi and prokaryotes [6,17,18]. Flowering plants have the highest number of *mTERF* genes among eukaryotes, and most annotated Arabidopsis *mTERFs* are targeted to mitochondria or chloroplasts [17,18]. To date, only five *mTERF* genes from plants have been characterized: *MOC1* (*mterf-like gene of Chlamydomonas 1*) from *Chlamydomonas reinhardtii* [19,20], and *SOLDAT10* (*SINGLET OXYGEN-LINKED DEATH ACTIVATOR 10*) [21], *BELAYA SMERT/RUGOSA2*(*BSM/RUG2*) [17,22], *MDA1* (*MTERF DEFECTIVE IN Arabidopsis1*) [23] and *SHOT1* (*suppressor of hot1-4 1*) [24] from Arabidopsis.

In the unicellular green alga *C. reinhardtii*, MOC1 binds specifically to an octanucleotide sequence within the mitochondrial rRNA-coding module *S3*, and a loss of MOC1 increases read-through transcription at the S3-binding site thereby causing elevated antisense RNA levels in the mutant *stm6*, suggesting that *MOC1* possesses the evolutionarily-conserved transcription termination activity as for *mTERF1* in human [20]. *SOLDAT10*, the first *mTERF* gene characterized in higher plants, is localized to chloroplasts and its loss reduces plant growth and pigmentation while complete inactivation of *SOLDAT10* is apparently lethal. The *soldat10* mutant has decreased levels of plastid-specific rRNAs and affects protein synthesis in plastids, which subsequently activates retrograde signaling to the nucleus and leads to overexpression of stress-related nuclear genes [21]. Another Arabidopsis *mTERF*, dubbed *MDA1*, is targeted to chloroplast, and *mda1* mutant exhibit altered chloroplast morphology and plant growth. Additionally, the *mda1* mutations enhance salt and osmotic stress tolerance and alter sugar responses during seedling establishment via perturbing abscisci acid (ABA) retrograde signaling [23]. *BSM* or *RUG2*, encoding a dual-targeted mTERF protein, has a broader function, and is essential for normal plant development. Its loss affects levels of transcripts in both mitochondria and chloroplasts [17,22]. *SHOT1* is characterized as one suppressor of Arabidopsis *hot1-4* mutant and resides in mitochondria. The *shot1* mutant can suppress the *hot1-4* heat-hypersensitive phenotype via changing mitochondrial function and increasing transcripts of other heat shock protein (HSP) genes. Expression alteration of other *HSPs* and redox-related genes in *shot1-2* are involved in retrograde signaling from mitochondria to nucleus [24]. Taken together, *mTERF* genes are required for organelle gene expression regulation and play important roles in plant growth, development and abiotic stress tolerance, at least in Arabidopsis and possibly other higher plants. However, little is known about the molecular mechanisms of *mTERF* that control transcription of the mitochondrial and chloroplastic genomes. More *mTERF* genes require characterization in diverse plants, especially crop plants.

In this report, 31 putative *mTERF* genes were identified in the maize genome. Most maize mTERF proteins are predicted to reside in mitochondria or plastids. Phylogenetic analysis of *mTERF* genes in maize, rice, and Arabidopsis indicates mitochondria- and plastid-targeting *mTERF* proteins form two divergent clades. Expression of *mTERF* genes are regulated in maize seedlings treated with light/dark, plant hormones and salts, showing their important roles in abiotic stress response. Our work will provide a basic biochemical characterization of maize *mTERF*s, paving the way for future functional studies and certainly contributing to improving knowledge of their role in plant biology.

Results

Identification and Nomenclature of *mTERF* Genes in Maize

Using keyword and homology searches, we obtained 90 maize mTERF proteins from the NCBI Protein Database, 26 maize *mTERF* unigenes from the NCBI Unigene Database, 26 *mTERF* cDNA sequences from the Maize Full Length cDNA Library Database, and 30 *mTERF* genes from maize genomes that were identical to maize *mTERF* genes identified by HMMER 3.0 [25] with mTERF PFAM file (PF02636) (Table S1). Sequence comparison for the above putative mTERF cDNAs and proteins indicated that all *mTERF* cDNA sequences and unigenes (except *Zm#S48278749*) were derived from 27 of 30 *mTERF* genes annotated in the maize genome and 87 of 90 maize mTERF proteins from GenBank were encoded by 26 of 30 *mTERF* genes identified in the maize genome; while the remaining three proteins and *Zm#S48278749* were assumed to be from two undiscovered *mTERF* genes in maize (Table S1). Two *mTERF* genes, *GRMZM2G426154* and *GRMZM2G175930*, were located in AC217966.3-Contig20 (forward strand) and AC205479.3-Contig45 (reverse strand), respectively, which were about 38 kb apart in the maize genome, and matched distinct parts of one *mTERF* cDNA (Figure S1). Therefore, *GRMZM2G426154* and *GRMZM2G175930* could be mis-annotated and could be two parts of one *mTERF* gene separated by unsequenced genomic gaps (Figure S1). Finally, 31 unique *mTERF* genes were identified in maize and designated *ZmTERF1–ZmTERF31* according to the order of these genes localizing in the chromosomes except for *ZmTERF30* and *-31* (Table S2); and *ZmTERF23* was the gene derived from *GRMZM2G426154* and *GRMZM2G175930* (Table 1). SMART [26] database searching revealed that each of the 31 maize *mTERF* gene products had at least one mTERF motif. *ZmTERF1–ZmTERF29* were selected for further study in this work since they had known genomic DNA sequences.

Our results showed that the maize, rice and Arabidopsis genomes encoded a similar number of mTERF proteins, which comprises 34 members identified in rice (*Oryza sativa* subsp. japonica) and 35 members in Arabidopsis [17,18]. RT-PCR assay was carried out to validate cDNA sequences of the *mTERF* genes with gene-specific primers listed in Table S3. Except for *ZmTERF4*, *-7* and *-20*, which had undetectable expression in B73 seedling leaves, most cDNA sequences of the 29 maize *mTERF* genes were obtained (Figure S2) and were identical to corresponding gene models annotated in MaizeGDB (http://www.maizegdb.org/). All identified *mTERF* genes encode proteins varying from 118 (ZmTERF7) to 649 aa (ZmTERF2), in which protein sizes of ZmTERF4, -7 and -20 were <150 aa (Table 1). Isoelectric points of most maize mTERF proteins were similar and >8.0, except for those of ZmTERF25 (6.48) and ZmTERF27 (5.70) (Table 1).

Phylogenetic Analysis and Classification of *ZmTERF* Genes

To evaluate the evolutionary relationships among the *mTERF* genes in maize, rice and Arabidopsis, we performed a phylogenetic analysis of the 98 mTERF protein sequences to construct an unrooted ML inference-based tree with RAxML [27]. The defined multiple alignment sequences for phylogenetic tree construction are found in File S1. The phylogenetic tree also included known *mTERF* genes identified in human and *C. reinhardtii*. The best ML scoring tree is shown in Figure 1. From the tree topology resembling that from Arabidopsis phylogenetic trees constructed previously [17] including lower plants, it was apparent that there were nine groups (groups I–IX) despite the low bootstrap values for deep nodes (Figure 1). Group IX contained 11 *mTERF* maize genes while the other groups had 1–5 members. It is noteworthy that all functionally identified Arabidopsis *mTERF* genes belonged to groups II, IV and VI with *SOLDAT10* (AT2G03050) in group II, *MDA1* (AT4G14605) and *BSM/RUG2* (AT4G02990) both in group IV, and *SHOT1* (AT3G60400) in group VI.

Relationship analysis of intragroup members with well-supported bootstrap values revealed that seven groups (group I–VII) included both proteins from eudicots and monocots (Figure 1). This suggested that nearly 80% of the groups here defined were already present in the common ancestor of both groups, and evolved before divergence of monocots and dicots. Therefore, these *mTERF* genes might play conservatively important roles in mediating expression of organellar genes as do corresponding characterized Arabidopsis *mTERF* genes. Group IX comprised

Table 1. Information of *mTERF* gene family in maize.

Gene Name[a]	Gene Locus[b]	Protein ID[c]	ORF(bp)[d]	Deduced protein[e]			Subcellular Location[f]	
				Length (aa)	MW(kDa)	pI	*in silico*	Proteomics[†]
ZmTERF1	GRMZM2G168665	DAA45475.1	912	303	32.53	10.44	C/C	C
ZmTERF2	GRMZM2G061542	DAA48102.1	1950	649	73.88	8.92	M/M	
ZmTERF3	GRMZM2G034217	ACG24064.1	1185	394	44.20	9.99	M/M	
ZmTERF4	GRMZM2G054517	DAA52215.1	651	216	24.24	9.78	−/−	
ZmTERF5	GRMZM2G159766	ACG24523.1	1203	400	43.67	10.00	M/M	
ZmTERF6	GRMZM2G170137	NP_001140442.1	645	214	23.65	8.93	−/C	C
ZmTERF7	GRMZM2G087679	This study	357	118	14.08	9.33	C/−	
ZmTERF8	GRMZM2G060114	ACG38629.1	1170	389	43.06	9.37	M/M	C
ZmTERF9	GRMZM2G130773	DAA41241.1	1527	508	57.95	8.98	M/M	C
ZmTERF10	GRMZM2G177019	NP_001144077.1	990	329	36.92	10.33	M/M	
ZmTERF11	GRMZM2G023257	NP_001151049.1	1167	388	42.98	9.93	M/M	
ZmTERF12	GRMZM2G113181	NP_001152615.1	1005	334	37.34	9.63	−/−	
ZmTERF13	GRMZM2G312806	NP_001141758.1	840	279	31.16	10.49	M/M	
ZmTERF14	GRMZM2G000610	NP_001145894.1	1470	489	54.68	8.09	M/M	
ZmTERF15	GRMZM2G119921	ACL52777.1	1188	395	44.62	9.30	M/M	
ZmTERF16	GRMZM2G395850	NP_001143033.1	1068	355	40.12	9.47	M/M	
ZmTERF17	GRMZM2G024550	ACF87053	1461	486	54.33	9.24	C/C	C
ZmTERF18	GRMZM2G017355	AFW73453.1	1176	391	42.83	9.31	M/C	C
ZmTERF19	GRMZM2G017429	NP_001169079.1	1173	390	42.43	9.11	ER/S	C
ZmTERF20	GRMZM2G158854	AFW73456.1	411	136	15.25	9.39	−/S	
ZmTERF21	GRMZM2G161146	DAA48102.1	1116	371	40.83	9.76	−/M	
ZmTERF22	GRMZM2G012999	AFW77957.1	1212	403	45.16	9.96	M/M	
ZmTERF23*	GRMZM2G426154	NP_001152167.1	1836	611	68.67	9.52	C/C	C
	GRMZM2G175930							
ZmTERF24	GRMZM2G142150	NP_001169565.1	1839	612	68.43	9.34	C/C	C
ZmTERF25	GRMZM2G062910	NP_001147866.1	1725	574	64.75	6.48	M/M	
ZmTERF26	GRMZM2G325350	NP_001130068.1	1155	384	42.90	9.47	M/M	
ZmTERF27	GRMZM2G029933	NP_001149660.1	1485	494	55.23	5.70	C/C	C
ZmTERF28	GRMZM2G068462	This study	510	169	18.58	9.94	−/−	
ZmTERF29	GRMZM2G157716	AFW88319.1	906	301	32.47	9.81	C/M	C
ZmTERF30		NP_001152154.1	900	290	33.25	9.81	ER/C	
ZmTERF31		ACG24302.1	999	332	37.71	9.12	−/−	

[a]Systematic nomenclature of maize mTERF genes according to the order of their location in maize chromosomes.
[b]Gene name annotated in MaizeGDB (http://www.maizegdb.org).
[c]NCBI accession number of the maize mTERF proteins; -, no identical protein found in NCBI Protein Database.
[d]Length of open reading frame (ORF) of maize *mTERF* genes in base pairs.
[e]Length (number of amino acids, aa), molecular weight(MW) (kilodaltons, kDa) and isoelectric point (pI) of the deduced proteins.
[f]Subcellular location of maize mTERF proteins predicted by Predotar [28] and (/) TargetP [29]. C, Chloroplast; M, Mitochondria; S, Secretion; ER, Endoplasmic Reticulum; -, none.
[†]The ZmTERF proteins were identified in maize plastid nucleoid or stroma in previous proteomic studies [31,32].
*Gene model of the maize *mTERFs* repredicted in this study instead of those annotated in MaizeGDB (http://www. maizegdb.org).

monocot-specific genes and its members were of similar protein length, except for ZmTERF4 and ZmTERF20. Another clade containing exclusive Arabidopsis mTERF proteins was found in the evolutionary tree and diverged from group VIII *mTERF* genes. This result suggests that group IX formed after divergence of monocots and dicots, and expansion of subgroups of group IX in monocots took place before divergence of rice and maize. However, we cannot exclude the possibility that rice-specific subgroups in group IX evolved after species formation of rice (Figure 1), because syntenic orthologs of several maize *mTERF* genes were not be found in rice genome (Table S2). Putative paralogous genes were identified with the criterion that two mTERF proteins were clustered pairwisely and supported by > 95% bootstrap values, leading to formation of three pairs of maize *mTERF* paralogs, including *ZmTERF15* vs. *ZmTERF16*, *ZmTERF4* vs. *ZmTERF5*, and *ZmTERF11* vs. *ZmTERF20*.

Prediction of Potential Subcellular Location for ZmTERF Proteins

Unlike human mTERF proteins which all reside in mitochondria, GFP fusion transient expression experiments showed that most members of Arabidopsis mTERF proteins were targeted to mitochondria or chloroplasts [17]. To investigate the subcellular location of ZmTERF proteins, we performed *in silico* prediction using TargetP [28] and Predotar [29]. Totally, 25 of 31 ZmTERF proteins were predicted to enter mitochondria or chloroplast in either TargetP or Predotar (Table 1), and three mTERF proteins ZmTERF19, -20 and -30 could be secretory proteins, while four mTERF proteins (ZmTERF4, -12, -28 and -31) were predicted to localize to any other location (Table 1). In this result, six maize mTERF proteins (ZmTERF6, -7, -18, -20, -21, and -29) were targeted to different organelles predicted by the two programs. In the six proteins, ZmTERF18 and ZmTERF29 proteins possibly have dual transit peptides for mitochondria and chloroplasts.

Mitochondrial and plastid proteomic database identification provide another way to survey the compartmentalization of the proteins of interest. Here we used the maize mTERF proteins to screen PPDB [30], and found that ten mTERF proteins (ZmTERF1, -8, -9, -17, -18, -19, -23, -24, -27 and -29) were detected in maize plastid nucleoid [31] and ZmTERF6 was observed in maize chloroplast stroma (Table 1 and Table S2) [32]. Of these mTERF proteins, ZmTERF8 and ZmTERF9 are not consistent with the prediction produced by TargetP [28] and Predotar [29] (Table 1). Subcellular targeting of rice mTERF proteins were also predicted using TargetP [28] and Predotar [29] (Table S3). By mapping the subcellular location information of plant mTERF proteins on the evolutionary tree (Figure 1), we found that the *mTERF* genes can group according to targeting predictions in the evolutionary tree (Figure 1). Intragroup mTERF proteins localize to the same organelles with several exceptions, such as ZmTERF9 in group I, ZmTERF13 in group II, and LOC_Os01g27690 in group IX. Groups I, II, IV and VII targeted to the chloroplast, and groups III, V, VI, VIII and IX targeted to the mitochondria. Surprisingly, there was a mitochondria-targeted subgroup in group IV, including AT2G44020, ZmTERF14 and LOC_Os02g54200.

Gene Structure and Conserved Motifs of *ZmTERF* Genes

Studying the structure of genes of interest is crucial in biology and can provide important clues concerning gene evolution. The exon–intron organization of maize *mTERF* genes was obtained by comparing the cDNA sequences of maize *mTERF* genes with the corresponding genomic DNA sequences. The maize *mTERF* cDNA sequences were derived from sequenced PCR products (Figure S2), which were amplified from B73 seedling leaf total RNA with gene-specific primers (Table S4) except for three genes *ZmTERF4*, *-7* and *-20* which were not detected in this study and of which cDNA sequences were extracted from corresponding gene models annotated in MaizeGDB (http://www.maizegdb.org/). The genomic DNA sequences were retrieved from MaizeGDB (http://www.maizegdb.org/). This result showed that quite a large number of *mTERF* genes (79%) were intron-free (Figure S3). Six *mTERF* genes containing introns belonged to groups I, III, IV and VII, of which only group IV had multiple members. *ZmTERF12* (group III) had a 5′-UE (Untranslated Exon) in 5′-UTR (UnTranslated Region) and had no intron in the coding sequence, while *ZmTERF6* (group VII) had multiple 5′-UEs in 5′-UTR and only one intron in the coding sequence. There were two different transcripts for *ZmTERF6*, nevertheless, the alternative splicing event occurs in 5′-UTR and had no effect on the coding sequence (Figure S3). Most intron-free *mTERF* groups contained multiple members, particularly group IX, implying that ancestors of these groups were evolved via retrotransposition.

Conserved mTERF motifs containing 32 aa residues have been characterized in human mTERF proteins and are thought to be DNA-binding modules [8]. To identify any possible known motifs in maize mTERF proteins, we carried out SMART and PFAM database searches and revealed that, except for mTERF motifs (approximately 38-aa SM00733 in SMART and 345-aa PF02536 in PFAM), no other known motifs were identified (Figure S4). This result suggested that these genes play conserved biological roles. For SM00733 mTERF motifs, maize mTERF proteins had 1–11 mTERF motifs. The intragroup members had similar protein architecture, for example group IX mTERF proteins had 4–6 mTERF motifs except for two putative pseudogenes, *ZmTERF7* and *ZmTERF20* (Figure S4), and group IV proteins with 9–11 mTERF motifs. To determine sequence features of maize mTERF motifs, 37 motif sequences within seven mTERF proteins from different groups were aligned with ClustalW 2.0 [33] and mTERF motif aa residues were graphically represented (Figure 2A). Three repeats of leucine zipper-like heptad X_3LX_3 identified in human mTERF proteins previously were also observed in maize mTERF motifs (Figure 2B). These results indicate that maize mTERF proteins harbor conserved mTERF motifs and these motifs contain conserved leucine resides like those in human mTERF motifs (Figure 2B) [8], implying plant mTERF proteins might possess similar structures and functions to human proteins.

To discover putative motifs shared among related proteins within the maize *mTERF* family, the MEME program [34] was used. Overall, 15 highly significant motifs (<1e-50) were mined among these 29 proteins, and designated as motifs 1–15 (Figure 3). Of these motifs, motifs 1, 3, 4 and 5 were common to all maize mTERF proteins. Sequence comparison with PFAM mTERF seed alignment indicated that motifs 1, 3, 4, 5, 9, 10 and 12 overlaid with the PFAM mTERF domain (PF02536) (Table 2). Motifs 2, 6, 7, 8, 13 and 14 were specific to mitochondria-targeting mTERF proteins, especially to the group IX mTERF proteins (Figure 3). It is possible that group-specific motifs contribute to distinct functions of these *mTERF* genes.

Higher Structure Analysis of ZmTERF Proteins

Human mTERF proteins regulate mitochondrial gene expression via binding to mitochondrial DNA (hsmTERF1–hsmTERF3) or RNA (hsmTERF4) [13–15,35]. The structure of three of four mTERF proteins have been determined by X-ray diffraction in human [5,16,36]. The mTERF protein structure shows they are modular proteins, and the mTERF module comprises two or three tandem α-helixes [36–38]. There are eight modules in hsmTERF1

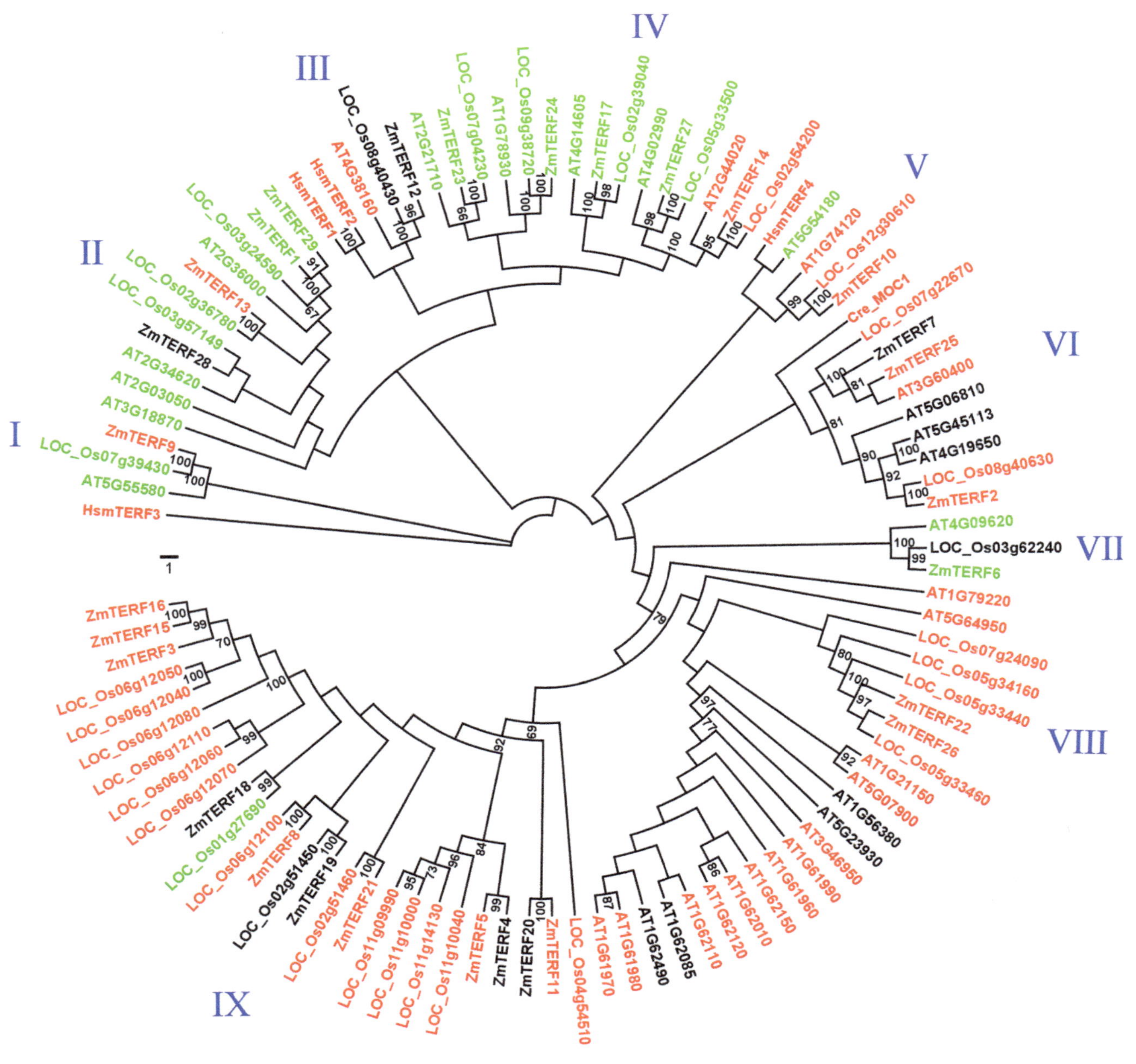

Figure 1. Phylogenetic analysis of maize and other known mTERF proteins. Phylogeny tree was constructed using RAxML with Arabidopsis, maize and rice mTERF protein sequences. The numbers associated with the branches indicate the bootstrap values (%) for 500 bootstrap replicates and only bootstrap values of >60% are shown. The scale bar represents the number of amino acid substitutions per site. Nine groups designated I–IX are shown outside. Subcellular localization of mTERF proteins is shown in red for mitochondria, in green for chloroplasts and in black for other places or a discrepancy in prediction by TargetP [28] and Predotar [29].

[37], seven in hsmTERF3 [36] and six in hsmTERF4 [38]. Human mTERF1 adopts a fold similar to HEAT [39,40] and PUM/PUF proteins [41] and forms a positively charged groove in the protein surface where dsDNA is docked [36,37]. The overall structure of mTERF proteins are very different from previously proposed models [8,11]; however, no structure has been resolved for plant mTERF proteins yet. Arabidopsis BSM/RUG2 protein was shown to bind to dsDNA in a nonspecific manner, and the structure of the protein was modeled based on homology to human mTERF1 [17].

Here, the structure of ZmTERF27 protein without N-terminal transit peptide, a maize mTERF protein homologous to hsmTERF1 and BSM/RUG2, was developed using I-TASSER program [42]. The 3D structure modeling of maize mTERF protein (Figure 2C) was similar to the structure of hsmTERF1. ZmTERF27 contains nine mTERF modules which differ from the mTERF motifs mined by SMART [26] (Figure 2D and Figure S5). ZmTERF27 protein structure can fold to form a positively charged groove (Figure 2E), suggesting that it could bind to dsDNA as does human mTERF1, and the 3D structure complex including ZmTERF27 and human mitochondrial DNA was developed using the TM-align program [43]. This result indicates that plant *mTERF* genes, or at least parts of them, play conserved roles in regulating organellar gene expression by DNA-binding activity similar to human *mTERF* genes. Compared with hsmTERF1, however, ZmTERF27 contained two excess domains in N- and C-termini, respectively (Figure 2). Even in the homologous region, the mTERF modules of ZmTERF27 cannot match those modules in hsmTERF1 very well due to unaligned aa residues (Figure 2 and Figure S5). These variations demonstrate that ZmTERF27 has specialized biological function in plants. Other maize mTERF proteins containing >4 mTERF motifs can also be modeled to form 3D structures like that of ZmTERF27 except for ZmTERF2 and -25 (data not shown).

Chromosomal Location and Gene Duplication Analysis of *ZmTERFs*

To investigate the genomic organization of *ZmTERF* genes, the positions of the *mTERF* loci on maize chromosomes were mapped. The distribution of 29 *ZmTERF* genes on nine of ten maize chromosomes was uneven (Figure 4). Chromosomes 1 and 5 have six and seven *ZmTERF* genes and only one *mTERF* gene was observed in chromosomes 6 and 9, respectively. Analysis of maize *mTERF* genes in SyMAP [44] and PLAZA [45] revealed four (i.e. two pairs) *mTERF* genes could be assigned to maize segmental duplication and five *mTERF* genes could be involved in tandem duplication (Figure 4). The five tandemly duplicated *mTERF* genes were localized on chromosomes 1 and 5, respectively, and fell into two clusters, one for *ZmTERF4* and *ZmTERF5* and the other for *ZmTERF18*, *-19* and *-20* (Figure 4). The duplicated genes were inconsistent with those identified in the phylogenetic tree because two *ZmTERF* genes (*ZmTERF11* and *ZmTERF20*) were not in homologous regions. All of these duplicated genes exhibited high sequence similarity, especially *ZmTERF15* and *ZmTERF16* which shared a 343-aa peptide. Three pairs of homologs involving two tandem duplication events and one segmental duplication event belonged to group IX, while two other pairs were of groups II and IV, respectively. It is worth noting that in three pairs of duplicated *mTERF* genes, *ZmTERF4*, *-7* and *-20* genes encoding truncated redundant peptides were silent in maize compared with their corresponding paralogous copies. These findings suggest tandem and segmental duplication events both contributed to the current complexes of the maize *mTERF* gene family unlike Arabidopsis *mTERF* genes which were duplicated mainly by tandem duplication events [17,18].

Analysis of *cis*-elements in Promoters of *ZmTERFs*

The *cis*-elements are important molecular switches involved in the transcriptional regulation of genes during plant growth and development, and abiotic stress responses [46–48]. The 2-kb promoter sequences upstream from the translation start codon of 23 maize *mTERF* genes were retrieved from maize genome sequences in MaizeSequence (http://www.maizesequence.org/index.html) and used for stress-related *cis*-element analysis in the PlantPAN server [49]. All of putative environment stimulus-responsive *cis*-elements were identified in *mTERF* gene promoters and divided into eight classes: including Auxin-responsive element (ARE), Light-regulated element (LRE), Drought/dehydration-responsive element (DRE), Gibberellic acid (GA) responsive element (GARE), ABA-responsive element (ABRE), Ethylene-responsive element (ERE), Anaerobically induced element (AIE) and Low-temperature responsive element (LTRE). Some *cis*-elements may respond to multiple plant hormones or environment stimuli – for instance LTRECOREATCOR15 (CCGAC) is induced by ABA, drought and low-temperature [48]. Only the *cis*-elements in positive strands were collected and displayed in Figure 5. The detailed information of stress-related *cis*-element sequences and annotation is found in Table S5, and the position and abundance of all *cis*-elements predicted to localize in promoter regions of maize *mTERF* genes are shown in Table S6.

All 23 *mTERF* promoters were highly enriched with LREs, and the average number of the LRE sequences in these promoters was 19.5, of which only five genes had <15 LRE boxes (Figure 5). Of 23 *mTERF* promoters, 17 were rich in DREs with >10 DREs per promoter, in which *ZmTERF25* had 25 DREs. There were nine *mTERF* promoters containing >5 ABREs per promoter – especially *ZmTERF19*, *-2* and *-3*, with 25, 14 and 14 ABREs, respectively. We also found that many EREs appeared in maize *mTERF* promoters, and eight *mTERF* promoters each contained > 5 EREs. In addition, *ZmTERF25* and *-19* contained 12 and six LTREs, respectively, in their promoters, and others had <4. The two *mTERF* genes also possessed multiple abiotic stress-responsive elements. There were 19 cold-, drought- and ABA-regulated elements in *ZmTERF25*; and 11 drought-, high-salt- and cold-responsive elements in *ZmTERF19* (Figure 5). Notably, six *mTERF* genes had 3–5 AIEs, suggesting these genes were probably induced by anaerobic conditions, such as waterlogging stress. Therefore, expression of maize *mTERF* genes would be regulated by various environmental factors.

Expression Pattern Analysis of Maize *mTERF* Genes in Different Tissues

The gene expression patterns can provide important clues for the gene function. Two approaches were applied to analyze the expression patterns of the maize *mTERF* genes in different tissues and organs, including expressed sequence tag (EST) profiles and microarray data.

For analysis of EST profiles, the maize ESTdb in NCBI (http://www.ncbi.nlm.nih.gov/dbEST/) was searched using BLASTN program. Most maize *mTERF* genes were expressed – except for *ZmTERF4*, *-7* and *-20* with no EST sequence detected (Table S7), providing strong indication for expression of most *mTERF* genes; However, the frequency and source of ESTs for *mTERF* genes varied, indicating their differential expression patterns. The EST sequences for *mTERF* genes were mainly from six single different tissues (root, shoot, ear embryo, endosperm and tassel) and several cDNA libraries derived from multiple mixed tissues. *ZmTERF6* and *-8* showed high expression in all tissues and organs used here, and several maize *mTERF* genes have tissue-specific expression patterns: *ZmTERF1* in embryo, and *ZmTERF11*, *-21* and *-29* in shoot. Additionally, EST hits for *ZmTERF13*, *-18*, *-19*, *-23* and *-28* were only from mixed tissues.

Expression patterns of maize *mTERF* genes were examined in different organs by estimating their transcript levels determined by Nimblegen maize microarray experiments encompassing 60 tissues at different developmental stages of maize: including germinating seed, primary root, whole seedling, stem and shoot apical meristem, internodes, cob, tassel and anthers, silk, leaf, husk and seed [50]. The detailed information for the tissues and organs was described by Sekhon *et al* [50]. The expression data for only 24 maize *mTERF* genes were collected, excluding *ZmTERF4*, *-7*, *-16*, *-20*, and *-28*. The signal values for all these maize *mTERF* genes are given in Table S8. The hierarchical clustering of microarray expression data using Cluster 3.0 [51] revealed expression patterns for 24 maize *mTERF* genes in diverse tissues and organs covering the whole life of maize, but no clearly distinct expression pattern groups emerged (Figure 6). The coefficient of variation (CV = sd/mean, where sd and mean represent the standard deviation and mean expression level of a gene across all

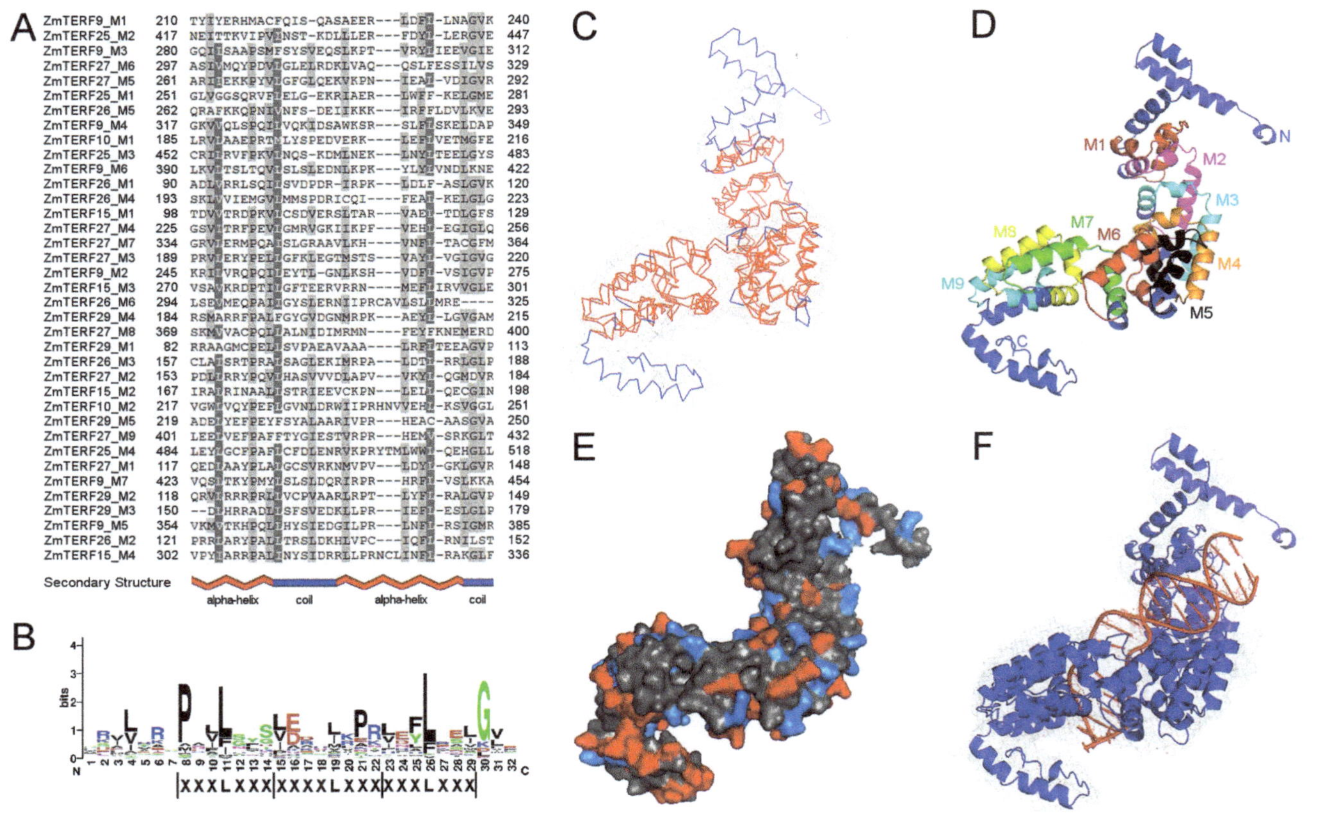

Figure 2. Structure of maize mTERF protein. (A) Multiple sequence alignment of maize mTERF motifs. Alignment of the conserved maize mTERF motifs was conducted with CLUSTALW2.0, and displayed using GeneDoc with the conserved residue shading mode. The conserved secondary structure was predicted by MINNOU [69] and shown under the alignment panel. (B) Sequence representation LOGO derived from multiple sequence alignment of the mTERF motifs and three leucine zipper-like X_3LX_3 repeats are highlighted. (C) Overlay between human mTERF1 model (PDB 3MVA) and ZmTERF27 structure developed by I-TASSER [42]. The 3D structures of HsmTERF1 and ZmTERF27 proteins are indicated by blue and red stricks, respectively. The matched regions are represented by double red stricks. (D) ZmTERF27 model is displayed in cartoon and nine mTERF motifs are indicated in different colors. (E) Electrostatic surface potential of ZmTERF27. Only basic amino acid residues (histidine, arginine and lysine) and acidic residues (aspartic and glutamic acids) were analyzed and are shown on the surface of ZmTERF27 model in light blue and red, respectively. (F) DNA-binding mode of ZmTERF27. The dsDNA (red) derived from human mitochondrial DNA was predicted to be docked in the groove formed by ZmTERF27 model (blue) using TM-align program [43].

the tissues, respectively) value of each *mTERF* gene was calculated to estimate the expression variation. This result showed that expressions of 18 *mTERF* genes were relatively stable across all tissues (CV<10%) (Table S8), implying that these genes were likely involved in basal metabolic or 'housekeeping' functions. In these genes, *ZmTERF2*, *-5*, *-6*, *-14*, *-15*, *-22*, *-25*, *-26* and *-27* were highly expressed throughout the 60-tissue panel and *ZmTERF1*, *-3* and *-13* genes were less expressed (Figure 6). The remaining eight stably expressed genes showed slight expression variation in certain tissues. *ZmTERF8* and *-10* showed similar expression patterns, with low expression in anther, root, pericarp and several types of leaves (Figure 6). *ZmTERF9* showed low expression in root and silk compared to other tissues. *ZmTERF24* was relatively less expressed in root, pericarp and germinating seed. *ZmTERF11* showed low expression in pericarp, silk, germinating seed and most types of leaves. The CV values of the other six genes (*ZmTERF12*, *-17*, *-18*, *-19*, *-23* and *-29*) were >10%, in which *ZmTERF17* and *-29* showed large expression variation with CV value >15%. The expression of *ZmTERF19* and *-29* was specific to leaf. *ZmTERF12* and *-17* were remarkably expressed in leaf and moderately expressed in embryo. The expression pattern of *ZmTERF23* is similar to that of *ZmTERF24*, and its expression level in leaf, endosperm, immature tassel and cob was higher than in other tissues or organs. Notably, *ZmTERF18* was lowest expressed over the whole life of maize plants. The chloroplast-targeting *mTERF* proteins, *ZmTERF6*, *-9*, *-23*, *-24*, *-27* and *-29* were highly expressed in leaves, especially *ZmTERF29* with leaf-specific expression, suggesting that these genes played important roles in chloroplast development and biogenesis.

To confirm the expression variations of maize *mTERF* genes, qRT-PCR was performed to measure the expression of eight *mTERF* genes in five different tissues, including mature seed, immature tassel, unpollinated silk, seedling leaf and root. In the eight genes, only *ZmTERF11* was targeted to mitochondria while the rest were all imported to chloroplasts. Of the senven chloroplast *ZmTERFs*, four *mTERF* genes (*ZmTERF19*, *-23*, *-27* and *-29*) exhibited similar expression patterns in the five tissues, and they were expressed predominantly in leaf. For other four mTERF genes, *ZmTERF6* was highly expressed in seed and leaf; *ZmTERF9* and *-24* showed high expression levels in seed, silk and leaf; and *ZmTERF11* was abundantly expressed in root (Figure S6). Additionally, all of the selected *ZmTERFs* showed lower expression

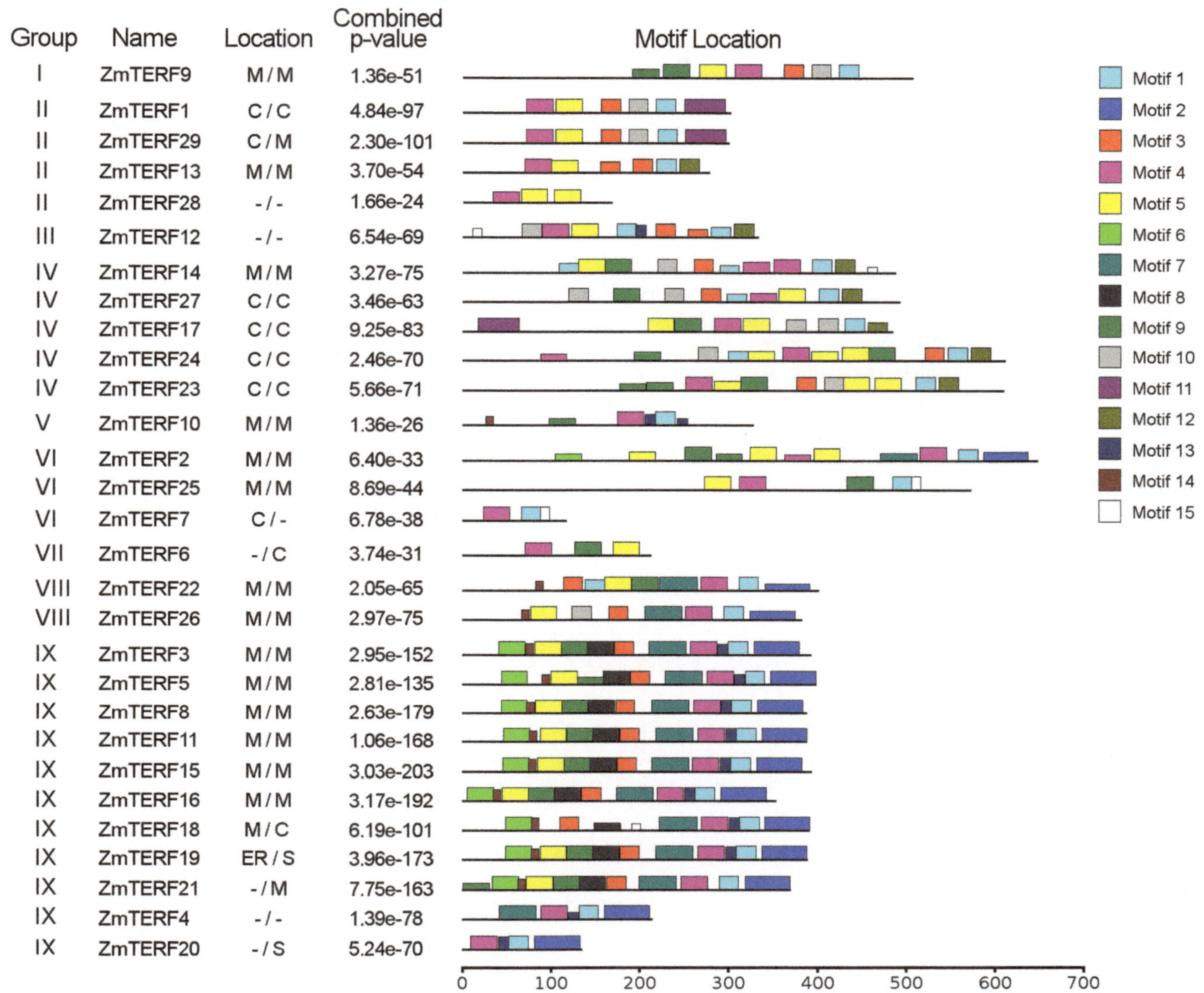

Figure 3. Schematic representation of putative conserved motifs identified MEME in maize mTERF proteins. Putative conserved motifs shared by maize mTERF proteins were mined in MEME program. Fifteen motifs are indicated by different colored boxes and the regular motif sequences are shown in the Table 2.

levels in tassel compared to other tissues. These results were consistent with the observation in macroarray data.

Expression pattern shifts of the duplicated paralogous genes can reflect the maintenance of duplicate genes through non-functionalization, subfunctionalization or neofunctionalization [52]. For duplicated *mTERF* genes, two pairs (segmentally duplicated *ZmTERF1* and *ZmTERF29*, and tandemly duplicated *ZmTERF18* and *ZmTERF19*) both showed divergent expression profiles for duplicate genes. *ZmTERF1* and *ZmTERF18* showed low expression levels in all detected tissues or organs; whereas their paralogs, *ZmTERF29* and *ZmTERF19*, were preferentially expressed in leaf. This result indicates the fate of two *ZmTERF* pairs could be described as neofunctionalization since the expression of one copy of the paralog had an obvious increase in leaf (Figure 6).

Expression Analysis of Maize *mTERF* Genes During Leaf Development

Corn is an important cereal and a typical C_4 plant. In addition, plant *mTERF* genes can regulate chloroplast gene expression. It is speculated that maize *mTERF* genes may be involved in chloroplast biogenesis and development, and C_4 photosynthesis occurring in the chloroplast. To validate this hypothesis, expression levels of maize *mTERF* genes during the development of C_4 photosynthesis were extracted from RNA-seq data deposited in the eFP browser [53]. The RNA-seq data were derived from a leaf continuous developmental gradient and mature bundle sheath and mesophyll cells [54]. The leaf gradient comprises four representative zones of the leaf blade, including a basal zone (Base: sink tissue where cell division is active, cell-fates are being determined and proplastid appears), a transitional zone (Ligule+4: undergoing the sink–source transition), a maturing zone (Ligule+9: with strong light-mediated development and formation of mature

Table 2. Multiple Em for Motif Elucidation (MEME) protein motifs detected in maize mTERF proteins.

Motif	Protein sequence (regular expression)*
1	[YE][LI][VA]E[FR]PA[LYF][LF]S[YF]SL[ED]RR[IL][VK]PR[HY]
2	**KGL[LV]NS[DN]RSF[FLY][STN]VAA[IL][GST][ED][ES]KFRS[KR][FY][VI] [HD]P[HY][KE][ED]SVP[GS]LA[AD]AYA[SA][AS]CAG[KE][VH][PQ]W**
3	[LI]LSY[DS][LIV]EKV[LV][KR]PN[LI][EA][FL]L[RQ][SE][CL]G
4	[ST]LG[VW]SEA[DE][VL]AR[AV]V[KS][KR]xPQ[ILV]LGFSEER[LIV]x[RK]
5	[FL]L[AS]ELG[LV]Sx[AK][ED][IL][AG][AK][VL][VLI][AR]RDPR[LV]LC[SA][SD][VI][DE]R
6	**F[AS][VA]EDYLVA[TAR]C[GH]LT[PQR][AE]QA[LA][KR][AV][SA]KH[LV]SHLK**
7	**P[ED][RS][VIL][QR]EA[LV]AR[AL]EE[FL]G[VIL][PQ][PR]G[ST][GQ][LM] FRH[AG][LV]x[ATV][VF]A[CIS]L[GS][PR]EK[IV]A[AS]K**
8	**R[HR][PS]S[FLV][VR][SL][KN][LV][EQ]F[WY][LI][SP][FLV][FL]GS[FLP][DE][RE][LV][LI][RQ]A[LV][RK]IN**
9	LAPR[VL]AEL[RL][DS]LG[LVF]SPSQ[IL]PR[LI]LxVxPxL[FL]
10	[ART][RKS][FY]PA[LV][LF][GT][LYC][SG][VL][DE][GK][NHK][LMI][RKV]P[KV][VAY][DEQ][YF]
11	P[LR][PG][AL][MQ]L[RP]P[GW][DER][AL][KS][FL][RS][ARS][TS][LS][DST] SC[VR][GS][SC][TM]L[PV][RS]R[RQ][SL][AP][LI][WC][HN] A[TQ][WS][VY][DA]DD[AL][ATW][AV]A
12	R[GN]Ix[CM]SLx[EW]ML[NT]C[NS]DEKFAER
13	**EFLIN[EV]VGLE[PV]**
14	**[SD][PA]S[KN][AP]DAV**
15	TML[RW]WLQEHG

*Mitochondrial mTERF protein-specific motifs are shown in bold, and the residues underlined overlap with the PFAM mTERF domain (PF02536).

chloroplast) and a mature zone (Tip-1: fully differentiated and active C_4 photosynthetic zone) [54]. RPKM (reads per kilobase per million mapped reads) values were determined for 25 maize *mTERF* genes in each of the zones and cells tested (Figure S7). In these genes, six genes showed higher expression levels with maximum RPKM >20 compared to *ZmTERF2*, *-3*, *-8*, *-9*, *-10*, *-13*, *-14*, *-15*, *-16*, *-21*, *-25* and *-26* with maximum RPKM<10. Of 25 maize *mTERF* genes, 16 were significantly differentially expressed along the leaf developmental gradient with CV >50% and *ZmTERF2*, *-10*, *-11*, *-12*, *-13*, *-15*, *-16*, *-21*, *-22*, *-26* and *-27* were down-regulated from the Base to Tip-1 zones. The result indicates that the *mTERF* genes were involved in leaf development and chloroplast biogenesis, especially required for early development of chloroplast. Ten *mTERF* genes (*ZmTERF2*, *-9*, *-10*, *-11*, *-13*, *-14*, *-15*, *-18*, *-25* and *-29*) were differentially expressed by >2-fold between bundle sheath and mesophyll cells. Notably, in the ten genes, only *ZmTERF29* was highly expressed with RPFM >20. These genes with different expression levels between bundle sheath and mesophyll cells are speculated to participate in coordination of chloroplast gene expression for C_4 photosynthesis.

Expression Profile of Maize *mTERF* Genes under Abiotic Stress

Previous reports on Arabidopsis *mTERF* genes have shown that loss of *mTERF* genes affect expression of stress-related genes [21,24], phytohormone signal tranduction [17], and salt and osmotic stress tolerance [23], as well as organellar gene expression regulation. It was therefore envisioned that the maize mTERF genes may also be influenced by environmental signals. To examine the expression patterns of maize *mTERF* genes under various conditions including abiotic stress, several representative maize *mTERF* genes were selected for quantitative RT-PCR analysis under light/dark, salt and phytohormone treatments. Seven *mTERF* genes (*ZmTERF8*, *-15*, *-19*, *-21*, *-23*, *-25* and *-27*) were investigated under light and dark conditions – all these genes were up-regulated under light with peaks at 4 h after illumination and conversely, they were all down-regulated in darkness (Figure 7). In these genes, two chloroplast-targeting genes, *ZmTERF23* and *-27*, were largely up-regulated under light; and were also as well as *ZmTERF8*, heavily down-regulated in darkness by >2-fold (Figure 7). Plastids will develop from etioplasts to chloroplasts during de-etiolation of maize seedlings under illumination and photosynthetic apparatus will be developed. The result implied that *mTERF* genes are generally required for chloroplast development and photosynthesis in maize.

Expression profiles of six maize mTERF genes at early stages (0–4 h) of response to salts (NaCl and $AlCl_3$) and phytohormones (ABA and NAA) were analyzed (Figure 8). Of these six genes, *ZmTERF12* and *ZmTERF28* gene products were targeted to neither chloroplast nor mitochondria, while the remaining four genes (*ZmTERF2*, *-5*, *-11* and *-13*) all resided in mitochondria, with *ZmTERF5* and *ZmTERF11* of group IX. The expression changes of six *ZmTERFs* in response to salts and plant hormones were all less than 2 folds except *ZmTERF28* which was remarkably up-regulated with four stimuli (Figure 8). Expression levels of *ZmTERF2*, *-5* and *-11* were not affected by NaCl, $AlCl_3$ and ABA treatments except those at certain time spots. For the *ZmTERFs* involved in stress response, *ZmTERF12* and *-13* showed similar expression patterns in response to NaCl and $AlCl_3$, in which they were up-regulated up to 1 h followed by suppression in expression. For ABA treatment, *ZmTERF12* was down-regulated at first one hour after treatment and it was up-regulated thereafter, while the expression of *ZmTERF13* was suppressed after ABA treatment. For NAA treatment, the expression of all *ZmTERF* genes tested here were slightly changed at 2–3 time spots (Figure 8). These results suggested that maize *mTERF* genes, at least for these *ZmTERF* genes investigated here, had a limited role in response of maize to stress conditions.

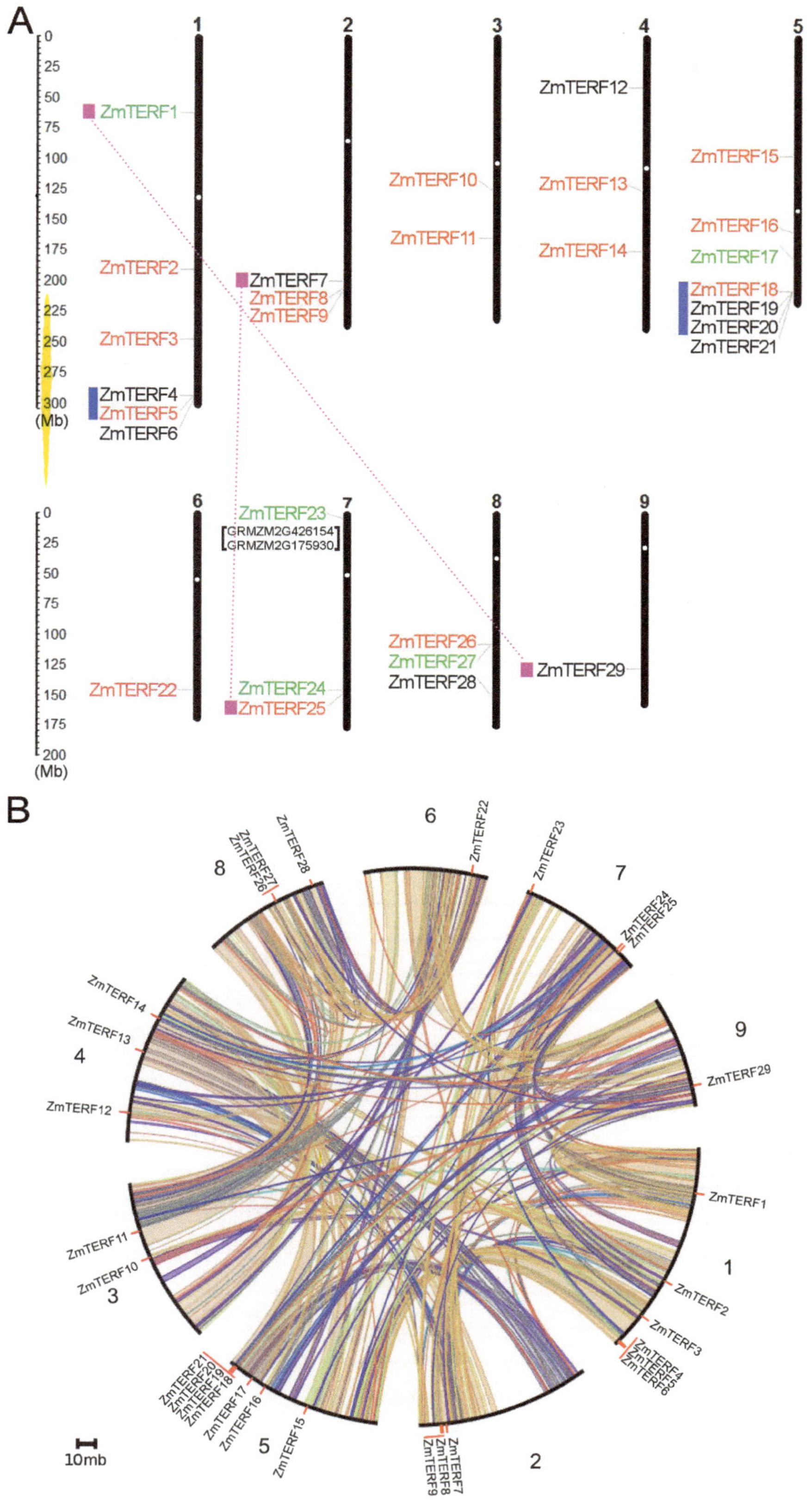

Figure 4. Chromosomal localization and gene duplication events of maize *mTERF* genes. (A) Distribution of *ZmTERF* genes in maize chromosomes. Twenty-nine *ZmTERF* genes were mapped on the nine maize chromosomes. Segmentally duplicated paralogs are connected by dotted magenta line, and tandemly duplicated homologs are marked by blue bars before the genes. The centromere of each chromosome is indicated with a white circle, and chromosome numbers are indicated at the top of each chromosome. (B) Duplicated blocks in maize chromosomes.

The circular image retrieved from PLAZA database [45] show inter-chromosome homologous regions connected by bands in different colors. The chromosome numbers and *mTERF* genes are indicated outside.

Discussion

To fully understand the role of *mTERF* genes in plants, it is therefore essential to characterize more *mTERF* genes in diverse plant species, particularly in crops. In this study, we report the systematic characterization of maize *mTERF* genes using bioinformatics and molecular biological approaches.

Evolution of *ZmTERF* Genes

We carried out comprehensive searches for putative *ZmTERF* genes and found 29 *mTERF* genes in the sequenced maize genome, including a spliced gene model by genomic gap, and two extra *mTERF* genes identified in the NCBI Protein Database and cDNA library (Table S1). The two *ZmTERF* genes could not be mapped onto maize chromosomes. The number of *ZmTERF* genes was slightly less than for rice and Arabidopsis despite maize having a much larger genome. Therefore, the number of *ZmTERF* genes would be likely underestimated due to unsequenced genomic gaps or mis-annotation of gene models in MaizeGDB (http://www.maizegdb.org/) [55]. However, flowering plants have more *mTERF* genes than lower plants and mammals [18]. Given the complicated regulation of organellar genome transcription in higher plants [2,3,56,57] and the conserved roles of Arabidopsis and *C. reinhardtii mTERF* genes in control of mitochondrial gene expression [5], expansion of the *mTERF* gene family in higher plants could be driven by complexity of organellar genomes.

Arabidopsis *mTERF* genes have been classified into eight groups based on phylogenetic relationships [17]. The phylogenetic tree including the *mTERF* proteins from maize, rice and Arabidopsis had similar topological structure to the tree constructed with *mTERF* genes of lower plants (Figure 1) [17]. Based on

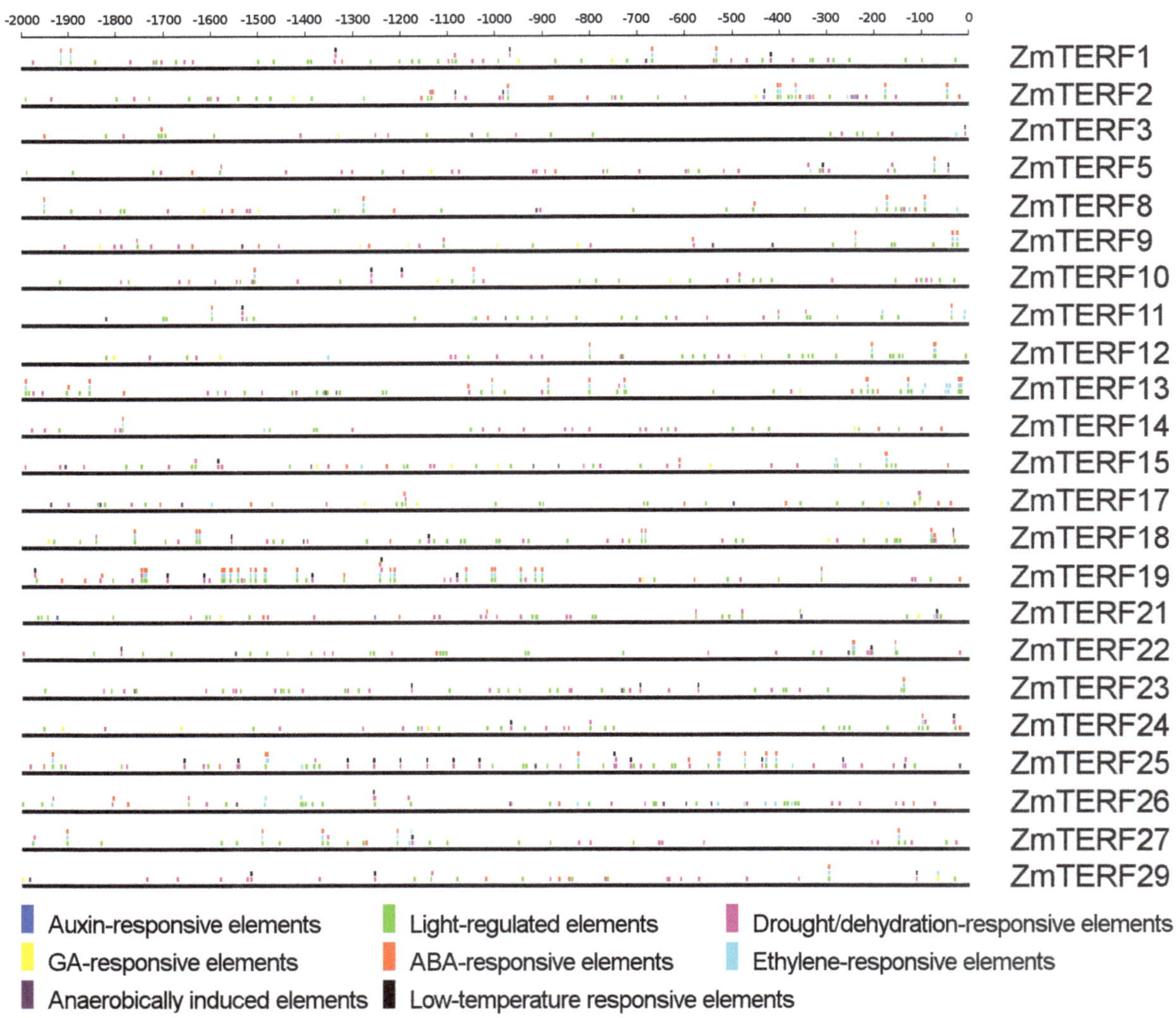

Figure 5. Phytohormone- and abiotic stress-responsive *cis*-elements in the promoter regions of maize *mTERF* genes. Putative LTRE, DRE, ARE, LRE, GARE, ABRE, ERE, AIRE and LTRE core sequences in the 2000-bp promoter regions of maize *mTERF* genes were predicted using PlantPAN [49]. The lines denote promoter sequences. Only the elements represented by different color boxes and located in the forward strand (sense strand of the gene) are indicated above the lines.

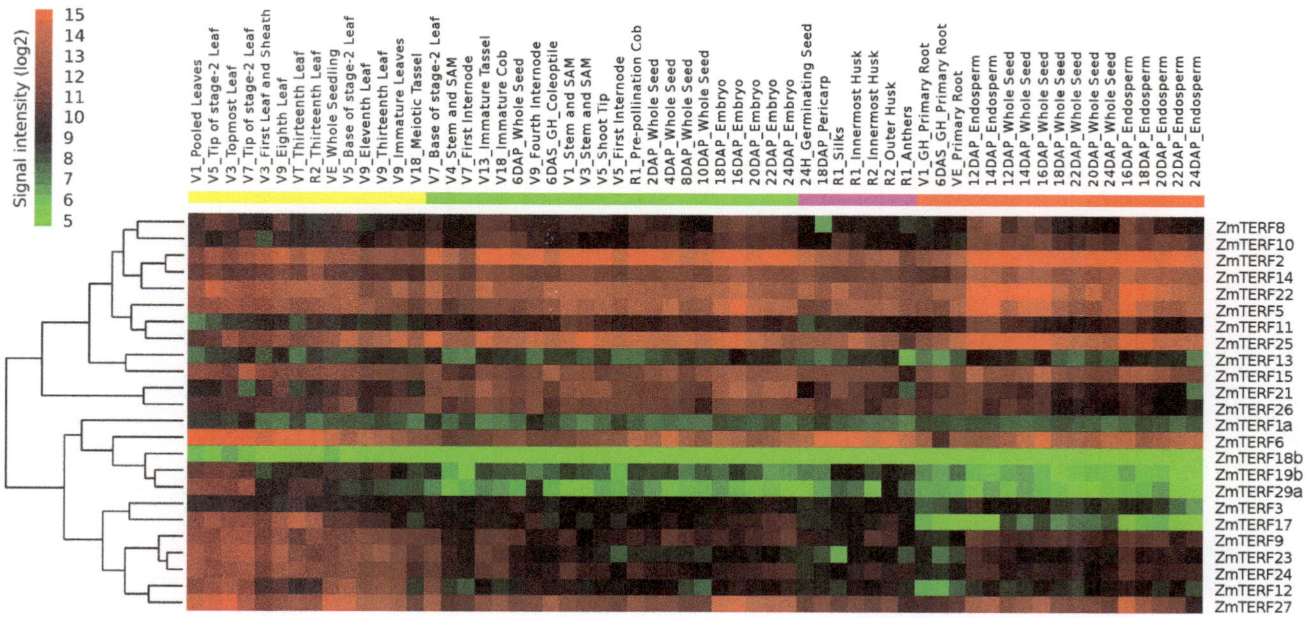

Figure 6. The expression profile of *mTERF* gene family in different tissues of maize. Dynamic expression profiles of maize mTERF genes in maize different tissues were derived from Nimblegen Maize Microarray data deposited in PLEXdb [72]. Relative expression level is denoted by intensity of red color (higher) or green color (lower). Two pairs of duplicated paralogs are marked by lowercase letters.

phylogenetic relationships, *ZmTERF* genes were divided into nine groups. All information about gene structure, motif distribution, subcellular location and protein size of *ZmTERF* genes support the conserved *ZmTERF* genes in the same groups shown in Figure 1. Group IX members shared group-specific motifs 2, 6, 7, 8 and 14, which confer novel functional characters compared with other groups. Only group IX was species-specific while the other groups were all composed of orthologs in various organisms used here, even including lower plants [17]. We conclude that group IX as well as VIII is newly generated before the divergence of angiosperms and have adopted a different evolutionary route from the other groups in plants; however, the other groups mainly containing chloroplast-targeting *mTERF* genes represent the ancestral family composition as supposed [17]. Sequence and evolutionary conservation leads to functional identification of Arabidopsis *mTERF* genes belonging to the ancestral groups (except for groups VIII and IX), since they are required for organellar gene expression as described for *hsmTERF*s.

It is largely accepted that the maize genome has undergone several rounds of genome duplication, including the latest whole genome duplication (WGD) about 5–12 million years ago (mya) after divergence from sorghum [55]. However, the WGD would not increase the amount of *ZmTERF* genes compared with sorghum, in which 38 members have been identified [18], suggesting that extension of the maize mTERF family cannot be explained by WGD alone. It is reported that nine Arabidopsis *mTERF* genes involved in tandem duplication were localized on chromosome 1 and formed a single cluster for these paralogs [18], whereas in the maize genome there were two tandemly duplicated *mTERF* gene clusters in chromosomes 1 and 5. The two clusters only contained five *ZmTERF* genes, representing 17% of *ZmTERF* genes, and are comparative with the four homologs generated by segmental duplication (Figure 4). Both segmental and tandem duplication are important for *mTERF* gene family expansion in maize, unlike for rice and Arabidopsis where tandem duplication played a predominant role in gene duplication [17,18]. Duplicated genes may have three different evolutionary fates nonfunctionalization, subfunctionalization or neofunctionalization [58]. Sequence homology and gene expression analysis revealed that *ZmTERF4*, *-7* and *-20* were pseudogenes encoding truncated *mTERF* peptides and showed no expression in microarray data (Figure 6) and RT-PCR assay (Figure S2). This result showed that the three genes underwent nonfunctionalization by degenerative mutations during maize evolution. Two pairs of duplicates (*ZmTERF1* and *-29*, and *ZmTERF18* and *-19*) were stably preserved since they differ in their functions – such as differential spatial and temporal gene expression (Figure 6). The duplicated genes showed divergent expression patterns and have undergone neofunctionalization.

Expression Divergence of *ZmTERF* Genes in Different Tissues or Organs

To establish organ and tissue-specific expression patterns of *ZmTERF* genes, NCBI ESTdb, available microarray and RNA-seq data were used. As shown in Figures 6, S5 and Table S7, most *ZmTERF* genes were actively transcribed as for Arabidopsis *mTERF* genes [18], and nine *ZmTERFs* were ubiquitously expressed in all organs and most tissues (Figure 6). In general, the temporal and tissue-specific expression of plastid-targeting *ZmTERF* genes was correlated with chloroplast development and photosynthesis proceeding, e.g. for *ZmTERF6*,*-9*, *-23*, *-24*, *-27* and *-29*. Expression divergence of *ZmTERF*s along the leaf developmental gradient and between mature bundle sheath and mesophyll cells were explored using RNA-seq data. The result reveals that quite a number of *ZmTERF*s showed differential expression levels among the four zones representing differentially developed chloroplasts, derived from developmental leaf, and ten *ZmTERF*s were differentially expressed between mature bundle sheath and mesophyll cells. Together, maize *mTERF* genes may have important roles in controlling organellar gene expression for chloroplast development and C_4 photosynthesis.

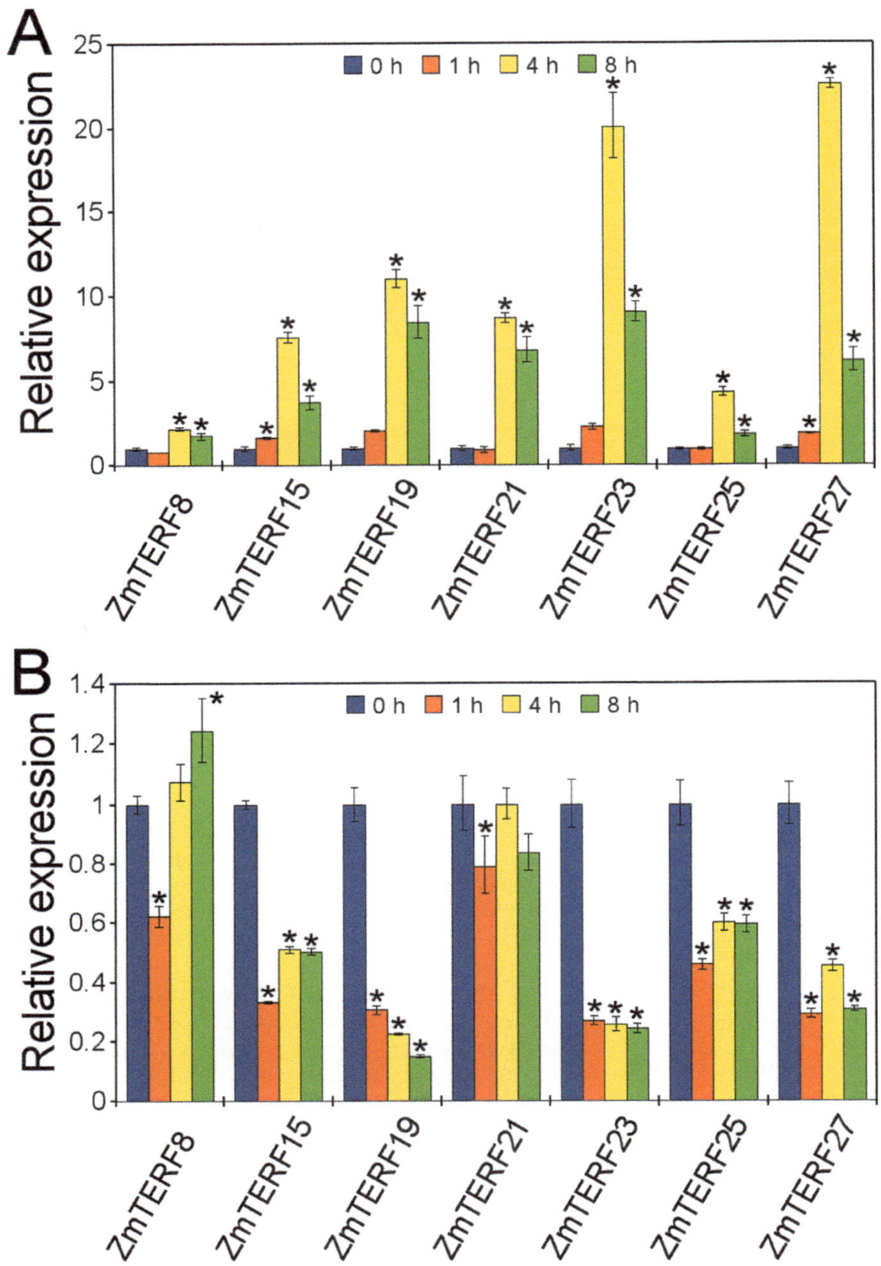

Figure 7. Differential expression of maize *mTERF* genes regulated by light/dark treatments. For the light treatment, total RNA was isolated from etiolated two-leaf seedling at 0, 1, 4 and 8 h after exposure to light (A). For the dark treatment, two-leaf seedlings cultured under 16 h light/8 h dark were used and placed in darkness after 24 h of continuous illumination (B). The expression levels were normalized against maize *Actin1* gene and the expression in the corresponding samples at 0 h was set as 1 using $2^{-\Delta\Delta CT}$ method [75]. Error bars represent the standard error of the mean. Asterisk (*) on top of error bar indicate the significant difference ($\alpha = 0.05$, n = 3) compared with control (0 h).

Arabidopsis *BSM/RUG2* is essential for chloroplast biogenesis and development [17]. T-DNA insertion in SOLDAT10 causes lethal, whereas point mutation of SOLDAT10 shows lighter green cotyledons [21]. *SOLDAT10* can abrogate $^1O_2$-mediated cell death of *flu* seedlings under illumination by perturbing chloroplast homeostasis and subsequently affecting the chloroplast-to-nucleus retrograde signaling pathway. The retrograde signals that adjust acclimation to light intensity are proposed to contribute to the signaling pathways that control photoperiodic acclimation of leaves [21]. Expression alteration of several *ZmTERF* genes in response to light and dark was investigated by qRT-PCR. Largely induced expression of chloroplast-targeting *ZmTERF23* and *-27* in response to light is understandable because these genes were probably requisite for transformation of plastid from etioplast to chloroplast during de-etiolation of maize seedling under illumination. The result is consistent with the presence of plentiful LREs in the corresponding promoters (Figure 5). Nevertheless, mitochondrial *ZmTERF* genes also show up-regulated expression in response to light. These genes might be required to optimize photosynthetic efficiency via coordinating the expression of

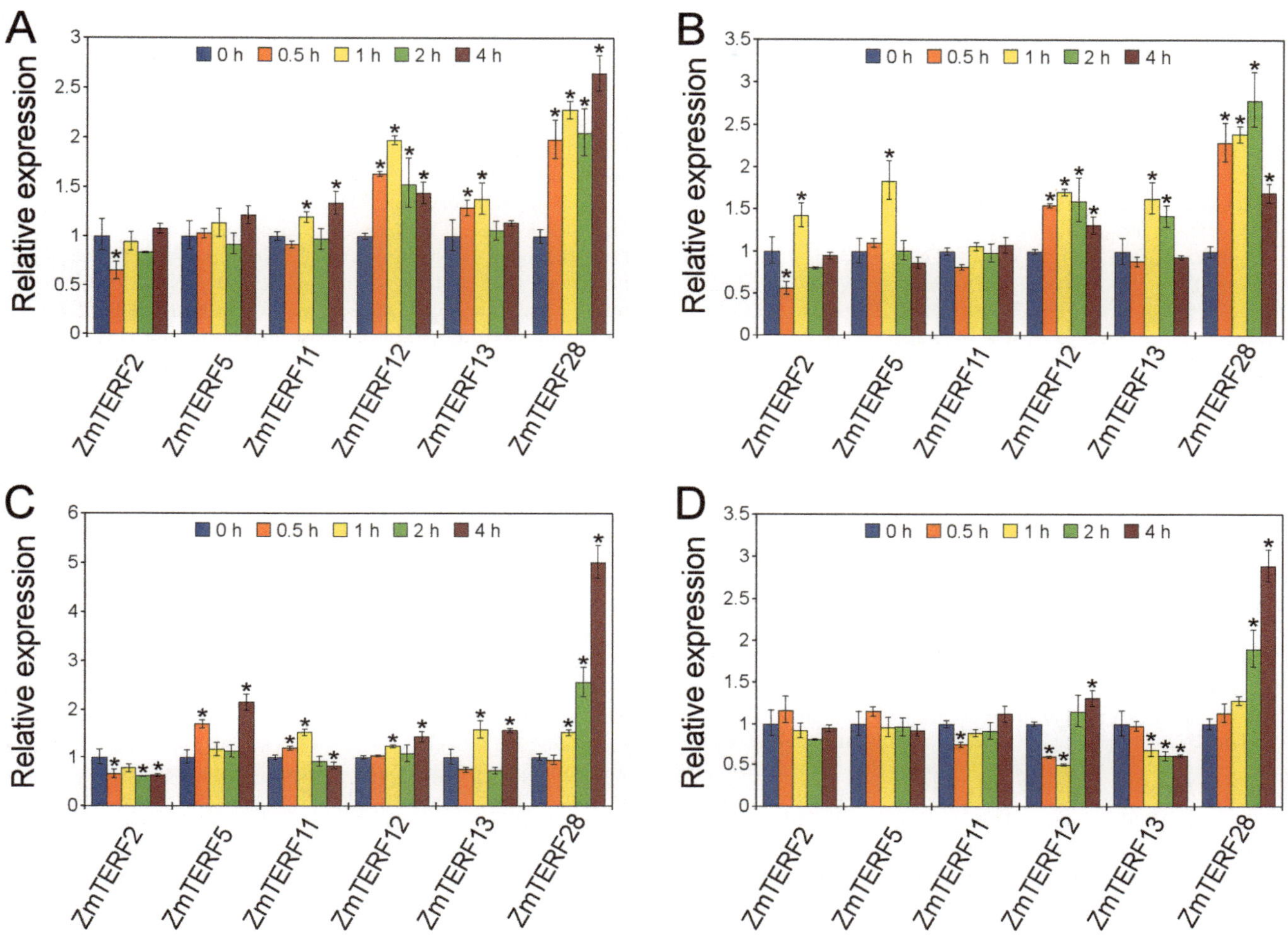

Figure 8. Quantitative RT-PCR analysis of maize *mTERF* gene expression under salts and phytohormones. Total RNA was extracted from two-leaf seedlings at 0.5, 1, 2 and 4 h after exposure to sodium chloride (A), aluminum chloride (B), NAA (C) and ABA (D), respectively. The expression levels were normalized against maize *Actin1* gene and the expression in the untreated control samples was set as 1 using $2^{-\Delta\Delta CT}$ method [75]. Error bars indicate the standard error of the mean. Asterisk (*) on top of error bar represent the significant difference ($\alpha = 0.05$, $n = 3$) compared with control (0 h).

mitochondrial genes and regulating the composition of the mitochondrial respiratory chain in the light, like *MOC1* in *C. reinhardtii* [19]. This hypothesis is also consistent with the down-regulated expression of these genes under dark treatment (Figure 7).

Selected *ZmTERF* genes show differential expression patterns under salt, ABA and NAA treatments. In Arabidopsis, most *mTERF* genes were heavily down-regulated in response to ABA, salt or mannitol [23]. In maize, however, *ZmTERF* genes showed different expression profiles to Arabidopsis *mTERF* genes, although several *ZmTERF* genes were down-regulated at certain periods after ABA treatment. This difference between Arabidopsis and maize *mTERF* genes might be due to the short-term response to ABA monitored in our study. Conversely, a few *ZmTERF* genes were up-regulated under salt treatments ($AlCl_3$ and NaCl) to a substantial extent (Figure 8). The result may reflect the differences in response of monocots and dicots to salts. All of *ZmTERF* genes checked here were up or down-regulated with NAA at more than one time spot (Figure 8). This expression alteration is not consistent with the fact there were no known GREs in the promoters of these *ZmTERF* genes, and thus other unidentified elements in the promoter regions could account for the NAA-induced expression.

Molecular Function of Plant *mTERF* Genes

In human, the biological roles of *mTERF* genes in controlling mitochondrial replication, transcription and translation have been well documented [10,14–16]. Detailed molecular characterizations have shown that hsmTERF1 binds to mitochondrial DNA as a monomer [11,37] and hsmTERF4 binds to mitochondrial rRNA within a protein complex containing NSUN4 [15,16]. In *C. reinhardtii*, MOC1 protein can bind to the mitochondrial DNA specifically as does hsmTERF1 [19]. The conserved transcriptional factor-like function shared by *mTERF* genes in *C. reinhardtii* and human is probably responsible for the similar mechanism for mitochondrial genome transcription by which the mitochondrial genomes of human and *C. reinhardtii* are transcribed bidirectionally to produce two long primary transcripts, which are further processed to yield the mature mRNAs [19]. Increased *mTERF* members in higher plants may account for complicated regulation of mitochondrial genome as mtDNA transcription regulators besides chloroplast genome. Arabidopsis *mTERF* genes are required for normal expression of organellar genes [17,21–24]. Moreover, plant mTERF proteins share conserved mTERF motif and similar 3D structure with human mTERF proteins (Figure 2) [17]. Arabidopsis mTERF protein, BSM/RUG2, has been shown to nonspecifically bind the chloroplast DNA segments [17]. It is envisaged that *mTERF* genes of higher plants have similar roles to *hsmTERFs* and *MOC1*. However, we cannot exclude the possibility that mTERF proteins of higher plants require auxiliary factors for specific DNA-binding because of existence of mTERF complex in chloroplast nucleoid [17], or these *mTERF* proteins have evolved the specialized property for RNA-binding, as has hsmTERF4 protein [15,16], since Arabidopsis *mTERF* genes show co-expression with pentatricopeptide repeat (PPR) genes and might regulate mitochondrial RNA metabolism [18].

To date, only four *mTERF* genes of flowering plants have been functionally identified by a reverse genetic approach and all were Arabidopsis *mTERF* genes. Furthermore, three of the four genes are targeted to chloroplast [17,21,23] while only one (*SHOT1*) is targeted to mitochondria [24]. No known mutations have been reported so far for Arabidopsis groups VIII and IX genes (Figure 1). It is possible that these genes are functionally redundant or their mutations are lethal or conditional. Further biochemical evidence for the function of plant *mTERF* genes is still required to address these hypotheses. The biological functions of maize *mTERF* genes will be interpreted through mutant analysis, which is now accessible in the UniformMu Resource (http://www.maizegdb.org/documentation/uniformmu/) and the Mu Insertion Database (http://teosinte.uoregon.edu/mu-illumina/). Recently, several chloroplast-targeting *mTERF* mutants showing abnormal development of chloroplasts were observed and reported in the Photosynthetic Mutant Library (PML) (http://pml.uoregon.edu/pml_table.php) in which *ZmTERF27* cloned by Hammani and Brakan very recently is orthologous to Arabidopsis *BSM/RUG2* and plays an important role in group II intron splicing in maize chloroplasts [59]. In Arabidopsis, absence of *BSM/RUG2* causes abortion of embryo development, while in maize, mutants of *ZmTERF27* (*Zm-mterf4-1* and *Zm-mterf4-3* in PML) only causes disturbed leaf development of seedlings after seed germination, showing that in maize the effect of *ZmTERF27* mutation on embryo development can be alleviated [59]. This result may be due to variation of malonyl-coA synthase in maize and Arabidopsis [17].

Conclusion

This is the first study of genome-scale analysis of *mTERF* genes in maize. We identifed 31 putative *ZmTERFs*, sorted them into nine groups based on their phylogenetic relationship, mined 15 conserved motifs, found contributions of gene duplications to the expansion of *mTERF* gene family in maize, and explored spatio-temporal and abiotic stress-responsive expression patterns of *ZmTERFs*. The results presented in this study provide basic information on maize *mTERF* genes and form the foundation for future functional studies of these proteins. Future functional analyses of maize *mTERF* genes identified in this study would eventually elucidate the biological mechanisms controlling organellar gene expression mediated by *ZmTERF* genes.

Materials and Methods

Identification of *mTERF* Genes in Maize

To obtain all putative *mTERF* family genes in maize, three approaches were used to mine maize expression and annotation databases. First, the keywords 'maize' and 'mTERF' were used to search the NCBI Protein database. Second, the reported sequences of the *mTERF1-mTERF4* genes in human [11,12,14,15], *C. reinhardtti MOC1* [19] and Arabidopsis *mTERFs* [22] were collected and used as query sequences to search the maize filtered-gene set (ZmB73_5b_FGS_translations.fasta downloaded from www.maizesequence.org) in local BLASTP program [60] with E-value<1e-10. Additional local TBLASTN [60] was performed to query the Maize Full Length cDNA database (downloaded from http://www.maizecdna.org/) [61] and Unigene database of NCBI (www.ncbi.nlm.nih.gov) with the above known mTERF proteins. Third, conserved mTERF domains were used to query the maize filtered-gene set with mTERF HMM file (PF02536) from the Pfam database [62] using the HMMER 3.0 package [25], and the mTERF proteins were collected in terms of default inclusion threshold. Unique maize mTERF proteins were collected by manually eliminating redundant repeats. The maize mTERF proteins were confirmed within SMART [26]. Information about chromosomal localization, coding sequence length and aa length was obtained for each gene from MaizeGDB (http://maizegdb.org/). The molecular weight and theoretical isoelectric point (pI) value of the maize mTERFs were investigated within Expasy online tools (http://expasy.org/tools/).

Gene Structure and Alternative Splicing

DNA and transcript sequences of maize *mTERF* genes were obtained from MaizeSequence (www.maizesequence.org) and MaizeGDB (http://maizegdb.org/) databases. There was more than one alternatively spliced transcript for most maize *mTERFs* annotated in the MaizeGDB database. Conserved gene structures of maize *mTERFs* were identified by comparing with their homologs in other higher plants in the CoGe program [63] and Phytozome v8.0 (http://www.phytozome.net). These conserved gene structures were drawn and displayed by Gene Structure Display Server (GSDS) [64].

To confirm the gene models of maize *mTERFs*, PCR amplification of maize *mTERFs* was performed with transcript specific primers (Table S2), and the DNA and cDNA derived from total RNA of B73 seedlings as a PCR template. PCR products were purified and sequenced in an ABI3730 sequencer (Sunny Bio., Shanghai). By comparing cDNA with DNA sequences, putative gene models were validated in the B73 inbred line.

Motif and Conserved Sequence Analysis

Pfam [62] and SMART [26] database searching was carried out to identify known conserved motifs in maize mTERF proteins, and the Multiple Em for Motif Elicitation (MEME) program v4.9.0 [34] was used to predict the potential motifs in the putative *mTERF* family gene sequences with the following parameters: zero or one per sequence for distribution of a single motif, 6≤ optimum width of each motif ≤50, and maximum number of motifs to find = 15. mTERF motifs in maize *mTERF* proteins were collected and clustered using the ClustalW 2.0 program [33], and graphical representation of aa residues was made using WebLogo [65].

Subcellular Location and Chromosomal Location of Maize *mTERFs*

TargetP [28] and Predotar [29] were used to predict subcellular location of maize mTERF proteins. Additionally, experimental proteomic data in the Plant Proteome DataBase (PPDB) for *Arabidopsis thaliana* and maize [30] were queried for validating the prediction results on the basis of appearance of mTERF proteins in specific organelle proteomes. Maize *mTERF* genes were mapped onto the corresponding maize chromosomes by identifying their chromosomal positions given in the MaizeGDB (http://maizegdb.org). Using MapChart software [66], the distribution of maize *mTERF* genes on chromosomes was drawn and modified manually with annotation information in MaizeGDB (http://maizegdb.org).

Synteny Analysis and Gene Duplication

To compare the genomic context of maize *mTERF* genes with that in other grass species, information on their patterns of microsynteny was retrieved from the database Phytozome v8.0 (www.phytozome.net). Orthologs of maize *mTERF* genes in other grasses were investigated in the CoGe database [63]. Duplication analysis of maize *mTERF* genes was carried out in the Synteny Mapping and Analysis Program (SyMAP) v4.0 [44] and WGMapping program in the PLAZA v2.5 platform [45]. Two *mTERFs* placed on the synteny blocks in maize genome were designated as segmental duplicated paralogs, and two *mTERFs* separated by five or fewer gene loci were regarded as tandem duplicated paralogs as described [67].

Sequence Alignment and Phylogenetic Tree Construction

A total of 98 aa sequences (Table S3) of the *mTERF* genes were aligned with the program MUSCLE [68]. The resulting alignment was manually optimized by removing unaligned residues in CINEMA 5 (http://aig.cs.man.ac.uk/research/utopia/cinema/cinema.php). A Maximum Likelihood (ML) phylogenetic tree was constructed using RAxML v7.2.8 [27] with the JTT aa substitution model. Bootstrap testing was performed with 500 re-sampling to search for the best tree. The ML phylogenetic tree was depicted by FigTree v1.4.0 (http://tree.bio.ed.ac.uk/software/-figtree/).

Secondary Structure and 3D Structure Analysis

To interpret the responsibility of conserved higher structure for roles of *mTERF* genes in plant organellar gene expression regulation, the secondary structures of mTERF proteins were analyzed using MINNOU [69], and their 3D structures were remodeled using the I-TASSER program [42] on the basis of crystal structure of previously resolved human mTERFs [16,36–37,70]. Superimposing 3D structures of maize mTERFs with that of human mTERFs was performed in the TM-align program [43]. PyMOL v1.6.0.0 [71] was used to display and analyze the tertiary structure of maize mTERFs.

Cis-Regulatory Elements in Promoter

Promoter sequences (2-kb upstream of the translation start codon) for maize mTERF genes were obtained from the MaizeSequence database (http://www.maizesequence.org/-index.html) and subjected to prediction of *cis*-acting regulatory DNA elements using PlantPAN [49] with the transcription factors selected from rice, maize and Arabidopsis. The regulatory DNA elements were displayed in Argo Genome Browser v1.0.26 (http://www.broadinstitute.org/annotation/argo/).

Gene Expression Profile Analysis

Spatio-temporal expression regulation of maize *mTERFs* was investigated using the microarray data (ZM37) for a genome-wide gene expression altas of the entire life cycle of B73 maize [50] from PLEXdb [72]. Robust Multi-array Average (RMA) normalized and $\log_2$-transformed expression data of maize *mTERFs* was retrieved from ZM37 and median expression values were loaded into Cluster 3.0 [51] for hierarchical clustering analysis. Clustered expression data of maize mTERFs were depicted using Java TreeView v1.1.5 [73]. eFP browser [53] was used to analyze the expression regulation of maize *mTERFs* during seedling leaf development as well as in bundle sheath and mesophyll cells.

Plant Stress Treatment

To verify the expression regulation of *mTERF* genes in maize under hormone and salt [sodium chloride (NaCl, 200 mM) and aluminum chloride ($AlCl_3$, 200 mM)] treatments, the two-leaf B73 seedlings were cultured in water under the following chemical treatments, ABA (100 μM) and 1-Naphthaleneacetic acid (NAA, a synthetic plant hormone in the auxin family) (100 μM) (Sigma-Aldrich, Shanghai), respectively. Samples were collected at 0.5, 1, 2 and 4 h after the above treatment, with three biological replicates per sample. To investigate effects of light on expression of maize *mTERFs*, etiolated B73 seedlings cultured in darkness were placed under continuous light, while normal-cultured B73 seedlings cultured under 16 h light/8 h dark were placed in darkness after 24 h of continuous light. Samples were collected at 1, 2, 4 and 8 h after light or dark treatment, with three biological replicates per sample. Total RNAs of collected samples were isolated with Trizol reagent (Invitrogen, USA) according to the manufacturer's instructions and quantified by NanoDrop 2000 spectrophotometer (ThermoFisher, USA) before cDNA synthesis.

Quantitative Real-Time PCR Analysis

For real-time PCR analysis, first-strand cDNA was synthesized from RNase-free DNase I (Fermentas, USA)-treated total RNA using Superscript II reverse transcriptase (Invitrogen, USA) according to the manufacturer's instructions. Real-time PCR was performed in an optical 96-well plate with a BioRad CFX96 Real-time PCR System (Bio-Rad Laboratories, Inc.) with the gene-specific primers (Table S9). Each reaction contained 12.5 μl of 2×SYBR Green Master Mix Reagent (Applied Biosystems, USA), 5 μl of diluted cDNA sample and 500 nM of each primer (Table S2) in a final volume of 25 μl. The thermal cycle used was as follows: 95°C for 5 min; 45 cycles of 95°C for 15 s and 60°C for 15 s. The maize *Actin1* gene (GenBank accession number, J01238) was used as the internal control [74]. The relative expression levels were determined using $2^{-\Delta\Delta CT}$ method as described previously [75]. The SAS v9.3 (SAS Institute Inc., USA) was used for the statistical analysis and the Dunnett's t test was used to compare the significant difference of all stress treatments against their controls.

Supporting Information

Figure S1 Correction of *ZmTERF23* gene model. (A) Multiple sequence alignment for *GRMZM2G426154*, *GRMZM2G175930* and *BT084020* was performed in Clustal Omega (http://www.ebi.ac.uk/Tools/msa/clustalo/). Identical nucleotides are denoted by asterisks under the alignment. (B) BAC contigs were rearranged to produce correct *ZmTERF23*.

Figure S2 Exon–intron organization of maize mTERF genes. The gene structures of maize mTERF genes were generated by comparing the transcript sequences with corresponding DNA sequences in GSDS [64]. Classification and subcellular localization information of maize mTERF genes shown in this picture were derived from Figure 1 and Table 1, respectively.

Figure S3 Schematic structure of maize mTERF proteins. mTERF motifs identified in SMART [26] are boxed in black. Light gray boxes denote mTERF motifs with lower reliability.

Figure S4 Validation of maize *mTERF* gene models by RT-PCR and DNA sequencing. DNA and cDNA derive from total RNA isolated from B73 seedling leaf were used as templates to amplify the maize *mTERF* genes with transcript-specific primers listed in. Each of maize mTERF genes was amplified and surveyed in 1% agarose gel followed by EB (ethidium bromide) staining.

Figure S5 Multiple sequence alignment of ZmTERF27 and its homologs in Arabidopsis, rice and human. MUSCLE program [68] was used to align the amino acid sequences of ZmTERF27, BSM/RUG2, LOC_Os05g33500 and HsmTERF1. The secondary structure of ZmTERF27 protein displayed under the aligned sequences was predicted in MINNOU [69]. mTERF motifs (M1–M9) are shown as black bars under their conserved residues. Putative mTERF structure modules with two or three α-helices are represented by red bars. Conserved arginine residues are highlighted by red asterisks.

Figure S6 Expression patterns of maize mTERF genes in different tissues analyzed by realtime RT-PCR. Total RNA was isolated from seedling leaf, root, unpollinated silk, mature seed and immature tassel of B73 maize. Three biological replicates per sample were conducted. The expression levels of each gene were normalized against *Actin1* gene using $2^{-\Delta Ct}$ methods [75]. Duncan's multiple range test (MRT) was used to statistically analyze the expression in different tissues and significant differences were marked by little letter on the top of bars ($\alpha = 0.05$, $n = 3$). Error bars indicate the standard error of the mean.

Figure S7 Expression levels of maize mTERF genes in B73 seedling leaf. Transcript abundance of maize mTERF genes in different developmental gradients and cells of B73 seedling leaf were retrieved from eFP browser (http://bar.utoronto.ca/efp_maize/cgi-bin/efpWeb.cgi) [53]. Base, base of the leaf; Ligule+4, −1 cm from ligule; Ligule+9, +4 cm from ligule; Tip_−1, −1 cm from tip; Tip_BS, bundle sheath cells of Tip_−1; Tip_ME, mesophyll cells of Tip_−1. Expression levels (RPKM) of maize *mTERF* genes are represented on y-axis.

Acknowledgments

We thank Liwei Wang and Pengcheng Yin for plant treatments and RNA isolation, and Dr. Jianyi Yang (University of Michigan) for his helpful suggestion on protein structure analysis.

Author Contributions

Conceived and designed the experiments: YXZ XBZ FZQ. Performed the experiments: YXZ XBZ MJC YRL JHZ HLZ FK. Analyzed the data: YXZ XBZ MJC FZQ. Wrote the manuscript: YXZ FZQ YLZ.

References

1. Gray MW, Burger G, Lang BF (1999) Mitochondrial evolution. Science 283: 1476–1481.
2. Binder S, Brennicke A (2003) Gene expression in plant mitochondria: transcriptional and post-transcriptional control. Philos Trans R Soc Lond B Biol Sci 358(1429): 181–188.
3. Liere K, Weihe A, Börner T (2011) The transcription machineries of plant mitochondria and chloroplasts: composition, function, and regulation. J Plant Physiol 168(12): 1345–1360.
4. Kühn K, Bohne AV, Liere K, Weihe A, Börner T (2007) Arabidopsis single-polypeptide RNA polymerases: accurate in vitro transcription of organellar genes. Plant Cell 19: 959–971.
5. Robles P, Micol JL, Quesada V (2012) Unveiling plant mTERF functions. Mol Plant 5(2): 294–296.
6. Linder T, Park CB, Asin-Cayuela J, Pellegrini M, Larsson NG, et al. (2005) A family of putative transcription termination factors shared amongst metazoans and plants. Curr Genet 48: 265–269.
7. Roberti M, Bruni F, Polosa PL, Gadaleta MN, Cantatore P (2006) The Drosophila termination factor DmTTF regulates in vivo mitochondrial transcription. Nucleic Acids Res 34(7): 2109–2116.
8. Roberti M, Bruni F, Loguercio Polosa P, Manzari C, Gadaleta MN, et al. (2006) MTERF3, the most conserved member of the mTERF family, is a modular factor involved in mitochondrial protein synthesis. Biochim Biophys Acta 1757(9–10): 1199–11206.
9. Roberti M, Polosa PL, Bruni F, Manzari C, Deceglie S, et al. (2009) The MTERF family proteins: Mitochondrial transcription regulators and beyond. Biochim Biophys Acta 1787(5): 303–311.
10. Kruse B, Narasimhan N, Attardi G (1989) Termination of transcription in human mitochondria: identification and purification of a DNA binding protein factor that promotes termination. Cell 58(2): 391–397.
11. Fernandez-Silva P, Martinez-Azorin F, Micol V, Attardi G (1997) The human mitochondrial transcription termination factor (mTERF) is a multizipper protein but binds to DNA as a monomer, with evidence pointing to intramolecular leucine zipper interactions. EMBO J 16(5): 1066–1079.
12. Pellegrini M, Asin-Cayuela J, Erdjument-Bromage H, Tempst P, Larsson NG, et al. (2009) MTERF2 is a nucleoid component in mammalian mitochondria. Biochim Biophys Acta 1787(5): 296–302.

13. Wenz T, Luca C, Torraco A, Moraes CT (2009) mTERF2 regulates oxidative phosphorylation by modulating mtDNA transcription. Cell Metab 9(6): 499–511.
14. Park CB, Asin-Cayuela J, Cámara Y, Shi Y, Pellegrini M, et al. (2007) MTERF3 is a negative regulator of mammalian mtDNA transcription. Cell 130(2): 273–285.
15. Cámara Y, Asin-Cayuela J, Park CB, Metodiev MD, Shi Y, et al. (2011) MTERF4 regulates translation by targeting the methyltransferase NSUN4 to the mammalian mitochondrial ribosome. Cell Metab 13(5): 527–539.
16. Spåhr H, Habermann B, Gustafsson CM, Larsson NG, Hallberg BM (2012) Structure of the human MTERF4-NSUN4 protein complex that regulates mitochondrial ribosome biogenesis. Proc Natl Acad Sci USA 109(38): 15253–15258.
17. Babiychuk E, Vandepoele K, Wissing J, Garcia-Diaz M, De Rycke R, et al. (2011) Plastid gene expression and plant development require a plastidic protein of the mitochondrial transcription termination factor family. Proc Natl Acad Sci USA 108(16): 6674–6679.
18. Kleine T (2012) Arabidopsis thaliana mTERF proteins: evolution and functional classification. Front Plant Sci 3: 233.
19. Schönfeld C, Wobbe L, Borgstädt R, Kienast A, Nixon PJ, et al. (2004) The nucleus-encoded protein MOC1 is essential for mitochondrial light acclimation in Chlamydomonas reinhardtii. J Biol Chem 279(48): 50366–50374.
20. Wobbe L, Nixon PJ (2013) The mTERF protein MOC1 terminates mitochondrial DNA transcription in the unicellular green alga Chlamydomonas reinhardtii. Nucleic Acids Res 41(13): 6553–6567.
21. Meskauskiene R, Würsch M, Laloi C, Vidi PA, Coll NS, et al. (2009) A mutation in the Arabidopsis mTERF-related plastid protein SOLDAT10 activates retrograde signaling and suppresses $^1O_2$-induced cell death. Plant J 60(3): 399–410.
22. Quesada V, Sarmiento-Mañús R, González-Bayón R, Hricová A, Pérez-Marcos R, et al. Arabidopsis RUGOSA2 encodes an mTERF family member required for mitochondrion, chloroplast and leaf development. (2011) Plant J 68: 738–753.
23. Robles P, Micol JL, Quesada V (2012) Arabidopsis MDA1, a nuclear-encoded protein, functions in chloroplast development and abiotic stress responses. PLoS One 7(8): e42924.
24. Kim M, Lee U, Small I, des Francs-Small CC, Vierling E (2012) Mutations in an Arabidopsis mitochondrial transcription termination factor-related protein enhance thermotolerance in the absence of the major molecular chaperone HSP101. Plant Cell 24(8): 3349–3365.
25. Eddy SR (1996) Hidden Markov models. Curr Opin Struct Biol 6(3): 361–365.
26. Letunic I, Doerks T, Bork P (2012) SMART 7: recent updates to the protein domain annotation resource. Nucleic Acids Res 40(Database issue): D302–D305.
27. Stamatakis A (2006) RAxML-VI-HPC: maximum likelihood-based phylogenetic analyses with thousands of taxa and mixed models. Bioinformatics 22(21): 2688–2690.
28. Emanuelsson O, Brunak S, von Heijne G, Nielsen H (2007) Locating proteins in the cell using TargetP, SignalP and related tools. Nat Protoc 2(4): 953–971.
29. Small I, Peeters N, Legeai F, Lurin C (2004) Predotar: A tool for rapidly screening proteomes for N-terminal targeting sequences. Proteomics 4(6): 1581–1590.
30. Sun Q, Zybailov B, Majeran W, Friso G, Olinares PD, et al. (2009) PPDB, the Plant Proteomics Database at Cornell. Nucleic Acids Res 37(Database issue): D969–D974.
31. Majeran W, Friso G, Asakura Y, Qu X, Huang M, et al. (2011) Nucleoid-enriched proteomes in developing plastids and chloroplasts from maize leaves: a new conceptual framework for nucleoid functions. Plant Physiol 158(1): 156–189.
32. Friso G, Majeran W, Huang M, Sun Q, van Wijk KJ (2010) Reconstruction of metabolic pathways, protein expression, and homeostasis machineries across maize bundle sheath and mesophyll chloroplasts: large-scale quantitative proteomics using the first maize genome assembly. Plant Physiol 152(3): 1219–1250.
33. Larkin MA, Blackshields G, Brown NP, Chenna R, McGettigan PA, et al. (2007) Clustal W and Clustal X version 2.0. Bioinformatics 23(21): 2947–2948.
34. Bailey TL, Boden M, Buske FA, Frith M, Grant CE, et al. (2009) MEME SUITE: tools for motif discovery and searching. Nucleic Acids Res 37(Web Server issue): W202–W208.
35. Martin M, Cho J, Cesare AJ, Griffith JD, Attardi G (2005) Termination factor-mediated DNA loop between termination and initiation sites drives mitochondrial rRNA synthesis. Cell 123(7): 1227–1240.
36. Spåhr H, Samuelsson T, Hällberg BM, Gustafsson CM (2010) Structure of mitochondrial transcription termination factor 3 reveals a novel nucleic acid-binding domain. Biochem Biophys Res Commun 397(3): 386–390.
37. Yakubovskaya E, Mejia E, Byrnes J, Hambardjieva E, Garcia-Diaz M (2010) Helix unwinding and base flipping enable human MTERF1 to terminate mitochondrial transcription. Cell 141(6): 982–993.
38. Yakubovskaya E, Guja KE, Mejia E, Castano S, Hambardjieva E, et al. (2012) Structure of the essential MTERF4: NSUN4 protein complex reveals how an MTERF protein collaborates to facilitate rRNA modification. Structure 20(11): 1940–1947.
39. Groves MR, Hanlon N, Turowski P, Hemmings BA, Barford D (1999) The structure of the protein phosphatase 2A PR65/A subunit reveals the conformation of its 15 tandemly repeated HEAT motifs. Cell 96(1): 99–110.
40. Sibanda BL, Chirgadze DY, Blundell TL (2010) Crystal structure of DNA-PKcs reveals a large open-ring cradle comprised of HEAT repeats. Nature 463: 118–121.
41. Edwards TA, Pyle SE, Wharton RP, Aggarwal AK (2001) Structure of Pumilio reveals similarity between RNA and peptide binding motifs. Cell 105(2): 281–289.
42. Zhang Y (2008) I-TASSER server for protein 3D structure prediction. BMC Bioinformatics 9: 40.
43. Zhang Y, Skolnick J (2005) TM-align: a protein structure alignment algorithm based on the TM-score. Nucleic Acids Res 33(7): 2302–2309.
44. Soderlund C, Bomhoff M, Nelson WM (2011) SyMAP v3.4: a turnkey synteny system with application to plant genomes. Nucleic Acids Res 39(10): e68.
45. Van Bel M, Proost S, Wischnitzki E, Movahedi S, Scheerlinck C, et al. (2012) Dissecting plant genomes with the PLAZA comparative genomics platform. Plant Physiol 158(2): 590–600.
46. Nakashima K, Ito Y, Yamaguchi-Shinozaki K (2009) Transcriptional regulatory networks in response to abiotic stresses in Arabidopsis and grasses. Plant Physiol 149(1): 88–95.
47. Khong G, Richaud F, Coudert Y, Pati PK, Santi C, et al. (2008) Modulating rice stress tolerance by transcription factors. Biotechnol Genet Eng Rev 25: 381–403.
48. Baker SS, Wilhelm KS, Thomashow MF (1994) The 5′-region of Arabidopsis thaliana cor15a has cis-acting elements that confer cold-, drought- and ABA-regulated gene expression. Plant Mol Biol 24(5): 701–713.
49. Chang WC, Lee TY, Huang HD, Huang HY, Pan RL (2008) PlantPAN: Plant promoter analysis navigator, for identifying combinatorial cis-regulatory elements with distance constraint in plant gene groups. BMC Genomics 9: 561.
50. Sekhon RS, Lin H, Childs KL, Hansey CN, Buell CR et al. (2011) Genome-wide atlas of transcription during maize development. Plant J 66(4): 553–563.
51. Eisen MB, Spellman PT, Brown PO, Botstein D (1998) Cluster analysis and display of genome-wide expression patterns. Proc Natl Acad Sci USA 95(25): 14863–14868.
52. Duarte JM, Cui L, Wall PK, Zhang Q, Zhang X, et al. (2006) Expression pattern shifts following duplication indicative of subfunctionalization and neofunctionalization in regulatory genes of Arabidopsis. Mol Biol Evol 23(2): 469–478.
53. Winter D, Vinegar B, Nahal H, Ammar R, Wilson GV, et al. (2007) An "Electronic Fluorescent Pictograph" browser for exploring and analyzing large-scale biological data sets. PLoS One 2(8): e718.
54. Li P, Ponnala L, Gandotra N, Wang L, Si Y, et al. (2010) The developmental dynamics of the maize leaf transcriptome. Nat Genet 42(12): 1060–1067.
55. Schnable PS, Ware D, Fulton RS, Stein JC, Wei F, et al. (2009) The B73 maize genome: complexity, diversity, and dynamics. Science 326: 1112–1115.
56. Forner J, Weber B, Thuss S, Wildum S, Binder S (2007) Mapping of mitochondrial mRNA termini in Arabidopsis thaliana: t-elements contribute to 5′ and 3′ end formation. Nucleic Acids Res 35: 3676–3692.
57. Gagliardi D, Binder S (2007) Expression of the plant mitochondrial genome. In D Logan, ed, Plant Mitochondria. Blackwell Publishing, Ames, IA, 50–95.
58. Lynch M, Conery JS (2000) The evolutionary fate and consequences of duplicate genes. Science 290: 1151–1155.
59. Hammani K, Barkan A (2014) An mTERF domain protein functions in group II intron splicing in maize chloroplasts. Nucleic Acids Res In press.
60. Altschul SF, Gish W, Miller W, Myers EW, Lipman DJ (1990) Basic local alignment search tool. J Mol Biol 215(3): 403–410.
61. Soderlund C, Descour A, Kudrna D, Bomhoff M, Boyd L, et al. (2009) Sequencing, mapping, and analysis of 27,455 maize full-length cDNAs. PLoS Genet 5(11): e1000740.
62. Punta M, Coggill PC, Eberhardt RY, Mistry J, Tate J, et al. (2012) The Pfam protein families database. Nucleic Acids Res 40(Database issue): D290–D301.
63. Schnable JC, Lyons E (2011) Comparative genomics with maize and other grasses: from genes to genomes! Maydica 56: 183–200.
64. Guo AY, Zhu QH, Chen X, Luo JC (2007) GSDS: a gene structure display server. Yi Chuan 29(8): 1023–1026.
65. Crooks GE, Hon G, Chandonia JM, Brenner SE (2004) WebLogo: a sequence logo generator. Genome Res 14(6): 1188–1190.
66. Voorrips RE (2002) MapChart: software for the graphical presentation of linkage maps and QTLs. J Hered 93(1): 77–78.
67. Peng X, Zhao Y, Cao J, Zhang W, Jiang H, et al. (2012) CCCH-type zinc finger family in maize: genome-wide identification, classification and expression profiling under abscisic acid and drought treatments. PLoS One 7(7): e40120.
68. Edgar RC (2004) MUSCLE: multiple sequence alignment with high accuracy and high throughput. Nucleic Acids Res 32(5): 1792–1797.
69. Cao B, Porollo A, Adamczak R, Jarrell M, Meller J (2006) Enhanced recognition of protein transmembrane domains with prediction-based structural profiles. Bioinformatics 22(3): 303–309.
70. Jiménez-Menéndez N, Fernández-Millán P, Rubio-Cosials A, Arnan C, Montoya J, et al. (2010) Human mitochondrial mTERF wraps around DNA through a left-handed superhelical tandem repeat. Nat Struct Mol Biol 17(7): 891–893.
71. Schrodinger L (2013) The PyMOL Molecular Graphics System, Version 1.6.0.0. Portland, OR, USA.

72. Dash S, Van Hemert J, Hong L, Wise RP, Dickerson JA (2012) PLEXdb: gene expression resources for plants and plant pathogens. Nucleic Acids Res 40(Database issue): D1194–D1201.
73. Saldanha AJ (2004) Java Treeview–extensible visualization of microarray data. Bioinformatics 20(17): 3246–3248.
74. Haring M, Offermann S, Danker T, Horst I, Peterhansel C, et al. (2007) Chromatin immunoprecipitation: optimization, quantitative analysis and data normalization. Plant Methods 3: 11.
75. Livak KJ, Schmittgen TD (2001) Analysis of relative gene expression data using real-time quantitative PCR and the $2^{-\Delta\Delta CT}$ method. Methods 25(4): 402–408.

5

Mitochondrial Genome Sequencing in Mesolithic North East Europe Unearths a New Sub-Clade within the Broadly Distributed Human Haplogroup C1

Clio Der Sarkissian[1*¤a], Paul Brotherton[1¤b], Oleg Balanovsky[2,3], Jennifer E. L. Templeton[1¤c], Bastien Llamas[1], Julien Soubrier[1], Vyacheslav Moiseyev[4], Valery Khartanovich[4], Alan Cooper[1], Wolfgang Haak[1*], The Genographic Consortium[¶]

1 Australian Centre for Ancient DNA, School of Earth and Environmental Sciences, University of Adelaide, Adelaide, South Australia, Australia, **2** Vavilov Institute for General Genetics, Russian Academy of Sciences, Moscow, Russia, **3** Research Centre for Medical Genetics, Russian Academy of Medical Sciences, Moscow, Russia, **4** Peter the Great Museum of Anthropology and Ethnography (Kunstkamera) RAS, St Petersburg, Russia

Abstract

The human mitochondrial haplogroup C1 has a broad global distribution but is extremely rare in Europe today. Recent ancient DNA evidence has demonstrated its presence in European Mesolithic individuals. Three individuals from the 7,500 year old Mesolithic site of Yuzhnyy Oleni Ostrov, Western Russia, could be assigned to haplogroup C1 based on mitochondrial hypervariable region I sequences. However, hypervariable region I data alone could not provide enough resolution to establish the phylogenetic relationship of these Mesolithic haplotypes with haplogroup C1 mitochondrial DNA sequences found today in populations of Europe, Asia and the Americas. In order to obtain high-resolution data and shed light on the origin of this European Mesolithic C1 haplotype, we target-enriched and sequenced the complete mitochondrial genome of one Yuzhnyy Oleni Ostrov C1 individual. The updated phylogeny of C1 haplogroups indicated that the Yuzhnyy Oleni Ostrov haplotype represents a new distinct clade, provisionally coined "C1f". We show that all three C1 carriers of Yuzhnyy Oleni Ostrov belong to this clade. No haplotype closely related to the C1f sequence could be found in the large current database of ancient and present-day mitochondrial genomes. Hence, we have discovered past human mitochondrial diversity that has not been observed in modern-day populations so far. The lack of positive matches in modern populations may be explained by under-sampling of rare modern C1 carriers or by demographic processes, population extinction or replacement, that may have impacted on populations of Northeast Europe since prehistoric times.

Editor: Alessandro Achilli, University of Perugia, Italy

Funding: This research was supported by The Genographic Project, which is supported by funding from the National Geographic Society, IBM, and the Waitt Family Foundation. PB was funded by the Australian Research Council Linkage Project LP0882622 awarded to AC, and the Discovery Projects DP1095782 and DP130102158 awarded to WH and BL. OB was funded by the RAS Programmes "Molecular and cell biology", "Gene pool dynamics" and RFBR grants 13-04-01711, Cover Letter 13-06-00670. The funders had no role in study design, data collection and analysis, decision to publish, or preparation of the manuscript.

Competing Interests: The authors declare having received funding from a commercial partner (IBM). This does not alter the authors' adherence to all the PLOS ONE policies on sharing data and materials.

* E-mail: clio.dersarkissian@gmail.com (CDS); wolfgang.haak@adelaide.edu.au (WH)

¤a Current address: Centre for Geogenetics, Natural History Museum of Denmark, University of Copenhagen, Copenhagen, Denmark
¤b Current address: Archaeogenetics Research Group, School of Applied Sciences, University of Huddersfield, Huddersfield, United Kingdom
¤c Current address: School of Biological Sciences, Flinders University, Adelaide, South Australia, Australia

¶Membership of The Genographic Consortium is provided in the Acknowledgments.

Introduction

Human mitochondrial haplogroup (hg) C is part of the non-African macro-haplogroup M. Most of the diversity of hg C is found today in indigenous populations of Asia and the Americas [1–2]. In northern Asia, hg C represents, together with hg D, more than half of the present-day mitochondrial (mtDNA) diversity [3]. Haplogroup Z, the sister-clade of hg C, has a broad distribution ranging from northern Scandinavia (in Saami) to central Asia, Siberia, northern China and Korea.

Phylogenetic analyses of complete mtDNA genomes revealed four major sub-clades of hg C, termed C1, C4, C5 and C7 (e.g., [3–7]). Of these, haplogroup C1 has one of the broadest distributions of all human mtDNA hgs in the world, ranging from Iceland to East Asia and the Americas. The C1 basal haplotype is defined by the hypervariable region I and II (HVR-I and HVR-II) motif: A16129G, T16187C, C16189T, G16230A, T16278C, T16298C, C16311T, T16325C, C16327T (HVR-I; numbering according to the Reconstructed Sapiens Reference Sequence RSRS; [8]) and C146T, C152T, C195T, A247G, A249d, 290–291d and T489C (HVR-II).

The phylogeny of hg C1 is structured into five distinct monophyletic sub-clades, C1a, C1b, C1c, C1d and C1e, which exhibit a clear geographical distribution pattern ([4], [7], [9–10]; Figure 1). Three of the C1 sub-clades (C1b, C1c and C1d) are restricted to Native American populations, although spread widely across the American continent [11–12]. It was proposed that these three Native American C1 sub-clades were among the ancestral founder lineages, along with hg A2, B2 and D1, which reached the Americas during the initial human colonisation of the continent [4–5], [7], [9]. The source population of this migration was assumed to be in eastern Asia, where most of the diversity of hg C is observed today, and where C1a, a sister clade of the American C1 clades, is found at low frequencies in diverse indigenous populations [9]. The peopling of the Americas was made possible by the Beringian ice-free land bridge that connected north-east Siberia and Alaska before (~30,000 years Before Present, BP) and after (~13,000 yBP) the last Ice Age [9], [13]. The place of origin of ancestral hg C1 was approximated in the Amur River region just south of Beringia (eastern Asia) on the basis of the current frequency distribution of hg C1 in Asia [9]. The last hg C1 clade to have been described, C1e, was only found recently in a few individuals in Iceland and was shown to be distinct from any of the previously defined Asian and American clades on the basis of seven coding region and three control region mutations [10].

In Europe, the dense and extensive sampling of the HVR-I diversity has revealed extremely low frequencies of hg C1, with very few haplotypes found in Germans [14], Canarians [15], Icelanders [16–17] and Bashkirs [18] (Figure 2). These sequences lack HVR-I Single Nucleotide Polymorphisms (SNPs) diagnostic of the sub-clades C1a (T16356C) and C1d (A16051G). However, a more detailed assignment of the European haplotypes into sub-haplogroups is limited by the low resolution provided by HVR-I and the lack of information from the coding region thus far. These limitations therefore impede the reconstruction of their precise phylogenetic placement, origin and relation to the Asian, American and Icelandic sister-clades.

Three hypotheses for the origins of the C1 lineages in Europe can be put forward [10], [16–17]. Hypothesis 1 proposes a recent genetic input from Asia into Europe during historical times. Historically, Central and East Europe experienced repeated influences from invading groups from the neighbouring Asian steppes, which could have introduced C1 into Europe. Well-documented examples include the Huns from Mongolia in the 4th–5th centuries Anno Domini (A.D.) and the Mongols in the 13th century A.D. [19]. However, the common Asian C1a clade is characterised by the HVR-I transition T16356C, which has not been found in any European C1 haplotype. In the case of a recent Asian ancestry, a reversal of the mutation at nucleotide position (np) 16356 in all the European sequences would be required. Hypothesis 2 assumes an American origin, where hg C1 would have reached Europe through admixture between Native Americans and Europeans. This gene flow may have occurred during and after the colonization of the New World by Europeans in the 15th century A.D., i.e. in post-Columbian times [16–17]. Alternatively, one explanation for the presence of the sub-clade C1e in Iceland was a pre-Columbian admixture between Native Americans and Icelandic Vikings, which are widely acknowledged to have built temporary pioneer settlements in the north-western coast of the Americas in the 10th century A.D. [10]. In accordance with the hypothesis of the American origin, few European C1 HVR-I sequences could belong to either the C1b or C1c American clades, as diagnostic SNPs for these two clades are located outside HVR-I [16–17]. Hypothesis 3, proposes that hg C1 has been present in Europe since prehistoric times in the light of the recent finding of hg C1 HVR-I haplotypes in three individuals of the 7,500-year-old Mesolithic site of Yuzhnyy Oleni Ostrov (individuals UZOO-7, UZOO-8, and UZOO-74), North West Russia (Figure 1; [20]). The classification of the corresponding mtDNA haplotype within hg C1 was previously determined by HVR-I sequencing (hg C1 defining mutation T16325C) as well as by typing informative SNPs in the coding region (hg C defining

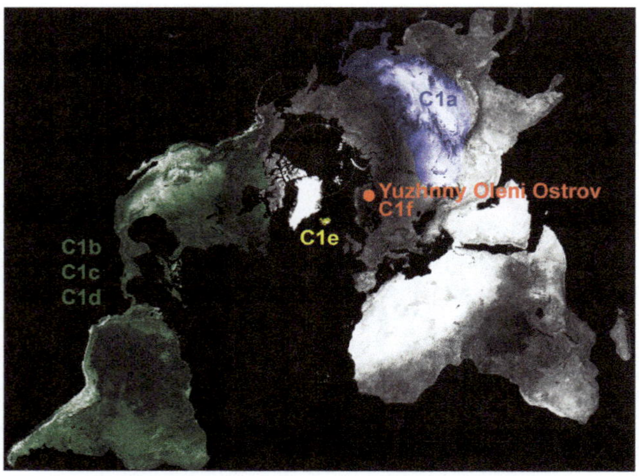

Figure 1. Approximate geographical distribution of the C1 sub-clades in modern and Mesolithic Yuzhnyy Oleni Ostrov populations.

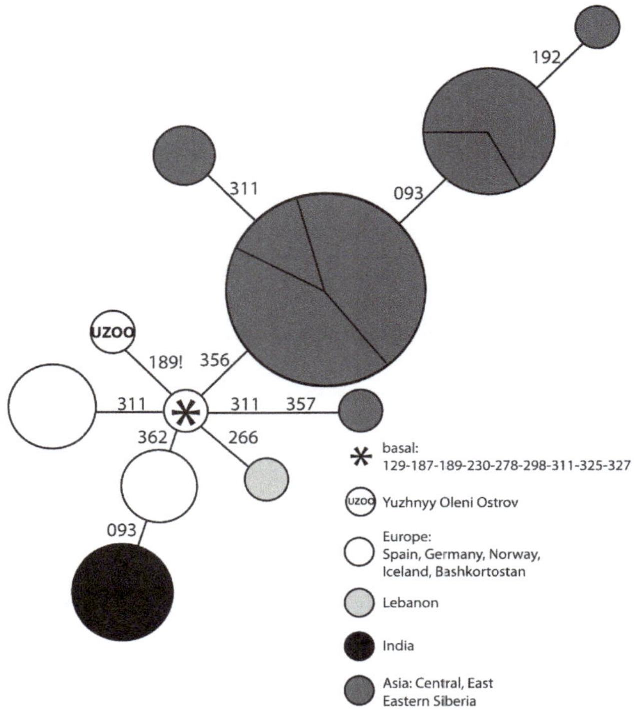

Figure 2. Network representation of C1 HVR-I sequences in Mesolithic Yuzhnyy Oleni Ostrov and modern Eurasian populations. Each haplotype is represented by a circle, the area of which is proportional to the number of individuals that were found to carry this haplotype in the literature. The haplotypes are colour-coded according to their geographical location: India (black), Asia (dark grey), Lebanon (light grey), and Europe (white). Each section of the circles represents individuals sampled from a same population. Mutations are all substitutions and are reported according to the Reconstructed Sapiens Reference Sequence minus 16000. The star represents the hypervariable region-I haplotype that characterizes the root of the C1 clade. The haplotype labeled 'UZOO' is the hypervariable region-I haplotype sequenced from individuals of the archaeological site of Yuzhnyy Oleni Ostrov. All the other haplotypes were found in modern populations.

mutation A13263G [20]), but the lack of resolution of the HVR-I sequence prevented establishing clear phylogenetic relationships with currently known hg C1 clades.

In this study, we sequenced the complete mtDNA genome of one of the three Mesolithic hg C1 carriers from the Yuzhnyy Oleni Ostrov archaeological site (individual UZOO-74) in order to shed further light on the population history of the Yuzhnyy Oleni Ostrov hunter-gatherers and to contribute to the characterisation of the mtDNA diversity, evolutionary history and phylogeography within hg C1.

To this day, complete mtDNA genome sequences from ancient specimens have successfully been determined for a Palaeo-Eskimo from Greenland [21], the 5,000-year-old Tyrolean Iceman [22], a 700-year-old individual from New Zealand [23], a Palaeolithic individual from Tianyan, China [24], as well as several Palaeolithic, Mesolithic and Neolithic individuals from Europe [25–28]. Most of these mtDNA genomes were obtained on high-throughput 'next-generation' sequencing platforms (e.g., [29]). In accordance with these recent studies, we first created a genomic library, which was subsequently enriched for mtDNA in two iterative rounds of hybridisation to in-house designed biotinylated DNA probes, following the protocol by [28]. The enriched DNA library was sequenced on an Ion Torrent PGM platform. We analysed the resulting mtDNA genome from the Yuzhnyy Oleni Ostrov specimen in the light of an updated phylogeny of all currently available hg C1 lineages. The resulting mtDNA genome sequence allowed us to identify a novel C1 sub-clade, coined "C1f", which fills a gap in the knowledge of the hg C1 distribution in West Eurasia.

Results

Ancient Mitochondrial Genome Sequencing

Our ancient DNA (aDNA) enrichment, followed by sequencing on an Ion Torrent PGM platform, allowed the unambiguous determination of 99.8% (16537 out of 16569 base pairs, bp) of the UZOO-74 mtDNA genome with 20,579 unique reads assembled to the RSRS at an average coverage of 68X (Table 1) and average read length of 55±14.5 bp. Indels, a well-defined homopolymer sequencing error, were observed in the resulting data set. However, adequate depth and coverage of the mtDNA genome sequence data prevented false-positive base calls.

Missing data comprised 32 consecutive bp, spanning nps 7525–7556 (Table 1). Other mtDNA genome sequences that we have generated following the same protocol have also exhibited a low coverage or dropout in exactly the same region [28]. Figure S1 shows that low GC content regions are characterised by a poor coverage; in particular in the region 7525–7556, GC content is only 25.0% compared to 44.4% GC for the whole mtDNA. It is therefore suspected that this region of the mtDNA genome is energetically sub-optimal (AT-rich) for two rounds of hybridisation and stringency washes in the ionic and temperature conditions used here [28], which may produce secondary DNA structures that adversely affect hybridisation-based DNA capture [30]. In addition, re-amplification of the enriched libraries was done with AmpliTaq Gold (Applied Biosystems; see [28]), a *Taq* polymerase known to be biased towards high GC content [31].

The ancient mtDNA haplotype of individual UZOO-74 differed from the RSRS at 58 nucleotide positions (Table 1). Among these, 51 substitutions define the sub-hg C1 in accordance with the current phylogeny (www.PhyloTree.org), including back mutations at T182C! and G11914A!, which are identical by state to the RSRS. In addition, UZOO-74 showed five private substitutions (G247A!, A8577G, A11605t, A12217G and T16189C!, including two additional back mutations in the hypervariable region). These five additional nucleotide differences were directly amplified and sequenced from two different extracts in order to verify whether they represented true private mutations defining a novel C1 sub-clade (Table S1). All five mutations have been confirmed by direct sequencing and were taken into account in further phylogenetic analysis of hg C1 sequences. Importantly, we did not observe any SNPs characteristic of other hgs, nor any mixed signals that could indicate systematic DNA degradation or DNA contamination from exogenous sources.

We analysed the pattern of nucleotide misincorporation at the 3′ and 5′-ends of the DNA fragments in order to assess whether the estimated age of the molecules reflects the age of the sample [32–33]. We observed a C-to-T substitution frequency of 22.4% at the 5′-end (Figure S2), which sits well with previous findings that suggested a correlation between this frequency and the age of the samples [34], whereby samples older than 500 years had a C-to-T substitution frequency >10%. The Bayesian statistical framework implemented in mapDamage v2.0.1 [33] also provided simulated posterior distribution of three parameters of the damage model: λ, probability of terminating in overhang; ∂D, probability of cytosine deamination in double strands; and ∂S, probability of cytosine deamination in single strands. The posterior distribution of these parameters all departed from 0 (λ: mean, 0.582; standard deviation, 0.021; ∂D: mean, 0.036; standard deviation, 0.001; ∂S: mean, 0.651; standard deviation, 0.041), in accordance with

Table 1. Positions and nucleotide changes in the Yuzhnyy Oleni Ostrov C1f haplotype when compared to the Reconstructed Sapiens Reference Sequence.

Coverage	99.8%
Missing bases	np 7525–7556
Average redundancy	68.5 (23.2 stdv)
Minimum redundancy	0
Maximum redundancy	126
Mutations	C146T, C152T, T182C!, C195T, **G247A!**, A249d, 290–291d, T489C, *522.AC*, A769G, A825t, A1018G, A2758G, C2885T, T3552a, T3594C, G4104A, T4312C, A4715G, G7146A, C7196a, T7256C, A7521G, T8468C, **A8577G**, G8584A, T8655C, A9545G, C10400T, T10664C, A10688G, C10810T, C10915T, **A11605t**, G11914A!, **A12217G**, G13105A, A13263G, G13276A, T13506C, T13650C, T14318C, T14783C, G15043A, G15301A, A15487t, A16129G, *A16183c*, T16187C, **T16189C!**, G16230A, T16278C, T16298C, C16311T, T16325C, C16327T, *C16519T*

Nucleotide changes in bold represent mutations in the Yuzhnyy Oleni Ostrov haplotype that are new within the C1 clade. Transitions are reported with upper case letters and transversions with lower case letters. "!" indicates a back mutation. np, nucleotide position. stdv, standard deviation.

the DNA sequences generated from the UZOO-74 individual arising from aDNA molecules, and not from contamination by more recent DNA molecules during post-excavation handling. The results of the DNA damage analyses support the authenticity of the aDNA data presented here.

The Sequence from Mesolithic Yuzhnyy Oleni Ostrov Defines a Novel Lineage within the C1 Phylogeny

Upon confirmation of the five novel mutations in the C1 mtDNA genome of individual UZOO-74, we genotyped the same SNPs in the two other C1 individuals UZOO-7 and UZOO-8 from Yuzhnyy Oleni Ostrov. Direct sequencing confirmed the presence of all five novel SNPs, suggesting that the three C1 individuals from Yuzhnyy Oleni Ostrov were maternally related. Their mtDNA genomes may be strictly identical, or they may display differences in the form of additional private SNPs at coding region positions that have not been sequenced in these remaining individuals. The precise nature of the genetic relationships between these individuals cannot be inferred from the archaeological and genetic data currently available.

A search against the public Phylotree database yielded no match for the newly sequenced C1 mtDNA haplotype in 16810 modern complete mtDNA genomes (entries, mtDNA tree build 15 (30 Sep 2012) on PhyloTree.org [35]). The five SNPs identified in these individuals, and among these, the three coding region mutations A8577G, A11605t and A12217G in particular, represent novel sub-clade defining mutations that have not been reported together within a single hg C1 haplotype before. We therefore assigned them to a distinct new clade, which we tentatively named "C1f" following the conventional nomenclature (Figure 3). The resulting phylogenetic reconstruction shows that clade C1 is now characterised by six monophyletic sub-clades: C1a, C1b, C1c, C1d, C1e and C1f. The tree topology suggests that the Eurasian C1 sub-clades, the East Asian C1a, the rare C1f branch from Yuzhnyy Oleni Ostrov and the Icelandic C1e split early from the most recent common ancestor of the C1 clades and evolved independently (Figure 3).

Discussion

Under-sampling of the Mitochondrial Genome Diversity

In the present study, we established that the hg C1 mtDNA genome sequence carried by the Mesolithic individuals of the Yuzhnyy Oleni Ostrov site in north-western Russia defines a new clade, C1f, within the hg C1 phylogeny. Because of the polytomous topology of the hg C1 tree, no direct phylogenetic relationship could be established between C1f and the other well geographically defined C1 clades. As a result, clear inferences regarding the origin and evolutionary history of the C1f clade will remain difficult to draw, unless future sequencing of complete mtDNA genomes uncovers sequences closely related to the C1f genome sequenced here.

The absence of a direct match with sequences in databases of complete mtDNA genomes could be explained by under-sampling of mtDNA genomes in modern human populations. The number of published modern-day *Homo sapiens* complete mtDNA genome sequences is still small compared to that of HVR-I sequences. As such it is not too surprising that studies regularly report the discovery of novel clades and lineages (e.g., hg C1e [10]; within hg C1d [7]). Furthermore, the geographical coverage of modern-day populations for complete mtDNA genome sequencing is still unequally distributed, and the sampling so far has focused either on few specific populations or on particular hgs (e.g., [36–38]). As a consequence, mtDNA genomes available from the literature can still only provide an incomplete yet biased picture of the full, extant mtDNA diversity.

Absence of Match for C1f in Asia

Asia, and more precisely Siberia, could be considered as potential places of origin for the C1f clade identified in the Mesolithic site of Yuzhnyy Oleni Ostrov. This hunter-gatherer group was indeed shown to exhibit mtDNA affinities with modern-day populations of western and southern Siberia, the Altai region, or Mongolia [20]. The hypothesis of an Asian origin for the C1f sub-clade is also supported by the fact that most of the diversity of hg C is found in present-day populations of East Eurasia [3]. Sequences closely related to hg C1f may persist in modern-day populations of East Eurasia but remain undetected to date, as mtDNA genomes for these populations have not been as densely and extensively sampled as, for example, European populations.

Absence of Match for C1f in Europe

Despite the dense sampling of mtDNA in modern-day populations of Europe, only a few hg C1 HVR-I and no hg C1f mtDNA genome sequences were detected. The close matches for the HVR-I sequence of C1f did not display the back mutation T16189C! (Figure 2) and hence, none matched the C1f HVR-I haplotype exactly. However, np 16189 has been described as one of the top five transitional hotspots in the human control region [39], and hence provides little phylogenetic discrimination power. It is possible that these European haplotypes belong to the C1f clade without harbouring the mutation at np 16189. Therefore, additional SNPs in the coding region are required to definitely rule out these Eurasian C1 haplotypes as potential members of the C1f clade, and potential persistence of hg C1f in Europe since the Mesolithic.

Extinction or Near-extinction of C1f due to Post-Mesolithic Population Dynamics

Low frequencies and a restricted distribution seem to have been characteristic of hg C1 already in Mesolithic times, as hg C1 could not be detected in any of the other European Mesolithic populations sampled for ancient mtDNA in Eurasia further west [20]: in central/eastern Europe [25], [27], [40], and in Scandinavia [41]. This suggests an under-sampling of Mesolithic populations for aDNA, mating isolation of the Yuzhnyy Oleni Ostrov population, and/or influences from Siberian populations that had not reached Central Europe. Because of its low frequency, the distribution of hg C1 is prone to be affected by demographic processes, such as genetic drift or population replacements that may have occurred since Mesolithic times. Eventually, hg C1 may have reached extremely low frequencies or have gone extinct, thus preventing it from being detected in present-day European populations. The effects of these population processes can be observed at the population level, as the Yuzhnyy Oleni Ostrov group, similarly to the other described Mesolithic populations of Europe, was indeed shown to exhibit little genetic continuity with present-day Europeans [20]. Significant dissimilarities have been shown between the mtDNA gene pool of European Mesolithic populations characterised by a low diversity and high frequencies of hg U sub-clades (U2, U4, U5 and U8 in particular), and the rather homogeneous mtDNA makeup of present-day Europeans, which arrived during the Neolithic transition and subsequent periods [20], [40–42].

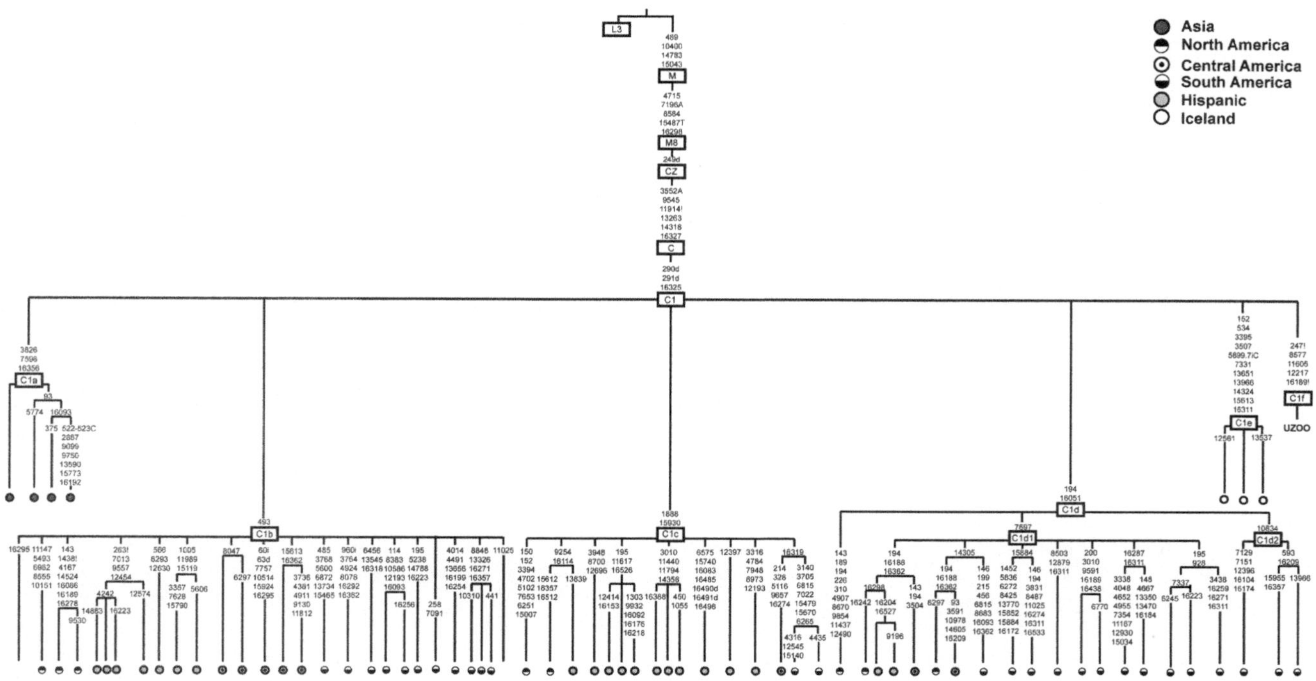

Figure 3. Median joining phylogenetic tree of haplogroup C1 complete mitochondrial genomes. A haplogroup sequence L3 sequence was chosen as the root of the tree. Mutations are reported according to the Reconstructed Sapiens Reference Sequence. "d" represents deletions. "i" represents insertions.

Absence of Match for C1f in the Americas

The Americas also remain under-sampled for complete mtDNA genomes and could be suggested as a potential geographical origin for the C1f lineage, as it has been for the Iceland-restricted C1e sub-clade [10]. For C1e, an American origin through mating of Viking explorers with Native American women sometime earlier than 300 years ago was proposed by [10]. Among other hypotheses including that of a European origin, an American origin was favoured on the basis that most of the hg C1 diversity is found on the American continent, despite the fact that no sequence belonging to hg C1e could be detected in the Americas (or anywhere else). This lack of match was explained by under-sampling of the American mtDNA genome diversity [10]. In any case, if admixture between Native Americans and Vikings did occur, it must have been limited, as no other American-specific lineage (e.g. hg A2, B2, D1, C1b, C1c, C1d) was detected in Iceland.

As for Mesolithic Europe, the possibility of a direct prehistoric genetic influence from the Americas is highly unlikely. However, in the eventuality that further sampling of complete mtDNA genomes in the Americas reveals the presence of additional haplotypes belonging to C1f, it would suggest an evolutionary history similar to that of mtDNA hg X2. Like hg C1, hg X2 displays relatively low frequencies albeit with a global distribution in the Northern hemisphere. For example, clade X2a was observed in Europe in the West, in the Near East, Europe, Central Asia, Siberia as well as North America [43]. One model for the present-day distribution of hg X2 suggests that clade X2a split early from the rest of the X2 lineages in the Near East, and reached east Siberia before participating in the second wave of migration into the Americas through admixture with Beringian populations [44]. A similar scenario involving an early split of the different C1 clades in Asia followed by their spread and subsequently isolated evolution could be considered as an explanation for the wide geographical distribution of hg C1 in general. However, this scenario currently lacks substantial support.

Similar Genetic Pre-history for the Icelandic-specific C1e and the Mesolithic C1f European Sub-clades

While the updated phylogeography of hg C1 does not allow defining the precise origins and divergence times of the C1f and C1e clades, the observation of C1f in Mesolithic Yuzhnyy Oleni Ostrov brings us to reconsider the hypotheses concerning the origins of C1e. Building on a hypothesis proposed by [10], we suggest that the Icelandic-specific C1e sub-clade could have had a recent origin in northern Europe rather than an American origin. This hypothesis is relevant with regard to the origins of the Icelandic population, as Iceland was discovered and first settled by Scandinavian Vikings around 1,130 years ago. Vikings raids extended as far from their homeland in Scandinavia as France, Spain and Sicily, but their main expansion range comprised western Russia, the Baltic region, Scandinavia, and the British Isles [16]. The study of the mtDNA diversity of present-day Icelanders identified that most of the Icelandic mtDNA lineages had Norse (from Scandinavia) or Gaelic origins (from the British Isles) and that the Icelandic gene pool had strongly been impacted by genetic drift [16–17,45].

Considering the Scandinavian origins of Icelanders and the identification of the sister clade C1f in Mesolithic North East Europe, it can be proposed that the Icelandic-specific C1e and C1f sub-clades might have both split from the common ancestors of the C1 lineages somewhere in Eurasia and later reached northern Europe during independent or similar migrations (before the Mesolithic for C1f). Therefore, the rare occurrence of the C1e and C1f sub-clades in Europe could be the result of their dilution within the pre-existing European mtDNA diversity when these lineages reached Europe. Of note, a contrasting pattern of elevated frequency and diversity was observed for the American C1 sister-clades (C1b, C1c and C1d): all three American sub-clades signal important population expansion during the initial peopling of the continent, which was void of human occupation and thus competing lineages. The distribution of the C1e sub-clade restricted to Iceland, associated with the presence of the novel sub-clade C1f in a region neighbouring the homeland of Vikings and clearly predating the Viking expansion, lends support to the hypothesis that hg C1e might have been brought in by the Vikings who first colonised Iceland. The presence of a novel sub-clade (C1f) closely related to the Icelandic-specific C1e sub-clade in a region neighbouring the homeland of Vikings and clearly predating the Viking expansion lends support to the hypothesis that hg C1e might have been brought in by the Vikings who first colonised Iceland. While the C1e sub-clade might have been preserved at detectable frequencies in the Icelandic population due to the effects of founder event, it most likely has gone extinct in the source population in northern Europe as a consequence of its low frequency. In contrast, due to the small size of the population through time, Icelandic mtDNA diversity has been greatly affected by genetic drift and increased rates of mtDNA haplotype extinctions [45]. As such, the C1e clade would be more likely to survive in the potential North European source population than in Iceland [45], but the extensive sampling of the Icelandic population makes it more likely to be detected there than anywhere else in North Europe. The potential long-term survival of C1 lineages in prehistoric Europe is highly relevant to the discussion about the prehistoric interactions between the ancestral populations of Europeans, Siberians and Native Americans. It is consistent with recently published genomic data from a 24,000 year-old Upper Paleolithic individual from Mal'ta, South Siberia [46]. Interestingly, this individual was shown to belong to the western Eurasian hg U, which was also the most frequent hg found in Yuzhnyy Oleni Ostrov Mesolithic individuals (64%) [20]. Genome-wide data from Upper Palaeolithic Mal'ta revealed affinities with both present-day western Eurasian and Native Americans, and further supports gene-flow between the ancestral populations of Europeans and Native Americans prior to the colonisation of the Americas [46]. The new C1f lineage thus bridges the geographic gap between the Icelandic, the Siberian and the Native American C1 lineages and argues for the presence of C1 lineages, albeit at low frequency, in prehistoric West Eurasia.

Materials and Methods

Ethics Statement

No specific permits were required for the described field studies.

Archaeological Samples

The three tooth samples analysed in this study were collected at the Mesolithic site of Yuzhnyy Oleni Ostrov, Onega Lake, Karelia, Russian Federation (61°30′N 35°45′E). These samples are under the custody of V.K. at the Peter the Great Museum of Anthropology and Ethnography (Kunstkamera) RAS, St Petersburg, Russian Federation, and were previously subjected to aDNA analyses in [20]. The three samples are identified as follows: UZOO-7 (MAE RAS collection number 5773–7, grave number 56), UZOO-8 (MAE RAS collection number 5773–8, grave number 57), and UZOO-74 (MAE RAS collection number 5773–74, grave number 114). Individuals UZOO-7 and UZOO-8 were found in adjacent graves (grave number 56 and grave number 57), whereas UZOO-74 was found in a grave located at the other end of the graveyard (grave number 114; see map in [47]).

Ancient DNA Extraction

Among the three Mesolithic individuals from Yuzhnyy Oleni Ostrov shown to carry hg C1 (UZOO-7, UZOO-8, UZOO-74) in [20], individual UZOO-74 was selected for mtDNA genome sequencing on the basis of its subjectively good preservation and robust performance in previous Polymerase Chain Reaction (PCR) amplification experiments [20]. DNA extractions followed established protocols as described previously [20].

Enrichment of Ancient Human Mitochondrial DNA

Ancient DNA extracts from specimens preserved in soil are expected to contain DNA molecules of various origins. In addition to the highly degraded DNA of the specimen under study, environmental, microbial DNA (bacteria and fungi), as well as from unidentified sources, has been shown to constitute a major proportion in the pool of DNA molecules present in aDNA extracts (e.g., [48]). The presence of a mixed population of DNA molecules from various organisms hampers the reliable sequencing of the DNA fragments of interest. Here, targeted enrichment of ancient human mtDNA of the extract for individual UZOO-74 was a crucial step prior to sequencing on the Ion Torrent PGM. Our aim was to increase the concentration of mtDNA fragments above the concentration threshold required for obtaining unambiguous sequencing at sufficient mtDNA genome coverage on an Ion Torrent PGM 316 chip. Ancient DNA libraries were enriched for human mtDNA using a hybridisation-based method described in [28]. All enrichment steps were performed twice for sample UZOO-74 to produce a 'second round enrichment' DNA library.

Sequencing on the Ion Torrent PGM

The enriched library DNA was prepared for Ion Torrent sequencing by re-amplification using Ion Torrent barcoded adapters (Adapter A: 5′-CCATCTCATCCCTGCGTGTCT-CCGACTCAGAAAAAGGTGTTGTTAGGAATGCGAGA-3′; Adapter B: 5′-CCTCTCTATGGGCAGTCGGTGATAGGA-TAGGTCGTTGCTGTGTA-3′). Eight reactions with a total volume of 25 µL were re-amplified using 1 µL of purified library DNA as template. Final reaction conditions comprised of 1x AmpliTaq Gold buffer II, 2.5 mM $MgCl_2$, 2.5 U AmpliTaq Gold (Applied Biosystems), 250 µM of each dNTP (Invitrogen), and 0.5 µM of each PCR primer. The thermocycling profile consisted of 94°C for 12 min, followed by 12 cycles of 30 s at 95°C, 30 s at 60°C and 45 s at 72°C, followed by a final 10 min at 72°C. The eight PCRs were pooled and purified using MiniElute spin columns (Qiagen), then eluted into 15 µL as per the manufacturer's instructions. The DNA was sized and quantified via gel electrophoresis against size markers (HyperLadder V, Bioline) and a Nanodrop 2000 (Thermo Scientific). Library DNA was size-selected above 120 bp and further purified, to remove adaptor dimers, using Qiagen's gel extraction purification kit following manufacturer's instructions.

Prior to sequencing, individual libraries were assessed for fragment size distribution and DNA concentration using a Bioanalyzer 2100 (Agilent Technologies) following manufacturer's instructions. The quantified indexed library DNA was pooled to an equimolar concentration alongside other samples. The pooled library DNA was adjusted to a final concentration of 10–15 pM prior to amplification (by emulsion PCR) and enriched for positive Ion Sphere Particles (ISPs) using the Ion Torrent One Touch System II (Life Technologies) and the Ion One Touch 200 template kit v2 DL (Life Technologies), following manufacturer's instructions. Templated ISPs were sequenced on a 316 micro-chip (up to 100 Mb of data) using the Ion Torrent Personal Genome Machine (PGM; Life Technologies) and the Ion PGM 200 sequencing kit v2 chemistry (Life Technologies) for 130 cycles (520 flows). After sequencing, the individual sequence reads were filtered within the PGM software to remove low quality and polyclonal sequences. Sequences matching the PGM 3′ adaptor were also automatically trimmed prior to bioinformatics analysis.

Bioinformatics and Sequence Analysis

Next generation sequencing data (Ion Torrent PGM platform) from the mtDNA capture was processed using an in-house customizable analytical pipeline based on available scripts. Reads were de-multiplexed by barcode, not allowing for any mismatch, using FASTX-Toolkit (http://hannonlab.cshl.edu/fastx_toolkit/index.html). Cutadapt v1.1 [49] was then used to trim adapters and filter the reads for quality and length (five successive rounds of trimming with 33.3% error rate allowed; minimum quality value of Phred 20; minimum and maximum length of 25 and 110 bp after trimming). The filtered reads were checked with FastQC (http://www.bioinformatics.bbsrc.ac.uk/projects/fastqc), before mapping against the RSRS [8] using TMAP v3.2.1 (https://github.com/nh13/TMAP) with the following options: -g 3 -M 3 -n 7 -v stage1–stage-keep-all map1–seed-length 12–seed-max-diff 4 stage2 map2–z-best 5 map3–max-seed-hits 10. Mapped reads with mapping quality below Phred 30 and all duplicates were removed using both Samtools v0.1.18 [50] and the MarkDuplicates option of Picard Tools v1.79 (http://picard.sourceforge.net). GC content of mapped reads was analysed using the CollectGcBiasMetrics tool of Picard Tools v1.79. Misincorporation patterns were assessed using MapDamage v2.0.1 [10–11] and default parameters. The resulting sequence assembly was analyzed using Geneious Pro (v5.6.2) software in order to build the consensus sequence and assign a haplotype following the latest nomenclature and phylogeny on PhyloTree.org (Build 15 (30 Sep 2012)).

Genomic coverage by unique reads was compared to the distribution of GC content across the mtDNA using a custom R script (see Figure S1). Coverage depth was directly estimated from the unique read mapping file by counting the number of reads covering each mtDNA position. We also calculated the average GC percentage in a sliding window of 55 nt centered around each mtDNA position, taking in account the circular nature of the molecule. The size of the window corresponds to the average size of the unique mapped reads (see the Results section and Figure S2).

SNP Confirmation by Direct Sequencing and Minisequencing

Selected regions of the mtDNA genome were sequenced independently via single and multiplex PCR amplification followed by direct sequencing (Table S1), and SNaPshot minisequencing following established PCR conditions [30] in order to:

1) verify the HVR-I sequence between np 15997 and 16410;
2) verify the 22 coding region hg diagnostic SNPs targeted by the GenoCoRe22 reaction;
3) interrogate the deletions at np 249, 290, 291, which are diagnostic of the CZ and C1 sub-haplogroups (Figure 3);
4) confirm the new SNPs identified here in the mtDNA genome sequence of specimen UZOO-74, but also to type these in the two other C1 carriers from Yuzhnyy Oleni Ostrov, UZOO-7 and UZOO-8.

Typing of coding-region SNPs using the GenoCore22 reaction was performed using the protocol described in [20].

Authentication of the Ancient mtDNA Sequence

The aDNA sequence data was validated using three lines of evidence: 1) monitoring of contamination, 2) reproducibility, 3) phylogenetic consistency.

1) Pre-PCR DNA work was carried out at the Australian Centre for Ancient DNA (ACAD), University of Adelaide, a purpose-built positive air pressure laboratory dedicated to aDNA studies, which is physically isolated from any molecular biology laboratory amplifying DNA. Routine decontamination of the laboratory surfaces and instruments involves exposure to UV radiation and thorough cleaning using DNA oxidants such as bleach, Decon (Decon labs) and Isopropanol. In order to protect the laboratory environment from human DNA, researchers wear protective clothes including body suit, a facemask, a face shield, gumboots, and three layers of surgical gloves that are changed on a regular basis. Obvious large-scale contamination within the laboratory or in the reagents were monitored and controlled by blank controls (one extraction blank for every five ancient samples and two PCR/GenoCoRe22 blank controls for every six reactions). In addition, no haplotype similar to any of those possessed by laboratory members was consistently amplified from aDNA extracts.
2) For all three Yuzhnyy Oleni Ostrov individuals, the HVR-I sequences and GenoCoRe22 profiles could be replicated from two samples extracted independently [20]. For UZOO-74, the same HVR-I sequence and coding region SNPs were obtained by direct sequencing/GenoCoRe22 minisequencing and sequencing the complete mtDNA genome. Private SNPs identified by complete mtDNA genome sequencing were also confirmed by direct sequencing.
3) The phylogenetic consistency of the combination of variable positions in the mtDNA genome was an additional indicator of the authenticity of the sequence. The fifty-one nucleotide differences with RSRS were 100% consistent with the phylogenetic position of UZOO-74 on the hg C1 branch. No deviation from this position could be detected. The combination of five additional mutations was found to be unique to the haplotype sequenced here. None of these additional mutations was found to define branches within the human mtDNA tree, thus providing little support for them arising from contamination. In addition, considering the multiple replications performed to type the SNPs of interest, jumping PCR is thought to have had little or no impact on the hg C1 haplotype presented here.

Phylogenetic Analysis of Haplogroup C1

In order to construct a phylogenetic network of the hg C1 HVR-I sequences displaying the C1 mutational pattern A16129G, T16187C, C16189T, G16230A, T16278C, T16298C, C16311T, T16325C, C16327T in Eurasia were gathered from the literature (Table S2). Sequences of whole mtDNA genomes belonging to hg C1 were also compiled from the online GenBank database on the basis of the list published in [10] (Table S2) and relevant new publications since. Sequences were corrected for known sequencing errors and ambiguities and, in particular, sequence length polymorphisms following recommendations in [39] and [51]. Mutations 309.1C, 315.1C, A16182c, A16183c, 16193.1C and C16519T were systematically ignored, as these positions are known to represent mutational hotspots and/or recurrent sequencing artefacts, which create reticulations in the phylogenetic analysis [35]. A tree was then constructed manually for complete mtDNA genome sequences on the basis of the tree constructed in [10] taking the latest mtDNA phylogeny into account (www.phylotree.org; [35]).

Divergence Time Estimate

The split of the six clades from the hg C1 root could theoretically be dated using mtDNA genome sequences. However, the use of molecular data for dating divergence times can be problematic. The most commonly used method in modern mtDNA studies is based on the calculation of the ρ statistic [52]. However, this method has been shown to produce inaccurate dates, especially when the genetic data was collected from populations with complex demographic histories (e.g., [53]). The tree topology of hg C1 suggests that this is likely to be the case for the six C1 clades. The tree obtained is indeed very imbalanced: the three American clades show strong signals of a recent expansion, contrary to the three clades in Asia, Iceland and Mesolithic Yuzhnyy Oleni Ostrov. Moreover, a limitation of molecular dating methods is the inaccuracy of the estimation of the human mtDNA substitution rate and the fact that they rely on the use of a constant mutation rate through time (e.g., [54]). A recent study could largely improve the accuracy of such substitution rate estimates by correcting for purifying selection but still rely on a fossil calibration point, i.e., the human/chimp split ~6.5 million years ago [55]. Based on this method, divergence time estimate were calculated for the modern sub-clades C1a, C1b and C1c at around 17,100 yBP (95% Confidence Interval: 12,000–22,500 yBP). The use of dated ancient mtDNA genome sequences to calibrate the molecular clock has so far not been considered by this method. However, recent studies have utilised ancient mtDNA genomes as tip calibrations in a Bayesian Markov Chain Monte Carlo framework using the program BEAST [56], which allows estimation of a substitution rate that is allowed to vary in time and can also take temporally heterogeneous mtDNA genome sequences into account [27–28]. However, the case of the polytomous hg C1 tree revealed problematic [7] it was recently demonstrated that a biased or incomplete representation of C1d lineages could lead to a much younger divergence time estimate (7,000–9,000 yBP; [7]). The addition of new data resulted in a revised divergence time to 18,700+/−1,400 yBP, in accordance with estimates for all other Pan-American haplogroups (15,000–18,000 yBP; [4], [7], [42], [57]). With only one C1f and very few Icelandic C1e mtDNA genomes, it is apparent that the sampling of the genetic diversity for these clades is not yet extensive enough to reliably calculate a divergence time estimates. Similarly, it seems also unreasonable to reconstruct the timing of the arrival of hg C1 lineages in Europe via coalescence age dating and/or determine whether all European C1 lineages reached Europe as part of the same migration as the Yuzhnyy Oleni Ostrov C1f branch or as part of other movements from the East.

Accession Numbers

The deep-sequencing data were deposited on the NCBI Sequence Read Archive, accession number: SRP033724.

Supporting Information

Figure S1 Coverage depth and GC content for the haplogroup C1f mitochondrial genome (individual UZOO-74). Mapping coverage of unique reads (in blue) is given per base. Local GC content (red) is shown for 55-bp intervals. The arrow indicate missing data between nucleotide positions 7277 and 7556, a region also characterized by low GC content.

Figure S2 Analysis of DNA damage patterns. A. Four upper plots: Frequencies of the A, C, G and T bases according to the nucleotide positions within the read (within the grey box) and outside the read (outside the grey box). Two lower plots: Frequency distribution of specific substitutions from the 5′-end (left) to the 3′-end (right) of the read sequence. B. Two upper plots: Read length distribution. Two lower plots: Observed cumulative frequency of C to T and G to A misincorporations. C. Observed frequencies of nucleotide misincorporation and simulated Bayesian posterior predictive intervals obtain from model fitting. D Simulated posterior distribution of model parameters: Lambda, probability of terminating in overhang (λ); DeltaD, probability of cytosine deamination in double strands (∂D); and DeltaS, probability of cytosine deamination in single strands (∂S).

Acknowledgments

We thank Marta Kasper for help with project logistics; Jeremy Austin, and Jessica Metcalf for helpful comments; Thomas Duffey for support with match searches. Members of The Genographic Project Consortium: Syama Adhikarla (Madurai Kamaraj University, Madurai, Tamil Nadu, India), Christina J. Adler (University of Adelaide, South Australia, Australia), Jaume Bertranpetit (Universitat Pompeu Fabra, Barcelona, Spain), Andrew C. Clarke (University of Otago, Dunedin, New Zealand), David Comas (Universitat Pompeu Fabra, Barcelona, Spain), Matthew C. Dulik (University of Pennsylvania, Philadelphia, Pennsylvania, United States), Jill B. Gaieski (University of Pennsylvania, Philadelphia, Pennsylvania, United States), Arun Kumar Ganesh Prasad (Madurai Kamaraj University, Madurai, Tamil Nadu, India), Marc Haber (Universitat Pompeu Fabra, Barcelona, Spain; Lebanese American University, Chouran, Beirut, Lebanon), Li Jin (Fudan University, Shanghai, China), Matthew E. Kaplan (University of Arizona, Tucson, Arizona, United States), Shilin Li (Fudan University, Shanghai, China), Begoña Martínez-Cruz (Universitat Pompeu Fabra, Barcelona, Spain), Elizabeth A. Matisoo-Smith (University of Otago, Dunedin, New Zealand), Nirav C. Merchant (University of Arizona, Tucson, Arizona, United States), R. John Mitchell (La Trobe University, Melbourne, Victoria, Australia), Amanda C. Owings (University of Pennsylvania, Philadelphia, Pennsylvania, United States), Laxmi Parida (IBM, Yorktown Heights, New York, United States), Ramasamy Pitchappan (Madurai Kamaraj University, Madurai, Tamil Nadu, India), Daniel E. Platt (IBM, Yorktown Heights, New York, United States), Lluis Quintana-Murci (Institut Pasteur, Paris, France), Colin Renfrew (University of Cambridge, Cambridge, United Kingdom), Daniela R. Lacerda (Universidade Federal de Minas Gerais, Belo Horizonte, Minas Gerais, Brazil), Ajay K. Royyuru (IBM, Yorktow Heights, New York, United States), Fabrício R. Santos (Universidade Federal de Minas Gerais, Belo Horizonte, Minas Gerais, Brazil), Theodore G. Schurr (University of Pennsylvania, Philadelphia, Pennsylvania, United States), Himla Soodyall (National Health Laboratory Service, Johannesburg, South Africa), David F. Soria Hernanz (National Geographic Society, Washington, District of Columbia, United States), Pandikumar Swamikrishnan (IBM, Somers, New York, United States), Chris Tyler-Smith (The Wellcome Trust Sanger Institute, Hinxton, United Kingdom), Arun Varatharajan Santhakumari (Madurai Kamaraj University, Madurai, Tamil Nadu, India), Pedro Paulo Vieira (Universidade Federal do Rio de Janeiro, Rio de Janeiro, Brazil), Miguel G. Vilar (University of Pennsylvania, Philadelphia, Pennsylvania, United States), R. Spencer Wells

(National Geographic Society, Washington, District of Columbia, United States), Pierre A. Zalloua (Lebanese American University, Chouran, Beirut, Lebanon), Janet S. Ziegle (Applied Biosystems, Foster City, California, United States).

Author Contributions

Conceived and designed the experiments: CDS WH PB BL AC. Performed the experiments: CDS WH PB BL JT. Analyzed the data: CDS WH PB BL JS. Contributed reagents/materials/analysis tools: CDS WH PB BL JS JT OB VM VK AC TGC. Wrote the paper: CDS WH AC.

References

1. Kong QP, Yao YG, Sun C, Bandelt HJ, Zhu CL, et al. (2003) Phylogeny of East Asian mitochondrial DNA lineages inferred from complete sequences. Am J Hum Genet 73: 671–676.
2. Metspalu M, Kivisild T, Bandelt HJ, Richards M, Villems R (2006) The pioneer settlement of modern humans in Asia. In: Bandelt HJ, Macaulay V, Richards M, editors. Human mitochondrial DNA and the evolution of Homo sapiens. Berlin: Springer-Verlag. 181–199.
3. Derenko M, Malyarchuk B, Grzybowski T, Denisova G, Rogalla U, et al. (2010) Origin and post-glacial dispersal of mitochondrial DNA haplogroups C and D in Northern Asia. PLoS One 5: e15214.
4. Tamm E, Kivisild T, Reidla M, Metspalu M, Smith DG, et al. (2007) Beringian standstill and spread of Native American founders. PLoS One 2(9): e829.
5. Achilli A, Perego UA, Bravi CM, Coble MD, Kong QP, et al. (2008) The phylogeny of the four pan-American MtDNA haplogroups: implications for evolutionary and disease studies. PLoS One 3(3): e1764.
6. Volodko N, Starikovskaya E, Mazunin I, Eltsov N, Naidenko P, et al. (2008) Mitochondrial genome diversity in arctic Siberians, with particular reference to the evolutionary history of Beringia and Pleistocenic peopling of the Americas. Am J Hum Genet 82: 1084–1100.
7. Perego UA, Angerhofer N, Pala M, Olivieri A, Lancioni H, et al. (2010) The initial peopling of the Americas: a growing number of founding mitochondrial genomes from Beringia. Genome Res 20(9): 1174–1179.
8. Behar DM, van Oven M, Rosset S, Metspalu M, Loogväli EL, et al. (2012) A "Copernican" reassessment of the human mitochondrial DNA tree from its root. Am J Hum Genet 90(4): 675–684.
9. Starikovskaya EB, Sukernik RI, Derbeneva OA, Volodko NV, Ruiz-Pesini E, et al. (2005) Mitochondrial DNA diversity in indigenous populations of the southern extent of Siberia, and the origins of Native American haplogroups. Ann Hum Genet 69: 67–89.
10. Ebenesersdóttir SS, Sigurðsson A, Sánchez-Quinto F, Lalueza-Fox C, Stefánsson K, et al. (2011) A new subclade of mtDNA haplogroup C1 found in Icelanders: evidence of pre-Columbian contact? Am J Phys Anthropol 144(1): 92–99.
11. Forster P, Harding R, Torroni A, Bandelt H (1996) Origin and evolution of Native American mtDNA variation: a reappraisal. Am J Hum Genet 59: 935–945.
12. Bandelt HJ, Herrnstadt C, Yao YG, Kong QP, Kivisild T, et al. (2003) Identification of Native American founder mtDNAs through the analysis of complete mtDNA sequences: some caveats. Ann Hum Genet 67(6): 512–524.
13. Shields G, Schmiechen A, Frazier B, Redd A, Voevoda M, et al. (1993) mtDNA sequences suggest a recent evolutionary divergence for Beringian and northern North American populations. Am J Hum Genet 53: 549–562.
14. Pfeiffer H, Brinkmann B, Hühne J, Rolf B, Morris AA, et al. (1999) Expanding the forensic German mitochondrial DNA control region database: genetic diversity as a function of sample size and microgeography. Int J Legal Med 112: 291–298.
15. Rando JC, Cabrera VM, Larruga JM, Hernandez M, Gonzalez AM, et al. (1999) Phylogeographic patterns of mtDNA reflecting the colonization of the Canary Islands. Ann Hum Genet 63: 413–428.
16. Helgason A, Sigurðardóttir S, Gulcher JR, Ward R, Stefánsson K (2000) mtDNA and the origin of the Icelanders: deciphering signals of recent population history. Am J Hum Genet 66: 999–1016.
17. Helgason A, Nicholson G, Stefánsson K, Donnelly P (2003) A reassessment of genetic diversity in Icelanders: strong evidence from multiple loci for relative homogeneity caused by genetic drift. Ann Hum Genet 67: 281–297.
18. Bermisheva MA, Kutuev IA, Spitsyn VA, Villems R, Batyrova AZ, et al. (2005) Analysis of mitochondrial DNA variation in the population of Oroks. Genetika 41: 78–84.
19. Grousset R (1970) The Empire of the Steppes: History of Central Asia. Rutgers University Press.
20. Der Sarkissian C, Balanovsky O, Brandt G, Khartanovich V, Buzhilova A, et al. (2013) Ancient DNA reveals prehistoric gene-flow from siberia in the complex human population history of North East Europe. PLoS Genet 9(2): e1003296.
21. Gilbert MT, Kivisild T, Grønnow B, Andersen PK, Metspalu E, et al. (2008) Paleo-Eskimo mtDNA genome reveals matrilineal discontinuity in Greenland. Science 320: 1787–1789.
22. Ermini L, Olivieri C, Rizzi E, Corti G, Bonnal R, et al. (2008) Complete mitochondrial genome sequence of the Tyrolean Iceman. Curr Biol 18: 1687–1693.
23. Knapp M, Horsburgh KA, Prost S, Stanton JA, Buckley HR, et al. (2012) Complete mitochondrial DNA genome sequences from the first New Zealanders. Proc Natl Acad Sci U S A 109(45): 18350–18354.
24. Fu Q, Rudan P, Pääbo S, Krause J (2012) Complete mitochondrial genomes reveal neolithic expansion into Europe. PLoS One (3): e32473.
25. Krause J, Briggs A, Kircher M, Maricic T, Zwyns N, et al. (2010) A complete mtDNA genome of an early modern human from Kostenki, Russia. Curr Biol 20: 231–236.
26. Sánchez-Quinto F, Schroeder H, Ramirez O, Avila-Arcos MC, Pybus M, et al. (2012) Genomic affinities of two 7,000-year-old Iberian hunter-gatherers. Curr Biol 22(16): 1494–1499.
27. Fu Q, Mittnik A, Johnson PL, Bos K, Lari M, et al. (2013) A revised timescale for human evolution based on ancient mitochondrial genomes. Curr Biol 23(7): 553–559.
28. Brotherton P, Haak W, Templeton J, Brandt G, Soubrier J, et al. (2013) Neolithic mitochondrial haplogroup H genomes and the genetic origins of Europeans. Nat Commun 4: 1764.
29. Margulies M, Egholm M, Altman WE, Attiya S, Bader JS, et al. (2005) Genome sequencing in microfabricated high-density picolitre reactors. Nature 437: 376–380.
30. Burrow AA, Marullo A, Holder LR, Wang Y-H (2010) Secondary structure formation and DNA instability at fragile site FRA16B. Nuc Acids Res 38: 2865–2877.
31. Dabney J, Meyer M (2012) Length and GC-biases during sequencing library amplification: a comparison of various polymerase-buffer systems with ancient and modern DNA sequencing libraries. Biotechniques 52: 87–94.
32. Ginolhac A, Rasmussen M, Gilbert MT, Willerslev E, Orlando L (2011) mapDamage: testing for damage patterns in ancient DNA sequences. Bioinformatics 27(15): 2153–2155.
33. Jónsson H, Ginolhac A, Schubert M, Johnson PL, Orlando L (2013) mapDamage2.0: fast approximate Bayesian estimates of ancient DNA damage parameters. Bioinformatics 29(13): 1682–1684.
34. Sawyer S, Krause J, Guschanski K, Savolainen V, Pääbo S (2012) Temporal patterns of nucleotide misincorporations and DNA fragmentation in ancient DNA. PLoS One 7(3): e34131.
35. van Oven M, Kayser M (2009) Updated comprehensive phylogenetic tree of global human mitochondrial DNA variation. Hum Mutat 30: E386–394. www.phylotree.org/.
36. Achilli A, Rengo C, Magri C, Battaglia V, Olivieri A, et al. (2004) The molecular dissection of mtDNA haplogroup H confirms that the Franco-Cantabrian glacial refuge was a major source for the European gene pool. Am J Hum Genet 75: 910–918.
37. Pereira L, Richards M, Goios A, Alonso A, Albarrán C, et al. (2005) High-resolution mtDNA evidence for the late-glacial resettlement of Europe from an Iberian refugium. Genome Res 15: 19–24.
38. Pala M, Olivieri A, Achilli A, Accetturo M, Metspalu E, et al. (2012) Mitochondrial DNA signals of late glacial recolonization of Europe from near eastern refugia. Am J Hum Genet 90: 915–24.
39. Bandelt H, Parson W (2008) Consistent treatment of length variants in the human mtDNA control region: a reappraisal. Int J Legal Med 122: 11–21.
40. Bramanti B, Thomas M, Haak W, Unterlaender M, Jores P, et al. (2009) Genetic discontinuity between local hunter-gatherers and central Europe's first farmers. Science 326: 137–140.
41. Malmström H, Gilbert M, Thomas M, Brandström M, Storå J, et al. (2009) Ancient DNA reveals lack of continuity between neolithic hunter-gatherers and contemporary Scandinavians. Curr Biol 19: 1758–1762.
42. Haak W, Balanovsky O, Sanchez JJ, Koshel S, Zaporozhchenko V, et al. (2010) Ancient DNA from European early Neolithic farmers reveals their near eastern affinities. PLoS Biol 8: e1000536.
43. Reidla M, Kivisild T, Metspalu E, Kaldma K, Tambets K, et al. (2003) Origin and diffusion of mtDNA haplogroup X. Am J Hum Genet 73: 1178–1190.
44. Perego UA, Achilli A, Angerhofer N, Accetturo M, Pala M, et al. (2009) Distinctive Paleo-Indian migration routes from Beringia marked by two rare mtDNA haplogroups. Curr Biol 19: 1–8.
45. Helgason A, Lalueza-Fox C, Ghosh S, Sigurethardóttir S, Sampietro ML, et al. (2009) Sequences from first settlers reveal rapid evolution in Icelandic mtDNA pool. PLoS Genet 5(1): e1000343.
46. Raghavan M, Skoglund P, Graf KE, Metspalu M, Albrechtsen A, et al. (2013) Upper Palaeolithic Siberian genome reveals dual ancestry of Native Americans. Nature. 2013 Nov 20.
47. O'Shea J, Zvelebil M (1984) Oleneostrovskii Mogilnik: Reconstructing Social and Economic Organisation of Prehistoric Hunter-Fishers in Northern Russia. J Anthropol Archaeol 1: 1–40.
48. Green RE, Krause J, Ptak SE, Briggs AW, Ronan MT, et al. (2006) Analysis of one million base pairs of Neanderthal DNA. Nature 444: 330–336.
49. Martin M (2011) Cutadapt removes adapter sequences from high-throughput sequencing reads. EMBnet journal 17(1): 10.
50. Li H, Handsaker B, Wysoker A, Fennell T, Ruan J, et al. (2009) The Sequence Alignment/Map format and SAMtools. Bioinformatics 25(16): 2078–2079.

51. Bandelt HJ (2006) Estimation of Mutation Rates and Coalescence Times: Some Caveats. In: Bandelt HJ, Macaulay V, Richards M. Mitochondrial DNA and the evolution of Homo sapiens. Berlin: Springer-Verlag. 64.
52. Saillard J, Forster P, Lynnerup N, Bandelt HJ, Nørby S (2000) mtDNA variation among Greenland Eskimos: the edge of the Beringian expansion. Am J Hum Genet 67(3): 718–726.
53. Cox MP (2008) Accuracy of molecular dating with the rho statistic: deviations from coalescent expectations under a range of demographic models. Hum Biol 80(4): 335–357.
54. Ho SY, Lanfear R, Bromham L, Phillips MJ, Soubrier J, et al. (2011) Time-dependent rates of molecular evolution. Mol Ecol 20(15): 3087–3101.
55. Soares P, Ermini L, Thomson N, Mormina M, Rito T, et al. (2009) Correcting for Purifying Selection: An Improved Human Mitochondrial Molecular Clock. Am J Hum Genet 84(6): 740–59.
56. Drummond A, Rambaut A (2007) BEAST: Bayesian evolutionary analysis by sampling trees. BMC Evol Biol 7: 214.
57. Fagundes N, Kanitz R, Bonatto S (2008) A reevaluation of the Native American mtDNA genome diversity and its bearing on the models of early colonization of Beringia. PLoS One 3: e3157.

6

Identification of the Maize Gravitropism Gene *lazy plant1* by a Transposon-Tagging Genome Resequencing Strategy

Thomas P. Howard III[1¤], Andrew P. Hayward[1], Anthony Tordillos[1], Christopher Fragoso[1], Maria A. Moreno[1], Joe Tohme[2], Albert P. Kausch[3], John P. Mottinger[3], Stephen L. Dellaporta[1*]

1 Department of Molecular, Cellular and Developmental Biology, Yale University, New Haven, Connecticut, United States of America, **2** International Center for Tropical Agriculture, Cali, Colombia, **3** Department of Cell and Molecular Biology, University of Rhode Island, Kingston, Rhode Island, United States of America

Abstract

Since their initial discovery, transposons have been widely used as mutagens for forward and reverse genetic screens in a range of organisms. The problems of high copy number and sequence divergence among related transposons have often limited the efficiency at which tagged genes can be identified. A method was developed to identity the locations of *Mutator (Mu)* transposons in the *Zea mays* genome using a simple enrichment method combined with genome resequencing to identify transposon junction fragments. The sequencing library was prepared from genomic DNA by digesting with a restriction enzyme that cuts within a perfectly conserved motif of the *Mu* terminal inverted repeats (TIR). Paired-end reads containing *Mu* TIR sequences were computationally identified and chromosomal sequences flanking the transposon were mapped to the maize reference genome. This method has been used to identify *Mu* insertions in a number of alleles and to isolate the previously unidentified *lazy plant1 (la1)* gene. The *la1* gene is required for the negatively gravitropic response of shoots and mutant plants lack the ability to sense gravity. Using bioinformatic and fluorescence microscopy approaches, we show that the *la1* gene encodes a cell membrane and nuclear localized protein. Our Mu-Taq method is readily adaptable to identify the genomic locations of any insertion of a known sequence in any organism using any sequencing platform.

Editor: Lewis Lukens, University of Guelph, Canada

Funding: This work was funded by a grant to SLD, MAM and JT from the National Science Foundation (0965420). The funders had no role in study design, data collection and analysis, decision to publish, or preparation of the manuscript.

Competing Interests: The authors have declared that no competing interests exist.

* E-mail: stephen.dellaporta@yale.edu

¤ Current address: Harvard Medical School, Boston, Massachusetts, United States of America

Introduction

The maize *Mutator* (*Mu*) is one of the most aggressively mobile transposon families yet characterized in any organism. *Mutator* lines were first described as a genetic system that increased the mutation rate by 30-fold [1]. The "Mutator" trait did not segregate according to a simple one-gene model, with nearly 100% of progeny between crosses of *Mutator* lines and non-*Mutator* lines exhibiting high mutation rates [1]. The first mutant allele to be cloned and characterized from a *Mutator* line contained a 1.4 kb insertion (later named *Mu1*) in the alcohol dehydrogenase (*adh1*) gene of maize [2]. The *Mu1* element had ~215 bp highly homologous terminal inverted repeats (TIRs), and was flanked by a 9 bp target sequence duplication at its site of insertion [3]. When Robertson's *Mu* lines were examined by Southern hybridization, plants were shown to possess between 10–70 copies of *Mu*-related sequences [4]. These early studies suggested that active transposable elements were the genetic basis for the high mutation rates found in Robertson's *Mutator* lines.

Continued characterization of *Mu* lines has elucidated a family of transposons that is both diverse and complex. More than a dozen different *Mu* elements are currently known (reviewed in [5]), including the autonomous *MuDR* element that encodes for an active transposase, MURA [6,7]. All active (mobile) *Mu* elements share highly similar TIRs but their internal regions can vary considerably in both size and sequence [reviewed in [5,8–11]]. MURA binds to a highly conserved region within the *Mu* TIR, promoting transposition both in *cis* and in *trans* for other *Mu* elements found in the plant genome [7]. In this way, the presence and activity of *MuDR* determines whether all non-autonomous *Mu* elements are mobile or immobile in the plant's genome. Other epigenetic factors, such as the cytosine methylation state of *Mu* elements, affect mobility of individual elements (reviewed in [5]).

A number of factors account for the exceptionally high mutation rates in *Mu* lines. First, active *Mu* lines can contain over 100 *Mu* elements per genome (reviewed in [9]), a number that can be maintained from generation to generation [12]. On average, each transposon is responsible for one new insertion event every generation, either non-conservatively through transposition or conservatively through duplication [13]. Studies show that *Mu* elements prefer to insert into low copy number (non-repetitive) DNA [14], and genome-wide analysis suggests that *Mu* elements preferentially insert into regions of the genome that are transcribed [15]. For instance, analysis of the *RescueMu* element (described below) shows that 69% of its insertion sites are located in putatively expressed genomic sequences [16] even though the maize genome appears to be made up of more than 80% repetitive sequences [17]. Finally, *Mu* elements exhibit little target site sequence specificity, allowing for essentially uniform genome-wide mutagenesis (reviewed in [9]). These studies show that *Mu* elements are abundant, highly active, and targeted towards genes throughout the entire genome, features that drive a forward mutation rate of up to 10^{-3} per locus per generation.

Given their robust mutagenic properties, *Mu* transposons have been employed in a number of gene cloning and functional genomics experiments. Originally, mutant alleles were recovered from *Mu* lines, and the corresponding gene cloned by their association with *Mu* sequences by standard transposon tagging methods [18–20]. The limitation with this approach was the high copy number and diversity of *Mu* elements, making the linkage association of any particular element with a mutant allele a difficult and time-consuming endeavor. In fact, using this approach, only a handful of genes had been cloned even decades after the molecular identification of *Mu* elements.

More efficient reverse genetic screens became feasible with the advent of PCR methods that incorporated a degenerate *Mu* primer that anneals to a TIR consensus sequence together with a gene-specific primer to clone *Mu*-induced mutant alleles [20,21]. The Maize-Targeted-Mutagenesis (MTM) project increased the scope of reverse genetic screens by producing a large population (>40,000) of *Mu* plants whose genomic DNA was multiplexed for screening for *Mu* insertions in specific gene sequences [22]. Since these methods required prior knowledge of a gene sequence for primer design, they did not necessarily provide a robust approach for forward genetic studies of *Mu*-induced mutant phenotypes.

Other approaches were developed to clone and sequence *Mu* flanking sequences (MFS) *en masse* without prior knowledge of the gene surrounding the insertion [23]. These "anonymous" methods would ligate adapters to restriction-digested genomic DNA, and then amplify DNA with a degenerate *Mu* TIR primer together with a primer specific to the adapter sequence. A variation of this method, called MuTAIL-PCR, was developed to achieve the same goal with fewer sample manipulations, incorporating nested degenerate *Mu* TIR primers together with random primers and programmed thermal alternations to gradually enrich for MFS [24]. With these technologies, numerous groups undertook large-scale MFS cloning and sequencing projects [15]. Chief among these was the "UniformMu" project, which generated extensive maize *Mu* populations and incorporated the ability to stabilize transpositions by epigenetically silencing *Mu* elements after a cycle of mutagenesis [25]. UniformMu studies identified almost 2,000 independent, stable insertions, creating, at the time, one of the most significant knockout resources in maize [26]. While these resources provide efficient reverse genetics methods of identifying insertions in known genes, they have not been widely employed as a method of conducting forward genetic screens.

The *Rescue Mutator* (*RescueMu*) project attempted to develop a facile transposon tagging strategy by introducing a recombinant *Mu* element containing a bacterial origin of replication (*ori*) and an antibiotic resistance gene (ampicillin) into maize harboring *MuDR* as a transposase source [16,27,28]. After rounds of transposition, genomic DNA was digested around *RescueMu*, the linear fragments were circularized by ligation, rescued in *E. coli*, and sequenced to identify the MFS. The main limitation of this method is that all *Mu* elements, including endogenous ones, are active in a *MuDR* line and any one could have caused a mutation of interest. Because of this, only a small subset of mutants derived from *RescueMu* lines was caused by the insertion of *RescueMu* and many mutations went uncharacterized at a molecular level. Nevertheless, this project, like MTM and UniformMu, contributed to the genetic resources in maize by providing an additional source of *Mu*-derived alleles.

The advent of next-generation sequencing (NGS) technologies has increased the throughput and efficiency of *Mutator* functional genomics. The UniformMu project now incorporates Illumina sequencing strategies with over 12,000 of the nearly 40,000 maize genes identified as *Mu*-tagged with 20,000 new insertions expected each year [29]. A recent method called Digestion-Ligation-Amplification (DLA) combines an adapter-ligation step together with NGS to efficiently identify *Mu* flanking genomic sequences [30]. More specifically, DLA employs a degenerate set of *Mu* TIR primers together with a single-stranded, blocked adapter to selectively amplify MFS fragments. A secondary PCR step is required to create a suitable fragment library for sequencing. The DLA method generates a large number of sequencing reads with 94% corresponding to MFS with exceptional depth of coverage. The method was used to identify 12/14 *Mu*-induced alleles of the *glossy4* gene [30]. An alternate NGS-based method, employing a biotinylated oligonucleotide corresponding to the *Mu* TIR, enriches for MFS reads by hybridization and selection from whole genome Illumina libraries [31]. This method has been shown to have an 80% success rate in identifying *Mu*-tagged alleles.

Both of these NGS-based methods are effective at characterizing *Mu* insertion sites, but require specific primers and multiple PCR steps. Here, we report on a method that is both efficient and cost-effective in identifying MFS using a standard sequencing library protocol with minor modifications. Our "Mu-Taq" method incorporates restriction-digested genomic DNA with standard Illumina library protocols for paired-end sequencing to enrich for MFS-containing reads. MFS reads were identified *in silico* with custom scripts and mapped to the maize reference genome to locate *Mu* insertion sites.

Using the Mu-Taq method, we identified several known *Mu*-induced mutations and isolated the previously uncharacterized gravitropism *lazy plant1* (*la1*) gene of maize. To confirm our identification of *la1*, we characterized a second, independent *la1* mutation arising from insertion of a TIR family CACTA transposon in a plant displaying the *lazy* phenotype. We further characterized the wildtype *La1* gene using bioinformatic and fluorescence microscopy approaches. The maize LA1 protein was found to localize to the cell membrane and nucleus, with nuclear localization dependent upon a predicted nuclear localization sequence (NLS).

Materials and Methods

Genetic Stocks

The sources of mutant alleles used in this study are listed in Table S1. Mutant germplasm was obtained from the Maize Genetics Cooperation Stock Center (http://maizecoop.cropsci.uiuc.edu/). Many of these alleles were first identified during D.S. Robertson's original *Mu* experiments and, in some cases, the molecular basis of the mutation was unknown. The "lazy plant" mutation arose as a segregating recessive mutation in an F2 family derived from *Mutator* lines grown at the University of Rhode Island Experiment Station (this study). This mutation was phenotypically similar to the *lazy plant1* mutation first described in 1931 [32] and later designated *la1-mu1*.

Library Construction

Genomic DNA was isolated as previously described [33]. Approximately 5 µg of DNA was digested by 30 units *Taq*$^{\alpha}$I restriction endonuclease (New England BioLabs, Inc.) according to the manufacturer's instructions. The enzyme was inactivated by phenol:chloroform (1:1) extraction, precipitated with two volumes of ethanol, and resuspended in 20 µl 10 mM Tris pH 8.0. *Taq*$^{\alpha}$I digestion leaves a 5′ overhang on the ends of fragments. One µg of *Taq*$^{\alpha}$I digested DNA was made blunt using the NEBNext End Repair Module (NEB) according to the manufacturer's instructions. Reactions were incubated at 20° for 30 min followed by 65° for 20 min. The reaction buffer was exchanged with 10 mM Tris pH 8.0 in a volume (~30 µl).

The concentrated end-repaired fragments were 3′-adenylated in a 50 µl reaction using the NEBNext dA-Tailing Module (NEB) according to the manufacturer's instructions. Reactions were incubated at 37° for 30 min followed by 65° for 20 min. The reaction buffer was exchanged with 10 mM Tris pH 8.0 and the adenylated fragments were ligated to Illumina paired-end adapters (Table S2) using the following conditions: 19.6 µl of adenylated fragments in a 30 µl reaction containing adapters (final concentration 0.33 µM), 3,000 units T4 DNA Ligase (NEB), and 1X T4 DNA Ligase Reaction Buffer (NEB) supplemented with 2 mM ATP (NEB). Reactions were incubated at 20° for 60 min followed by 65° for 15 min. Unligated adapters were removed using the QIAquick PCR Purification Kit (Qiagen) according to the manufacturer's protocol. Samples were eluted in 30 µl elution buffer prewarmed to 65°.

The adapted fragments were amplified by PCR using modified Illumina PCR primers (Table S3). Triplicate 50 µl reactions were performed for each allele. Reactions contained 1X Phusion High-Fidelity PCR Master Mix with GC Buffer (NEB) or 1X KAPA HiFi HotStart ReadyMix (Kapa Biosystems) (preferred), 60 nM of each primer, and ~50 ng adapted DNA. Cycling instructions were as follows: for Phusion: 98° (2 min); 12 cycles of 98° (10 sec), 65° (30 sec), 72° (30 sec); 72° (5 min); for KAPA: 95° (4 min); 12 cycles of 98° (20 sec), 65° (15 sec), 72° (15 sec); 72° (5 min). Triplicate samples were pooled and concentrated using Microcon YM-30 Centrifugal Filter Devices (Millipore) as described. Pooled samples were subjected to gel electrophoresis and 200–800 bp fragments were excised and purified using Freeze 'N Squeeze DNA Gel Extraction Spin Columns (Bio-Rad) according to the manufacturer's protocol. The samples' buffer was exchanged to 1X TE and concentrated to a minimal volume.

Barcoded Adapter Design and Preparation

In these experiments, custom bar-coded Illumina adapters were employed for library preparation. Custom adapters incorporated a unique four-base "barcode" sequence (see Table S2 for adapter sequences). The barcode sequence was sequenced in the first four nucleotides in each of the paired-end reads followed by the thymine added in the adenylation reaction. Note, however, that our method incorporates ligating adapters to adenylated library fragments. Therefore, standard Illumina paired-end adapters, with or without indexing, can be substituted in lieu of custom barcoded primers without any modifications to the protocol.

Multiplexing and qPCR Library Normalization

The gel-extracted libraries were subjected to qPCR analysis in order to normalize the amount of DNA from each library in the multiplexed sample. The qPCR reaction employed primers (Table S3) that anneal to the extension of the adapters that was added to fragments during the PCR amplification step. The qPCR step estimates the relative concentration of only those fragments with suitable adapters at both ends for sequencing. Triplicate 20 µl reactions were performed containing 1X KAPA SYBR FAST Universal qPCR Master Mix (Kapa Biosystems), 200 nM of each primer, and 1 µl (<1 ng) DNA. Cycling conditions were as follows: 95° (3 min); 40 cycles of 95° (3 sec), 60° (30 sec); ramp to 95° over 5 min; in an Eppendorf Mastercycler ep realplex real-time PCR system, with Ct values calculated using the CalQPlex algorithm. Relative concentrations of each sample were calculated from the mean of triplicate Ct values using a ΔCt method ($2^{Ct1-Ct2}$). Samples were multiplexed according to their calculated relative concentrations. The multiplexed pool was then concentrated to a minimal volume using the Microcon columns as described. The pool was paired-end sequenced (2×75 nt) using an Illumina HiSeq 2000 at the W.M. Keck Foundation's Yale Center for Genome Analysis (YCGA).

Data Analysis

Raw image files were processed by the YCGA using the Illumina CASAVA computational pipeline for base calling and quality score determination. The paired-end fastq reads were demultiplexed and parsed into separate fastq files according to the four-bp terminal barcode using a custom Java script. This algorithm also removed the first five bases of the read, which represents the barcode along with the thymine introduced in the adenylation and adapter ligation steps. In some cases, the parsed fastq files were mapped against the reference maize genome using Burrows-Wheeler algorithm [34]. The parsed fastq files were then analyzed using a second custom script, which identified and parsed reads that contain the *Mu* TIR tag on one or the other end (read 1 or read 2). The script also extracted the sequences of each MFS (31 bases adjacent to the *Mu* TIR tag) into a separate fasta sequence file. The MFS were mapped using the BLASTn algorithm [35] against the *Zea mays* B73 reference genome v2 (maizesequence.org) [17]. BLAST output files were imported into Microsoft Excel for all subsequent analyses. All primary sequence files, extracted MFS, and scripts are provided in File S1.

Localization Studies

Fusion constructs. Plasmid pYU2973 (*La1*:Citrine) was derived from plant expression vector pPZP200b and contained the full-length *La1* CDS with the stop codon removed in frame with the citrine CDS under control of the single CaMV 35S promoter with TEV leader and the 35S terminator (construction details available upon request). Plasmid pYU2972 (La1ΔNLS:Citrine) contained *La1* CDS with an in-frame deletion of 381 bases encompassing amino acids 263–389 of the LA1 protein, and was otherwise identical to pYU2973. The La1ΔNLS:Citrine deletion arose from spontaneous PCR amplification error. The single CaMV 35S promoter was amplified from plasmid pDPG165. The TEV leader sequence and 35S terminator were amplified from plasmid pRTL2.

Transient protein expression and confocal microscopy. *Agrobacterium*-mediated transient expression was performed in *N. benthamiana* leaves as previously described [36]. Briefly, *Agrobacterium* GV2260 containing pYU2972 or pYU2973 was grown overnight, pelleted, and resuspended in infiltration medium containing 10 mM MgCl2, 10 mM 2-Morpholinoethanesulfonic acid and 200 µM acetosyringone. Strains were induced at room temperature for 4 hours followed by vacuum infiltration into 4–5 week old *N. benthamiana* leaves at OD_{600} 1.2. Live tissue imaging was performed on a Zeiss LSM510 META confocal microscope (Carl Zeiss) using a 40× C-Apochromat water immersion objective lens. Tissue samples were cut from *N. benthamiana* leaves at approximately 42 hours post infiltration. The 514 nm laser line of an argon laser with appropriate emission filters was used to image citrine and chloroplasts.

Results

Identification of a Conserved Motif in *Mu* Elements

We sought to devise a simple and efficient NGS-based method for identifying transposon-induced mutations by virtue of their association with a conserved terminal inverted repeat (TIR) sequence. Current enrichment methods for transposon sequences suffer from limitations, such as complex and costly library preparation, the use of degenerate primers, or assumptions about sequence conservation in *Mu* TIRs. Often, these methods are designed for a single transposon family, or a single sequencing platform, or both, and are not flexible and adaptable.

To address these limitations, we tested a simple transposon enrichment method based on digestion of genomic DNA with a restriction enzyme that recognizes a conserved sequence motif in its TIR element. The enrichment strategy was to digest genomic DNA with an appropriate restriction enzyme as to define one end of the genomic fragments in the TIR element. Given the high read coverage of current instrumentation –200 million or more paired-end reads – these TIR-containing fragments could be readily identified by computational methods and mapped to a reference genome.

In this study, the transposon of interest was Robertson's *Mutator*, a diverse family of transposable elements in maize that share conserved TIRs [reviewed in [5]]. Illumina libraries were constructed using restriction-digested genomic DNA. The choice of the restriction enzyme was crucial since it must recognize a highly conserved or invariant motif in all TIRs found in active *Mu* elements to enrich for these fragments and to maximize the chances that all, or nearly all, *Mu* Flanking Sequences (MFS) were represented in the library. Furthermore, the restriction site needed to be an optimal distance from the end of the TIR to enable identification of TIRs and to map flanking sequences. In the Illumina sequencing protocol, for instance, only 75–100 nt of sequence are generated from one or both ends of library fragments [37]. Therefore, if one end of the library fragment is fixed within the TIR, then the chosen site must be located so as to generate sufficient TIR sequence (*Mu* tag) for identification, as well as sufficient flanking genomic sequence (MFS) for mapping to a reference genomic location. Paired-end read technology was chosen so that either read of a pair could be examined for a TIR tag. Moreover, if necessary, the paired, non-TIR end could be later used for resolution or confirmation of the mapping location of any individual MFS.

An additional requirement is that the restriction site be frequent enough to generate fragment sizes appropriate for NGS methods. For instance, Illumina library fragments are typically between 200 and 800 bp [38]. An enzyme with a recognition site of N bp will cut once every 4^N bp [excluding degenerate recognition and assuming 50% G-C genome composition]. In our protocol, the library preparation steps add an additional 130 bp of adapter sequences to every fragment. For example, an enzyme with a 4 bp non-degenerate recognition site would generate a library consisting of fragments normally distributed around 386 bp, an ideal size for Illumina-based sequencing methods. Lastly, it would be preferable that the enzyme not be inhibited by CG or CNG methylation since plant genomes contain high levels of cytosine methylation at these sites [39,40]. Hence, a methylation-sensitive enzyme will be blocked from digesting a significant fraction of its recognition sites that are methylated, increasing the average fragment size in the library. Taken together, we sought to identify a cytosine methylation-insensitive restriction enzyme that recognizes a conserved 4 bp motif located approximately 20–50 bp from the end of *Mu* TIRs. Depending on its exact location, Illumina paired-end sequencing (2×75 nt) would yield a suitable 20–50 nt TIR tag for fragment identification and a suitable 20–50 nt of MFS for reference mapping.

To identify a suitable restriction enzyme, a ClustalW [41] alignment was performed on 70 bp of TIR sequences from all known active and potentially active (mobile) *Mu* elements (Figure 1). Of the enzyme options available in the *Mu* TIR (Table S4), *Taq*$^{\alpha}$I and *Mnl*I fit the criteria we sought. *Taq*$^{\alpha}$I was chosen because it recognizes a perfectly conserved 4 bp motif (TCGA), located 37–40 bp from the TIR end, and this enzyme was cytosine-methylation insensitive (NEB product information). Using this enzyme with barcoded adapters, we would generate enough data to identify a TIR tag of 39 bp and a MFS tag of 31 bp, ample enough sequence data to unambiguously identify *Mu* TIRs and to map MFS to the maize genome (Figure 2). [Note: using standard Illumina adapters, *Taq*$^{\alpha}$I digests will generate a TIR sequence of 39 bp and a MFS sequence of 36 bp.].

Identification of Mu Flanking Sequences

The basic strategy was to digest genomic DNA with *Taq*$^{\alpha}$I, construct Illumina libraries, and pair-end sequence to identify MFS for mapping to the B73 reference genome (Figure 2). Genomic DNA samples were isolated from plants homozygous for several mutant alleles (Table S1) and digested using *Taq*$^{\alpha}$I. The DNA fragments were prepared for Illumina sequencing by end-repair, adenylation, adapter ligation, and library amplification (see Materials and Methods). Each genomic library was ligated to a specific barcoded adapter for later identification. After qPCR normalization, the individual libraries were multiplexed and sequenced using the Illumina HiSeq 2000. The resultant sequencing data were demultiplexed and parsed by barcodes. Each parsed file was analyzed to identify reads containing a *Mu* TIR tag on one end, and its adjacent MFS extracted and parsed as fasta files. MFS were mapped to the B73 reference genome (version 2) using the BWA [34] and Bowtie 2 [42,43].

The results of this analysis are summarized in Table 1. The average number of reads per sample was ~14,500,000 reads, with a range from 8,955,215 to 25,196,240 reads. This range indicated that the qPCR step was fairly effective at normalizing the individual libraries in the multiplexed pool. Given that 150 bp of sequence are generated for each read, the average number of bases generated per sample was ~2.15 Gb, which, if the reads were distributed evenly, is approximately 1X coverage of the entire maize genome [17]. Since we are only sampling a subset of the genome that is directly adjacent to a *Taq*$^{\alpha}$I recognition site, however, the depth of coverage in these regions should be much higher.

To assess coverage, all paired-end reads from one library, *la1-mu1* were mapped to the maize B73 reference genome using the BWA algorithm [34]. These results showed a uniform distribution across the genome (Figure 3A). A closer look at the read coverage in a 1 MB subinterval (Figure 3B) showed a relatively even distribution of reads. Given that some *Taq*$^{\alpha}$I fragments created through digestion of the genome will be either too small or too large to be sequenced and that some areas will not have a recognition site at all, it is not surprising that certain areas in the interval were not covered.

Next, the number of TIRs and associated MFS detected in each sample was analyzed. On average, 281, with a range of 144 to 506, were found in each genomic library (Table 1). Some of these MFS were likely independent reads from the same insertion, or the results of PCR replication. Because the paired end reads were fixed at restriction sites, it was not possible to distinguish between these two possibilities. Approximately 50% of the total MFS

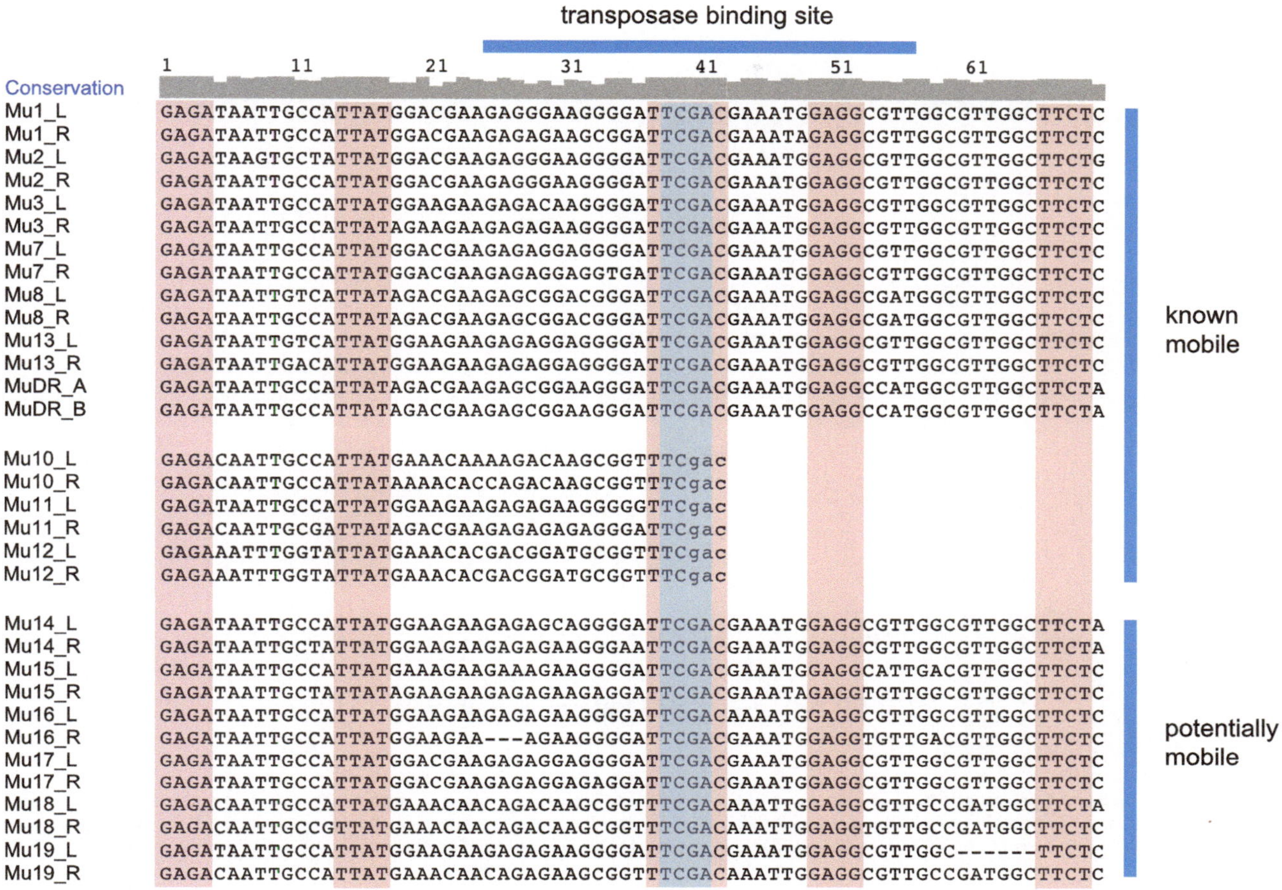

Figure 1. Alignment of *Mu* Terminal Inverted Repeats (TIRs). ClustalW was used to align all known active and potentially active *Mu* elements. Strings of four or more consecutive bases that are entirely conserved among all elements are shaded in red. The conserved *Taq*$^{\alpha}$I site is shaded in blue. *Mu4, 5,* and *6* are inactive and are not included [reviewed in [5]]. *Mu9* is MuDR. Only the first 39 bp of the *Mu10, 11,* and *12* TIRs have been sequenced [62]. However, the primer used to amplify the TIR ended in a 3′ GTC, allowing for the assumption that the sequence continues as GAC (shown as small case) and that the *Taq*$^{\alpha}$I site remains intact. Of the most recently discovered *Mu* elements (13–19), only *Mu13* has been confirmed to actively move and create new mutations [58]. TIR sequences obtained from [3,58,62–68].

represent unique reads. Given that there were between 50 and 100 *Mu* elements per genome [reviewed in [10]] this number of unique TIRs (twice the number of expected insertions) was in the range expected for a genome containing active *Mu* elements. As long as there is not a strong bias towards amplifying *Mu* TIR fragments, PCR replicates should alter read depth evenly throughout the genome. If we were to assume the number of *Mu* TIRs was high (e.g. 200), the 39 bp *Mu* TIR tag that we search for would make up 0.000312% of the genome. Nonetheless, in our data, an average of 0.00204% of sequencing reads contained a *Mu* TIR and MFS, representing 6.5-fold enrichment for *Mu* TIR fragments.

Analysis of Mutant Alleles

In total, we analyzed MFS from 13 different plant genomes. Eleven of the 13 genomes harbored a mutation in a known gene, while the remaining two genomes harbored mutations in previously unidentified genes, *lazy plant1* and *silkless1*. Each of these 13 mutations was derived from *Mutator* lines of maize (see Table S1 for alleles and sources) and four of these were previously known to be caused by an insertion of a *Mu* element. The molecular basis of the remaining nine alleles had not been previously characterized at the molecular level. Seven of these nine alleles showed mutability in the presence of *Mutator* activity, a likely indicator of Mu insertion in the gene.

The locations of MFS found in each of these genomes were examined for those mapping to the genomic location of the affected gene. In the case of the four known *Mu*-induced mutations, our analysis identified *Mu* insertions in three of these four genes. In the previously uncharacterized alleles, we identified a *Mu* insertion in four out of nine alleles. These included the two previously uncharacterized genes, *lazy plant1* (discussed below) and *silkless1* (manuscript in preparation) both derived from active *Mutator* stocks.

The *lazy plant1* (*la1*) maize mutation results in a prostrate growth pattern (Figure 4A). Interestingly, mutant plants do not have any type of structural abnormality but instead lack the drive to grow upwards, leading to their appropriately named "lazy" phenotype [44–46]. Charles Darwin was one of the first scientists to document that plant shoots show negative gravitropism, *i.e.* grow in the opposite direction as gravity [47]. The *lazy plant1* mutants have been shown to continue growing in whichever direction they are pointed in, regardless of the direction of gravity [44]. Without the perception of gravity, the maturing plant grows prostrate (Figure 4A). These phenotypes suggest that the wild type gene is required for the process of negative gravitropism in the shoot.

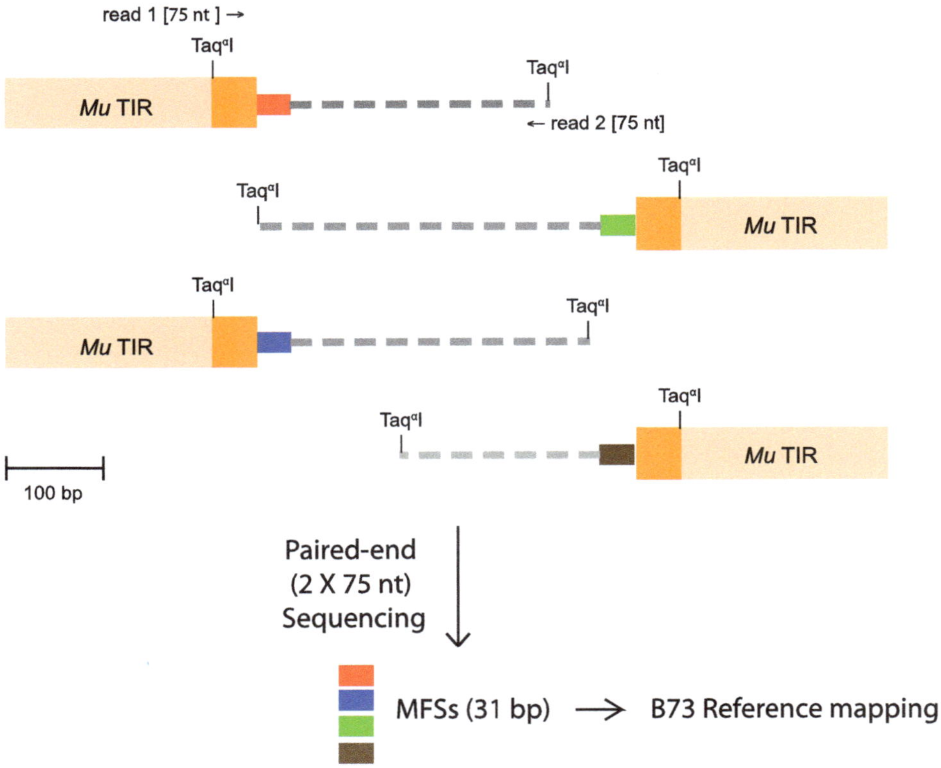

Figure 2. ***Mu-Taq*$^{\alpha}$I Library Construction.** Digesting genomic DNA with *Taq*$^{\alpha}$I creates a library of fragments. Fragments containing a *Mu*-MFS junction will all contain a degenerate 39 nt *Mu* TIR tag along with 31–35 nt of flanking genomic sequence at one end of the 2×75 nt paired-end read. These fragments are computationally identified and the *Mu* flanking sequence (MFS) is mapped to the maize reference genome.

The *la1* gene of maize had not been cloned at the time of this study but its chromosomal position had been genetically mapped to chromosome 4 (Maize IBM2 2008 Neighbors map, www.maizegdb.org). In 2009, a recessive "lazy" mutation was found segregating in an F2 population derived from an active Robertson's *Mutator* line. This mutation represented an ideal

Table 1. Summary of Sequenced Alleles.

Allele	Mu insertion[1]	Total Number of Paired-End Reads	Number of MFS Reads (total/unique)	Gene Identified (this study)
a1-mum1	Unknown[1]	21,746,017	506/163	No
a1-mum2	Yes	12,486,199	328/132	No
a2-mum2	Unknown[2]	11,122,765	230/120	No
a2-mum4	Unknown[2]	12,301,174	331/146	Yes
bz1-mum4::Mu1	Unknown[2]	9,824,747	263/167	No
bz1-mum9	Yes	9,100,676	228/101	Yes
bz2-mVW4::MuDR	Yes	8,955,215	224/120	Yes
c2-mum1	Unknown[2]	14,066,613	374/172	Yes
wx1-mum1	Unknown[2]	11,681,516	370/171	No
wx1-mum2	Unknown[3]	16,439,709	275/130	No
wx1-mum5::Mu8	Yes	18,280,957	217/100	Yes
la1-mu1 (replicate 1)	Unknown	25,196,240	280/161	Yes
la1-mu1 (replicate 2)	Unknown	10,754,601	144/105	Yes
sk1-mu1	Unknown	14,327,049	168/141	Yes

[1]Allele shown to contain a *Mu* element by molecular analysis.
[2]Uncharacterized at the molecular level but likely caused by *Mu* insertion based on mutable phenotype dependent on *Mutator* activity.
[3]Atypical mutability pattern. May not contain *Mu* element.

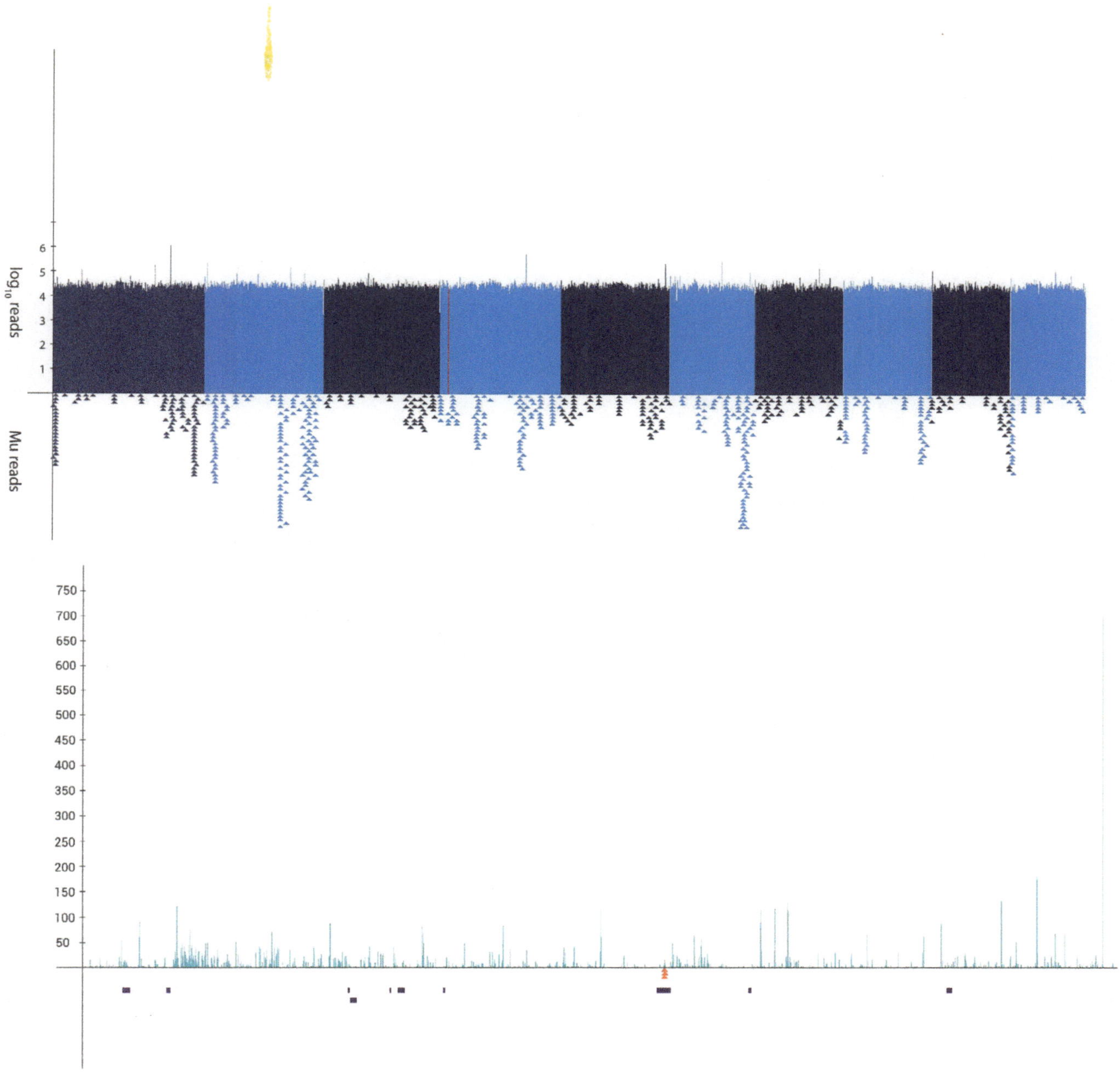

Figure 3. Manhattan plot of *la1-mu1 Taq*[a]I sequencing. (**A**) Manhattan plot showing the distribution of reads from *la1-mu1* genomic DNA mapped throughout the B73 genome. Alternating colors represent each of the ten maize chromosomes. Each x-axis pixel represents a bin of 1 Mb and the logarithmic y-axis denotes the number of reads mapping to each bin. The red line represents the known genetic map position for the *la1* reference mutation. Each triangle below the plot represents the approximate location of mapped MFS. (**B**) Expanded Manhattan plot of a 1 Mb interval corresponding to the approximate map position of *la1*. Same as top with each x-axis pixel representing a bin of 1 kb. Filtered genes in the 1 Mb interval are shown as black rectangles. MFS mapping to this interval are shown as inverted red triangles.

situation for testing our method since any MFS in a gene mapping to the genetic location of *la1* in these plants would potentially represent a candidate *La1* gene. The Mu-Taq strategy was applied to two biological replicates of *la1* plants. The libraries returned 25,196,240 reads with 280 MFS and 10,754,601 reads with 144 MFS, respectively (shown diagrammatically in Figure 3A). In both libraries, a single MFS mapped within a few MB of the predicted *la1* map location (Figure 3B). This MFS, identical in both replicates, mapped to a predicted exon within the GRMZM2G135019 gene (*la1-mu1;* Figure 4C) located on chromosome 4 at position 17,977,557 to 17,984,094, an estimated 4 Mb from the genetic map position of *la1*. BLASTn analysis of the predicted cDNA sequence returned no putative conserved domains but did identify an *Oryza sativa* (rice) gene (Os11g0490600) with significant homology (E value = 8e^{-68}). In rice, mutant alleles of Os11g0490600 display a gravitropic phenotype similar to the maize "lazy" phenotype [48]. Moreover, the maize and rice genes are located in a region of extensive synteny [49] suggesting that they are orthologous. BLASTp [50] alignment of the maize and rice proteins indicates that they share 60% identity and 67% similarity (Figure 4D). These results indicate that GRMZM2G135019, identified by virtue of its

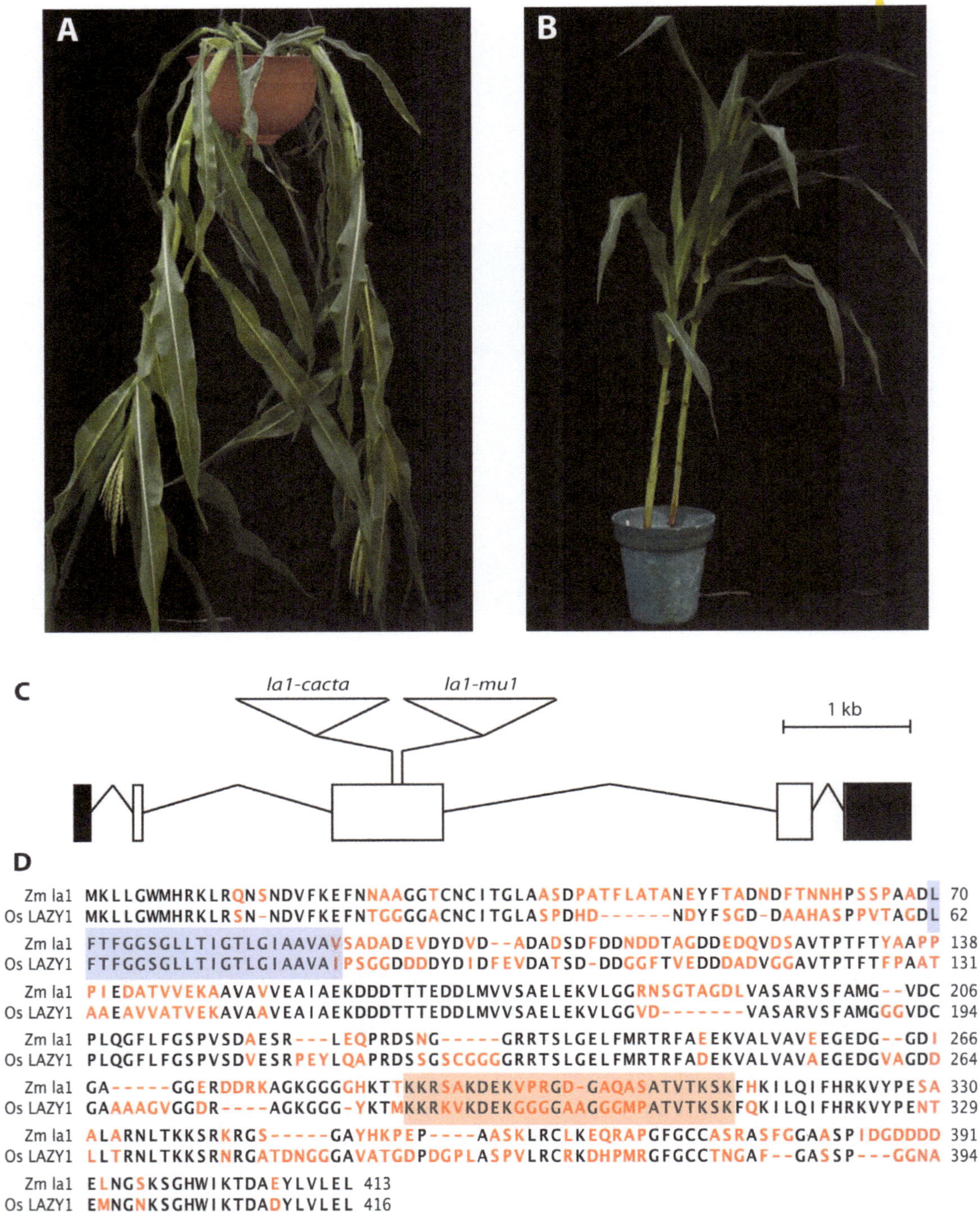

Figure 4. Identification of *la1*. The *la1* mutant plants (**A**) exhibit prostrate growth habit caused by a lack of a negative gravitropic response. Wild type maize plants (**B**) purposefully grow away from gravity (negative gravitropism). (**C**) Gene structure of *la1* with the site of the *Mu* insertion in the *la1-mu1* mutant allele and the CACTA family insertion in the *la1-cacta* allele. Exons are shown as empty boxes and UTRs are shown as filled boxes. The *Mu* insertion has not been characterized and is not to scale. (**D**) Amino acid ClustalW alignment of the maize and rice predicted LAZY1 proteins. The pair shares 60% identity. The predicted transmembrane domain is shaded in blue, the predicted NLS domain is shaded in orange.

association with a linked MFS found through Mu-Taq analysis, is a strong candidate for the *lazy plant1* gene of maize.

Confirmation of the identification of the *La1* gene was obtained by characterizing an independent *la1* mutation. Primers derived from the GRMZM2G135019 gene were used to amplify and sequence an independent *la1* mutant alleles obtained from the Maize Stock Center. Sequencing of this allele revealed disruption of GRMZM2G135019 by the insertion of a previously uncharacterized TIR family CACTA transposable element (*la1-cacta*; Figure 4C). Complementation test of this allele with the *la1-reference* allele (Maize Genetics Cooperative Stock Center) provide further confirmation that GRMZM2G135019 is indeed the *la1* gene.

Characterization of the Maize *La1* Gene

The maize *La1* was predicted to encode a protein of 413 amino acids of unknown function with a molecular weight of 43.6 kD (Figure 4D). Prediction of transmembrane helices using both HMMTOP [51,52] and TMpred [53] indicated a strong likelihood of the presence of a transmembrane helix at position 70–91 (blue box in Figure 4D). Strong homology was also seen at an NLS region previously described in rice *LAZY1* at position 288–313 (orange box in Figure 4D). The subcellular localization of maize LA1 was determined with a fluorescent-tagged LA1 protein (*La1*:Citrine). LA1 localized to the cell membrane and nucleus during transient expression in *N. benthamiana* epidermal and mesophyll cells (Figure 5A, 5B). The LA1-Citrine fusion protein was predicted to be larger than the threshold for cytoplasm-to-nucleus diffusion through the nuclear pores ($< \sim 50$ KDa), and LA1-Citrine was not detected in the cytoplasm, suggesting that nonspecific nuclear localization was unlikely. Western blotting also revealed no evidence of smaller cleavage products containing citrine that could cause artifactual nuclear signal (Figure 5D). To confirm that the putative maize LA1 NLS domain was required for nuclear localization, a deletion encompassing the NLS (amino-acid residues 263–389) was generated (*La1*ΔNLS:Citrine). Deletion of the NLS in LA1ΔNLS-Citrine abrogated nuclear localization (Figure 5C). These results suggest that full-length LA1 protein is present at both the cell membrane and nucleus, though its molecular function remains unclear.

Discussion

The Mu-Taq method identified three out of four alleles known to be caused by *Mu* insertions and four out of nine alleles (including *la1* and *sk1*) that were potentially caused by insertion of a *Mu* element. This efficiency of known alleles (75%) is comparable to that of previously published methods such as DLA (86%) and the biotinylated oligonucleotide method (80%). The overall efficiency of the method, however, was lower for previously uncharacterized alleles. For example, it is likely that many of these unknown alleles actually contain *Mu* insertions since their mutable phenotype is dependent on *Mutator* activity. This would therefore place a lower limit on the efficiency at 54%. From analysis of bulk read data, we detect over 100 unique MFS in a plant known to harbor *Mu* elements (Table 1). This is above the number detected per genome by other methods and in the range predicted from the average number of *Mu* elements in a genome. Given that the *a1-mum2* allele is confirmed to have a *Mu* insertion [54] and that other alleles showing mutability are likely to be *Mu*-induced, certainly a fraction of MFS have gone undetected. Based on the number of unique MFS identified in each genome the depth of sequencing is probably below saturation and additional sequencing may lead to the identification of some of these alleles. Therefore, the efficiency of our method at detecting a *Mu* insertion is between the estimates of 54–75% [54].

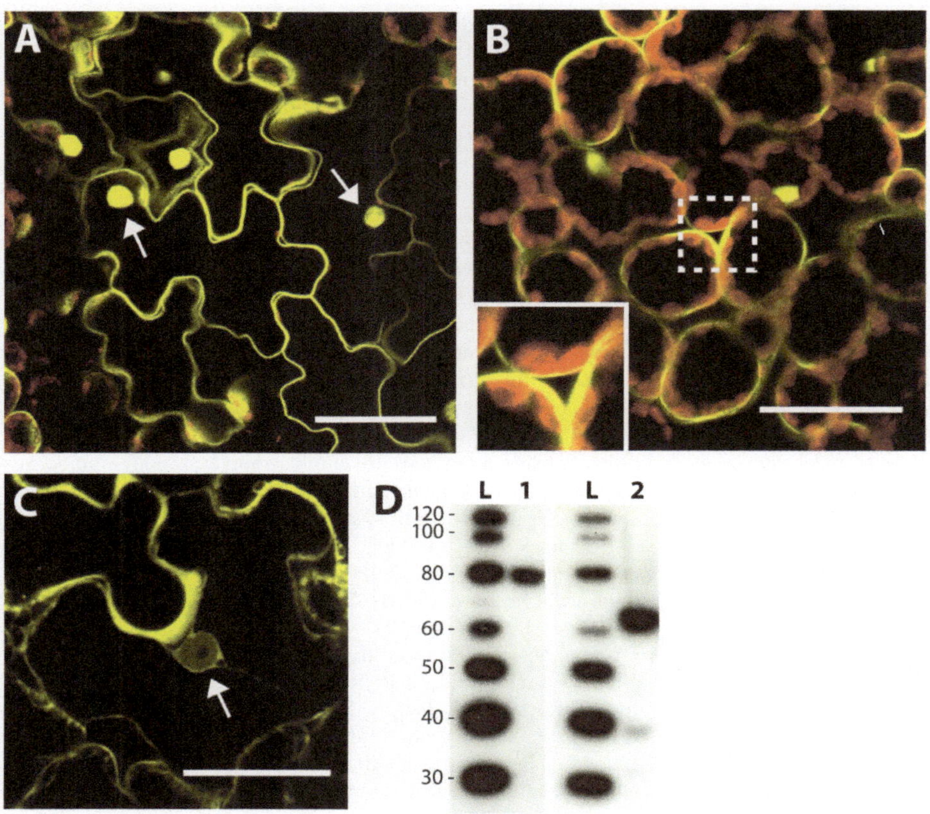

Figure 5. Localization of LA1. LA1-Citrine localizes to the nucleus (arrows) and cell membrane in (A) *N. benthamiana* epidermal cells and (B) *N. benthamiana* mesophyll cells. Inset shows detail of the mesophyll cell membrane. (C) Deletion of the putative NLS in LA1ΔNLS-Citrine abrogates nuclear localization. Scale bars are 20 uM. (D) Western blotting confirms expression of full length LA1-Citrine (lane 1) and LA1ΔNLS-Citrine (lane 2).

Besides depth of sequencing, there are many possible reasons for not detecting a MFS when it is present in the genome. Some of these reasons are technical in nature. One obvious reason would be the absence of the conserved Taq^{α}I site in the TIR. Based on the conservation of Taq^{α}I in all known mobile and potentially mobile elements, we believe that this possibility is unlikely. Another reason is that any individual Taq^{α}I fragment generated during restriction digestion may be too large for efficient cluster formation, estimated to be greater than 800 bp. While the average size of a Taq^{α}I fragment, based on a random distribution of bases and an AT-GC content of 50%, would be 256 bp, the actual size will vary according to local and global base composition and a quasi-random distribution of sites. Also, given that 130 bp of adapter sequence are added during library construction, it is possible that some Taq^{α}I fragments will exceed the size necessary to generate efficient clusters during the sequencing process. Nevertheless, this is predicted to be a relatively minor fraction of Taq^{α}I fragments. A third possible reason for missing an individual MFS is the problem of Taq^{α}I digesting too close to the end of the TIR sequence to identify a MFS. Since we are mapping MFS using the adjacent 31 nt, the frequency of smaller size fragments is not expected to be significant. In the specific case of *Mu* elements and Taq^{α}I, the terminal two bases of the TIR are part of the Taq^{α}I recognition site. Therefore, on average, one out of sixteen TIR ends would reconstitute the full Taq^{α}I site and be cut directly after the TIR. These fragments would have no detectable MFS.

It is also possible that the map position of an individual MFS is not accurate due to mapping errors or reference genome limitations. Our current analysis includes only the single, most significant mapped hit. Given that variation exists between the maize lines used in this study and the inbred B73 reference genome as well as the relatively high Illumina sequencing error rate, some MFS may not have been properly mapped. In the short term, reducing the stringency of possible matches and mapping both ends of the paired-end reads may resolve this issue. A longer-term solution would be to continue to characterize and catalog the variation found in maize lines. Nevertheless, for these issues to be relevant, both ends of the transposon would have to suffer from one or more of these problems to cause the element to be completely absent from a particular MFS dataset.

Another issue with the efficiency of MFS detection is depth of read coverage and absence of coverage for some areas of the genome. As previously mentioned in cases where the gene went undetected in the MFS reads, it is possible that read depth was insufficient to detect all MFS sequences. In our experiments we achieve an overall depth of coverage of approximately 1X. However, since only fragments adjacent to a Taq^{α}I sites were sequenced, our effective genome size is considerably smaller and our effective depth of coverage much higher for these regions. Our estimate of 6.5X enrichment for *Mu* TIR fragments should enhance the effectiveness of this method to detect individual MFS. Evidence that depth of coverage may not be a significant issue is that the allele with the lowest number of total paired-end reads (*bz2-mVW4*) was detected, but the allele with the second highest number of total paired-end reads (*a1-mum1*) was not. Therefore, increasing read depth may not be the only solution sought to resolve the issue of MFS detection.

One biological reason for failing to identify an MFS-associated gene is that the mutant allele may not contain a *Mu* element. Since *Mu* is mobile, it may no longer be associated with a particular allele. *Mu* elements frequently insert and excise imperfectly, causing loss-of-function alleles even though the *Mu* element is no longer present. In our study, nine alleles derived from *Mutator* lines were not previously characterized at the molecular level to confirm *Mu* insertions in the gene. Nevertheless, the mutable phenotype associated with many of these alleles makes it likely that a Mu insertion is the causative agent. Yet, only four of these nine alleles were found to be associated with a particular MFS. It has been well documented that many mutations recovered from *Mu* lines do not contain heritable *Mu* insertions. For instance, at the *liguleless1* locus of maize, two alleles derived from *Mutator* stocks did not contain *Mu* insertions [55], while at the *tasselseed1* locus, only two out of four alleles derived from *Mutator* lines contained *Mu* insertions [56]. If the alleles used in this study did not contain a *Mu* element, this would explain the failure to detect at least some of these genes in the dataset. At this point, the molecular basis of the mutations in the alleles that were not identified need to be further investigated to determine whether or not they actually contain *Mu* elements with conserved TIR sequences. This additional data would allow us to accurately assess the effectiveness of our protocol.

Comparison to Other Methods

Three other methods have been reported to massively identify MFS using next-generation sequencing. Though successful, these methods rely on imperfectly conserved sequences in the *Mu* TIR (Figure 1). The DLA method [30] uses a set of degenerate primers to amplify out of the *Mu* TIR to create a library of MFS. Certain *Mu* elements are likely to be selectively amplified because of differential annealing efficiency of the various primers. The biotinylated oligonucleotide method [31] relies on annealing over a larger range of the TIR, expanding the possibility that divergent sequences of certain *Mu* elements limit their capture and sequencing. A recent method called Mu-seq incorporates a single 23 base, 12-fold-degenerate *Mu* TIR primer to target fragment amplification [57]. In comparison, our method did not rely on degenerate *Mu* primers but rather on anchoring reads to a highly conserved of a 4 bp Taq^{α}I site, which is located in the MURA transposase binding site. This site has been perfectly conserved in all active *Mu* elements characterized to date. In addition, our method would identify MFS from any unknown *Mu* elements as they likely contain the conserved TIR Taq^{α}I site. Seven new *Mu* elements were recently identified [58] and it seems likely that other *Mu* elements have yet to be identified. In our method, the identification of TIR tags was incorporated into an algorithm that could be easily adapted to include additional TIR elements and to modulate the stringency and flexibility of TIR identification.

One of the main advantages of the Mu-Taq strategy is its simplicity. Standard Illumina reagents and adapters can be used for libraries and sequencing with little protocol modification. Other published methods are cumbersome and require numerous steps to prepare sequencing libraries or complicated post-library enrichment methods. For instance, the DLA method requires a dideoxynucleotide blocking step as well as an extra PCR amplification, introducing biases and read redundancy into the libraries. The biotinylated oligonucleotide method requires two rounds of hybridization as well as an extra amplification, also introducing amplification bias and read redundancy. Mu-seq employs three rounds of PCR amplification and special adapters. In contrast, our Mu-Taq method requires only the typical steps – end-repair, adenylation, and ligation – used during standard Illumina library construction. Next-generation library protocols also require a size fractionation step. In this method, instead of the standard random DNA fragmentation by sonication or nebulization, a simple restriction digestion step was employed as both a size fractionation and enrichment step. Once end-repaired and adenylated, these fragments can be ligated to standard Illumina adapters and amplified in a single low cycle PCR reaction.

Subsequently, the Mu-Taq method incorporates a custom bioinformatics pipeline to identify *Mu* TIR sequences to extract MFS and standard algorithms to map their corresponding MFS.

While other NGS methods generate a much higher percentage of reads that correspond to MFS, their exceptional coverage depth is unnecessary for some applications such as single gene identification. Even though in our method only 0.00204% of reads correspond to MFS, mutations in eight independent plant genomes were identified from reads derived from a single Illumina HiSeq lane. With the ever-improving efficiency of NGS platforms, an increase in read depth will become even more superfluous for regular gene cloning applications. While larger MFS screening projects may wish to identify more alleles in a single run using an MFS amplification protocol, our method is ideal for identifying single or multiple *Mu*-tagged genes of interest.

Adaptability of Mu-Taq to Other Transposon Families

The Mu-Taq method is readily adaptable to other transposon systems, and indeed any type of known DNA insertion, in organisms besides plants. The only requirement is the presence of a suitable and highly conserved restriction site close to the end of the insertion. The "Dynamic TIR Finder" algorithm has been written to accept a Fasta file of any set of TIR tags and to parse out all corresponding transposon flanking sequence reads. Therefore, in principle, this method is readily adaptable to any species, and to any family of transposable elements with conserved terminal elements. In addition, the method should be readily adaptable to other sequencing platforms. In this way, the Mu-Taq method described here represents a general and flexible strategy for identifying transposon-tagged genes in the ever-progressing NGS field.

Application of Mu-Taq Cloning to Identify and Characterize Maize *la1*

The application of the Mu-Taq method effectively identified the maize *la1* gravitropism gene in two biological replicates of an allele derived from a *Mutator* line. We were able to further identify a second mutant *la1* allele that failed to complement the *la1-reference* that contained a previously uncharacterized CACTA transposon insertion. We show that maize *La1* shares significant homology to its recently identified rice ortholog [48]. Concurrent with this study, Dong et al. (2013) have also reported on the cloning of the maize *La1* gene using a Mutator-derived allele [59]. Interestingly, the mapping and cloning strategy employed required a time- and labor-intensive effort involving several generations and nearly 1300 plants. In our study, from mutant to candidate gene identification required only a few weeks at a substantial savings in time and expense.

Both our study and Dong et al. [59] show that GRMZM2G135019 corresponds to the maize *La1* gene and LA1 protein localizes to the nucleus and cell membrane, a localization profile previously reported for both the rice and *Arabidopsis La1* orthologs [48,60]. Using sequence homology between rice and maize *La1*, we confirm a putative NLS required for nuclear localization of maize LA1. The rice *LAZY1* gene has been shown to affect gravitropism through asymmetric auxin distribution in the developing plant [48,61]. Similar perturbations of auxin transport have been reported in maize *la1* plants [59], and the *la1* phenotype was shown to be auxin dependent in the original investigations into the nature of the mutant in the 1930s [45]. Even though the effector (auxin) is known, there still remain significant questions regarding how *la1* actually senses gravity. Although trafficking of membrane-bound LA1 to the nucleus could hypothetically mediate auxin signaling, we are unable to detect shuttling of LA1 by confocal microscopy. Furthermore, a recent study showed that deletion of the AtLAZY1 NLS did not prevent rescue of the *atlazy1* mutant phenotype in Arabidopsis, leaving the functional relevance of nuclear LA1 unclear [60]. More complete understanding of LA1 activity will certainly be a productive area of future research on gravitropism.

Acknowledgments

The authors wish to thank Yingchun Tong, Jorge Duitama, Christopher Bolick, and Christopher Heffelfinger for technical assistance and suggestions to the manuscript.

Author Contributions

Conceived and designed the experiments: TPH APH AT JPM SLD TPH APK AT JPM SLD. Performed the experiments: TPH APH AT CF MAM JPM TPH APK AT CF MAM JPM. Analyzed the data: TPH APH AT CF MAM JPM SLD TPH APK AT CF MAM JPM SLD. Contributed reagents/materials/analysis tools: TPH APH AT TPH APK AT JT. Wrote the paper: TPH APH SLD TPH APK SLD.

References

1. Robertson DS (1978) Characterization of a mutator system in maize. Mutat Res 51: 21–28.
2. Bennetzen JL, Swanson J, Taylor WC, Freeling M (1984) DNA insertion in the first intron of maize Adh1 affects message levels: cloning of progenitor and mutant Adh1 alleles. Proc Natl Acad Sci U S A 81: 4125–4128.
3. Barker RF, Thompson DV, Talbot DR, Swanson J, Bennetzen JL (1984) Nucleotide sequence of the maize transposable element Mul. Nucleic Acids Res 12: 5955–5967.
4. Bennetzen JL (1984) Transposable element Mu1 is found in multiple copies only in Robertson's Mutator maize lines. J Mol Appl Genet 2: 519–524.
5. Walbot V, Rudenko GN (2002) *MuDR/Mu* Transposable Elements of Maize. In: Craig NL, Craigie R, Gellert M, Lambowitz AM, editors. Mobile DNA II. Washington, D.C.: ASM Press. 533–564.
6. Chomet P, Lisch D, Hardeman KJ, Chandler VL, Freeling M (1991) Identification of a regulatory transposon that controls the Mutator transposable element system in maize. Genetics 129: 261–270.
7. Benito MI, Walbot V (1997) Characterization of the maize Mutator transposable element MURA transposase as a DNA-binding protein. Mol Cell Biol 17: 5165–5175.
8. Chandler VL, Hardeman KJ (1992) The Mu elements of Zea mays. Adv Genet 30: 77–122.
9. Bennetzen JL, Springer PS, Cresse AD, Hendrickx M (1993) Specificity and regulation of the *mutator* transposable element system in maize. Critical Reviews in Plant Sciences 12: 57–95.
10. Bennetzen JL (1996) The Mutator transposable element system of maize. Curr Top Microbiol Immunol 204: 195–229.
11. Lisch D (2002) Mutator transposons. Trends Plant Sci 7: 498–504.

12. Walbot V, Warren C (1988) Regulation of Mu element copy number in maize lines with an active or inactive Mutator transposable element system. Mol Gen Genet 211: 27–34.
13. Alleman M, Freeling M (1986) The Mu transposable elements of maize: evidence for transposition and copy number regulation during development. Genetics 112: 107–119.
14. Cresse AD, Hulbert SH, Brown WE, Lucas JR, Bennetzen JL (1995) Mu1-related transposable elements of maize preferentially insert into low copy number DNA. Genetics 140: 315–324.
15. Hanley S, Edwards D, Stevenson D, Haines S, Hegarty M, et al. (2000) Identification of transposon-tagged genes by the random sequencing of Mutator-tagged DNA fragments from Zea mays. Plant J 23: 557–566.
16. Raizada MN, Nan GL, Walbot V (2001) Somatic and germinal mobility of the RescueMu transposon in transgenic maize. Plant Cell 13: 1587–1608.
17. Schnable PS, Ware D, Fulton RS, Stein JC, Wei F, et al. (2009) The B73 maize genome: complexity, diversity, and dynamics. Science 326: 1112–1115.
18. O'Reilly C, Shepherd NS, Pereira A, Schwarz-Sommer Z, Bertram I, et al. (1985) Molecular cloning of the a1 locus of Zea mays using the transposable elements En and Mu1. EMBO J 4: 877–882.
19. McLaughlin M, Walbot V (1987) Cloning of a mutable bz2 allele of maize by transposon tagging and differential hybridization. Genetics 117: 771–776.
20. Bensen RJ, Johal GS, Crane VC, Tossberg JT, Schnable PS, et al. (1995) Cloning and characterization of the maize An1 gene. Plant Cell 7: 75–84.
21. Das L, Martienssen R (1995) Site-selected transposon mutagenesis at the hcf106 locus in maize. Plant Cell 7: 287–294.
22. May BP, Liu H, Vollbrecht E, Senior L, Rabinowicz PD, et al. (2003) Maize-targeted mutagenesis: A knockout resource for maize. Proc Natl Acad Sci U S A 100: 11541–11546.
23. Frey M, Stettner C, Gierl A (1998) A general method for gene isolation in tagging approaches: amplification of insertion mutagenised sites (AIMS). The Plant Journal 12: 717–721.
24. Settles AM, Latshaw S, McCarty DR (2004) Molecular analysis of high-copy insertion sites in maize. Nucleic Acids Res 32: e54.
25. McCarty DR, Settles AM, Suzuki M, Tan BC, Latshaw S, et al. (2005) Steady-state transposon mutagenesis in inbred maize. Plant J 44: 52–61.
26. Settles AM, Holding DR, Tan BC, Latshaw SP, Liu J, et al. (2007) Sequence-indexed mutations in maize using the UniformMu transposon-tagging population. BMC Genomics 8: 116.
27. Raizada MN (2003) RescueMu protocols for maize functional genomics. Methods Mol Biol 236: 37–58.
28. Fernandes J, Dong Q, Schneider B, Morrow DJ, Nan GL, et al. (2004) Genome-wide mutagenesis of Zea mays L. using RescueMu transposons. Genome Biol 5: R82.
29. Project TUM (2011) Project Documentation & Protocols: UniformMu Transposon Resource.
30. Liu S, Dietrich CR, Schnable PS (2009) DLA-based strategies for cloning insertion mutants: cloning the gl4 locus of maize using Mu transposon tagged alleles. Genetics 183: 1215–1225.
31. Williams-Carrier R, Stiffler N, Belcher S, Kroeger T, Stern DB, et al. (2010) Use of Illumina sequencing to identify transposon insertions underlying mutant phenotypes in high-copy Mutator lines of maize. Plant J 63: 167–177.
32. Jenkins M, Gerhardt F (1931) A gene influencing the composition of the culm in maize. Iowa Ag Exp Sta Research Bull 138: 121–151.
33. Chen J, Dellaporta S (1993) Urea-based plant DNA miniprep. In: Freeling M, Walbot V, editors. The Maize Handbook. New York: Springer-Verlag. 526–527.
34. Li H, Durbin R (2009) Fast and accurate short read alignment with Burrows-Wheeler transform. Bioinformatics 25: 1754–1760.
35. Altschul SF, Gish W, Miller W, Myer EW, Lipman DJ (1990) Basic local alignment search tool. J Mol Biol 215: 403–410.
36. Hayward A, Padmanabhan M, Dinesh-Kumar SP (2011) Virus-induced gene silencing in nicotiana benthamiana and other plant species. Methods Mol Biol 678: 55–63.
37. Bentley DR, Balasubramanian S, Swerdlow HP, Smith GP, Milton J, et al. (2008) Accurate whole human genome sequencing using reversible terminator chemistry. Nature 456: 53–59.
38. Illumina (2010) Genomic Sequencing.
39. Palmer LE, Rabinowicz PD, O'Shaughnessy AL, Balija VS, Nascimento LU, et al. (2003) Maize genome sequencing by methylation filtration. Science 302: 2115–2117.
40. Zhao X, Chai Y, Liu B (2007) Epigenetic inheritance and variation of DNA methylation level and pattern in maize intra-specific hybrids. Plant Science 172: 930–938.
41. Thompson JD, Higgins DG, Gibson TJ (1994) CLUSTAL W: improving the sensitivity of progressive multiple sequence alignment through sequence weighting, position-specific gap penalties and weight matrix choice. Nucleic Acids Res 22: 4673–4680.
42. Langmead B, Trapnell C, Pop M, Salzberg SL (2009) Ultrafast and memory-efficient alignment of short DNA sequences to the human genome. Genome Biol 10: R25.
43. Langmead B, Salzberg SL (2012) Fast gapped-read alignment with Bowtie 2. Nat Methods 9: 357–359.
44. van Overbeek J (1936) "Lazy," an a-geotropic form of maize. The Journal of Heredity 29: 339–341.
45. van Overbeek J (1938) "Laziness" in maize due to abnormal distribution of growth hormone. The Journal of Heredity 29: 339–341.
46. Shafer J (1939) Physiology of Lazy Corn. Botanical Gazette 101: 66–80.
47. Darwin C, Darwin F (1880) The Power of Movement in Plants. London: John Murray.
48. Li P, Wang Y, Qian Q, Fu Z, Wang M, et al. (2007) LAZY1 controls rice shoot gravitropism through regulating polar auxin transport. Cell Res 17: 402–410.
49. Soderlund C, Bomhoff M, Nelson WM (2011) SyMAP v3.4: a turnkey synteny system with application to plant genomes. Nucleic Acids Res 39: e68.
50. Altschul SF, Madden TL, Schaffer AA, Zhang J, Zhang Z, et al. (1997) Gapped BLAST and PSI-BLAST: a new generation of protein database search programs. Nucleic Acids Res 25: 3389–3402.
51. Tusnady GE, Simon I (1998) Principles governing amino acid composition of integral membrane proteins: application to topology prediction. J Mol Biol 283: 489–506.
52. Tusnady GE, Simon I (2001) The HMMTOP transmembrane topology prediction server. Bioinformatics 17: 849–850.
53. Hofmann K, Stoffel W (1993) TMbase- A database of membrane spanning proteins segments. The Journal of Biological Chemistry 374: 166.
54. Pooma W, Gersos C, Grotewold E (2002) Transposon insertions in the promoter of the Zea mays a1 gene differentially affect transcription by the Myb factors P and C1. Genetics 161: 793–801.
55. Moreno MA, Harper LC, Krueger RW, Dellaporta SL, Freeling M (1997) *liguleless1* encodes a nuclear-localized protein required for induction of ligules and auricles during maize leaf organogenesis. Genes Dev 11: 616–628.
56. Acosta IF, Laparra H, Romero SP, Schmelz E, Hamberg M, et al. (2009) tasselseed1 is a lipoxygenase affecting jasmonic acid signaling in sex determination of maize. Science 323: 262–265.
57. McCarty DR, Latshaw S, Wu S, Suzuki M, Hunter CT, et al. (2013) Mu-seq: Sequence-Based Mapping and Identification of Transposon Induced Mutations. PLoS One 8: e77172.
58. Tan BC, Chen Z, Shen Y, Zhang Y, Lai J, et al. (2011) Identification of an active new mutator transposable element in maize. G3 (Bethesda) 1: 293–302.
59. Dong Z, Jiang C, Chen X, Zhang T, Ding L, et al. (2013) Maize LAZY1 Mediates Shoot Gravitropism and Inflorescence Development through Regulating Auxin Transport, Auxin Signaling, and Light Response. Plant Physiol 163: 1306–1322.
60. Yoshihara T, Spalding EP, Iino M (2013) AtLAZY1 is a signaling component required for gravitropism of the Arabidopsis thaliana inflorescence. Plant J 74: 267–279.
61. Yoshihara T, Iino M (2007) Identification of the gravitropism-related rice gene LAZY1 and elucidation of LAZY1-dependent and -independent gravity signaling pathways. Plant Cell Physiol 48: 678–688.
62. Dietrich CR, Cui F, Packila ML, Li J, Ashlock DA, et al. (2002) Maize Mu transposons are targeted to the 5′ untranslated region of the gl8 gene and sequences flanking Mu target-site duplications exhibit nonrandom nucleotide composition throughout the genome. Genetics 160: 697–716.
63. Taylor LP, Walbot V (1987) Isolation and characterization of a 1.7-kb transposable element from a mutator line of maize. Genetics 117: 297–307.
64. Chen CH, Oishi KK, Kloeckener-Gruissem B, Freeling M (1987) Organ-specific expression of maize Adh1 is altered after a Mu transposon insertion. Genetics 116: 469–477.
65. Talbert LE, Patterson GI, Chandler VL (1989) Mu transposable elements are structurally diverse and distributed throughout the genus Zea. J Mol Evol 29: 28–39.
66. Schnable PS, Peterson PA, Saedler H (1989) The bz-rcy allele of the Cy transposable element system of Zea mays contains a Mu-like element insertion. Mol Gen Genet 217: 459–463.
67. Fleenor D, Spell M, Robertson D, Wessler S (1990) Nucleotide sequence of the maize Mutator element, Mu8. Nucleic Acids Res 18: 6725.
68. James MG, Scanlon MJ, Qin M, Robertson DS, Myers AM (1993) DNA sequence and transcript analysis of transposon MuA2, a regulator of Mutator transposable element activity in maize. Plant Mol Biol 21: 1181–1185.

7

Genome-Wide Investigation and Expression Analyses of WD40 Protein Family in the Model Plant Foxtail Millet (*Setaria italica* L.)

Awdhesh Kumar Mishra, Mehanathan Muthamilarasan, Yusuf Khan, Swarup Kumar Parida, Manoj Prasad*

National Institute of Plant Genome Research, Aruna Asaf Ali Marg, New Delhi

Abstract

WD40 proteins play a crucial role in diverse protein-protein interactions by acting as scaffolding molecules and thus assisting in the proper activity of proteins. Hence, systematic characterization and expression profiling of these *WD40* genes in foxtail millet would enable us to understand the networks of WD40 proteins and their biological processes and gene functions. In the present study, a genome-wide survey was conducted and 225 potential *WD40* genes were identified. Phylogenetic analysis categorized the WD40 proteins into 5 distinct sub-families (I–V). Gene Ontology annotation revealed the biological roles of the WD40 proteins along with its cellular components and molecular functions. *In silico* comparative mapping with sorghum, maize and rice demonstrated the orthologous relationships and chromosomal rearrangements including duplication, inversion and deletion of *WD40* genes. Estimation of synonymous and non-synonymous substitution rates revealed its evolutionary significance in terms of gene-duplication and divergence. Expression profiling against abiotic stresses provided novel insights into specific and/or overlapping expression patterns of *SiWD40* genes. Homology modeling enabled three-dimensional structure prediction was performed to understand the molecular functions of WD40 proteins. Although, recent findings had shown the importance of WD40 domains in acting as hubs for cellular networks during many biological processes, it has invited a lesser research attention unlike other common domains. Being a most promiscuous interactors, WD40 domains are versatile in mediating critical cellular functions and hence this genome-wide study especially in the model crop foxtail millet would serve as a blue-print for functional characterization of WD40s in millets and bioenergy grass species. In addition, the present analyses would also assist the research community in choosing the candidate WD40s for comprehensive studies towards crop improvement of millets and biofuel grasses.

Editor: Girdhar K. Pandey, University of Delhi South Campus, India

Funding: The authors work in this area was supported by the core grant of NIPGR. AKM and MM acknowledge the award of Senior Research Fellowship and Junior Research Fellowship from Council of Scientific and Industrial Research and University Grants Commission, New Delhi, India, respectively. The funders had no role in study design, data collection and analysis, decision to publish, or preparation of the manuscript.

Competing Interests: The authors declare that Manoj Prasad is serving as Academic Editor of PLOS ONE. This does not alter the authors' adherence to all the PLOS ONE policies on sharing data and materials.

* E-mail: manoj_prasad@nipgr.ac.in

Introduction

Foxtail millet [*Setaria italica* (L.) P. Beauv.], the second largest cultivated millet species in the world, possesses several salient attributes such as small genome (~515 Mb; 2n = 2x = 18), relatively lower repetitive DNA, short life-cycle, inbreeding nature and is closely-related to several bioenergy grasses [1,2]. These features along with its potential abiotic stress tolerance have accentuated this crop as an experimental model system for examining the architectural traits, evolutionary genomics and physiological aspects of C4 panicoid grass crops [2–4]. Hence, considering its significance, the US Department of Energy - Joint Genome Institute and the Beijing Genomics Institute, China had sequenced the genome and the draft sequence was released in 2012 [5,6]. Consequently, the availability of foxtail millet sequence information encouraged the scientific research community to decipher its structural and functional genomics, thus ultimately assisting in crop improvement and ensuring food security [7]. In this regard, we had also reported substantial findings in the aspects of both structural [8–13], and functional genomics [14–22] in the model crop, foxtail millet.

In our earlier study, we identified and characterized a differentially expressed transcript encoding for WD40 protein from a salinity and dehydration stress-induced subtractive cDNA library [20]. Being the first report, we showed a putative regulation of *SiWD40* expression by dehydration responsive elements (DRE) during abiotic stress [20]. WD40 proteins were identified to play a crucial role in diverse protein-protein interactions by acting as scaffolding molecules and thus assisting the proper activity of the proteins [23]. Structurally, the WD40 domain is characterized by the presence of several copies of WD40 repeats with each repeat containing 44–60 residue units. Each unit includes a glycinehistidine (GH) dipeptide about 11–24 residues from its N terminus and terminates with Trp-Asp (WD) doublet residues at the C-terminus [24,25]. Each of the repeat folds into four-stranded anti-parallel β-sheet and is proposed to originate from intragenic duplication and recombination events and diversify during evolution [25,26]. A subset of WD40 proteins have been named as DWD [Damaged DNA binding (DDB) WD40] based on their interaction with DDB1 and CULLIN4 (CUL4) [27]. CUL4–DDB1 ubiquitin E3 ligases use DWD proteins as molecular adaptors for substrate recognition, and modulate multiple biological processes through ubiquitin-dependent proteolysis such as DNA- repair mechanism caused by UV-damage and histone methylation (post-translational modification). These proteins contain 16 conserved amino acids within the WD40 repeats, called "DWD box" [28,29].

Considering the importance of deciphering the molecular networks, biological processes and gene functions of WD40 proteins, genome-wide investigations have been conducted in *Arabidopsis* [30] and rice [31], but no report was available in foxtail millet till date. Hence, this is the first comprehensive report on genome-wide survey, expression profiling and evolutionary analysis of WD40 proteins in foxtail millet (internally annotated as 'SiWD40'). We have identified about 225 *SiWD40* genes spanning the nine chromosomes of foxtail millet and classified them into five classes. Sequence comparison of *SiWD40* genes within themselves and with other grasses like sorghum, maize and rice facilitated the study on presence and distribution of paralogous and orthologous *WD40* genes between the grasses. These experimental outcomes have paved a way for further comparative genomic and phylogenetic analyses of WD40 proteins among members of grass family. Subsequently, quantitative real-time PCR (qRT-PCR)-based gene expression profiling showed the temporal and stress-specific expression pattern of candidate *SiWD40* genes. Homology modeling enabled three-dimensional structure prediction was then performed, which would facilitate studies on understanding its molecular function. Positively, this first report will serve as a solid base for functional genomic studies including further molecular characterization of WD40 genes towards various stress responses in foxtail millet.

Results and Discussion

Identification of Novel SiWD40 Members in *Setaria italica*

In order to identify the *SiWD40* genes in *Setaria italica*, the characteristic eukaryotes domain sequence of WD40 (GECKXVLXGHTSTVTCVAFSPDGPLLASGSRDGTI-KIWD) was generated by hmmemit from HMM profile (PF00400). The BLASTP analysis was performed using this sequence as a query in PHYTOZOME, with a threshold E value of ≤10. This identified a total of 321 sequences and the removal of different transcripts of the same gene identified 225 putative *SiWD40* genes (Table S1). Further, the presence of WD40 domain was confirmed by SMART and Pfam searching. Both search outputs showed the presence of WD40 domain in all the 225 *SiWD40* genes. For convenience, the 225 *SiWD40* genes were named from SiWD001 to SiWD225 according to the order of their chromosomal locations.

Except for the presence of a conserved WD40 domain, the SiWD40 genes vary substantially in the size and sequences of their encoded proteins, and their physicochemical properties (Table S1). The location of the WD40 domain within the protein also differs. The length of SiWD40 proteins varied from 98 to 3518 amino acids. EXPASY analysis suggested that the SiWD40 protein sequences had large variations in isoelectric point (pI) values (ranging from 4.54 to 9.69) and molecular weight (ranging from10.866 kDa to 390.606 kDa; Table S1). The characteristic features of SiWD40 protein sequences were summarized in Table S1.

Chromosomal Distribution and Structure of *SiWD40*

In silico mapping of *SiWD40*s on chromosomes indicated an uneven distribution of the genes on all the 9 chromosomes of foxtail millet (Figure 1). Among all, chromosome 9 contains the highest number of *SiWD40*s [45 (20%)], while lesser number genes were distributed on chromosome 8 [8 (~3.5%)] (Figure 1). The exact position (in bp) of each *SiWD40* on foxtail millet chromosome is given in Table S1. Pattern of their distribution on individual chromosomes also revealed certain physical regions with a relatively higher accumulation of *SiWD40* gene clusters. For example, *SiWD40* genes located on chromosomes 3 and 7 appear to be congregate at the upper end and lower end of the arms, respectively (Figure 1). Recently, Zhang et al. [5] reported the occurrence of whole-genome duplication in foxtail millet similar to other grasses ~70 million years ago (Mya). Hence, the presence of such large number of *SiWD40* genes in foxtail millet indicates the amplification of this gene family during the course of evolution. In all, 12 (~5%) *SiWD40* genes were found to be tandem repeats with a maximum of six intervening genes separating the tandem repeats (Figure 1). The distance between these genes ranged from 6.2 kb to 32.2 kb. In the whole foxtail millet genome, 6688 (~19%) genes are segmentally duplicated. Among the *SiWD40* genes, 32 (~14%) were found to be segmentally duplicated (Figure 2).

Investigation of *SiWD40* gene structures revealed highly diverse distribution of intronic regions (from 0 to 29 in numbers) amid the exonic sequences, signifying considerable evolutionary changes that have occurred in the foxtail millet genome (Figure S1). The shortest *SiWD40* gene was merely 461 bp (*SiWD084*) whereas the longest one was identified as *SiWD006* with ~ 23.5 kb genomic sequence (Table S1). This suggests that the evolution of these genes might have progressed immediately through some gene duplications or by integration into genomic region after reverse transcription [21,32,33].

Phylogenetic Classification of SiWD40s and Identification of Domain Conservation

A phylogenetic tree was constructed with 223 SiWD40 proteins by neighbour-joining (NJ) method. SiWD063 and SiWD216 being small sequences were excluded from alignment and phylogenetic tree construction. The phylogenetic analysis categorized all the SiWD40s into five discrete groups (Cluster I to V) comprising of 25, 48, 08, 11, and 131 proteins, respectively (Figure 3). Since a good number of the internal branches were observed to have high bootstrap values, it clearly shows the derivation of statistically reliable pairs of possible homologous proteins sharing similar functions from a common ancestor.

Further, the 225 SiWD40 proteins were classified into 12 subfamilies according to their domain compositions (Figure 4). About 146 members with only WD40 domain were categorized in subfamily A. Besides WD40 domain, SiWD40 proteins contained several other known functional domains and were classified into the following subfamilies. Four members containing the zinc finger domain were identified as subfamily B; Six members containing the Beige/BEACH domain were identified as subfamily C; Two members with breast carcinoma amplified sequence 3 (BCAS3) were identified as D subfamily; E subfamily (11 members) had LisH domain; F subfamily (7 members) had histone-binding protein RBBP4 or subunit C of CAF1 complex domains before WD40 repeats; G subfamily (3 members) had protein kinase domain or HEAT repeat; Eight members with the Coatomer WD associated region (WDAD) or Coatomer (COPI) alpha subunit C-terminus were identified as H subfamily; I subfamily (5 members) contained F-BOX and U-BOX; J subfamily (9 members) contained NLE (NUC) domain N terminal toWD40 repeats;

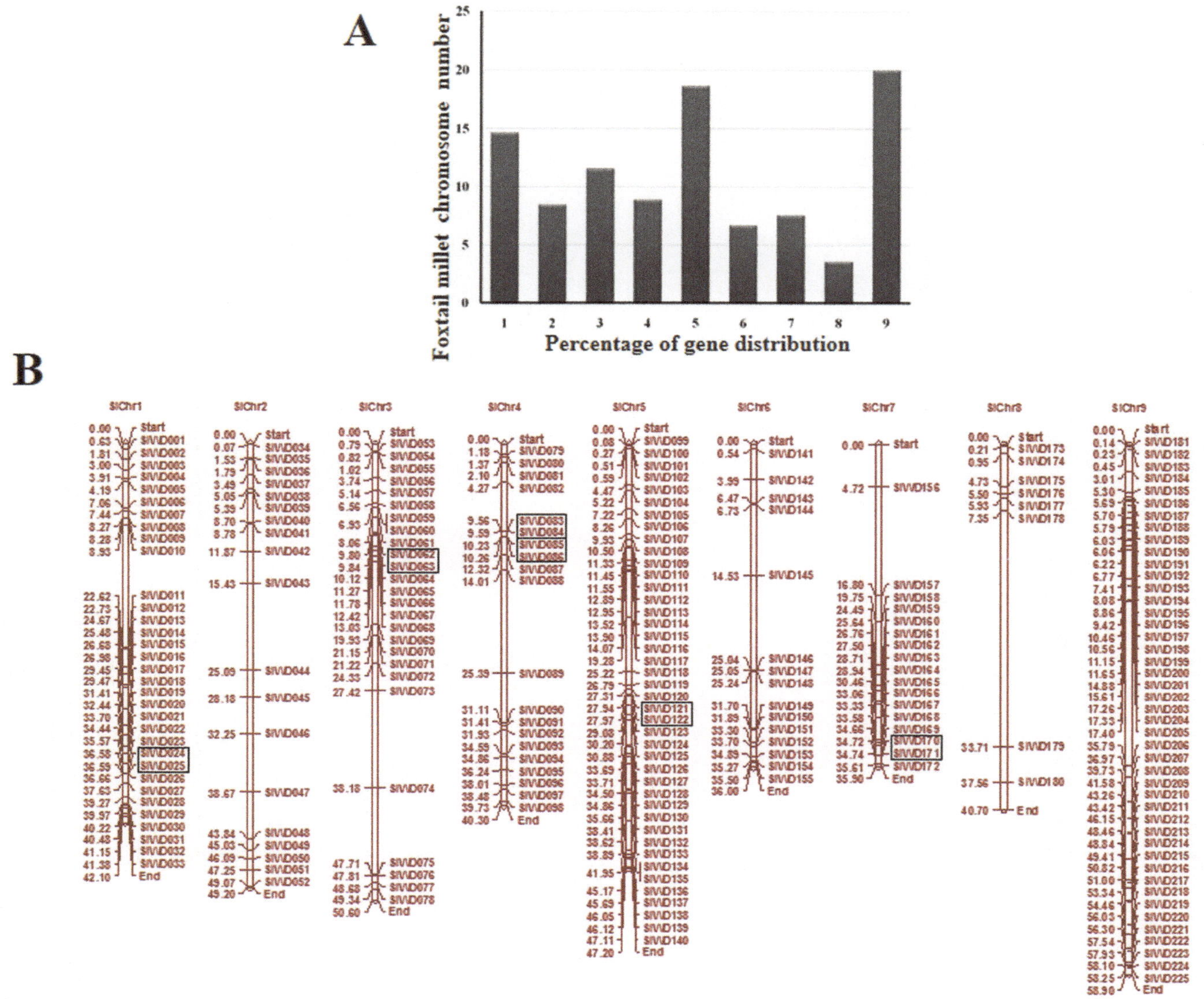

Figure 1. Distribution of 225 *SiWD40* genes onto nine foxtail millet chromosomes. (A) Percentage of *SiWD40* genes on each the foxtail millet chromosome to show their distribution abundance. (B) Graphical (scaled) representation of physical locations for each *SiWD40* gene on foxtail millet chromosomes (numbered 1–9). Tandem duplicated genes on a particular chromosome are depicted by black boxes. Chromosomal distances are given in Mb.

Utp12, Utp13, Utp15, and Utp21; Six member of UTP containing domain were identified as subfamily K; L subfamily (21 members) contained other domains including TUP1-like,IIPc, DENN, Cyclophilin and domains with unknown function (Figure 4). The members of HBRBBP4 domain containing SiWD40 proteins are found in one cluster in the subgroup Vc (Figure 4). Interestingly, 97 out of 225 SiWD40 were identified as DWD proteins. Further, these 97 DWD proteins possess 116 DWD domains, of which 82 had one DWD domain, 11 had two domains and four had three domains. Thus, a diverse domain variation and conservations were evidenced and such conservation or variation between the proteins specifies the functional equivalence or diversification, respectively, with respect to the various aspects of biological functions [34].

Gene Ontology Annotation

The GO slim analysis performed using Blast2GO showed the putative participation of SiWD40 proteins in diverse biological processes (Figure 5; Table S2). Out of 225 SiWD40 proteins, annotation could not be performed for 49 sequences and the results for the rest of 176 SiWD40s were defined in 26 categories of biological processes. The analysis showed that, predominant SiWD40 proteins were involved in response to primary metabolic process [75 (~43%)], followed by cellular metabolic processes [68 (~39%)]. Noteworthy, about 42 (~24%) SiWD40 were evidenced to participate in response to stress stimulus. This highlights the putative association of SiWD40 proteins in stress tolerance behaviour of foxtail millet (Figure 5). In case of molecular functions, about 76 (~43%) SiWD40 proteins were shown to participate in small molecule binding which concords with the molecular role of WD40 proteins in assisting protein-protein

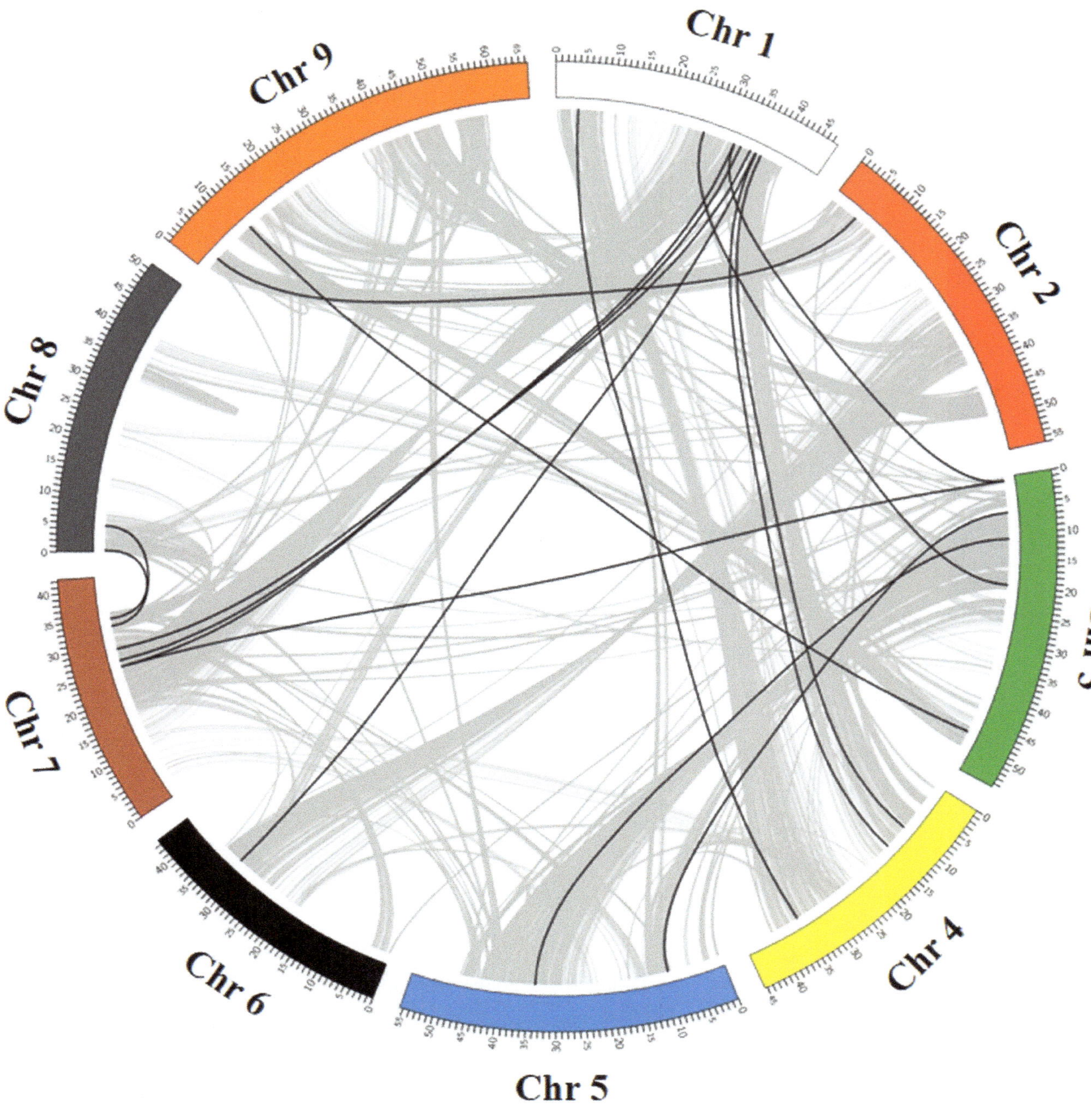

Figure 2. Distribution of segmentally duplicated *SiWD40* genes on foxtail millet chromosomes. Grey lines indicate collinear blocks in whole foxtail millet genome, and black lines indicate duplicated *SiWD40* gene pairs.

interactions. Cellular localization prediction showed that predominant [144 (~82%)] SiWD40 proteins are localized in the cell part, of which 60 (~42%) are nuclear localized (Figure 5; Table S2). This agrees with the experimental findings reported earlier [20,35]. Further, Blast2GO was performed to draw a connection between the domain composition of the families/sub families and the functional classes, but there were no correlation observed.

Promoter Analysis and miRNA Targets of *SiWD40* Genes

To support the functional predictions of the 42 stress-related *WD40* genes in foxtail millet, a comprehensive promoter analysis was performed. For this purpose, promoters and their regulatory elements were identified in DNA sequences (~2 kb upstream of their putative start codons) using PlantPAN (Table S3). The analysis identified *cis*-acting regulatory elements (CARE) in the upstream DNA sequences that are involved in regulation of gene expression under stress conditions. The data might indicate a major role for the identified stress-related *WD40* genes in regulating their gene expression in response to different stresses in foxtail millet. Further, putative microRNAs (miRNA) targeting the *SiWD40* genes were also identified using psRNATarget server. It showed that about eight *SiWD40* genes were targeted by *Setaria*

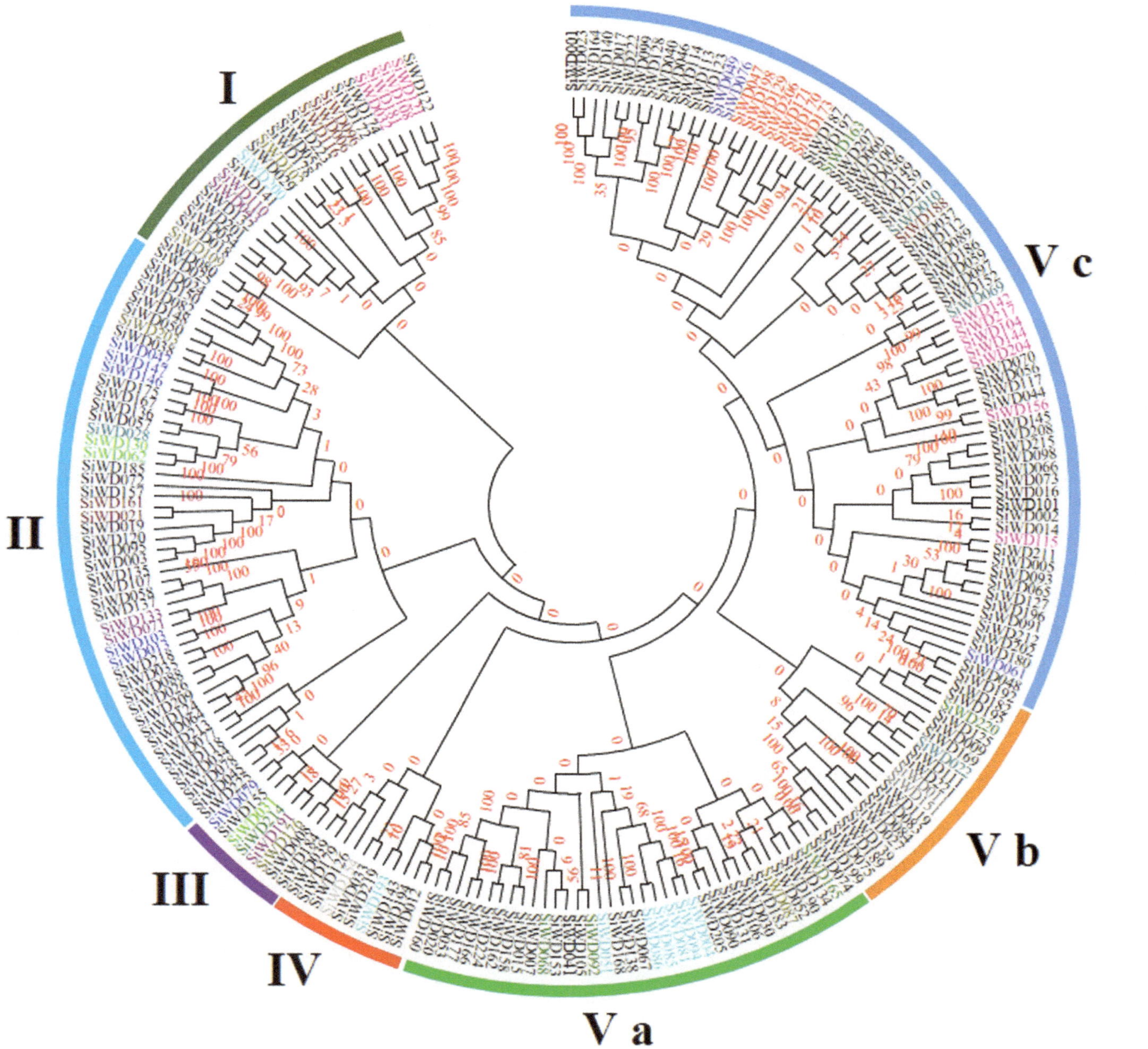

Figure 3. Phylogenetic relationships of foxtail millet WD40 proteins. The sequences were aligned by CLUSTALW at MEGA5 and the unrooted phylogenetic tree was deduced by neighbor-joining method with 1000 bootstrap replicates. The evolutionary distances were computed using p-distance method. The bootstrap values are shown at the nodes. The tree was divided into five phylogenetic cluster designated as I to V. The members of the SiWD40 were distinctly coloured to represent respective WD40 subfamilies.

italica miRNAs (Table S4). These miRNAs identified in the present study would assist in deciphering the post-transcriptional control of gene regulation during physiological and stress-induced cellular responses.

Orthologous Relationships of *WD40* Genes between Foxtail Millet and other Grass Species

To derive comparative mapping-based orthologous relationships of *SiWD40*, the physically mapped *WD40* genes were compared with those in the chromosomes of other related grass genomes namely, sorghum, maize and rice (Table 1; Figure S2). Of the identified 225 *SiWD40* protein-encoding genes in foxtail millet, the specific orthologous relationships could be derived on an average for ~ 83.6% proteins. Maximum orthology of *SiWD40* genes annotated on the foxtail millet chromosomes was obtained with sorghum (86.2%) followed by rice (82.7%). The close evolutionary relationships would be the plausible reason for the extensive gene-level synteny shared between foxtail millet, sorghum and maize [5,6,21]. Interestingly, most of *SiWD40* genes revealed syntenic bias towards particular chromosomes of rice, maize and sorghum. For instance, the *SiWD40* genes on foxtail

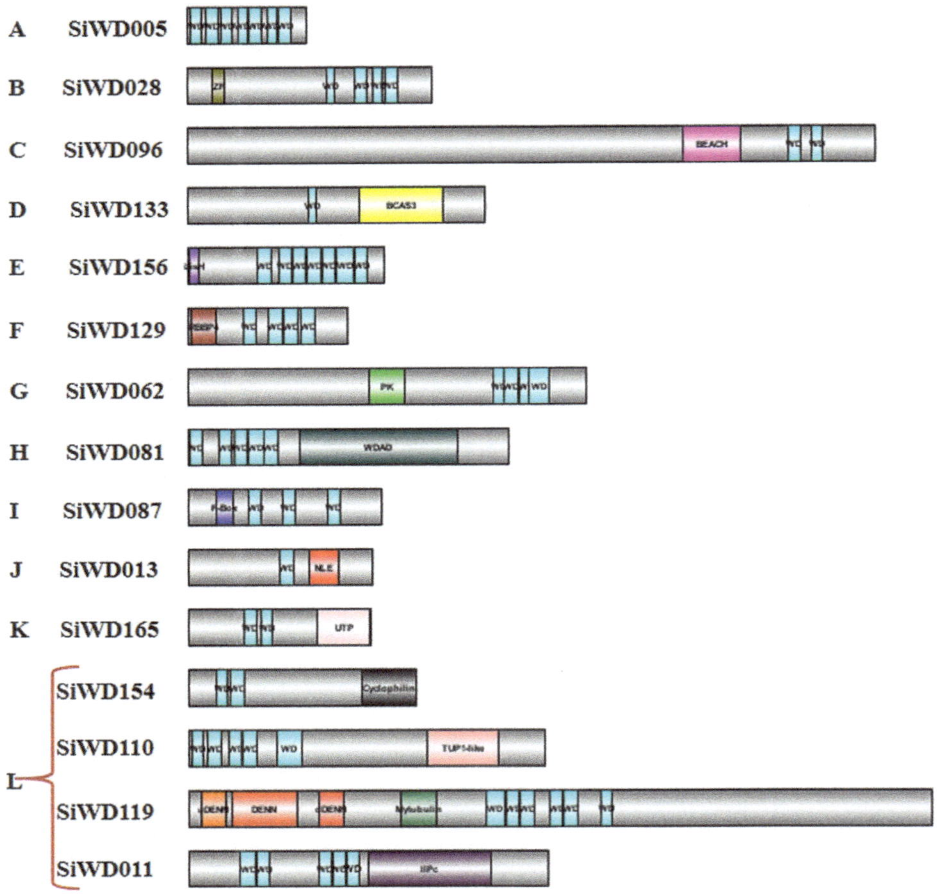

Figure 4. Structure of representative SiWD40 proteins from each subfamily. The protein structure is based on the presence of WD40 and other additional domains as identified by SMART and pfam. This categorizes the SiWD40 in 12 Subfamilies (A–L).

millet chromosome 1 showed 93% orthology and colinearity with sorghum chromosome 4 and rice chromosome 2 (90%) (Table 1; Figure S2). The *SiWD40* genes mapped on foxtail millet chromosome 9 showed inter-chromosomal inversions with rice chromosome 3 (72.7%) and maize chromosome 1 (65%), while colinearity with sorghum chromosome 1 (85.4%). Like-wise the *SiWD40* genes mapped on foxtail millet chromosome 5 revealed collinear relationships with rice chromosome 1 (82.5%) and sorghum chromosome 3 (92%) and inverted relationship with maize chromosome 3 (63.6%). The results indicated that the chromosomal rearrangements like duplication and inversion were predominant in shaping the distribution and organization of *WD40* genes in foxtail millet, rice, maize and sorghum genomes. The comparative mapping information provides a useful preface for understanding the evolutionary process of *WD40* genes among grasses involving the foxtail millet genome. Further, this study would be useful in selecting candidate *WD40* genes from foxtail millet and utilize them in genetic enhancement of other related grass family members.

Duplication and Divergence Rate of the *SiWD40* Genes

Multiple copies of genes in a gene family possibly evolve due to evolutionary events like whole genome tandem and segmental duplications. Such gene duplication has been documented in several plant transcription factor (TF) gene families such as MYB, F-box as well as in NAC [21,36,37]. We thus explored the effect of Darwinian positive selection in duplication and divergence of *WD40* genes. To interpret this, the ratios of non-synonymous (Ka) versus synonymous (Ks) substitution rate (Ka/Ks) were estimated for six tandem and 15 segmentally duplicated gene-pairs as well as between orthologous gene-pairs of *SiWD40* with those of rice (186-pairs), maize (183) and sorghum (194). The ratios of Ka/Ks for tandemly duplicated gene-pairs ranged from 0.09 to 0.15 with an average of 0.12 (Table S5), whereas Ka/Ks for segmentally duplicated gene-pairs ranging from 0.11 to 0.20 with an average of 0.13 (Table S6). It suggested that the duplicated *SiWD40* genes are under strong purifying selection pressure since their Ka/Ks ratios estimated as <1. Additionally, the duplication event of these tandemly and segmentally duplicated genes may be estimated to have occurred around 25–27 and 18–22 Mya, respectively (Figure 6). Among the orthologous gene-pairs of *SiWD40* with those of other grass species, the average Ka/Ks value was maximum between rice and foxtail millet (0.55) and least for sorghum-foxtail millet gene-pairs (0.23; Table S7). The relatively higher rate of synonymous substitution between rice and foxtail millet *WD40* genes indicated their earlier divergence around 33–44 Mya from foxtail millet as compared to sorghum and maize *WD40* genes (Figure 6). Remarkably, the *WD40* gene-pairs between sorghum and foxtail millet (average Ka/Ks = 0.23) appear to have undergone extensive intense purifying selection in comparison to foxtail millet-maize (Ka/Ks = 0.30) and foxtail millet-rice (Ka/Ks = 0.55) *WD40* genes (Table S7). This conforms

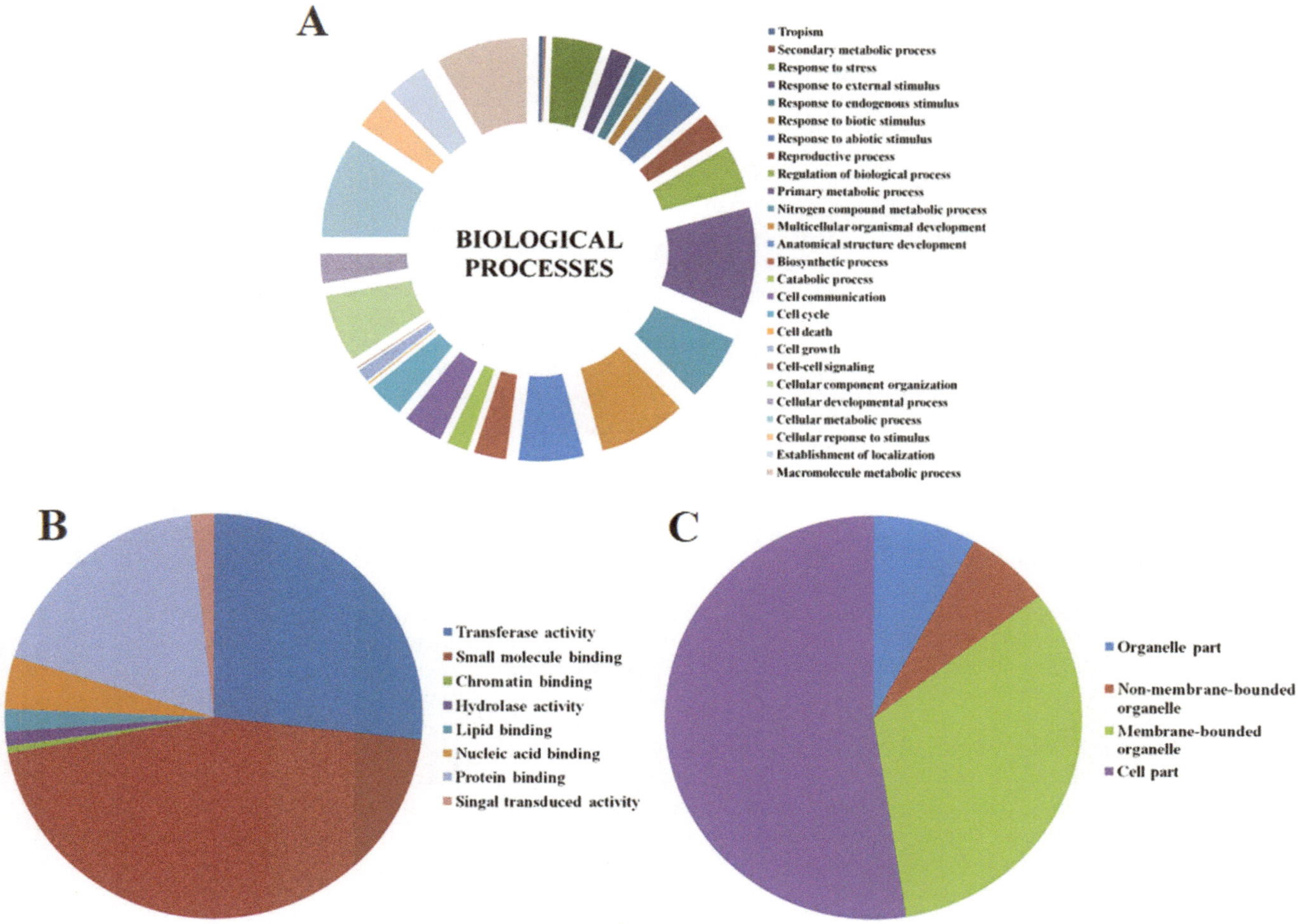

Figure 5. Gene Ontology (GO) distributions for the SiWD40 protein. The Blast2Go program defines the gene ontology under three categories, (A) biological processes, (B) molecular functions and (C) cellular component.

to their recent time of divergence around 16–21 Mya. The estimation of tandem and segmental duplication time (average of 22 Mya) of foxtail millet *WD40* genes in between the divergence time of foxtail millet-rice (37.7 Mya) and foxtail millet-maize (20.8 Mya) and foxtail millet-sorghum (19.2 Mya) orthologous *WD40* gene-pairs are comparable to evolutionary studies involving the protein-coding genes annotated from the recently released draft genome sequences of foxtail millet [5]. Interestingly, the *SiWD40* gene-pairs showing segmental and tandem duplication events are under similar evolutionary pressure (Ka/Ks = 0.12) of which, the segmentally duplicated genes revealed much recent duplication events (average 18.5 Mya) in contrast to tandemly

Table 1. A summary of comparative mapping of foxtail millet *SiWD40* genes on sorghum, maize and rice.

Setaria italica	*Sorghum bicolor*	*Zea mays*	*Oryza sativa*
Chr1	Chr4 (92.85%)	Chr5 (50%), Chr4 (28.57%)	Chr2 (90%)
Chr2	Chr2 (83.33%)	Chr7 (75%)	Chr7 (58%), Chr9 (25%)
Chr3	Chr9 (50%)	Chr8 (30%), Chr6 (25%)	Chr5 (45%), Chr12 (20%)
Chr4	Chr10 (86.67%)	Chr9 (46.67%), Chr6 (26.67%)	Chr6 (80%)
Chr5	Chr3 (92%)	Chr3 (63.6%)	Chr1 (82.5%)
Chr6	Chr7 (84.61%)	Chr1 (28.57%), Chr4 (28.57%)	Chr8 (78.57%)
Chr7	Chr6 (50%), Chr8 (42.85%)	Chr10 (66.67%)	Chr4 (38.4%), Chr12 (23%)
Chr8	Chr5 (100%)	Chr3 (60%), Chr2 (40%)	Chr11 (62.5%)
Chr9	Chr1 (85.36%)	Chr1 (65%)	Chr3 (72.7%)

duplicated gene-pairs (average 25.4 Mya) and orthologous foxtail millet-sorghum gene-pairs (19.2 Mya). It overall suggests that the segmental and tandem duplication events including the divergence events of *SiWD40* genes from other grass species have played a predominant role in evolution for shaping such gene family in foxtail millet.

In silico Tissue-specific Expression Profiling of *SiWD40*

Heat map generated for examining the tissue-specific expression showed a differential transcript abundance of 225 *SiWD40* genes in 4 major tissues namely root, leaf, stem and spica (Figure S3). About 87 genes (~39%) showed higher expression in all the four tissues and conversely, 37 (~16%) were found to be low expressed in all the four tissues (Figure S3). Comparing the expression of all the 225 *SiWD40* showed a relatively higher expression of SiWD024 and SiWD065 in all the tissues. Some of the *SiWD40s* also showed tissue-specific expression, such as *SiWD158* expressed only in root, *SiWD063* in leaf, and *SiWD023, SiWD108* and *SiWD162* express specifically in spica. The tissue-specific expression profiling of *SiWD40s* would facilitate the combinatorial usage of *SiWD40s* in transcriptional regulation of different tissues, whereas ubiquitously expressed *SiWD40s* might regulate the transcription of a broad set of genes. This heatmap data also enables the overexpression studies of *SiWD40s* across the tissues to impart stress tolerance in both foxtail millet and related crop species.

SiWD40 Expression Profiles of during Abiotic Stresses and Homology Modeling

Gene expression patterns can offer crucial indications for determining the gene function. Considering the potential abiotic stress tolerance characteristic of foxtail millet, we studied the expression pattern of *WD40* genes during dehydration, salinity, abscisic acid (ABA) and cold stress. About 13 candidate genes were chosen for quantitative expression analysis based on the GO annotation (possessing roles in abiotic stress stimuli) and representing all the sub-families. The expression pattern of the candidate genes in response to dehydration, salinity, ABA and cold stress during 0, 1, 3, 6, 12, 24 and 48 h durations of treatments was examined (Figure 7A–D). In summary, qRT-PCR analyses showed that all the candidate *SiWD40* genes have incurred variations in their expression patterns in response to one or more stresses in course of the experimentations. Higher expression of *SiWD40* genes were evidenced at 12th hr during dehydration stress and at 6th hr during salinity stress (Figure 7A-6B). During ABA treatment, higher number of genes was evidenced to be expressed at 3rd hr (Figure 7C) while higher expressions of *SiWD40* genes was observed at 24th hr during cold stress (Figure 7D). Noteworthy, *SiWD063* was found to be highly expressed in all the four stresses. Further, *SiWD028, SiWD037, SiWD063* and *SiWD182* were found to be highly expressed during dehydration stress, whereas *SiWD63, SiWD106, SiWD144* and *SiWD202* were upregulated during salinity stress. In ABA stress, *SiWD063* and *SiWD182* were evidenced to be highly expressed. Cold stress showed higher expression of *SiWD37, SiWD63* and *SiWD195*. This variability in gene expression patterns implies that *SiWD40*s may regulate a complex network of pathways to perform different physiological functions for acclimatizing towards multiple challenges. Since no reports were available on the study of WD40 expression patterns during stress, this comprehensive expression profile would invoke investigations on the role of WD40 in imparting stress tolerance.

Three dimensional protein models were constructed by sequence similarity searching the PDB database using BLASTP.

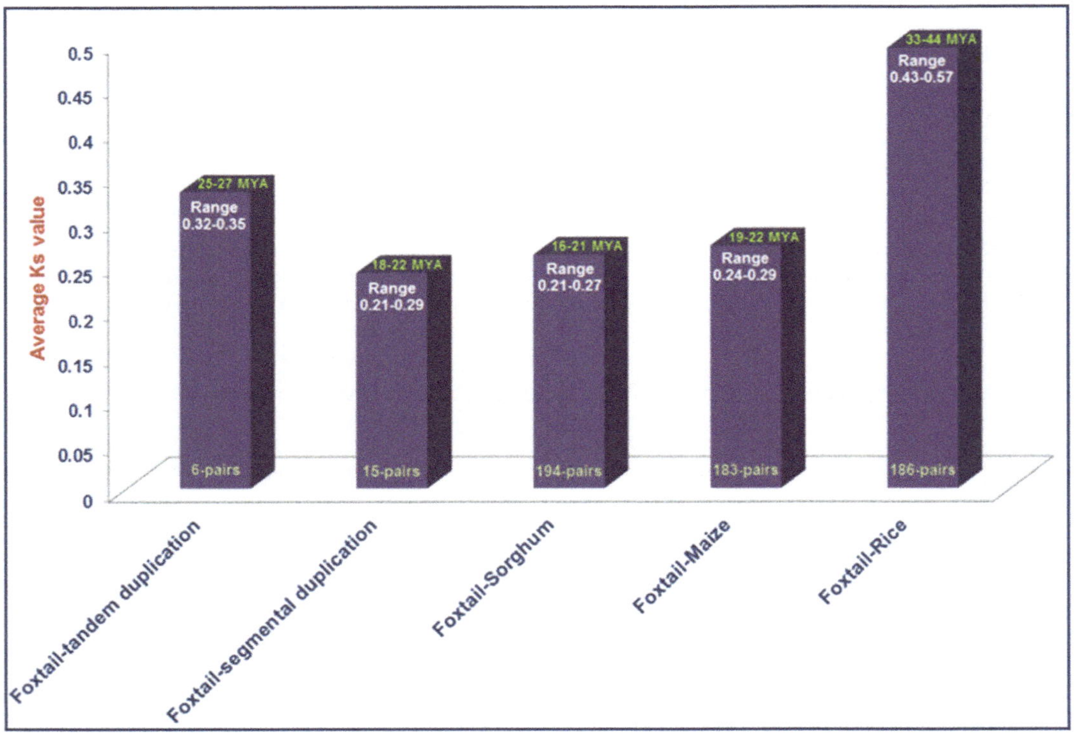

Figure 6. Time of duplication and divergence (MYA) based on synonymous substitution rate (Ks) estimated using duplicated *SiWD40* gene pairs of foxtail millet and orthologous *SiWD40* gene pairs between foxtail millet and rice or maize or sorghum.

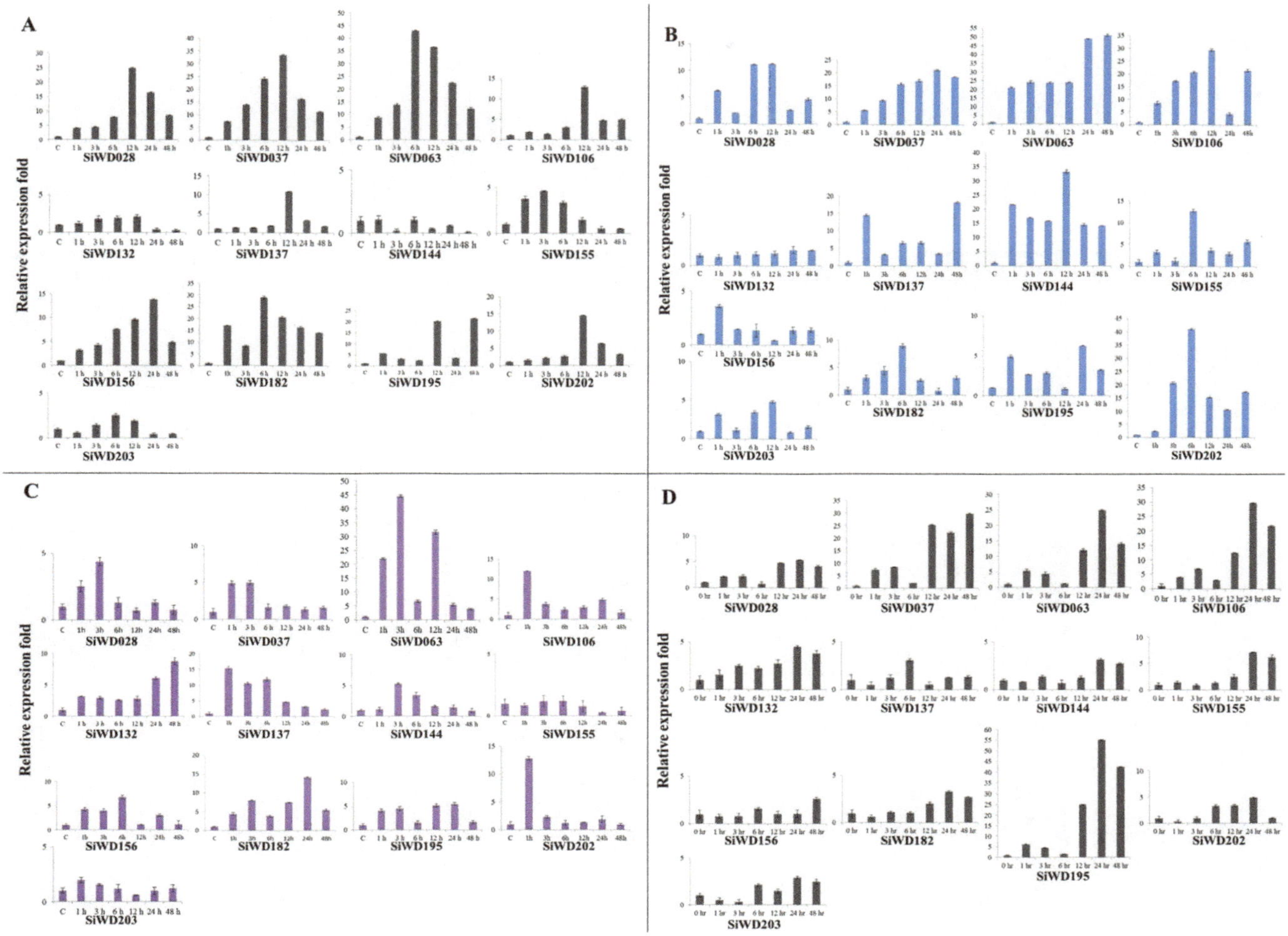

Figure 7. The relative expression ratio of 13 candidate *SiWD40* genes analyzed using qRT-PCR under (A) dehydration stress, (B) salinity stress, (C) ABA treatment (D) Cold stress for 0, 1, 3, 6, 12, 24 and 48 h. The relative expression ratio of each gene was calculated relative to its expression in control sample (0 h). *Act2* was used as an internal control to normalize the data. The error bars representing standard deviation were calculated based on three technical replicates for each biological duplicates.

Twenty four proteins having higher homology were selected and Phyre2 was used to predict the homology modeling (Figure 8). Noticeably, these 24 proteins represent diverse WD40s, in terms of repeats and domains (Table S9). Phyre2 uses the alignment of hidden Markov models via HMM-HMM search [38] to significantly improve the accuracy of alignment and detection rate. The intensive mode of Phyre 2 uses the multi-template modeling for higher accuracy. Furthermore it integrates a new *ab initio* folding simulation termed as Poing [39] to model regions of proteins with no noticeable homology to known structures. The protein structure of all the 24 SiWD40 are modelled at >90% confidence and the percentage residue varied from 81 to 100 (Figure 8, Dataset S1). The secondary structure predominantly comprised of *β* - sheets and coils, with rare occurrence of α - helices (Figure 8). Hence all the predicted protein structures are considered highly reliable and this offers a preliminary basis for understanding the molecular function of SiWD40 proteins.

Conclusions

The WD-repeat proteins possess seven WD40-repeat motifs, with the conserved core of the repeat containing 44 to 60 residues that terminates with Trp and Asp. The repeats form a β - propeller fold, allowing formation of a highly stable structure that coordinates the interactions with several other proteins [40]. Hence, its role is deemed imperative in protein-protein interactions and our recent identification on the role of WD40 proteins in abiotic stress tolerance in foxtail millet [20] had motivated us to conduct a genome-wide survey in this model crop. In summary, a total of 225 *SiWD40* genes were found to be present in foxtail millet genome. The variations in the lengths and genomic structure of SiWD40s support the great deal of complexity that has evolved within this gene family. Noteworthy, the *SiWD40* genes shared high orthology with their counter-parts in sorghum and maize supporting their close evolutionary relationship. Further, for the first time, we had showed a preliminary expression profiling of some *SiWD40* genes influenced by several environmental stimuli, including dehydration, salinity, ABA treatment and cold stress. We have also described the structure of 24 SiWD40 proteins which would expedite the investigation of its molecular functions. Hence, this report would be useful for the millet research community in selecting candidate genes for functional studies of WD40 members in foxtail millet, and other millets and bioenergy grasses.

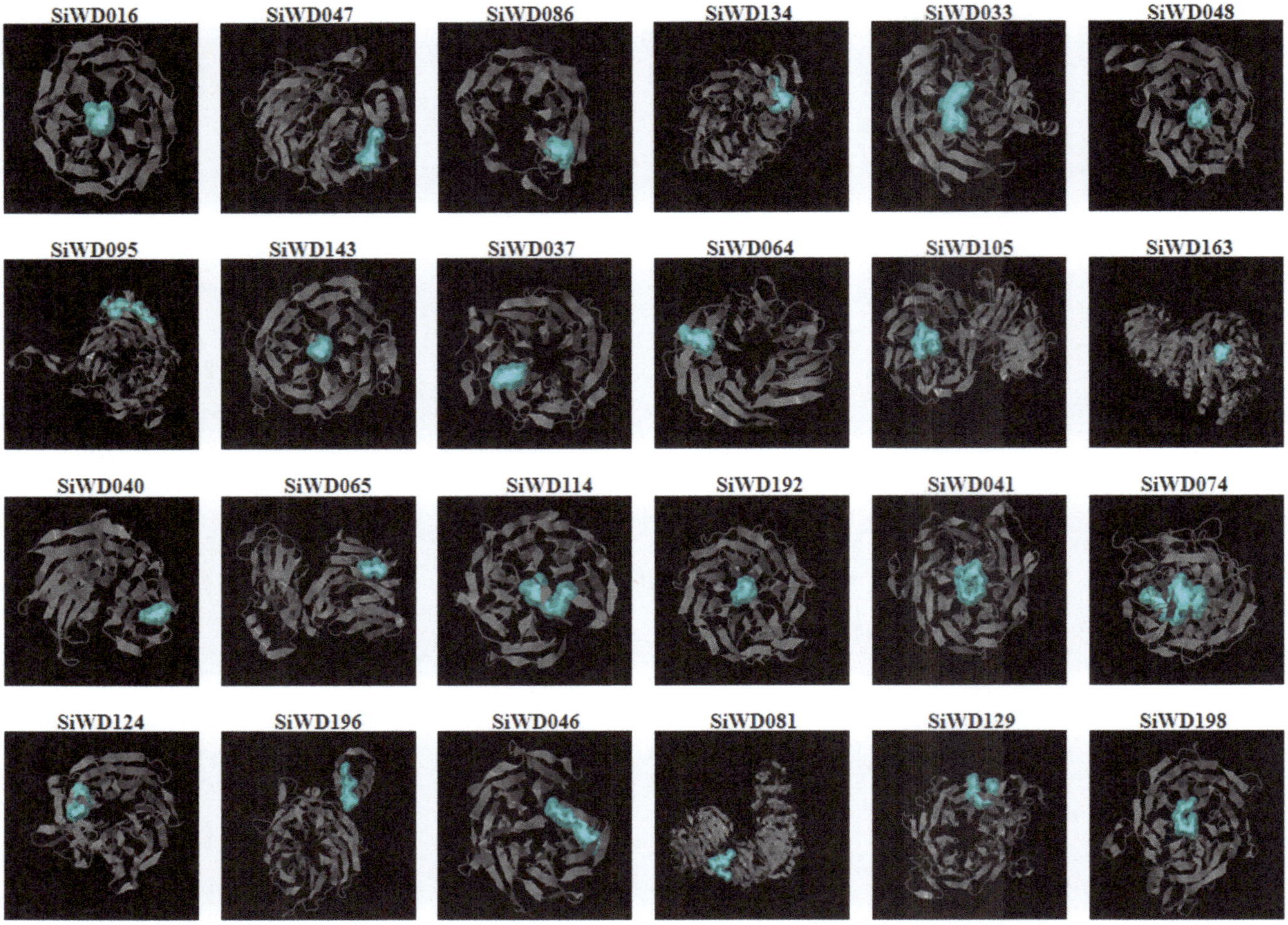

Figure 8. Predicated structures of SiWD40 proteins. The structure of 24 SiWD40 proteins with >90% confidence level were shown. Active sites are highlighted in blue colour.

Materials and Methods

Retrieval and Identification of WD40 Genes in *Setaria italica*

The Hidden Markov Model (HMM) profile of the WD40 domain (PF00400) retrieved from Pfam v27.0 (http://Pfam.sanger.ac.uk/) was queried against the PHYTOZOME v8.0 database (www.phytozome.net/) of *Setaria italica.* All hits with expected values less than 1.0 were retrieved and redundant sequences were removed using BLASTclust v2.17 (http://toolkit.tuebingen.mpg.de/blastclust). Each non-redundant sequence was checked manually for the presence of the conserved WD40 domain by executing SMART (http://smart.embl-heidelberg.de/) [41] and Pfam searches.

Physical Mapping, Gene Structure Prediction and Estimation of Genomic Distribution

Physical mapping of the genes encoding *SiWD40* onto the foxtail millet genome was performed by conducting BLASTP search of respective sequences against the PHYTOZOME database using default settings. Subsequently the genes were plotted onto the nine chromosomes according to their ascending order of physical position (bp), from the short arm telomere to the long arm telomere and ultimately the map was displayed using MapChart [42]. Since tandem and segmental duplication events that have occurred in the genome would plausibly result in the expansion of gene family, we investigated the mechanisms involved in the expansion of WD40 members in foxtail millet. The method of Plant Genome Duplication Database was used to identify segmental duplications [43]. Precisely, BLASTP search was performed against the complete peptide sequences of *Setaria italica* and the first 5 matches with E-value <1e-05 were identified as potential anchors. Collinear blocks were evaluated by MCScan v0.8 and alignments with an E value <1e-5 were considered as significant matches [44,45]. The segmental duplication was finally visualized using Circos 0.55 (http://circos.ca) [46]. Tandem duplications were characterized as adjacent genes of same sub-family located within the same or neighbouring intergenic region [45]. The exon-intron positioning of the genes were determined using Gene structure display server (gsds.cbi.pku.edu.cn/) [47] by comparing the full-length cDNA or predicted coding sequence (CDS) of *SiWD40* with their corresponding genomic sequence.

Phylogenetic Analysis and Gene Ontology (GO) Annotation

The amino acid sequences of SiWD40 were imported into MEGA5 [48] and multiple sequence alignments were performed using ClustalW with a gap open penalty of 10 and a gap extension penalty of 0.1 [49]. The alignment file was then subjected to create an unrooted phylogenetic tree based on the neighbor-joining method [50] and after bootstrap analysis for 1000 replicates, the final tree was generated. The functional annotation of SiWD40 sequences and the analysis of annotation data were performed using Blast2GO (http://www.blast2go.com) [51]. The amino acid sequences of SiWD40 were imported into Blast2GO program to execute three steps viz, (i) BLASTp against the non-redundant protein database of NCBI, (ii) mapping and retrieval of GO terms associated with the BLAST results, and (iii) annotation of GO terms associated with each query to relate the sequences to known protein function. The program provides the output defining three categories of GO classification namely biological processes, cellular components and molecular functions.

Analysis of Promoter and miRNA Targets

The upstream sequences (~2000 bp) of each identified *SiWD40* gene were retrieved from the PHYTOZOME (http://phytozome.net/). The upstream sequences were analyzed for the identification of regulatory cis-elements important for gene expression under stress conditions using PlantPAN [52]. Further, from our database of *Setaria italica* miRNAs (unpublished data) putative miRNAs targeting the SiWD40 genes were identified using psRNATarget [53].

Comparative Physical Mapping of SiWD40 Proteins between *S. italica* and other Grass Species

The amino acid sequences of physically mapped SiWD40 protein-encoding genes spanning the nine foxtail millet chromosomes were BLASTP searched against peptide sequences of sorghum, maize and rice (http://gramene.org/; www.phytozome.net) to infer orthologous relationship among the chromosomes of foxtail millet and the other three grass species. Reciprocal BLAST has also been performed to ensure the unique relationship between the orthologous genes. BLAST hits with E-value ≤1e-5 and at least 80% homology were considered significant. The comparative orthologous relationships of *WD40* genes among foxtail millet, rice, sorghum and maize chromosomes were finally visualized using MapChart [42].

Estimation of Synonymous and Non-synonymous Substitution Rates

The amino acid sequences duplicated protein-encoding WD40 genes as well as orthologous gene-pairs between foxtail millet and rice, maize and sorghum were aligned using ClustalW based multiple sequence alignment tool. The CODEML program in PAML interface tool of PAL2NAL (http://www.bork.embl.de/pal2nal/) [54], was used to estimate the synonymous (Ks) and non-synonymous (Ka) substitution rates by aligning the amino acid sequences and their respective original cDNA sequences of *SiWD40* genes. Time (million years ago, Mya) of duplication and divergence of each *SiWD40* genes were estimated using a synonymous mutation rate of λ substitutions per synonymous site per year, as $T = Ks/2\lambda$ ($\lambda = 6.5\times10^{-9}$) [55,56].

Expression Profiling using RNA-seq Data

To elucidate the tissue-specific expression profile of SiWD40 genes, the *Setaria italica* Illumina RNA-HiSeq reads from 4 tissues namely spica, stem, leaf and root were retrieved from European Nucleotide Archive [SRX128226 (spica); SRX128225 (stem); SRX128224 (leaf); SRX128223 (root)] [57]. The RNA-seq data was then filtered by NGS toolkit [58] to remove low quality reads and was mapped onto the gene sequences of *Setaria italica* by CLC Genomics Workbench v.4.7.1 (http://www.clcbio.com/genomics). The number of reads mapped was normalized by RPKM (reads per kilobase per million) method. The heat map showing tissue specific expression was generated on the RPKM value for each gene in all the tissue samples using TIGR MultiExperiment Viewer (MeV4) software package [59,60].

Plant Materials and Stress Treatments

Seeds of foxtail millet cv. Prasad known for its abiotic stress tolerance were procured from National Bureau of Plant Genetic Resources (NBPGR), Hyderabad, India and grown in a plant growth chamber (PGC-6L; Percival Scientific Inc., USA) at 28±1°C day/23±1°C night with 70±5% relative humidity and photoperiod of 14 h. For stress treatments, 21-day-old seedlings were exposed to 250 mM NaCl (salinity), 20% PEG 6000 (dehydration), 150 µM abscisic acid (ABA) and incubation at 4°C (cold) for 1 h, 3 h, 6 h, 12 h, 24 h and 48 h. Unstressed plants were maintained as controls. After the treatments, seedlings were immediately frozen in liquid nitrogen and stored at −80°C until RNA isolation. The above experiments were repeated thrice to ensure precision and reproducibility.

RNA Extraction and Quantitative Real-time PCR Analysis

Total RNA was isolated by following the procedure described by Longeman et al. [61] and treated with RNase-free DNase I (50 U/µl; Fermentas, USA) for removing DNA contamination. The quality and purity of the preparations were determined at OD_{260}:OD_{280} nm absorption ratio (1.8–2.0) and the integrity of the preparations was determined by resolving in 1.2% agarose gel containing formaldehyde. About 1 µg total RNA was reverse transcribed to first strand cDNA using random primers by Protoscript M-MuLV RT (New England Biolabs, USA) following manufacturer's instructions [21]. The qRT-PCR primers were designed using Primer Express 3.0 software (PE Applied Biosystems, USA) with default parameters (Table S8). qRT-PCR was carried out in three technical replicate for each biological duplicate by one step real time PCR system of Applied Biosytems (USA). The PCR mixtures and reactions were used as described previously by Kumar et al.[21] Melting curve analysis (60 to 95°C after 40 cycles) and agarose gel electrophoresis were performed to check amplification specificity for absence of multiple amplicons or primer dimers [22]. A constitutive *Act2* gene-based primer was used as endogenous control. The amount of transcript accumulated for *SiWD40* genes normalized to the internal control *Act2* were analyzed using $2^{-\Delta\Delta Ct}$ method cDNA synthesis. The PCR efficiency which is dependent on the assay, performance of the master mix and quality of sample, was calculated as: Efficiency = $10^{(-1/slope)} - 1$ by the software itself (Applied Biosystems).

Homology Modeling of SiWD40 Proteins

All the SiWD40 proteins were searched against the Protein Data Bank (PDB) [62] by BLASTP (with the default parameters) to identify the best template having similar sequence and known three-dimensional structure (Table S9). The data was fed in Phyre2 (Protein Homology/AnalogY Recognition Engine; http://www.sbg.bio.ic.ac.uk/phyre2) for predicting the protein structure by homology modeling under 'intensive' mode [63]. For active site prediction, the PDB code was submitted to Q-SiteFinder [64].

Supporting Information

Figure S1 Gene structures of 225 *SiWD40* proteins. Exons and introns are represented by green boxes and black lines, respectively.

Figure S2 Comparative physical mapping revealed high degree of orthologous relationships of *SiWD40* genes located on nine chromosomes of foxtail millet with (A) sorghum, (B) maize and (C) rice.

Figure S3 Heat map representation of *SiWD40* genes across different tissues. The Illumina RNA-seq data were re-analyzed and the heat map was generated. Bar at the top represents $\log_2$ transformed values, thereby values −2.0, 2.0 and 4.0 represent low, intermediate and high expression, respectively.

Table S1 A catalog of 225 *Setaria italica* WD40 proteins.

Table S2 Blast2GO annotation details of SiWD40 protein sequences.

Table S3 Characteristics of the promoter region of 42 stress-related *SiWD40* genes.

Table S4 Summary of putative miRNA targeting the *SiWD40* genes.

Table S5 The Ka/Ks ratios and estimated divergence time for tandemly duplicated SiWD40 proteins.

Table S6 The Ka/Ks ratios and estimated divergence time for segmentally duplicated SiWD40 proteins.

Table S7 The Ka/Ks ratios and estimated divergence time for orthologous WD proteins between foxtail millet, rice, sorghum and maize.

Table S8 List of primers used in quantitative real time-PCR expression analysis of 13 SiWD40 genes.

Table S9 Characteristics of 24 candidate SiWD40 proteins chosen for homology modeling.

Dataset S1 Compilation of PDB files used for homology modeling of SiWD40 proteins.

Acknowledgments

Grateful thanks are due to the Director, National Institute of Plant Genome Research (NIPGR), New Delhi, India for providing facilities. The authors also thank Mr. Venkata Suresh B, NIPGR for his timely assistance.

Author Contributions

Conceived and designed the experiments: MP. Performed the experiments: AKM MM YK SKP. Analyzed the data: AKM MM SKP MP. Contributed reagents/materials/analysis tools: MP. Wrote the paper: AKM MM SKP MP.

References

1. Li P, Brutnell TP (2011) *Setaria viridis* and *Setaria italica*, model genetic systems for Panicoid grasses. J Exp Bot 62: 3031–3037.
2. Lata C, Gupta S, Prasad M (2013) Foxtail millet, a model crop for genetic and genomic studies in bioenergy grasses. Crit Rev Biotechnol 33: 328–343.
3. Lata C, Prasad M (2013) *Setaria* genome sequencing: an overview. J Plant Biochem Biotechnol 22: 257–260.
4. Doust AN, Kellogg EA, Devos KM, Bennetzen JL (2009) Foxtail millet, a sequence-driven grass model system. Plant Physiol 149: 137–141.
5. Zhang G, Liu X, Quan Z, Cheng S, Xu X, et al. (2012) Genome sequence of foxtail millet (*Setaria italica*) provides insights into grass evolution and biofuel potential. Nature Biotechnol 30: 549–554.
6. Bennetzen JL, Schmutz J, Wang H, Percifield R, Hawkins J, et al. (2012) Reference genome sequence of the model plant *Setaria*. Nature Biotechnol 30: 555–561.
7. Muthamilarasan M, Theriappan P, Prasad M (2013) Recent advances in crop genomics for ensuring food security. Curr Sci 105: 155–158.
8. Gupta S, Kumari K, Das J, Lata C, Puranik S, et al. (2011) Development and utilization of novel intron length polymorphic markers in foxtail millet [*Setaria italica* (L.) P. Beauv.]. Genome 54: 586–602.
9. Gupta S, Kumari K, Sahu PP, Vidapu S, Prasad M (2012) Sequence based novel genomic microsatellite markers for robust genotyping purposes in foxtail millet [*Setaria italica* (L.) P. Beauv.]. Plant Cell Rep 31: 323–337.
10. Pandey G, Misra G, Kumari K, Gupta S, Parida SK, et al. (2013) Genome-wide development and use of microsatellite markers for large-scale genotyping applications in foxtail millet [*Setaria italica* (L.)]. DNA Res 20: 197–207.
11. Gupta S, Kumari K, Muthamilarasan M, Subramanian A, Prasad M (2013) Development and utilization of novel SSRs in foxtail millet [*Setaria italica* (L.) P. Beauv.]. Plant Breed doi: 10.1111/pbr.12070.
12. Kumari K, Muthamilarasan M, Misra G, Gupta S, Subramanian A, et al. (2013) Development of eSSR-markers in *Setaria italica* and their applicability in studying genetic diversity, cross-transferability and comparative mapping in millet and non-millet species. PLoS ONE 8: e67742.
13. Muthamilarasan M, Venkata Suresh B, Pandey G, Kumari K, Parida SK, et al. (2013) Development of 5123 intron-length polymorphic markers for large-scale genotyping applications in foxtail millet. DNA Res doi: 10.1093/dnares/dst039.
14. Jayaraman A, Puranik S, Rai NK, Vidapu S, Sahu PP, et al. (2008) cDNA-AFLP analysis reveals differential gene expression in response to salt stress in foxtail millet (*Setaria italica* L.). Mol Biotechnol 40: 241–251.
15. Lata C, Sahu PP, Prasad M (2010) Comparative transcriptome analysis of differentially expressed genes in foxtail millet (*Setaria italica* L.) during dehydration stress. Biochem Biophy Res Commun 393: 720–727.
16. Puranik S, Bahadur RP, Srivastava PS, Prasad M (2011) Molecular cloning and characterization of a membrane associated NAC family gene, SiNAC from foxtail millet [*Setaria italica* (L.) P. Beauv.]. Mol Biotechnol 49: 138–150.
17. Lata C, Bhutty S, Bahadur RP, Majee M, Prasad M (2011) Association of an SNP in a novel DREB2-like gene SiDREB2 with stress tolerance in foxtail millet [*Setaria italica* (L.)]. J Exp Bot 62: 3387–3401.
18. Lata C, Jha S, Dixit V, Sreenivasulu N, Prasad M (2011) Differential antioxidative responses to dehydration-induced oxidative stress in core set of foxtail millet cultivars [*Setaria italica* (L.)]. Protoplasma 248: 817–828.
19. Puranik S, Kumar K, Srivastava PS, Prasad M (2011) Electrophoretic Mobility Shift Assay reveals a novel recognition sequence for *Setaria italica* NAC protein. Plant Signal Behav 6: 1588–1590.
20. Mishra AK, Puranik S, Bahadur RP, Prasad M (2012) The DNA-binding activity of an AP2 protein is involved in transcriptional regulation of a stress-responsive gene, SiWD40, in foxtail millet. Genomics 100: 252–263.
21. Puranik S, Sahu PP, Mandal SN, B VS, Parida SK, et al. (2013) Comprehensive genome-wide survey, genomic constitution and expression profiling of the NAC transcription factor family in foxtail millet (*Setaria italica* L.). PLoS ONE 8: e64594.
22. Kumar K, Muthamilarasan M, Prasad M (2013) Reference genes for quantitative Real-time PCR analysis in the model plant foxtail millet (*Setaria italica* L.) subjected to abiotic stress conditions. Plant Cell Tiss Organ Cult 115: 13–22.
23. Mishra AK, Puranik S, Prasad M: (2012) Structure and regulatory networks of WD40 protein in plants. J Plant Biochem Biotechnol 21: 32–39.
24. Neer EJ, Schmidt CJ, Nambudripad R, Smith TF (1994) The ancient regulatory-protein family of WD-repeat proteins. Nature 371: 297–300.
25. Smith TF, Gaitatzes C, Saxena K, Neer EJ (1999) The WD repeat: a common architecture for diverse functions. Trends Biochem Sci 24: 181–185.
26. Andrade MA, Perez-Iratxeta C, Ponting CP (2001) Protein repeats: structures, functions, and evolution. J Struct Biol 134: 117–131.
27. Lee JH, Terzaghi W, Gusmaroli G, Charron JB, Yoon HJ, et al. (2008) Characterization of *Arabidopsis* and rice DWD proteins and their roles as substrate receptors for CUL4-RING E3 ubiquitin ligases. Plant Cell 20: 152–167.
28. Angers S, Li T, Yi X, MacCoss MJ, Moon RT, Zheng N (2006) Molecular architecture and assembly of the DDB1-CUL4A ubiquitin ligase machinery. Nature 443: 590–593.
29. Hua Z, Vierstra RD (2011) The cullin-RING ubiquitin-protein ligases. Annu Rev Plant Biol 62: 299–334.

30. Van Nocker S, Ludwig P (2003) The WD-repeat protein superfamily in Arabidopsis: conservation and divergence in structure and function. BMC Genomics 4: 50.
31. Ouyang Y, Huang X, Lu Z, Yao J (2012) Genomic survey, expression profile and co-expression network analysis of *OsWD40* family in rice. BMC Genomics 13: 100.
32. Lecharny A, Boudet N, Gy I, Aubourg S, Kreis M (2003) Introns in, introns out in plant gene families: a genomic approach of the dynamics of gene structure. J Struct Funct Genomics 3: 111–116.
33. Jain M, Khurana P, Tyagi AK, Khurana JP (2008) Genome-wide analysis of intronless genes in rice and *Arabidopsis*. Funct Integr Genomics 8: 69–78.
34. Puranik S, Sahu PP, Srivastava PS, Prasad M (2012) NAC proteins: regulation and role in stress tolerance. Trends Plant Sci 17: 1360–1385.
35. Han Z, Guo L, Wang H, Shen Y, Deng XW, Chai J (2006) Structural basis for the specific recognition of methylated histone H3 lysine 4 by the WD-40 protein WDR5. Mol Cell 22: 137–144.
36. Jain M, Nijhawan A, Arora R, Agarwal P, Ray S, et al. (2007) F-Box Proteins in Rice. genome-wide analysis, classification, temporal and spatial gene expression during panicle and seed development, and regulation by light and abiotic stress. Plant Physiol 143: 1467–1483.
37. Cannon SB, Mitra A, Baumgarten A, Young ND, May G (2004) The roles of segmental and tandem gene duplication in the evolution of large gene families in *Arabidopsis thaliana*. BMC Plant Biol 4: 10.
38. Söding J (2005) Protein homology detection by HMM-HMM comparison. Bioinformatics 21: 951–960.
39. Jefferys BR, Kelley LA, Sternberg MJE (2010) Protein folding requires crowd control in a simulated cell. J Mol Biol 397: 1329–1338.
40. Stirnimann CU, Petsalaki E, Russell RB, Müller CW (2010) WD40 proteins propel cellular networks. Trends Biochem Sci 35: 565–574.
41. Letunic I, Doerks T, Bork P (2012) SMART 7: recent updates to the protein domain annotation resource. Nucleic Acids Res 40: D302–D305.
42. Voorrips RE (2002) MapChart: software for the graphical presentation of linkage maps and QTLs. J Hered 93: 77–78.
43. Tang H, Bowers JE, Wang X, Ming R, Alam M, et al. (2008) Synteny and collinearity in plant genomes. Science 320: 486–488.
44. Du D, Zhang Q, Cheng T, Pan H, Yang W, Sun L (2012) Genome-wide identification and analysis of late embryogenesis abundant (LEA) genes in *Prunus mume*. Mol Biol Rep 40: 1937–1946.
45. Shiu S-H, Bleecker AB (2003) Expansion of the Receptor-Like Kinase/Pelle Gene Family and Receptor-Like Proteins in *Arabidopsis*. Plant Physiol 132: 530–543.
46. Krzywinski M, Schein J, Birol I, Connors J, Gascoyne R, et al. (2009) Circos: an information aesthetic for comparative genomics. Genome Res 19: 1639–1645.
47. Guo AY, Zhu QH, Chen X, Luo JC (2007) GSDS: a gene structure display server. Yi Chuan 29: 1023–1026.
48. Tamura K, Peterson D, Peterson N, Stecher G, Nei M, et al. (2011) MEGA5: Molecular evolutionary genetics analysis using maximum likelihood, evolutionary distance, and maximum parsimony methods. Mol Biol Evol 28: 2731–2739.
49. Thompson JD, Gibson TJ, Plewniak F, Jeanmougin F, Higgins DG (1997) The CLUSTAL_X windows interface: Flexible strategies for multiple sequence alignment aided by quality analysis tools. Nucleic Acids Res 25: 4876–4882.
50. Saitou N, Nei M (1987) The neighbor-joining method: a new method for reconstructing phylogenetic trees. Mol Biol Evol 4: 406–425.
51. Conesa A, Götz S (2008) Blast2GO: a comprehensive suite for functional analysis in plant genomics. Int J Plant Genomics 2008: 619832.
52. Chang WC, Lee TY, Huang HD, Huang HY, Pan RL (2008) PlantPAN: Plant Promoter Analysis Navigator, for identifying combinatorial cis-regulatory elements with distance constraint in plant gene group. BMC Genomics 9: 561.
53. Dai X, Zhao PX (2011) psRNATarget: a plant small RNA target analysis server. Nucleic Acids Res 39: W155–159.
54. Suyama M, Torrents D, Bork P (2006) PAL2NAL: robust conversion of protein sequence alignments into the corresponding codon alignments. Nucleic Acids Res 34: W609–W612.
55. Lynch M, Conery JS (2000) The evolutionary fate and consequences of duplicate genes. Science 290: 1151–1155.
56. Yang Z, Gu S, Wang X, Li W, Tang Z, Xu C (2008) Molecular evolution of the cpp-like gene family in plants: insights from comparative genomics of *Arabidopsis* and rice. J Mol Evol 67: 266–277.
57. Cochrane G, Alako B, Amid C, Bower L, Cerdeño-Tárraga A, et al. (2013) Facing growth in the European Nucleotide Archive. Nucleic Acids Res 41: D30–D35.
58. Patel RK, Jain M (2012) NGS QC Toolkit: A toolkit for quality control of next generation sequencing data. PLoS ONE 7: e30619.
59. Saeed AI, Sharov V, White J, Li J, Liang W, et al. (2003) TM4: a free, open-source system for microarray data management and analysis. Biotechniques 34: 374–378.
60. Saeed AI, Bhagabati NK, Braisted JC, Liang W, Sharov V, et al. (2006) TM4 microarray software suite. Methods Enzymol 411: 134–193.
61. Longeman J, Schell J, Willmitzer L (1987) Improved method for the isolation of RNA from plant tissues. Anal Biochem 163: 16–20.
62. Berman HM, Westbrook J, Feng Z, Gilliland G, Bhat TN, et al. (2000) The protein data bank. Nucleic Acids Res 28: 235–242.
63. Kelley LA, Sternberg MJE (2009) Protein structure prediction on the Web: a case study using the Phyre server. Nature Protocols 4: 363–371.
64. Laurie AT, Jackson RM (2005) Q-SiteFinder: an energy-based method for the prediction of protein-ligand binding sites. Bioinformatics 21: 1908–1916.

The Updated Phylogenies of the Phasianidae Based on Combined Data of Nuclear and Mitochondrial DNA

Yong-Yi Shen[1,2]*, Kun Dai[3], Xue Cao[1,6], Robert W. Murphy[1,4], Xue-Juan Shen[2], Ya-Ping Zhang[1,5]*

1 State Key Laboratory of Genetic Resources and Evolution, Kunming Institute of Zoology, the Chinese Academy of Sciences, Kunming, China, **2** Joint Influenza Research Centre (SUMC/HKU), Shantou University Medical College, Shantou, China, **3** Xinjiang Institute of Ecology and Geography, the Chinese Academy of Sciences, Urumqi, China, **4** Department of Natural History, Royal Ontario Museum, Toronto, Canada, **5** Laboratory for Conservation and Utilization of Bio-resources, Yunnan University, Kunming, China, **6** University of the Chinese Academy of Sciences, Beijing, China

Abstract

The phylogenetic relationships of species in the Phasianidae, Order Galliformes, are the object of intensive study. However, convergent morphological evolution and rapid species radiation result in much ambiguity in the group. Further, matrilineal (mtDNA) genealogies conflict with trees based on nuclear DNA retrotransposable elements. Herein, we analyze 39 nearly complete mitochondrial genomes (three new) and up to seven nuclear DNA segments. We combine these multiple unlinked, more informative genetic markers to infer historical relationships of the major groups of phasianids. The nuclear DNA tree is largely congruent with the tree derived from mt genomes. However, branching orders of mt/nuclear trees largely conflict with those based on retrotransposons. For example, *Gallus/Bambusicola/Francolinus* forms the sister-group of *Coturnix/Alectoris* in the nuclear/mtDNA trees, yet the tree based on retrotransposable elements roots the former at the base of the tree and not with the latter. Further, while peafowls cluster with *Gallus/Coturnix* in the mt tree, they root at the base of the phasianids following *Gallus* in the tree based on retrotransposable elements. The conflicting branch orders in nuclear/mtDNA and retrotransposons-based trees in our study reveal the complex topology of the Phasianidae.

Editor: Yidong Bai, University of Texas Health Science Center at San Antonio, United States of America

Funding: This work was supported by National Natural Science Foundation of China (30621092). R. W. Murphy was supported by a Visiting Professorship for Senior International Scientists from the Chinese Academy of Sciences and manuscript preparation was supported by a Natural Sciences and Engineering Research Council of Canada Discovery Grant A3148. The funders had no role in study design, data collection and analysis, decision to publish, or preparation of the manuscript.

Competing Interests: The authors have declared that no competing interests exist.

* E-mail: shenyy@mail.kiz.ac.cn (YYS); zhangyp@mail.kiz.ac.cn (YPZ)

Introduction

Rapid species radiations often result in ambiguous phylogenetic relationships because too little time is available to accrue and fix shared derived character states. This may manifest itself, in part, as incomplete lineage sorting. In such cases, different datasets, such as genes, may resolve conflicting suites of relationships. The Phasianidae, one of four families in the Galliformes, typifies this problem. Rapid radiation and convergent morphological evolution confound the resolution of relationships for many pheasants and partridges. Although the family has been the target of much phylogenetic research [1–17], not surprisingly, many unsolved nodes and much conflict remain.

Most previous molecular studies of phasianids analyze either one or a few mitochondrial (mt) genes [3,9,10,12], a single nuclear gene [1,18], or a combination of mt and a few nuclear gene sequences [2,7]. Employing complete mt genomes, Shen et al. [19] resolve a well-supported topology. The topology (Figure 1A) is largely congruent with previous molecular studies based on mt genes and nuclear segments [2,7]. However, this tree strongly conflicts with that based on retrotransposable elements (Figure 1B) [6,17,20]. For example, in the matrilineal genealogy, *Gallus* forms the sister-group of *Coturnix*, while the tree based on retrotransposable elements roots *Gallus* at the base of the phasianids, and *Coturnix* is the sister-group of the gallopheasants. Further, while *Pavo* is the sister-group of *Gallus/Coturnix* in the mt tree, it roots at the base of the phasianids following *Gallus* in the tree based on retrotransposable elements. Trees based on unclear genes exhibit many ambiguous nodes (Figure 1C).

The branching order in trees based on the mt genome conflicts with those derived from nuclear retrotransposons. This incongruence requires a reassessment of the phylogeny of the Phasianidae. Mitochondrial DNA markers reflect the matrilineal genealogy only; they do not provide information on paternal contributions. For retrotransposons, only a few phylogenetic approaches use indels as characters. Few genomes are available from which to design conservative retrotransposon primers for phasianids; only the chicken and a limited number of other avian genomes are available. This situation makes it difficult to obtain a sufficient number of phylogenetically informative characters. Considering the shortcomings of the mt and retrotransposon approaches, herein we report the sequencing of up to seven independent nuclear segments for 20 species and the complete mt genomes of three phasianids. We obtain other mt genomes from GenBank (Table S1 in File S1) and then combine these unlinked markers for the major groups of the Phasianidae to infer phylogenetic relationships.

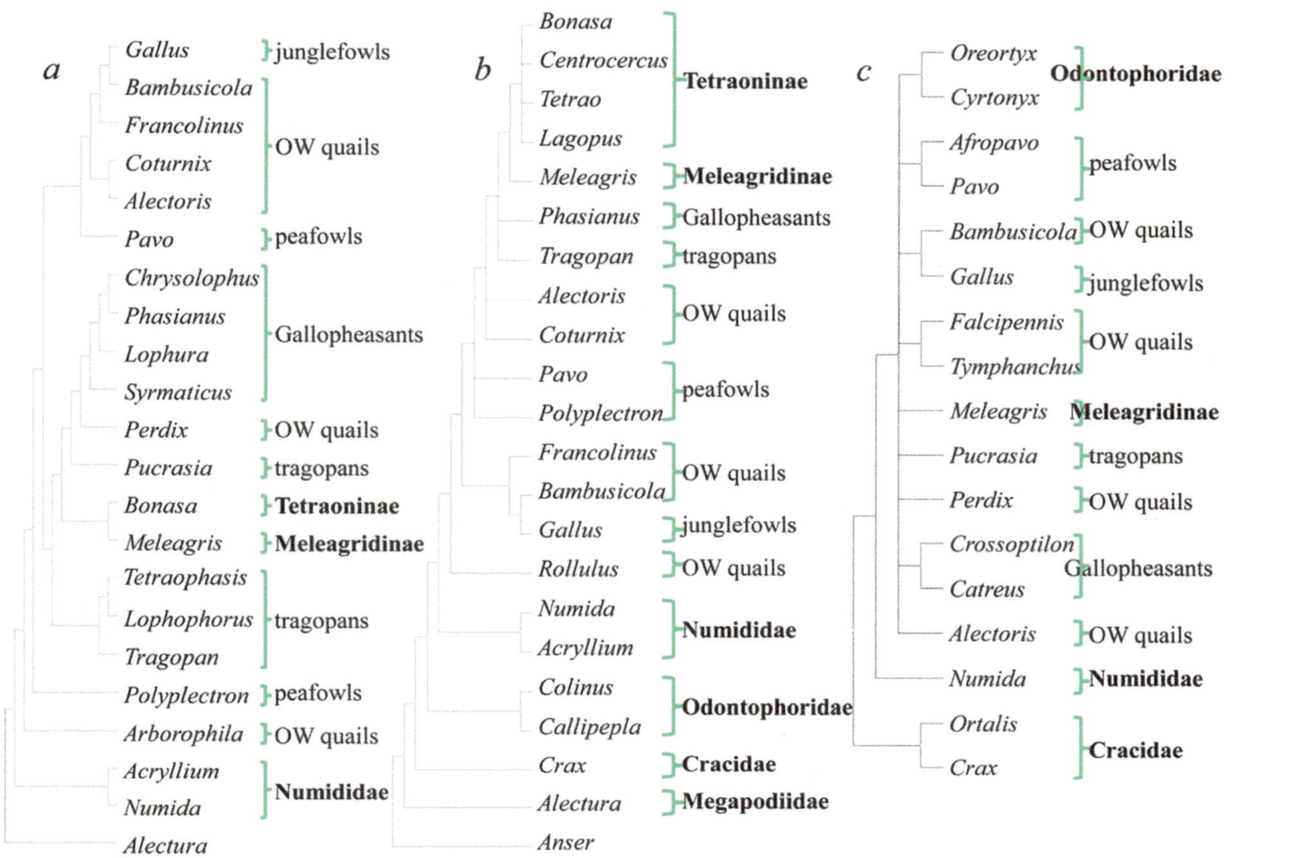

Figure 1. Phylogenetic hypotheses from the mitochondrial (mt) genome and retrotransposable elements for the Phasianidae. (a) Topology based on mt genomes (Shen *et al.* 2010); (b) topology based on insertion events of CR1 retrotransposable nuclear DNA elements [6,17,20]; (C) topology based on nuclear DNA segments (Crowe et al., 2006).

Materials and Methods

Specimens Sampling

The Animal Use Ethics Committee of the Kunming Institute of Zoology, the Chinese Academy of Sciences approved the study. *Argusianus argus*, *Crossoptilon crossoptilon* and *Ithaginis cruentus* were used for mt genome sequencing. A total of 23 species were used for nuclear gene sequencing (Table S1 File S1). Feather samples of *Argusianus argus* were provided by Beijing Zoo and the Museum of the Kunming Institute of Zoology provided muscle tissue for all other samples. Additional complete mt genomes and nuclear segments were obtained from GenBank (Table S1 in File S1).

DNA Extraction, PCR Amplification, and Sequencing

Total genomic DNA was extracted using standard 3-step phenol/chloroform extraction methods [21]. For mitochondrial genomes, primers were described in our previous study [19]. For seven nuclear segments (*BDNF*, *CMOS*, *FIB4*, *NGFB*, *NTF3*, *OVOG*, and *ZENK*), primers were described in Table S2 in File S1. PCR amplifications were conducted in a 50 µl volume containing 5 µl of 10×reaction buffer, 0.2 mM dNTPs, 0.2 µM each primer, 1.5 U Taq DNA polymerase (TaKaRa Biosystems), and approximately 2 ng total DNA. PCR amplifications were carried out using the following parameters: 95°C 4 min, 20 cycles of denaturation at 94°C for 1 min, annealing at 60–50°C (1 min; 0.5°C/cycle), extension at 72°C for 1 min, and finally 15 cycles of 94°C 1 min, 50°C 1 min, 72°C 1 min. PCR products were cleaned using Watson RCR Purification Kits (Watson BioTechnologies, Shanghai). PCR products were sequenced at least three times in both directions on an ABI 3730 Sequencer (Applied Biosystems, Foster, CA, USA) using the ABI PRISM BigDye Terminator v3.0 sequencing kit. DNA sequences were edited using DNAstar Seqman software (DNASTAR Inc., Madison, WI, USA). The newly determined sequences were deposited in GenBank (GenBank accession numbers: JQ713766–JQ713768; JQ713656–JQ713765).

Phylogenetic Reconstruction

The nucleotide sequence data sets were initially aligned using ClustalX 1.81 [22] with default parameters. The combined and individual 13 mitochondrial protein coding genes, and the combined data of seven nuclear segments were analyzed separately using maximum likelihood (ML) implemented in PAUP* 4.0b10 [23]. Modeltest 3.7 [24] was used to select the preferred models of evolution under the Akaike Information Criterion. ML heuristic searches used TBR branch swapping executed in 100 replicates with the selected models. Because heuristic searches in PAUP* were very slow, we used two additional fast ML-based inference packages using 1,000 replicates each: RAxML [25] and PHYML [26]. Because their topologies were identical, and only a few bootstrap values slightly differed, we only presented trees with bootstrap values from PAUP*. Bayesian inference (BI) was performed using MrBayes 3.1.2 [27]. The analyses used models estimated with Modeltest 3.7 under AIC.

Two separate runs were performed with four Markov chains. Each run was conducted with 3×10^6 generations and sampled every 100 generations. When the log-likelihood scores were found to stabilize, a consensus tree was calculated after omitting the first 25% trees as burn-in. In all these topology reconstruction, *Alectura lathami* was set as the outgroup according a previous study [19].

Results

Phylogenetic Analyses of the Mitochondrial DNA Dataset

We evaluated 39 mt genomes including those from GenBank. The 13 protein-coding genes consisted of 11,359 aligned nucleotide positions and the best-fit model of evolution was GTR+I+G. ML and BI analyses involving equal weight for each position resolved a single, robust tree (Figure 2A). Eight lineages, each with very high BSPs and BPPs, were resolved as follows: Group 1, *Arborophila*; Group 2, *Tragopan*, *Lophophorus*, and *Tetraophasis*; Group 3, *Chrysolophus*, *Phasianus*, *Lophura*, and *Syrmaticus*; Group 4, *Perdix*; Group 5, *Pucrasia*; Group 6, *Gallus*, *Bambusicola*, and *Francolinus*; Group 7, *Coturnix* and *Alectoris*; and Group 8, the peacocks.

Three additional weighting strategies were applied to the analysis of combined 13 protein-coding genes to avoid possible bias of nucleotide composition and saturation: (1) excluding the 3rd codon positions (Figure S1), (2) recoding the 3rd codon position nucleotides to two-state categories, R (purine) and Y (pyrimidine) (Figure S2), and (3) recoding the 1st and the 3rd codon position nucleotides to RY categories (Figure S3). The major topologies based on these weighting strategies were the same as evaluating all positions equally (Figure 2A).

Individual mt gene trees (Figure S4) were largely congruent with the mt genomic tree. As expected because of a low number of potentially phylogenetically informative characters for individual genes, nodes were supported by lower BSPs.

Phylogenetic Analyses of Nuclear Dataset

Segments of seven nuclear genes (*BDNF*, *CMOS*, *FIB4*, *NGFB*, *NTF3*, *OVOG*, and *ZENK*) were sequenced (Table S1 in File S1). Combined, these data consisted of 4,604 nucleotide positions. The best-fit model of nucleotide substitution was TrN+I+G. ML and BI analyses of the combined data resolved a single tree (Figure 2B). The nuclear tree was largely congruent with mitochondrial tree although some nodes conflicted. For example, *Ithagins cruentus* rooted at the base of Gallopheasants/tragopans in mt tree but the nuclear tree resolved it as the sister group of *Pucrasia*. The position

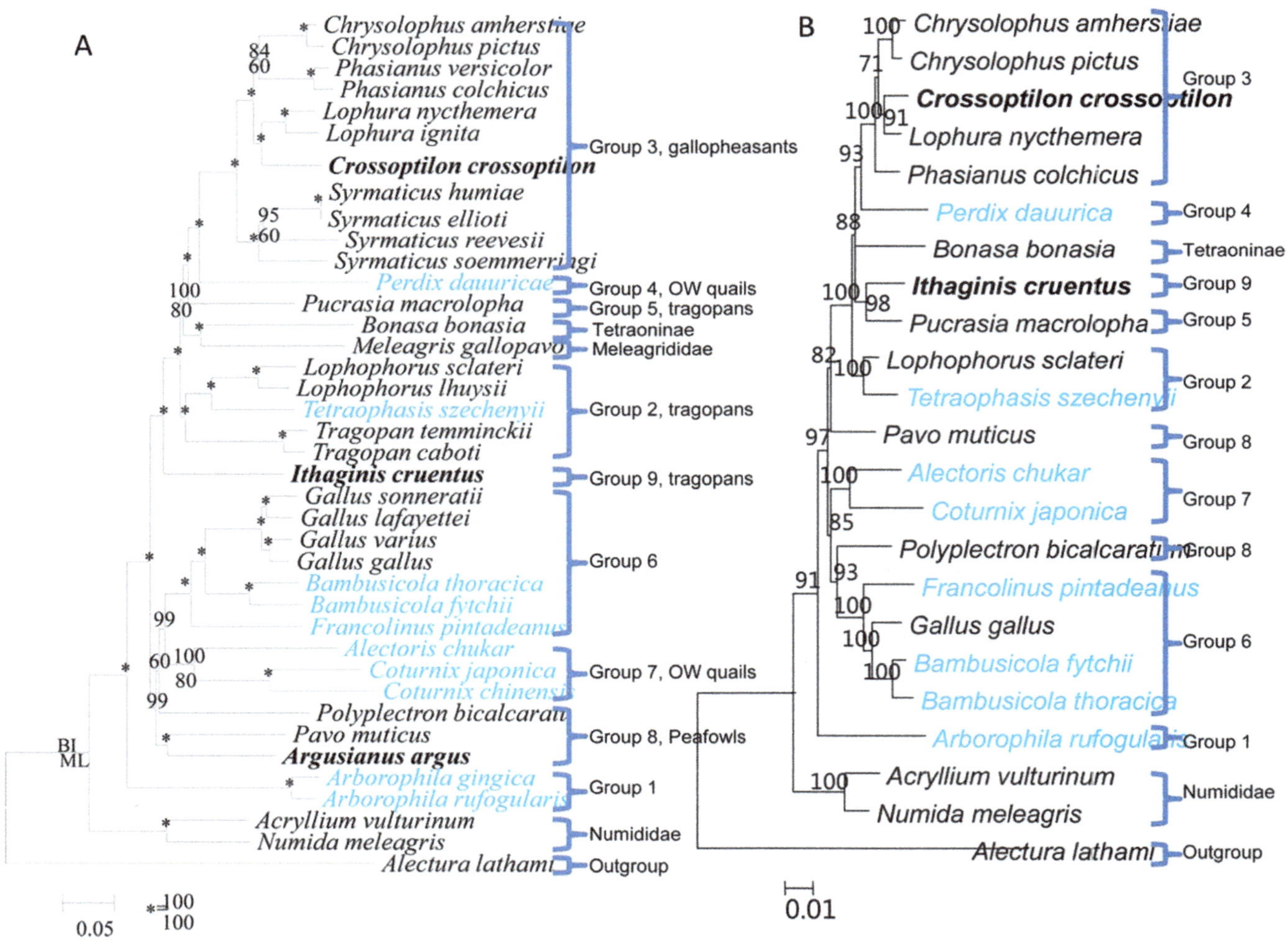

Figure 2. Phylogenetic hypotheses based on the complete mitochondrial genomes and nuclear segments for the Phasianidae. (a) Mt genomes; newly sequenced mt genomes denoted in bold. (b) Nuclear segments. Bayesian posterior probabilities >70%, and maximum likelihood bootstrap proportions >50% were indicated on the branches.

of peafowls was unstable in nuclear tree, while in mt tree they rooted at the base of *Gallus*/Old World quails.

Phylogenetic Analyses of the Combined Mitochondrial and Nuclear DNA Dataset

The mt and nuclear datasets shared 23 species. The combined dataset consisted of 15,972 aligned nucleotide positions. The best-fit model of nucleotide substitution was GTR+I+G. The topology (Figure S5) based on the combined dataset was nearly identical to that of mt tree (Figure 2A); they differed in the position of *Phasianus*. The mt genome tree clustered *Phasianus* with *Chrysolophus* with low support values (BPP = 84, and BSP = 60). In turn, this group clustered as the sister group of *Lophura*/*Crossoptilon* (BPP = 100, and BSP = 100). Trees based on the combined dataset clusterd *Chrysolophus* with *Lophura*/*Crossoptilon* (BPP = 99), then with *Phasianus* (BPP = 100). Trees derived from combined dataset were largely congruent with the nuclear gene phylogeny. However, a few conflicts occurred. For example, the nuclear gene tree resolved *Ithagins* and *Pucrasia* as sister taxa, while in combined and mt genome trees separated them far apart.

Discussion

Our mt genome tree depicts *Arborophila* as the sister-group to all other phasianids plus the Meleagrididae and Tetraonidae. *Tetraophasis* clusters independently with *Lophophorus* and their sister-group is *Tragopan*; *Perdix* and *Arborophila* do not cluster with other partridges. The non-monophyly of the pheasants and partridges is more common than not and this resolution involves a strongly supported association of *Gallus*, *Bambusicola*, and *Francolinus*. The previous matrilineal genealogy did not cluster *Polyplectron* with *Pavo* [16,19]. Herein, we add a new peacock–*Argusianus argus*. *Polyplectron* forms the sister-group of *Gallus*/*Bambusicola*/*Coturnix* albeit with relatively low support. This group clusters with other peacocks (*Pavo* and *Argusianus*). The phylogenetic position of *Polyplectron* remains unstable. More data involving new peacocks may further resolve this group's position.

We add the nearly complete mt genome of *Crossoptilon crossoptilon* and analyses involving this species clusters it with *Lophura* (BPP = 100; BSP = 100). Analyses of the new mt genome of *Ithaginis cruentus* strongly unites it with the gallopheasants/*Perdix*/tragopans/Tetraoninae/Meleagrididae (BPP = 100; BSP = 100). *Tragopan*, *Ithaginis*, *Pucrasia*, *Lophophorus* form the tragopans tribe [28]. Our resolution of a sister relationship for *Tragopan* and *Lophophorus* is well supported, but *Ithaginis* and *Pucrasia* do not cluster with them. The mt genome tree also does not support the morphological and behavioral placement of *Ithaginis* as the sister-group of New World and Old World quails [4]. Previous molecular studies did not conclusively resolve the phyletic position of *Ithaginis* [2,10].

The mt genomes provide a greater abundance of information, thus, have a greater likelihood of fully resolving a tree than individual protein-coding genes (Figure S4). In our analyses, almost all nodes receive very high support. Thus, the rapid rate of mutation renders the mt genome phylogenetically informative at the levels of genera and species for the phasianids.

Reliance on mt data for phylogenetic reconstruction may be fraught with problems. Functionally, in most animals the mt genome serves as a single, large genetic locus and it provides a matrilineal perspective only on the evolutionary history of a group [29,30]. Paternal contributions are not considered. Thus, mtDNA data alone are often inadequate for macroevolutionary phylogenetic analyses, especially in the face of complex evolutionary scenarios such as gene introgression, hybridization, and/or selection [31]. Our seven nuclear segments address this concern. BSPs tend to be lower in nuclear tree compared to the mt tree. The relatively slow rate of mutation rate of nuclear DNA compared to mtDNA generally results in relatively poorly resolved nuclear gene trees.

Previous studies based on a single nuclear gene failed to solve many nodes [1,18]. In contrast, our multi-gene analyses resolve many nodes with very high levels of support. This result indicates that additional informative sites greatly help to resolve ambiguous relationships. Our nuclear phylogeny is largely congruent with trees derived from mt genomes. However, our nuclear tree resolves *Ithagins* and *Pucrasia* as sister groups, but the mt tree depicts divergent relationships. The position of *Ithagins* was ambiguous in the previous morphological-behavioral parsimony cladogram of Dyke et al. (2003) and molecular studies [2,9,10]. The position of *Ithagins* received high BSPs in our mt and nuclear trees, yet further explorations into the nature of conflicting trees is necessary. The positions of peafowl are unstable in nuclear tree. *Pavo* and *Polyplectron* do not cluster together, including in our mt tree.

The tree based on combined mitochondrial and nuclear dataset results a well-supported tree (Figure S5). In mt genome tree (Figure 2A), the position of genus *Phasianus* is not well supported; it clusters with *Chrysolophus* with low support values (BPP = 84; BSP = 60). In contrast, the trees of the combined datasets cluster *Phasianus* with (*Chrysolophus*, (*Lophura*,*Crossoptilon*), and with strong support. The positions of peafowls are unstable in the mt genome, nuclear data, and combined data trees. More species of peafowl and additional markers may resolve the position of the peafowl.

Retrotransposons-based trees (Figure 1B) strongly conflict with nuclear/mtDNA trees (Figure 2) at the level of genus. For example, in the latter case, *Gallus*/*Bambusicola*/*Francolinus* forms the sister-group of *Coturnix*/*Alectoris*, while the former analyses root *Gallus*/*Bambusicola*/*Francolinus* at the base of the phasianids, and *Coturnix*/*Alectoris* formed the sister-group of gallopheasants/tragopans. Further, while peafowls cluster with *Gallus*/*Coturnix* in the mt tree, they root at the base of the phasianids following *Gallus* in the tree based on retrotransposable elements.

Retrotransposon data often consist of insertion/deletion (indel) events. Only a few phylogenetic approaches use indels as characters. Most researchers either delete them or treat the gaps as missing data. Indels cannot resolve relationships of clades branching off the focal clade–the lineage leading to the species in which the markers are originally identified–either before or after the insertion event [6]. Further, few genomes are available for identifying retrotransposon markers. These consist of the chicken and a limited number of other avian genomes. This paucity not only limits the design of conservative PCR primers for the target group, it also limits identification of an adequate number of informative characters. Conclusions based upon a few markers may be lead to inaccurate findings [17,32]. Thus, retrotransposons appear to be severely limited in their ability to resolve relationships at the hierarchical levels of genus and species, especially in cases of rapid radiations of species. This may explain the conflicting branching orders.

In conclusion, we combine mt genomes and segments of seven nuclear genes to reassess the phylogenetic relationships of phasianids. These multiple unlinked and informative genetic markers provide an updated topology. Our nuclear gene phylogeny is largely congruent with trees derived from mt genomes. However, our mt and nuclear topology largely conflict retrotransposons-based trees.

Supporting Information

Figure S1 Bayesian inference analyses of 13 mt genes that excluding the 3rd codon position.

Figure S2 Bayesian inference analyses of 13 mt genes that recoding the 3rd codon position nucleotides to two-state categories, R (purine) and Y (pyrimidine).

Figure S3 Bayesian inference analyses of 13 mt genes that recoding the 1st and the 3rd codon position nucleotides to two-state categories, R (purine) and Y (pyrimidine).

Figure S4 Bayesian inference analyses of individual mt genes and control region (CR). Each run was conducted with 5,000,000 generations and sampled every 100 generations. Bayesian Posterior Probabilities >70% were indicated on the branches. (A) *12S*, 1,036 aligned sites; (B) *16S*, 1,702 aligned sites;(C) *ATP6*, 681 aligned sites; (D) *ATP8*, 165 aligned sites; (E) *CoxI*, 1,548 aligned sites; (F) *CoxII*, 681 aligned sites; (G) *CoxIII*, 783 aligned sites; (H) CR, 1,352 aligned sites; (I) *ND1*, 972 aligned sites; (J) *ND2*, 1,038 aligned sites; (K) *ND3*, 348 aligned sites; (L) *ND4*, 1,377 aligned sites; (M) *ND4L*, 291 aligned sites; (N) *ND5*, 1,818 aligned sites; (O) *ND6*, 519 aligned sites; (P) *CytB*, 1,137 aligned sites.

Figure S5 Bayesian phylogenetic tree based on the combined dataset of 13 mt protein-coding genes and seven nuclear segments.

File S1 This file contains Table S1 and Table S2. Table S1, Source of sequence data for mitochondrial genomes and nuclear segments. Table S2, List of primers used in this study of the Phasianidae.

Author Contributions

Conceived and designed the experiments: YYS YPZ. Performed the experiments: YYS KD XC. Analyzed the data: YYS XJS. Contributed reagents/materials/analysis tools: YPZ. Wrote the paper: YYS YPZ RWM.

References

1. Armstrong MH, Braun EL, Kimball RT (2001) Phylogenetic utility of avian ovomucoid intron G: A comparison of nuclear and mitochondrial phylogenies in Galliformes. Auk 118: 799–804.
2. Crowe TM, Bowie RCK, Bloomer P, Mandiwana TG, Hedderson TAJ, et al. (2006) Phylogenetics, biogeography and classification of, and character evolution in, gamebirds (Aves: Galliformes): effects of character exclusion, data partitioning and missing data. Cladistics 22: 495–532.
3. Dimcheff DE, Drovetski SV, Mindell DP (2002) Phylogeny of Tetraoninae and other galliform birds using mitochondrial 12S and *ND2* genes. Mol Phylogenet Evol 24: 203–215.
4. Dyke GJ, Gulas BE, Crowe TM (2003) Suprageneric relationships of galliform birds (Aves, Galliformes): a cladistic analysis of morphological characters. Zool J Linn Soc 137: 227–244.
5. Fumihito A, Miyake T, Takada M, Ohno S, Kondo N (1995) The genetic link between the chinese bamboo partridge (*Bambusicola thoracica*) and the chicken and junglefowls of the genus *Gallus*. Proc Natl Acad Sci USA 92: 11053–11056.
6. Kaiser VB, van Tuinen M, Ellegren H (2007) Insertion events of CR1 retrotransposable elements elucidate the phylogenetic branching order in galliform birds. Mol Biol Evol 24: 338–347.
7. Kimball RT, Braun EL (2008) A multigene phylogeny of Galliformes supports a single origin of erectile ability in non-feathered facial traits. J Avian Biol 39: 438–445.
8. Kimball RT, Braun EL, Ligon JD (1997) Resolution of the phylogenetic position of the Congo peafowl, *Afropavo congensis*: a biogeographic and evolutionary enigma. Proc R Soc Lond B 264: 1517–1523.
9. Kimball RT, Braun EL, Ligon JD, Lucchini V, Randi E (2001) A molecular phylogeny of the peacock-pheasants (Galliformes: *Polyplectron* spp.) indicates loss and reduction of ornamental traits and display behaviours. Biol J Linn Soc 73: 187–198.
10. Kimball RT, Braun EL, Zwartjes PW, Crowe TM, Ligon JD (1999) A molecular phylogeny of the pheasants and partridges suggests that these lineages are not monophyletic. Mol Phylogenet Evol 11: 38–54.
11. Meng Y, Dai B, Ran J, Li J, Yue B (2008) Phylogenetic position of the genus *Tetraophasis* (Aves, Galliformes, Phasianidae) as inferred from mitochondrial and nuclear sequences. Biochem Syst Ecol 36: 626–637.
12. Pereira SL, Baker AJ (2006) A molecular timescale for galliform birds accounting for uncertainty in time estimates and heterogeneity of rates of DNA substitutions across lineages and sites. Mol Phylogenet Evol 38: 499–509.
13. Randi E, Lucchini V, Armijo-Prewitt T, Kimball RT, Braun EL, et al. (2000) Mitochondrial DNA phylogeny and speciation in the tragopans. Auk 117: 1003–1015.
14. van Tuinen M, Hedges SB (2004) The effect of external and internal fossil calibrations on the avian evolutionary timescale. J Paleont 78: 45–50.
15. Kan XZ, Yang JK, Li XF, Chen L, Lei ZP, et al. (2010) Phylogeny of major lineages of galliform birds (Aves: Galliformes) based on complete mitochondrial genomes. Genet Mol Res 9: 1625–1633.
16. Bonilla AJ, Braun EL, Kimball RT (2010) Comparative molecular evolution and phylogenetic utility of 3′-UTRs and introns in Galliformes. Mol Phylogenet Evol 56: 536–542.
17. Liu Z, He L, Yuan H, Yue B, Li J (2012) CR1 retroposons provide a new insight into the phylogeny of Phasianidae species (Aves: Galliformes). Gene 502: 125–132.
18. Smith EJ, Shi L, Tu ZJ (2005) *Gallus gallus* aggrecan gene-based phylogenetic analysis of selected avian taxonomic groups. Genetica 124: 23–32.
19. Shen YY, Liang L, Sun YB, Yue BS, Yang XJ, et al. (2010) A mitogenomic perspective on the ancient, rapid radiation in the Galliformes with an emphasis on the Phasianidae. BMC Evol Biol 10: 132.
20. Kriegs JO, Matzke A, Churakov G, Kuritzin A, Mayr G, et al. (2007) Waves of genomic hitchhikers shed light on the evolution of gamebirds (Aves: Galliformes). BMC Evol Biol 7: 190–200.
21. Sambrook J, Fritsch EF, Maniatis T (1989) Molecular cloning: a laboratory mannual. New York: Cold Spring Harbor Laboratory Press.
22. Thompson JD, Gibson TJ, Plewniak F, Jeanmougin F, Higgins DG (1997) The Clustal X windows interface: flexible strategies for multiple sequence alignment aided by quality analysis tools. Versión 1.8. Nucleic Acids Res 24: 4876–4882.
23. Swofford DL (2003) PAUP* 4.06. 10: Phylogenetic analysis using parsimony. Sunderland (MA): Sinauer Associates.
24. Posada D, Buckley TR (2004) Model selection and model averaging in phylogenetics: advantages of akaike information criterion and Bayesian approaches over likelihood ratio tests. Syst Biol 53: 793–808.
25. Stamatakis A (2006) RAxML-VI-HPC: maximum likelihood-based phylogenetic analyses with thousands of taxa and mixed models. Bioinformatics 22: 2688–2690.
26. Guindon S, Gascuel O (2003) A simple, fast, and accurate algorithm to estimate large phylogenies by maximum likelihood. Syst Biol 52: 696–704.
27. Ronquist F, Huelsenbeck JP (2003) MrBayes 3: Bayesian phylogenetic inference under mixed models. Bioinformatics 19: 1572–1574.
28. Johnsgard PA (1986) The pheasants of the world: Oxford University Press New York.
29. Zhang DX, Hewitt GM (2003) Nuclear DNA analyses in genetic studies of populations: practice, problems and prospects. Mol Ecol 12: 563–584.
30. Ballard JWO, Whitlock MC (2004) The incomplete natural history of mitochondria. Mol Ecol 13: 729–744.
31. Chan KMA, Levin SA, Gavrilets S (2005) Leaky prezygotic isolation and porous genomes: rapid introgression of maternally inherited DNA. Evolution 59: 720–729.
32. Han KL, Braun EL, Kimball RT, Reddy S, Bowie RCK, et al. (2011) Are transposable element insertions homoplasy free?: An examination using the avian tree of life. Syst Biol 60: 375–386.

9

Cross-Family Translational Genomics of Abiotic Stress-Responsive Genes between Arabidopsis and *Medicago truncatula*

Daejin Hyung[1]☯, Chaeyoung Lee[2]☯, Jin-Hyun Kim[2], Dongwoon Yoo[3], Young-Su Seo[4], Soon-Chun Jeong[5], Jai-Heon Lee[3], Youngsoo Chung[3], Ki-Hong Jung[6], Douglas R. Cook[7], Hong-kyu Choi[3]*

1 Department of Computer Science, Dong-A University, Busan, Republic of Korea, **2** Department of Medical Bioscience, Dong-A University, Busan, Republic of Korea, **3** Department of Genetic Engineering, Dong-A University, Busan, Republic of Korea, **4** Department of Microbiology, Busan National University, Busan, Republic of Korea, **5** Bio-Evaluation Center, Korea Research Institute of Bioscience and Biotechnology, Cheongwon, Republic of Korea, **6** Department of Plant Molecular Systems Biotechnology & Graduate School of Biotechnology, Kyunghee University, Yongin, Republic of Korea, **7** Department of Plant Pathology, University of California Davis, Davis, California, United States of America

Abstract

Cross-species translation of genomic information may play a pivotal role in applying biological knowledge gained from relatively simple model system to other less studied, but related, genomes. The information of abiotic stress (ABS)-responsive genes in Arabidopsis was identified and translated into the legume model system, *Medicago truncatula*. Various data resources, such as TAIR/AtGI DB, expression profiles and literatures, were used to build a genome-wide list of ABS genes. tBlastX/BlastP similarity search tools and manual inspection of alignments were used to identify orthologous genes between the two genomes. A total of 1,377 genes were finally collected and classified into 18 functional criteria of gene ontology (GO). The data analysis according to the expression cues showed that there was substantial level of interaction among three major types (i.e., drought, salinity and cold stress) of abiotic stresses. In an attempt to translate the ABS genes between these two species, genomic locations for each gene were mapped using an in-house-developed comparative analysis platform. The comparative analysis revealed that fragmental colinearity, represented by only 37 synteny blocks, existed between Arabidopsis and *M. truncatula*. Based on the combination of E-value and alignment remarks, estimated translation rate was 60.2% for this cross-family translation. As a prelude of the functional comparative genomic approaches, in-silico gene network/interactome analyses were conducted to predict key components in the ABS responses, and one of the sub-networks was integrated with corresponding comparative map. The results demonstrated that core members of the sub-network were well aligned with previously reported ABS regulatory networks. Taken together, the results indicate that network-based integrative approaches of comparative and functional genomics are important to interpret and translate genomic information for complex traits such as abiotic stresses.

Editor: Valerie de Crécy-Lagard, University of Florida, United States of America

Funding: This work was conducted within the framework of the Next-Generation BioGreen21 Program (PJ009089, PJ008124), Rural Development Administration and the Basic Science Research Program through the National Research Foundation of Korea (NRF), funded by the Ministry of Education, Science and Technology (NRF-2010-0023585), Republic of Korea. The funders had no role in study design, data collection and analysis, decision to publish, or preparation of manuscript.

Competing Interests: The authors have declared that no competing interests exist.

* E-mail: hkchoi@dau.ac.kr

☯ These authors contributed equally to this work.

Introduction

Model systems may play a central role in translating genomic information from one species to another. A key rationale of sequencing the model species with enormous expenses is that knowledge gained from relatively small genome can be transferred to a related, but commonly larger and more complex, species and even across different families. Such cross-species translation is largely based on precisely identifying orthologous genes with shared evolutionary origins.

In plants, Arabidopsis is the first species whose genome was fully sequenced and contains perhaps the most comprehensive context of genomic information than any other model systems, and thereby can serve as the central model system for the translational genomics approaches. Comparative genomics should be the most prevalent analysis tool used for purposes of translating genomic information across different species, and works based upon the assumption that genomic contents and orders are, to some extent, conserved among different, but related, species despite of various levels of evolutionary changes. This assumption has been validated many times with respect to conservation of genomic structures and gene functions between orthologous loci. Naturally, the ability to transfer genomic information between different species depends on evolutionary distances and the nature of genomic change in particular regions. In the past, the translation of genomic information had, with limited capability, been conducted using common RFLP probes [1] and EST-based gene markers [2,3]. Presently, it is becoming more direct, but even greater complex, due to strikingly rapid advancement of sequencing technologies (i.e., next generation sequencing or NGS) and resulting availability of genomic information. This technological advance leads the genomic science to a situation where the capability of bioinformatic analysis is increasingly emphasized.

Since the TAIR project (http://www.arabidopsis.org) first started in 1999, Arabidopsis has been playing a pivotal role, largely due to its well curated genomic information and functional annotation, in translational genomic studies within the Brassicaceae, including a broad range of related crop species such as *Brassica oleracea* (C genome, n = 9) [4], *B. rapa* (A genome, n = 10) and amphidiploid *B. napus* (AC genome, n = 19) [5]. Arabidopsis genomic information was also used to compare its genomic structure with other fully sequenced eudicot species including grape [6], poplar and papaya [7,8], and even with far more distant monocot species, such as rice [9] and maize [10]. It was shown that Arabidopsis genome had only scant synteny with relatively distant monocot species [9,10]. However, coexpression-based comparative network analysis of orthologous genes predicted between Arabidopsis and rice showed considerable level (approximately 77%) of conserved coexpression patterns [11], suggesting that cross-species translation of conserved functional genomic context would be more meaningful rather than simple structural genomic comparison.

The legume family includes a broad span of species, with roughly 20,000 species and 700 genera [12]. The family consists largely of three subfamilies: the Papilionoideae, with approximately 70% of species, which is a relatively recent-diverged group and includes most crop legumes; the Mimosoideae, with some 15%; and the Caesalpionoideae, known to contain the most ancient legume taxa, with the remainder. The potential of translational genomic analyses in the legume family and its application for the crop improvement may be particularly promising than any other plant groups, because well-defined model systems, such as *Medicago truncatula* and soybean, and a large number of domesticated legume species are available. More importantly, whole genome sequencing of five legume species, including *M. truncatula* [13], *Lotus japonicus* [14], *Glycine max* [15], *Phaseolus vulgaris* (unpublished, genome data available at Phytozome; www.phytozome.net) and *Cajanus cajan* [16], has recently been completed. At the dawn of the whole genome sequencing completion of these legume species, translational genomic approaches were suggested for the purposes of DNA marker development, positional cloning followed by gene discovery, and other omics studies with ultimate goals of legume crop improvement [17,18]. As an example of the application, 5460 unigenes, which were identified as orthologs in *M. truncatula*, *L. japonicas* and soybean, were newly mapped on Pea (*Pisum sativum*) genome, and hence leading to comparative analysis of these four legume genomes [19].

Of these five legume species, *M. truncatula* should be a nodal legume model system that may have a potential as the translational interface to accelerate crop improvements in legumes, because of its evolutionary proximity and frequently found genome conservation with other economically important legumes [17]. Nevertheless, a functional genomic translation has yet not been conducted between these two representative model systems, Arabidopsis and *M. truncatula*. Therefore, as the first step toward cross-family translation of genomic information, it is expected that *M. truncatula*, due to its relatively better established genomic data than any other sequenced legume species and its diploid nature, will make it anchoring resource to translate genomic information from Arabidopsis, which is the representative model system of the Brassicaceae, into the legumes.

These two families, the Brassicaceae and Fabaceae, in which contain significant portion of agro-economically important crops, are known to have diverged very long time ago, approximately 125-136 million years ago (MYA) [20], and presumably have experienced diverse scenario of evolutionary events (Figure 1). In addition to the ancient divergence time, even for model species with relatively small genomes, cross-species translation of genomic information is not malleable largely due to the complexity of genomic structures accumulated throughout the evolutionary history of given species.

In this study, we attempted to identify abiotic stress (ABS)-responsive genes, mainly using Arabidopsis as a central species for the translation of genomic information, and to transfer the genomic information into *M. truncatula*, as a anchoring model system within the legume family. Here, we report corresponding results, including genome-wide collection of ABS-related genes, estimation of cross-family translation rate, identification of key components in ABS response and genomic locus translation through comparative analysis. The main goal of this study is to provide researchers with basic information on genome-widely collected ABS-related genes towards the improvement of crop quality in ABS tolerance in the future.

Materials and Methods

Genome-wide screening of ABS-related genes and identification of cross-family orthologous genes

To search for the ABS-responsive genes, we employed different types of available data resources including literature, gene expression profiles, Arabidopsis database (TAIR at http://www.arabidopsis.org) and Arabidopsis Gene Index (AtGI; http://compbio.dfci.harvard. edu/tgi/cgi-bin/tgi/gimain.pl?gudb = arab). Genes associated with three major and the most important stresses (drought, salinity and cold stress) were screened in this study. In the events Arabidopsis DBs were used for the search, different key words relevant to the abiotic stresses were used to search ABS genes, and each of gene descriptions were investigated to confirm its functional relevance to the abiotic stresses. The genes collected from multiple data resources were combined into a single Excell worksheet and organized, basically, according to their locus IDs with other related information, such as annotations, expression cues, references and other related information (see Table S1).

For the first step of the translational analysis, the whole genome information for two model species, TAIR ver10 (http://www.arabidopsis.org) and Mt ver3.5 (*M. truncatula* HapMap Project; http://www.medicagohapmap.org/genome), was downloaded into our own server system. To identify orthologous genes between *M. truncatula* and Arabidopsis, we used in-house batch tBlastX/BlastP homolog search tools with the selected ABS-responsive gene set. Final decision on the orthology of pair-wise aligned genes was carefully made by taking integrated knowledge into consideration of the E-values, percent identities, cumulated total length of alignment and finally visual inspection for the quality of individual alignments.

Genomic locus translation and comparative mapping

Comparative genomic mapping was conducted with the orthologous genes predicted between Arabidopsis and *M. truncatula*. Genomic locus positioning of each gene was made by using fully sequenced genomic information of Arabidopsis and *M. truncatula*. For purposes of the comparative analysis, we developed an in-house platform for the comparative mapping. Accordingly, ABS gene database was constructed to store and use the ABS gene information to produce the corresponding comparative maps with

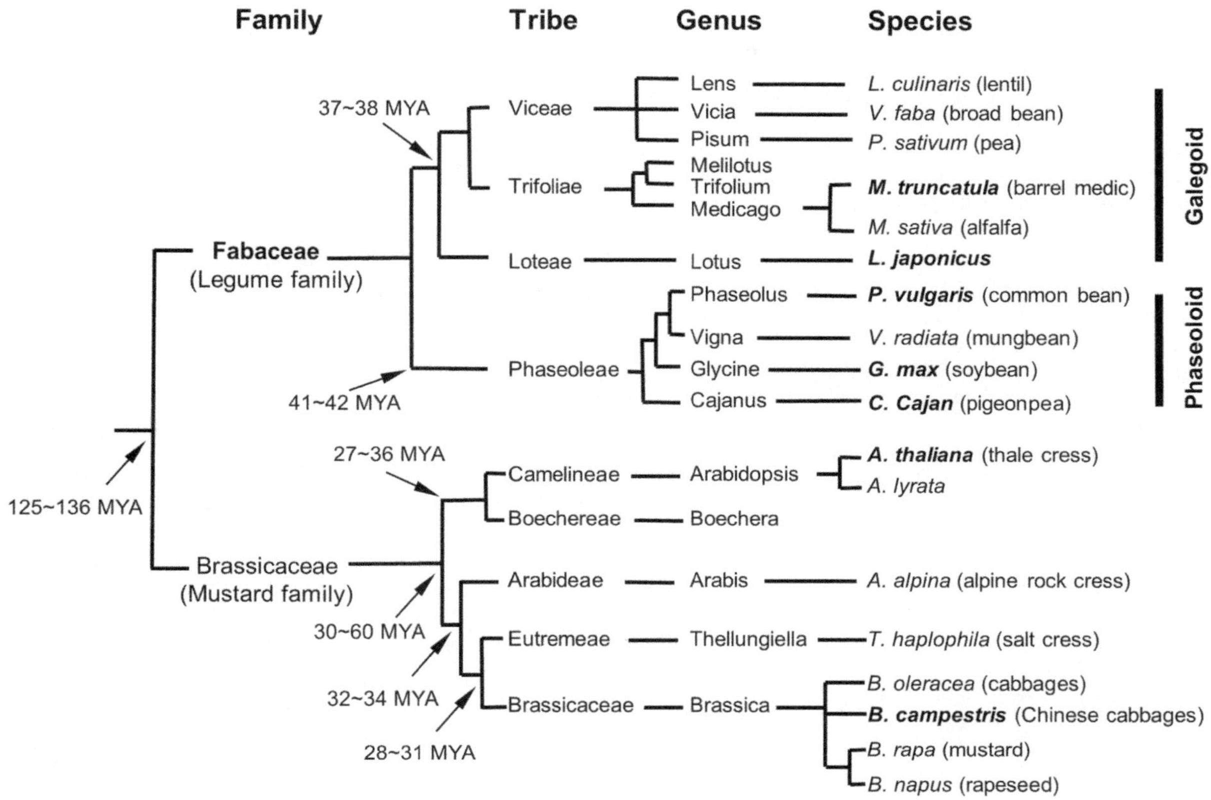

Figure 1. Taxonomic relationship of major crop and model species within two families, the Fabaceae and Brassicaceae. Whole genome-sequenced species are highlighted in bold italic. Most crop legumes occur in galegoid and phaseoloid clades. Known divergence time is denoted by arrows. MYA, million years ago.

selected options, for example chromosome number selection and gene ontology criteria, for the analysis (the manuscript for the DB and analysis tool development will be prepared separately). Predicted orthologs were interlinked with other relevant information, such as functional description of genes, locus ID and expression cues, within the tentative orthologous gene database (TOGDB). To properly describe functional aspects of entire gene set, eighteen different gene ontology (GO) criteria were selected by considering appropriate level of GO hierarchies.

Synteny blocks were determined by following criteria: (1) each block must contain at least more than four consecutively collinear genes, (2) these collinear genes within the synteny block should have consistently increasing or decreasing locus IDs, (3) distance between neighboring genes should be less than 2Mb. Finally, the synteny blocks were determined by re-examining the orthologous relationships of genes in each block.

Gene network and interactome analysis

To extract functional meaning of ABS-responsive genes and to connect this information with comparative genomic data, we employed a combination of tools and database including AraNet (http://www.functionalnet.org/aranet) [21] and CytoScape [22]. General gene network analysis pipeline was as follows. First, we used the web-based AraNet for the gene network analysis according to eighteen GO functional groups, and each of which gave ROC (Receiver Operating Characteristic) curve. Based on the ROC analyses, 12 GO criteria were further analyzed to select seed gene set. Initially, 125 seed genes were selected based on the evidence code (http://functionalnet.org/aranet/evidence_code.txt) and they were further analyzed against remaining GO groups, resulting in a total of 240 seed genes. The 240 gene set was further used to obtain additional protein-to-protein interaction (PPI) network data by integrating published PPI information [23,24]. After removing overlapped portion of these two data sets, integrated PPI network was constructed, which was composed of 2,406 protein nodes and 6325 direct physical interactions. Sub-networks predicted to be associated with the abiotic stresses were further analyzed and selected from the integrated network by extracting gene and PPI networks shared between these two datasets. We further elaborated the selected sub-networks using the Cytoscape program and combined them with the information on seed genes and their corresponding phenotypes.

Results

Collection and identification of abiotic stress-responsive genes

To translate ABS-responsive genes across legume species, we used Arabidopsis as the central anchor model system, because of its richest contents of gene information among other plant model systems, to connect corresponding information with *M. truncatula*, the representative model legume species. In addition, dought stress-responsive genes identified by Affymetrix DNA microarray experiment in Grapevine (*Vitis vinifera*) [25] were translated into Arabidopsis orthologous genes, with the intent of initially centralizing all corresponding information into a single model system. TAIR database and AtGI were used as major resources for identification of ABS-related genes. Promoter-targeted data were also included [26]. As a result, a total of 1377 genes (1.4K gene set) associated with abiotic stresses were finally collected (Table S1) and classified according to the gene ontology (GO) biological process criteria. Eighteen GO terms were carefully selected to best describe the functional criteria of collected gene set, including cell homeostasis, regulation of transcription, signal transduction and so forth (Figure 2). Of these GO criteria, it is noticeable that stress/defense-responsive genes (GO6) took a relatively large portion, approximately 13.4%, of entire GO-based classification of the genes. In addition, transcription factors (GO3, 196 genes) and signal transduction-involved proteins (GO4, 110 genes), whose functions generally act at higher level of regulation in cellular process and may be collectively called transcriptional regulator proteins, occupied 22.2% in total, which is a largest portion amongst all the GO groups. This group of genes may act as central modulators in adaptation under ever-changing environmental conditions and the result indicates that many of them should significantly be involved in responses to the abiotic stresses. The MYB family proteins (55 genes) predominated the GO3 followed by zinc finger (31 genes) and NAC (NAM/ATAF/CUC, 14 genes) transcription factor family proteins (Table S1 & S2). Many of them contained genes previously well-known as responsive to abiotic stresses, of which, for example, included DREB/CBF (dehydration-responsive element-binding/C-repeat and DRE binding factor, 13 genes), AREB or ABF/ABI (ABA-responsive element-binding/ABA-insensitive protein, 5 genes), ANAC019 (At1g52890; positive regulator of ABA signaling) [27], ICE1 (At3g26744; inducer of CBF expression 1) [28] and many other genes (Table S1).

To see how much portion of genes operate within the interface of different types of abiotic stresses or share their pathways, the 1.4K gene set was analyzed according to their expression cues (Figure 3). Out of three representative abiotic stresses, drought assumed the largest portion with a total of 811 genes followed by salt (434 genes) and cold stresses (300 genes). 128 Genes in the miscellaneous category include some indistinct, but potentially implicated with ABS response, descriptions for the expression cues to be classified into any of three criteria, such as "ABA-responsive" and "ROS-induced", implying that they would be reclassified into any distinct groups in the future. It is noteworthy that a total of 185 genes appeared to respond against more than two different abiotic stresses. Of these genes, 112 genes seemed to respond in all the three different types of stresses, indicating that those genes and/or proteins involved in ABS responses share their pathways in certain interactive manner.

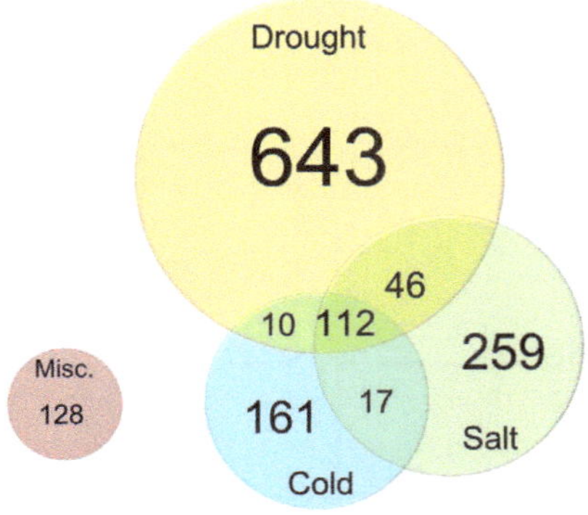

Figure 3. Distribution of ABS-responsive genes among three representative abiotic stresses, drought, cold and salinity. Expression cues for individual genes were determined by experimental platforms of relevant literature and the description of corresponding genes in the Arabidopsis database. Genes in the miscellaneous category usually had broad or ambiguous functional descriptions, such as abiotic stresses, ROS-responsive and ABA-induced.

A total of twenty resources were employed to make the genome-wide collection of ABS-related genes, including 2 databases and 18 research articles (Table S1). Within the 1.4K gene set, 188 genes were hit by more than two references, which may implicate their functional involvement in ABS responses. These genes were further organized according to their GO (Figure S1). Interestingly, genes belonging to three GO classes (i.e., GO3/4/6), collectively called proteins associated with expressional regulation (GO3 and GO4) in response to abiotic stresses (GO6), occupied almost the half portion (86 genes, 45.8%) of the multi-hit genes (Figure S1, Table S3), which provide a major portion, but not all, of genes involved in both upstream and downstream of the ABS-responsive gene networks. This gene set appears to contain many functionally important proteins, for example RD29A/LTI78 (At5g52310;

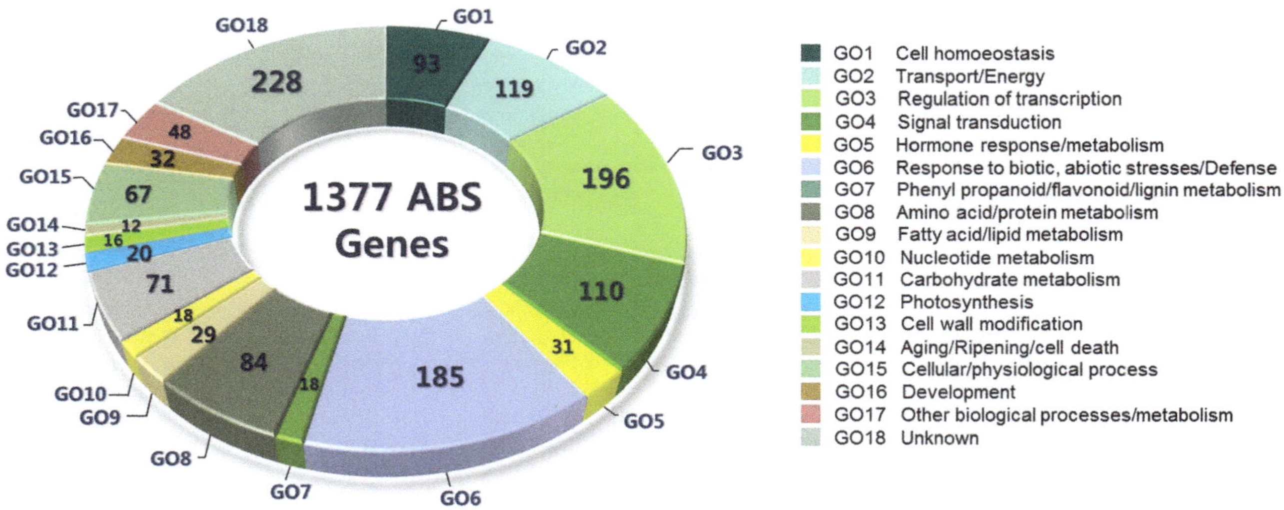

Figure 2. Classification of ABS-related genes according to 18 biological process gene ontology (GO) criteria.

responsive to dehydration/low temperature-induced; hit by 10 references), RD26/ANAC072 (At4g27410; hit by 8 references), RD20 (At2g33380), COR15a and COR15b (At2g42540 and At2g42530, respectively; cold-regulated proteins) and many others. Of those 86 genes, 22 genes were shown to be induced by all three abiotic stresses, in some cases even by pathogen infection (Table S3), reconfirming the fact, as already shown in the Figure 3, that drought/cold/salt stresses dynamically interact with each other and potentially even with pathogenic processes.

Identification of cross-family orthologous genes and translational efficiency of genomic information

Accuracy of identifying orthologs within the evolutionary context is crucial in translating genomic information between different species. For purposes of cross-family transfer of genomic information, we used a combination of parameters and tools by taking an integrative consideration of E-value and bit-score (i.e., a total length of corresponding alignment) of the tBlastX/BlastP-derived alignments. Final decision for the cross-species orthology was carefully made by scrutinizing resulting alignments along with investigator's comments. Firstly, we divided the translated genes according to the E-values, because it is a fundamental parameter of the first priority to be considered for the orthologous gene prediction, with regular intervals (Figure 4A). Of these, 352 genes (25.6%) with the value of $E = 0$ took the largest portion followed by 307 genes with $0<E<E^{-100}$, 123 genes with $E^{-100}<E<E^{-80}$ and 150 genes with $E^{-80}<E<E^{-60}$ of E value intervals. Taken together, a total of 932 genes (67.7%), although other genes above these E value intervals were not necessarily excluded from the ortholog criteria, showed reasonably significant E value-based prediction of orthology. Occasionally, E value-based identification of orthologs, in some sense, by an automatic prediction through bioinformatics data processing might be prone to errors mainly due to the complexity of genome, especially such as protein families resulting from gene duplications and homeologous chromosomes by polyploidization. Thus, we employed an additional means to ensure more accurate identification of orthologs by remarking and/or giving comments on actual tBlastX alignment outcomes (see legend of Figure 4 for the detailed information of remarks). The remarks fall into nine criteria (Figure 4). With the alignment remarks, a total of 829 genes (60.2%) were predicted as tentative orthologs above the level of FS (fairly specific) homolog search criterion (data not shown). To more accurately confirm the orthology of homologous candidates, genes with the value of $<E^{-30}$ and remarks of UC (uncertain)/NA (not available) were removed and reanalyzed according to the alignment remarks, thereby resulting in a total of 1134 genes (82.4% of 1.4K gene set; Figure 4B). Of these, a total of 797 genes (70.3% of 1134 genes; 57.9% of 1377 genes) were determined potentially to be orthologous between these two model systems, which seems reasonable for the cross-family prediction of orthologous genes. In addition, it is noteworthy that the removal of genes with less significant E-values substantially contributed to increasing the probability, which increased from 60.2% to 70.3%, of finding orthologous counterparts between these two species (Figure 4B). These analyses also showed that the removal of genes with the E-values of low significance did not greatly affected the orthology prediction of genes above FS level of the remarks, by which only 32 genes (3.8% of 829 genes) were excluded, while below this remark level, 78 genes (18.8% of 415 genes without UC/NA criteria) were removed (Figure 4B). This result indicates that proper use of the E-value criteria is obviously useful for the translation of genomic information, but it has limitations to the accuracy in identifying orthologous genes. Therefore, it is recommended that a combination of approaches, including the E-value, total length of alignment and up-close investigation of individual alignments, are necessary to ensure cross-species translation of genomic information. The calculated GO-based translational efficiencies appeared various according to different GOs, ranging approximately from 60% (GO 12 & 16) to 96% (GO11) (data not shown). However, it looked obvious that low translation rates were always seen in the case where the gene numbers of corresponding GO were low (for example GO12/16), which may be more sensitively affected by counting predicted orthologs. Taken together, this implicates that translation of genomic information is not significantly affected by different functional criteria of genes.

Genomic positional translation and integration of the comparative map with a gene sub-network

Comparative genomic analysis is an essential and efficient tool for translating genomic contexts of one species into other different, but taxonomically related, species. Comparative translation of genomic information is especially useful to save the research resources, not only because it enables to gain insights into genome-wide relationships among multiple genomes under comparison, but also because the translated information can be effectively applied to other organisms of researcher's interest. For purposes of interconnecting genomic information between Arabidopsis and *M. truncatula*, which represent two plant families of agricultural importance, loci of each predicted orthologous genes were positioned, based on the fully sequenced genomic locus information, in both model genomes. An in-house user-friendly platform for the comparative analysis was developed (relevant manuscript being prepared separately) and used to produce the comparative maps (Figure 5 and Figure S2).

Each of inferred orthologous gene loci were translated between the two genomes and analyzed for syntenic relationships. To avoid excessive complexity of resulting comparative maps, we conducted the analyses by translating locus information from single chromosome of one species to entire chromosomes of another compared species. Figure S2 A&B demonstrate Arabidopsis-centered comparison and Medicago-centered analysis, respectively. With 1377 gene loci, only 4.2% of 33,198 Arabidopsis annotated gene models [7], associated potentially with the ABS responses, we couldn't find extensive synteny blocks between these two species. The comparative analysis revealed only fragmental colinearity with a total of 37 synteny blocks (Figure S2 C, Table S4). The segmental syntenic relationship could readily be expected because two plant families of the Fabaceae and Brassicaceae, to which two model systems belong, diverged at the ancient time of approximately 125~136 MYA (Figure 1) and consequently their genomes should have experienced a diverse array of evolutionary events. The numbers of synteny blocks on each Arabidopsis and *M. truncatula* chromosome ranged from 5 to 9 and from 2 to 9, respectively (Figure S2 A&B), while no synteny blocks were found in MtChr6 (Figure S2B). To see how genomic locations of ABS genes were correlated with their functional specialization, we counted ABS genes in each chromosome according to different GO criteria (Figure 5 and Table S5). Based on this analysis, although the data is not fully supportive, it seems, in general, unlikely that there is a significant correlation between the conserved synteny and functional organization of ABS-responsive genes within certain genomic regions. However, we could observe that relatively higher number of ABS genes resided in AtChr 1&5 and MtChr 4&5, which might implicate these genomic regions were relatively better equipped with ABS responses. Of these four chromosomes, interestingly, MtChr5 was most populated with ABS genes and

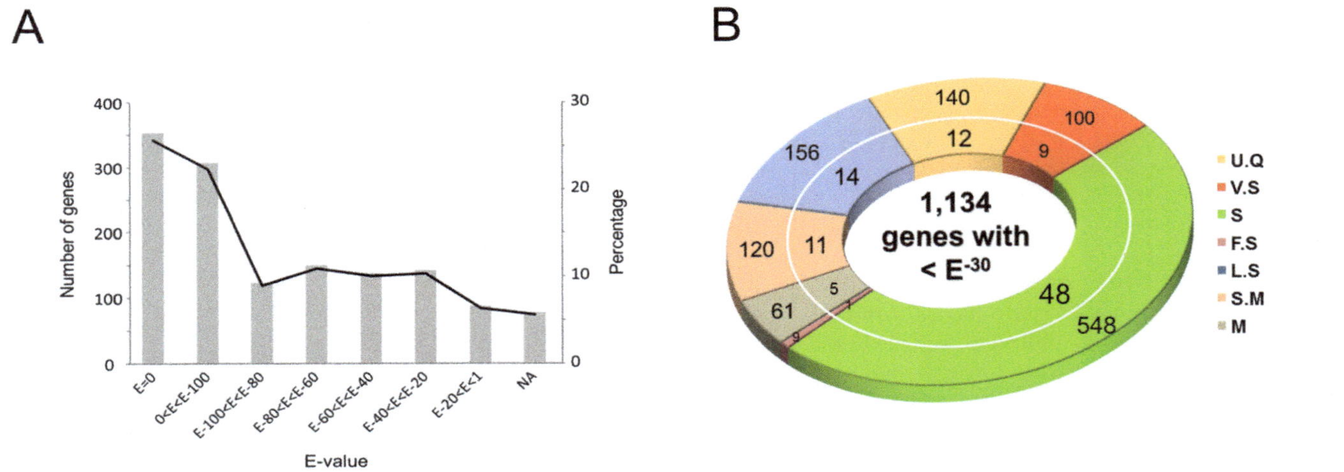

Figure 4. Cross-family translational efficiency of ABS-responsive gens between Arabidopsis and *M. truncatula*. A. Prediction of cross-species orthologous genes according to the Expect values. Bars indicate the number of genes, while the solid lines the percentage of translated genes. B. Distribution of translated genes with the E-value of $<E^{-30}$ according to visual inspection of the tBlastX alignments: UQ, unique; VS, very specific; S, specific; FS, fairly specific; LS, less specific; SM, specific but multiple homologs; M, multiple homologs. The numbers in outer circle denote the number of genes, while the numbers of inner circle denote corresponding percentage.

showed the highest counts of ABS genes in the GO3/4/6 (Table S5), which are relatively more important functional criteria in the aspect of ABS responsiveness. On the other hand, it is interesting to note that any single synteny block was not found in the MtChr6, in which 52 ABS genes, relatively smaller number than in other chromosomes, were contained. This observation seems consistent with previous reports on comparative analysis done with limited number of genetic markers [29] and recent *M. truncatula* whole genome analysis [13] that MtChr6 is rich in heterochromatic DNA and relatively poor in expressed genes [12]. Furthermore, it has been known that the majority of NBS-LRR resistance analogous genes were localized as clusters in the MtChr6 [30], indicating that the MtChr6 is functionally specialized for the resistance mechanism against pathogen infections. Likewise, functionally specialized genomic regions for the ABS responses might exist and remain to be further elucidated with more extended data in the future.

A total of eight tandem duplication, two in Arabidopsis (ERD6 and glutathione S-transferase genes) and six in *M. truncatula* (ubiquitin-conjugating enzyme, sarcosine oxidase and six more), were found within the 37 synteny blocks (for detail see Table S4). On the other hand, genome-wide analyses of ABS genes resulted in identification of 49 tandemly duplicated genomic sites. Of these, 24 sites were tandemly duplicated only in Arabidopsis, 13 only in *M. truncatula* and 8 in both genomes (Table S6). It appears that genes tandemly duplicated in both genomes include mainly effector proteins by which exert their functions to recover the balance in homeostasis, with sufficient copy numbers, against the stresses. For example, these included glutathione S-transferase, plasma membrane intrinsic proteins (RD28), cytochrome P450 monooxygenase and many others (for more detail see Table S6). Interestingly, five copies of Na^+/H^+-exchanger, which is an important protein in regulating osmotic homeostasis, are shown to be tandemely duplicated in both MtChr4 and MtChr5.

Function of genes involved in certain biological responses and/or metabolisms can be better triangulated when component genes are considered within the context of functional gene networks. Subsequently, as a prelude of the comparative functional genomic analysis, we attempted to integrate one of gene sub-networks, in which all interactions among genes were supported by physical protein-to-protein interactions [23], with corresponding comparative map (Figure 6).

The sub-network presented in Figure 6 elucidates a central ABA-mediated pathway in response to abiotic stresses, in which harbors mostly known core components of the pathway from the upstream, for example soluble ABA receptors (PYR/PYL/RCAR: pyrabactin resistance/PYR-like protein/regulatory component of ABA receptor), to the downstream, for example, two potassium channels (KT1 & KT2-3). Also, the network demonstrates correlations and/or connections between these components, including a series of cellular processes occurring upon stress such as perception, signal transduction, transcriptional regulation and expression of effector genes. The sub-network is composed of a total of 36 genes and can be further classified three groups: (i) perception and signaling-related sub-network; three ABA receptors, two SnRK2s (subfamily 2 snf1-related kinase), five PP2Cs (type 2C protein phosphatase); (ii) calcium-mediated signal transduction; one CaM (calmodulin), two CBLs (calcineurin B-like proteins) and nine CIPKs (CBL-interacting serine/threonine protein kinase); (iii) transcriptional regulation; six ABRE/ABF bZIP transcription factors and three AFPs (ABI five binding proteins, also known as Ninja-family protein, by which act as a negative regulator of bZIP TFs). In addition, five downstream effector genes are connected with the network: two potassium channels (KT1 & KT2-3), catalase (CAT2), H^+-ATPase (HA2) and 1-amiocyclopropane-1-carboxylase synthase (ACS6; involved in ethylene synthesis) (Table 1, Table S7).

There have recently been two important breakthroughs in ABA signaling model: (i) the discovery of soluble ABA receptor [31,32]; (ii) the identification of PP2C-SnRK complex, negative and positive regulators of ABA receptor respectively, as a downstream component of ABA receptor complex [33,34]. Due to these two important findings in 2009, ABA signaling model has been dramatically figured out that the PYR/PYL/RCAR-PP2C-SnRK2 complex plays the central role in response to abiotic stresses, such as drought and salinity [35]. In Arabidopsis, a total of 14 PYR/PYL/RCAR ABA receptor homologs have been identified [35]. Of these, three PYR/PYL/RCAR proteins

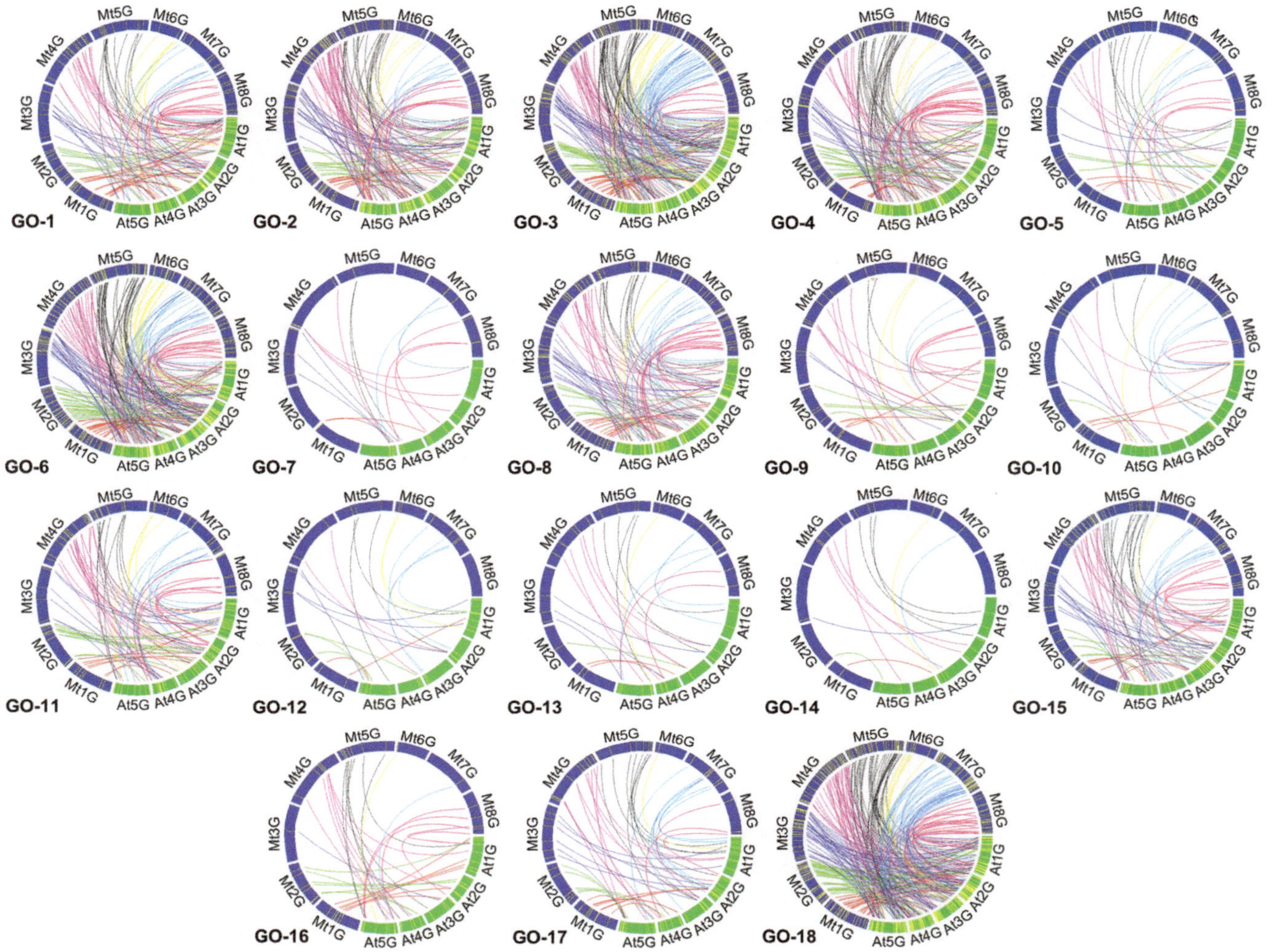

Figure 5. Cross-family locus translation of ABS-responsive genes through comparative analyses according to gene ontologies. Eight different colors, for *M. truncatula* chromosomes, were used to distinguish different chromosomal locations; Mt1G-red, Mt2G-green, Mt3G-deep blue, Mt4G-pale pink, Mt5G-black, Mt6G-yellow, Mt7G-sky blue, Mt8G-deep pink. At, *A. thaliana*; Mt, *M. truncatula*.

(At2g40330, At2g38310, At4g17870) are involved as members of the sub-network, in which the ABA receptors are shown to interact with three PP2Cs (At1g72770; known as HAB1 for homolog to ABI1, At3g11410, At4g26080; known as ABI1) and two SnRKs (At3g50500, At4g33950; also known as OST1 for open stomata 1) (Figure 6). This tripartite ABA signaling core complex is shown to associate with the calcium-mediated regulation of gene expression. With regard to this, CBL and CIPK interactions play a major role in the downstream signaling. In Arabidopsis, 10 CBLs and 26 CIPKs constitute a complicated interaction network with some extent of functional redundancy [36]. Of these, three Ca^{2+} sensors and eight CIPKs participated in the sub-network, in which a diverse array of interactions between sensors such as CBLs and CIPKs were apparent (Figure 6) and well documented in genetic studies of plant's responses to the salinity, for example SOS (salt overly sensitive) pathway [37]. The same research group reported that SOS2 (a type of CIPKs) forms a complex with SOS3 (a type of CBL proteins), and lead to activation of SOS1 (a Na^+/H^+ antiporter) important for ion homeostasis and salt tolerance. In this SOS pathway, ABI2 (a type of PP2C) act as a negative regulator by deactivating SOS2 or dephosphorylating SOS1 [38]. The network seems to fairly well represent the CBL-CIPK-PP2C interactions among genes involved in the Ca^{2+}-mediated signaling, although their modes of action, such as activation and deactivation, are not denoted. Within the network, it seems likely that the secondary signal of calcium ion is transduced mainly through CPK32, a calcium-dependent protein kinase, to AREB/ABF bZIP transcription factors. It was experimentally verified that AtCPK32 is involved in Ca^{2+}-mediated transduction of ABS signals by interacting with ABF4, a transcriptional regulator [39]. In addition to ABF4, five more transcription factors (i.e., ABF2/3, AREB3 and ABI3/5, which are all bZIP TFs except for ABI3) constitute positive regulatory part of the TF sub-network. In contrast, three negative regulators, which are named as AFP (i.e., ABI five binding proteins), take parts in the transcriptional regulatory network (Figure 6). It is currently believed that AFPs are needed to fine-tune stress responses by interacting with ABI5 and/or ABI5-related bZIP TFs [40].

On the other hand, ion transporters play crucial roles especially in maintaining ion homeostasis through osmoregulation. Among others, potassium ion is the most important and abundant cation, and subsequently its corresponding channels or transporters may

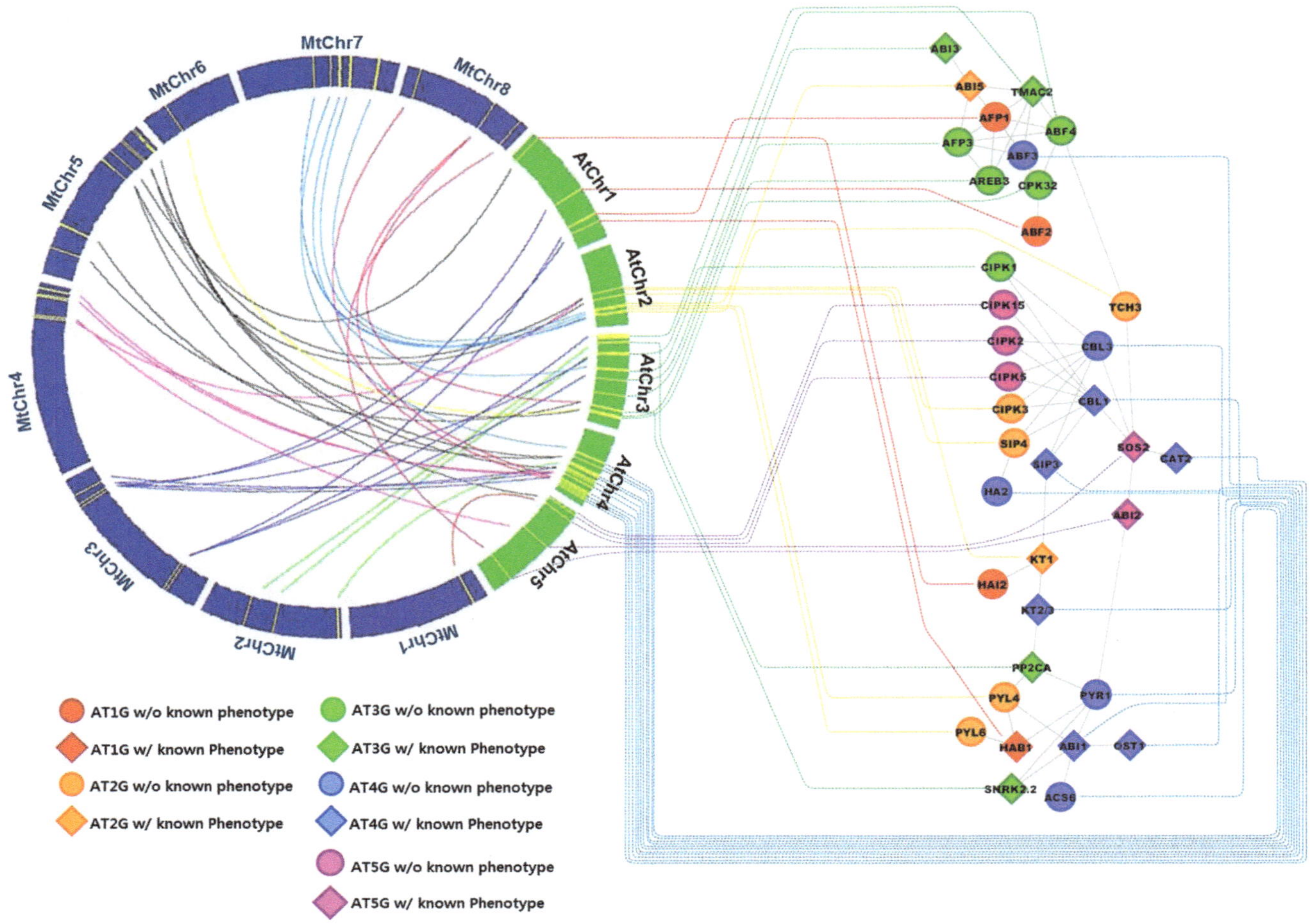

Figure 6. An example of integrative analysis for a core ABS regulatory sub-network and corresponding locus-translated comparative map. The sub-network was analyzed and reconstructed based on the Arabidopsis protein-to-protein interactome data. Availability of phenotypic information (for detail, see Table S7) and chromosome locations are denoted by a combination of different colors and shapes. Loci of individual genes within the sub-network are connected to corresponding positions on the Arabidopsis chromosomes. Detailed information of sub-network component genes is available in Table 1.

essentially affect the responses to abiotic stresses. Two potassium channels/transporters (KTs: At2g26650, At4g22200) are involved in the gene network in conjunction with the CBL-CIPK signaling pathway (Figure 6). Interestingly, it has recently been suggested that CBLs and CIPKs may compete with each other to bind with the channel and act in antagonistic manner, in which CIPK is a positive modulator and vice versa, to modulate KT's activity [41]. It has also recently been validated that HA2 (At1g 07430; a ABA-induced and seed specific PP2C, also known as HONSU) directly interacts with KT1 and regulates seed dormancy by acting as a negative modulator of ABA signaling [42].

Taken together, component genes within the chosen PPI-supported subnetwork appear to be well aligned not only with previously known functional studies, but also with interactome data [23], offering a nice case study of network-based translational genomic approaches. This type of genomic translation integrated with functional gene networks may facilitate more accurate transfer of genomic information between different species and can be applied to systematic functional studies of genes.

Discussion

Whole genome sequences and complete gene sets within the genomic contexts, which have recently become available in many plant species, can provide broad insights into how genes and genomic structures are organized and have changed during the processes of adaptation to fit ever changing environment. Therefore, it is expected that the translational genomics, if one properly utilizes such approaches, may play critical roles, based on genome-wide understanding, to leverage some key areas, such as basic plant biology and development of more efficient selection processes for molecular breeding, ultimately towards the crop improvement. Genomic responses to abiotic stress involve, in both quantitative and qualitative senses, complex traits, which are determined by interactions between multiple genes. In this study, we intended to establish and provide a genome-wide list of ABS-responsive genes for the purpose of building up a basic resource for the cross-species translational genomics, between Arabidopsis and *M. truncatula*, and further gaining insight into key components governing the ABS-responses in plants.

Validation of homologous gene's orthology is the prerequisite for the comparative genomic translation, because multi-copy

Table 1. List of genes involved in responses to abiotic stresses within the subnetwork.

	OrthologousLocusID		E-value	GeneSymbol	Annotation	Phenotype	Reference
	Arabidopsis	*M.truncatula*					
AtChr1	At1g07430	Mt5g080680	9.00E-108	HAI2	HighlyABA-inducedPP2Cgene2	NA	
	At1g45249	Mt3g101780	2.00E-56	ABF2	ABA-insensitive5-likeprotein	NA	
	At1g69260	Mt3g025970	1.00E-32	AFP1	ABIfivebindingprotein	NA	
	At1g72770	Mt3g104710	1.00E-118	HAB1	Proteinphosphatase2C16;homologytoABI1	Yes	N/A
AtChr2	At2g26650	Mt4g113530	0	KT1	PotassiumchannelAKT1	Yes	[52]
	At2g26980	Mt5g088350	0	CIPK3	CBL-interactingserine/threonine-proteinkinase3	NA	
	At2g30360	Mt5g075100	4.00E-154	SIP4	CBL-interactingserine/threonine-proteinkinase11	NA	
	At2g36270	Mt7g104480	3.00E-93	ABI5	ABAinsensitive5	Yes	[53]
	At2g38310	Mt7g070050	2.00E-73	PYL4	AbscisicacidreceptorPYL4	NA	
	At2g40330	Mt3g071740	3.00E-60	PYL6	AbscisicacidreceptorPYL6	NA	
	At2g41100	Mt7g087610	2.00E-51	TCH3	Calmodulin-likeprotein12	NA	
AtChr3	At3g02140	Mt3g025970	7.00E-07	TMAC2	Ninja-familyproteinAFP4;ABIfivebindingprotein4	Yes	[40]
	At3g11410	Mt2g066110	0	PP2CA	Proteinphosphatase2C37	Yes	N/A
	At3g17510	Mt2g045470	2.00E-159	CIPK1	CBL-interactingserine/threonine-proteinkinase1	NA	
	At3g19290	Mt3g101780	1.00E-67	ABF4	ABA-insensitive5-likeprotein7	NA	
	At3g24650	Mt7g083700	3.00E-36	ABI3	B3domain-containingtranscriptionfactorABI3	Yes	[54,55]
	At3g29575	Mt3g025970	8.00E-31	AFP3	Ninja-familyproteinAFP3	NA	
	At3g50500	Mt8g079560	0	SnRK2.2	Serine/threonine-proteinkinaseSRK2D	Yes	[56]
		Mt5g018050	0				
	At3g56850	Mt6g026800	3.00E-84	AREB3	ABA-insensitive5-likeprotein2	NA	
	At3g57530	Mt5g089320	2.00E-177	CPK32	Calcium-dependentproteinkinase32	NA	
AtChr4	At4g11280	Mt7g079080	0	ACS6	1-aminocyclopropane-1-carboxylatesynthase6	NA	
	At4g17615	Mt5g096420	1.00E-88	CBL1	CalcineurinB-likeprotein1	Yes	[57,58]
	At4g17870	Mt5g030500	2.00E-78	PYR1	AbscisicacidreceptorPYR1	NA	
	At4g22200	Mt2g006870	0	KT2.3	PotassiumchannelAKT2.3	Yes	[59]
	At4g26080	Mt3g104710	2.00E-112	ABI1	ABAinsensitive1	Yes	[60,61]
	At4g26570	Mt5g096420	2.00E-136	CBL3	CalcineurinB-likeprotein3	NA	
	At4g30190	Mt4g127710	0	HA2	H(+)-ATPase2	NA	
		Mt3g108800	0				
	At4g30960	Mt4g131060	0	SIP3	CBL-interactingserine/threonine-proteinkinase6	Yes	[62]
	At4g33950	Mt8g079560	0	OST1	Serine/threonine-proteinkinaseSRK2E	Yes	[50,63]
	At4g34000	Mt3g101780	6.00E-73	ABF3	ABA-insensitive5-likeprotein6	NA	
	At4g35090	-	-	CAT2	Catalase2	Yes	[64,65]
AtChr5	At5g01810	-	-	CIPK15	CBL-interactingserine/threonine-proteinkinase15	NA	

Table 1. Cont.

OrthologousLocusID		E-value	GeneSymbol	Annotation	Phenotype	Reference
Arabidopsis	*M.truncatula*					
At5g07070	Mt5g075060	0	CIPK2	CBL-interactingserine/threonine-proteinkinase2	NA	
At5g10930	Mt1g013700	0	CIPK5	CBL-interactingserine/threonine-proteinkinase5	NA	
At5g35410	Mt4g114670	0	SOS2	CBL-interactingserine/threonine-proteinkinase24	Yes	[66]
At5g57050	Mt8g102550	1.00E-78	ABI2	Proteinphosphatase2C77	Yes	[61]

genes, for example family proteins, may frequently lead to erroneous transfer of genomic information. However, accurate prediction of the gene's orthology between different species is not straightforward largely due to complex history of speciation and duplication. A substantial effort has been made to transfer orthologous functional annotation of genes between different genomes. For example, EnsemblCompara Gene Trees [43], specialized only for vertebrates, and InParanoid [44,45], developed mainly for eukaryotic species, employed phylogenetic approaches to resolve the complexity of ortholog identification, using TreeBeST method and two-pass BLAST, respectively. Particularly in plants, Arabidopsis played a central role in organizing functional annotation of genes in association with the Gene Ontology Consortium, and each of annotations were defined by 11 evidence codes, for typical examples ISS for "inferred from sequence similarity" and IPI for "inferred from physical interaction" [46]. The functional annotations of the gene in corresponding TAIR database (http://www.arabidopsis.org) is being continuously updated by integrating experimental data and literature-based information. Phytozome (http://www.phytozome.net) is one of the representative comparative genomics platforms for plants [47]. The database contains information of 25 completed genomic sequences and employs the common BLAST or BLAT (BLAST-like alignment tool) to perform the similarity searches. In contrast, GreenPhylDB (http://greenphyl.cirad.fr) provides comparative analysis platform specialized only for two representative plant species, Arabidopsis and rice, with the emphasis on phylogenomic approach for the family proteins [48]. On the other hand, Abrouk et al. [7] emphasized three components to ensure orthologous relationships between genes: (i) comparison of inter-specific coding sequence, (ii) similarity based on cumulative identity percentage (CIP), (iii) relative length of the sequences represented by cumulative alignment length percentage (CALP).

Although we did not develop or use an automated ortholog detection tool equipped with mathematically programmed algorithms, we tried to meet above orthology criteria, and subsequently orthologous genes were finally determined by inspecting individual alignments derived from the tBlastX/BlastP with highly trained human eyes and integrated knowledge on similarity searches. We also used our own multi-layered denotation on the sequence alignments from "unique" (for clear orthologous relationship; i.e., one-to-one relationship) to "multiple" (for multi-gene family and/or obscure orthology) (Figure 4). We imagined that the denotation would be especially useful when fully sequenced genomes, because they represented entirety of genomic context, were compared to translate the genomic information. Taken together, reliable range of translation rate was estimated approximately 60% (Figure 4) between Arabidopsis and *M. truncatula*. This translation rate is seemingly low, but probably reasonable if one considers their divergence time of 125~136 MYA (Figure 1). This estimation could be applied when researchers intend to transfer genomic information between two plant species with similar divergence time.

In nature, plants are simultaneously exposed to multiple stresses, including abiotic and biotic stresses, and the stresses often interact with each other. Although many genes known to be involved in the ABS responses have been identified and elucidated within the context of gene interaction network [49], little has, in reality, been known about how these genes are interwoven within the network and interact between different abiotic stresses. Thus it would be critical to discover key cross-stress interplayers towards full understanding of orchestrated networks composed of ABS genes. In this study, a total of 185 (13.4% of 1.4K gene set) genes were identified as potential candidates for such interplaying components

(Figure 3), although they could not be further analyzed towards the end of interactome analysis due to lack of information. However, we expect that the area or number of genes within the interacting interface of these three abiotic stresses would probably increase as more data and experimental evidences accumulate. Intriguingly, several recent reports suggest that the ABA, previously known as a central plant hormone regulating ABS responses, also play important roles in the interaction between abiotic stresses and pathogenesis [49]. One interesting example was shown that ABA was also involved in the defense response by closing stomata upon pathogen attack [50], thereby implicating that ABA plays its role against pathogen infection in more direct way than previously expected. As seen, a broader array of interactions between multiple stresses, including ABS and even biotic stresses, is becoming an emerging topic in this field of researches. Therefore, further analyses towards this direction are in great demand to shed more light on comprehensive understanding of network-based ABS responses, even with biotic stress responses.

Because ABS responses are QTL traits, extremely complex and thereby numerous genes are involved, the list of 1377 genes collected in this study seems large enough to explain full scenario of responses against abiotic stresses. However, the gene set is probably heterogeneous, because some portion of genes are perhaps be involved in general physiological responses, which means that individual genes may have differential significance in their functions. Various functions of genes commonly constitute networks within the hierarchical context at different levels on which they mainly act in the courses of reactions against given external stimuli. Thus, it is important to triangulate key genes from the central networks within which govern corresponding responses. In order for the comparatively translated genomic data to be more valuable, functional genomic data need to be integrated. For purposes of integrating these two different types of data, gene network and/or interactome analyses with the full set of 1.4K genes are currently under way (relevant manuscript will be prepared separately).

In addition to the functional aspect of the sub-network (Figure 6), we could observe some features of genomic organization of the sub-network genes. For example, we could find a couple of regions in which genes are localized in relatively closer proximity: (i) CIPK2/5/15 in terminal region of AtChr5, (ii) CPK32, AREB3 and SnRK2.2 in AtChr3, (iii) PYL4/6, TCH3 and ABI5 in AtChr2, (iv) CIPK3, SIP4 and KT1 in AtChr2 (Figure 6 & Table 1). It would be interesting to see that the AtChr4 is most populated or organized with genes from the broadest functional criteria of the gene network, which might be suggestive of structural organization, to some extent, for the functional specialization in ABS responses. However, the idea seems premature, remain elusive and need to be further assessed with more integrated data of genomic organization in conjunction with more extended gene network analyses.

As exemplified in Figure 6, the sub-network for ABA-triggered stress responses, although reconstructed only by using bioinformatic tools and/or resources, appeared to be well aligned with previously reported protein-to-protein interaction data [23,49], and may play a central role towards further expansion of the network by integrating with other pathways, for example ethylene pathway (e.g., ACS6 in the sub-network) and ROS pathway (e.g., CAT2 in the sub-network). The subnetwork-associated comparative translation of genomic information presented in this study (Figure 6) was only a small example as a prelude towards a genome scale functional comparative analysis. The 1.4K gene set-wide gene/interactome network analysis is currently underway with necessary update of data and/or information, and the results will be integrated with structural genomic data, in the near future, with an ultimate goal of gaining broader and deeper insights towards understanding the whole picture of complex interactions of genes involved in abiotic stresses. This way, one would be able to better identify key components associated with tolerance mechanisms against ABS-derived challenges and can apply the gained knowledge for the molecular breeding programs. However, species-to-species translation of genomic information may not be such a straightforward process, where the situation is more crucial for complex traits such as the abiotic stresses, that one need to take a careful caution to enhance accuracy and/or quality of the information transfer. In leguminous plants, for example, such genomic translation was tried to discover core genes responsive to salt stress, and successfully done using a combination of transcriptomic and metabolomics data with six Lotus species [51]. In this particular example, we need to notice that they utilized translational genomic approaches within only closely related model and crop species from the same genus. On the other hand, even cross eudicot-to-monocot genomic translation was conducted with 4,630 orthologous gene pairs identified between Arabidopsis and rice [11]. Corresponding results showed highly conserved coexpression pattern of 77%, consequently indicating a broad range of applicability of translational genomics, when properly applied to species of interest.

Translational genomics, from a practical point of view, seems likely still in its infancy. It still has considerable obstacles to be overcome, such as development of more accurate ortholog finding algorithms, comprehensive curation of functional genomic data and reliable integration of large scale omics data. Despite all these obstacles, it seems most likely that translational genomics should be a strong tool to overcome the barriers in translating genomic context among different species. Towards this direction, the data gained from translation of genomic information could be utilized to identify trait-associated genes, and subsequently exploit marker discovery and development for the purpose of molecular breeding.

Since ABS-related genomic information has been translated into the legume model system *M. truncatula* in this study, it would readily become possible to conduct pan-legume translation of genomic information on ABS genes across fully sequenced legume species. Until now, five legume species have been fully sequenced and their genomic information is available in public domains. This situation may provide a great opportunity for translating functional discoveries in model systems into many other legume crops, which are relatively poor in genomic information, through breeding programs and molecular biotechnology. Furthermore, similar approaches could be applied in other functional criteria, which are of great biological importance in legumes, such as symbiotic nitrogen fixation, seed maturation and physiology, and secondary metabolisms. However, the efficiency and/or predictability of the cross-species translational genomic applications remain to be further refined so that individual genes and/or gene networks would functionally work properly for the complex QTL-associated traits such as the abiotic stresses. Nevertheless, we anticipate that our study would provide rich information on ABS-responsive genes and potentially contribute to identifying key genes and developing trait-associated molecular genetic markers towards the improvement of important crop quality related to the abiotic stress tolerance.

Supporting Information

Figure S1 GO-based classification of genes hit by multiple resources. Three GO classes (GO3/4/6), which occupy largest portion and are associated with transcriptional regulation and response to abiotic stresses, are demonstrated separately using pie chart.

Figure S2 Locus translation and synteny analysis between Arabidopsis and *M. truncatula*. A. Comparative analyses of single Arabidopsis chromosome-to-*M. truncatula* whole genome. Each of five Arabidopsis chromosomes was separately, to avoid excessive complexity, analyzed with eight chromosomes of *M. truncatula*. Corresponding syntenic comparative analyses are shown side-by-side. B. Comparative analyses of single *M. truncatula* chromosome-to-Arabidopsis whole genome. Each of eight chromosomes of *M. truncatula* was separately analyzed with five Arabidopsis chromosomes. Corresponding syntenic comparative analyses are shown side-by-side. C. Integrated comparative map and synteny analysis of translated ABS-responsive gene loci.

Author Contributions

Conceived and designed the experiments: HC. Performed the experiments: DH CL JK DY. Analyzed the data: DH CL YS KJ. Wrote the paper: HC DH CL. Edited and commented the manuscript: DRC SJ JL YC.

References

1. Gale MD, Devos KM (1998) Plant comparative genetics after 10 years. Science 282: 656–659.
2. Choi H-K, Mun J-H, Kim D-J, Zhu H, Baek J-M, et al. (2004) Estimating genome conservation between crop and model legume species. Proc Natl Acad Sci USA 101: 15289–15294.
3. Choi H-K, Luckow MA, Doyle J, Cook DR (2006) Development of nuclear gene-derived molecular markers linked to legume genetic maps. Mol Gen Genom 276: 56–70.
4. Town CD, Cheung F, Maiti R, Crabtree J, Haas B, et al. (2006) Comparative genomics of Brassica oleracea and Arabidopsis thaliana reveal gene loss, fragmentation, and dispersal after polyploidy. Plant Cell 18: 1348–1359.
5. Wang J, Lydiate DJ, Parkin IAP, Falentin C, Delourme R, et al. (2011) Integration of linkage maps for the amphidiploid Brassica napus and comparative mapping with Arabidopsis and Brassica rapa. BMC Genomics 12: 101.
6. The French-Italian Public Consortium for Grapevine Genome Characterization (2007) The grapevine genome sequence suggests ancestral hexaploidization in major angiosperm phyla. Nature 449: 463–468.
7. Abrouk M, Murat F, Pont C, Messing J, Jackson S, et al. (2010) Paleogenomics of plants: synteny-based modeling of extinct ancestors. Trends Plant Sci 15: 479–487.
8. Lyons E, Pedersen B, Kane J, Alam M, Ming R, et al. (2008) Finding and comparing syntenic regions among Arabidopsis and the outgroups papaya, poplar, and grape: CoGe with Rocids. Plant Physiol 148: 1772–1781.
9. Liu H, Sachidanandam R, Stein L (2001) Comparative genomics between rice and Arabidopsis shows scant collinearity in gene order. Genome Res 11: 2020–2026.
10. Brendel V, Kurtz S, Walbot V (2002) Comparative genomics of Arabidopsis and maize: prospects and limitations. Genome Biol 3: 1005.1–1005.6.
11. Movahedi S, Van de Peer Y, Vandepoele K (2011) Comparative network analysis reveals that tissue specificity and gene function are important factors influencing the mode of expression evolution in Arabidopsis and rice. Plant Physiol 156: 1316–1330.
12. Doyle JJ, Luckow MA (2003) The rest of iceberg: legume diversity and evolution in a phylogenetic context. Plant Physiol131: 900–910.
13. Young ND, Debelle F, Oldroyd GED, Geurts R, Cannon SB, et al. (2011) The Medicago genome provides insight into the evolution of rhizobial symbioses. Nature 480: 520–524.
14. Sato S, Nakamura Y, Kaneko T, Asamizu E, Kato T, et al. (2008) Genome structure of the legume, *Lotus japonicus*. DNA Research 15: 227–239.
15. Schmutz J, Cannon SB, Schlueter J, Ma J, Mitros T, et al. (2010) Genome sequence of the palaeopolyploid soybean. Nature 463: 178–183.
16. Varshney RK, Chen W, Li Y, Bharti AK, Saxena RK, et al. (2012) Draft genome sequence of pigeonpea (*Cajanus cajan*), an orphan legume crop of resource-poor farmers. Nature Biotech 30: 83–89.
17. Young ND, Udvardi M (2009) Translating *Medicago truncatula* genomics to crop legumes. Curr Op Plant Biol 12: 193–201.
18. Cannon SB, May GD, Jackson SA (2009) Three sequenced legume genomes and many crop species: rich opportunities for translational genomics. Plant Physiol 151: 970–977.
19. Bordat A, Savois V, Nicolas M, Salse J, Chauveau A (2011) Translational genomics in legumes allowed placing in silico 5460 unigenes on the pea functional map and identified candidate genes in Pisum sativum L. Genes Genom Genet1: 93–103.
20. Hirsch AM, Lum MR, Downie JA (2001) What makes the Rhizobia-legume symbiosis so special? Plant Physiol127: 1484–1492.
21. Lee I, Ambaru B, Thakkar P, Marcotte EM, Rhee SY (2010) Rational association of genes with traits using a genome-scale gene network for Arabidopsis thaliana. Nature Biotech 28: 149–156.
22. Cline MS, Smoot M, Cerami E, Kuchinsky A, Landys N, et al. (2007) Integration of biological networks and gene expression data using Cytoscape. Nature Protocols 2: 2366–2382.
23. Arabidopsis Interactome Mapping Consortium (2011) Evidence for network evolution in an Arabidopsis interactome map. Science 333: 601–607.
24. Mukhtar MS, Carvunis A-R, Dreze M, Epple P, Steinbrenner J, et al. (2011) Independently evolved virulence effectors converge onto hubs in a plant immune system network. Science 333: 596–601.
25. Choi H-K, Iandolino A, Goes da Silva F, Cook DR (2013) Water deficit modulates the response of *Vitis vinifera* the the Pierce's disease pathogen *Xylella fastidiosa*. Mol Plant-Microbe Interact 26: 643–657.
26. Zhang W, Ruan J, Ho TD, You Y-S, Yu T, et al. (2005) Cis-regulatory element based targeted gene finding: genome-wide identification of abscisic acid- and abiotic stress-responsive genes in *Arabidopsis thaliana*, Bioinformatics 21:3074–3081.
27. Jensen MK, Kjaersgaard T, Nielsen MM, Galberg P, Peterson K, et al. (2010) The *Arabidopsis thaliana* NAC transcription facto family: structure-function relationships and determinants of ANAC019 stress signaling. Biochem. J426: 183-196.
28. Chinnusamy V, Ohta M, Kanrar S, Lee BH, Hong X, et al. (2003) ICE1: a regulator of cold-induced transcriptome and freezing tolerance in Arabidopsis. Genes Dev 17: 1043–1054.
29. Zhu H, Kim D-J, Baek J-M, Choi H-K, Ellis LC, et al. (2003) Syntenic relationships between *Medicago truncatula* and Arabidopsis reveal extensive divergence of genome organization. Plant Physiol 131: 1018–1026.
30. Zhu H, Cannon SB, Young ND, Cook DR (2002) Phylogeny and genomic organization of the TIR and non-TIR NBS-LRR resistance gene family in *Medicago truncatula*. Mol Plant-Microbe Interact 15: 529–539.
31. Ma Y, Szostkiewicz I, Korte A, Moes D, Yang Y, et al. (2009) Regulators of PP2C phosphatase activity functionas abscisic acid sensors. Science 324: 1064–1068.
32. Park S, Fung P, Nishimura N, Jensen DR, Fujii H, et al. (2009) Abscisic acid inhibits type 2C protein phosphatase via the PYR/PYL family of SYART proteins. Science 324: 1068–1071.
33. Umezawa T, Sugiyama N, Mizoguchi M, Hayashi S, Myouga F, et al. (2009) Type 2C protein phosphatases directlyregulate abscisic acid-activated protein kinases in Arabidopsis. Proc Natl Aacd Sci USA 106: 17588–17593.
34. Vlad F, Rubio S, Rodrigues A, Sirichandra C, Belin C, et al. (2009) Protein phosphatases 2C regulate the activation of the Snf1-related kinase OST1 by abscisic acid in Arabidopsis. Plant Cell 21: 3170–3184.

35. Umezawa T, Nakashima K, Miyakawa T, Kuromori T, Tanokura M, et al. (2010) Molecular basis of the core regulatory network in ABA responses: sensing, signaling and transport. . Plant Cell Physiol51: 1821–1839.
36. Wang Y, Wu W-H (2013) Potassium transport and signaling in higher plants. Annu Rev Plant Biol 64: 451–476.
37. Guo Y, Halfter U, Ishitani M, Zhu J-K (2001) Molecular characterization of functional domains in the protein kinase SOS2 that is required for plant salt tolerance. Plant Cell 13: 1383–1399.
38. Ohta M, Guo Y, Halftr U, Zhu J-K (2003) A novel domain in the protein kinase SOS2 mediates interaction with the protein phosphatase 2C ABI2. Proc Natl Acad Sci USA 100: 11771–11776.
39. Choi H-I, Park H-J, Park JH, Kim S, Im M-Y, et al. (2005) Arabidopsis calcium-dependent protein kinase AtCPK32 interact with ABF4, a transcriptional regulator of abscisic acid-responsive gene expression, and modulates its activity. Plant Physiol 139: 1750–1761.
40. Garcia ME, Lynch T, Peeters J, Snowden C, Finkelstein R (2008) A small plant-specific protein family of ABI five binding proteins (AFPs) regulates stress response in germinating Arabidopsis seeds and seedlings. Plant Mol Biol 67: 643–658.
41. Ren XL, Qi GN, Feng HQ, Zhao S, Zhao SS, et al. (2013) Calcineurin B-like protein CBL10 directly interact with AKT1 and modulates K^+ homeostasis in Arabidopsis. Plant J 74: 258–266.
42. Kim W, Lee Y, Park J, Lee N, Choi G (2013) HONSU, a protein phosphatase 2C, regulates seed dormancy by inhibiting ABA signaling in Arabidopsis. Plant Cell Physiol 54: 555–572.
43. Vilella AJ, Severin J, Ureta-Vidal A, Heng L, Durbin R, et al. (2009) EnsemblCompara GeneTrees: complete, duplication-aware phylogenetic trees in vertebrates. Genome Res 19: 327–335.
44. O'Brien KP, Remm M, Sonnhammer ELL (2005) Inparanoid: a comprehensive database of eukaryotic orthologs. Nu Acids Res 33: D476–D480.
45. Östuland G, Schmitt T, Forslund K, Köstler T, Messina DN, et al. (2010) InParanoid 7: new algorithms and tools for eukaryotic orthology analysis. Nu Acids Res 38: D196–D203.
46. Berardini TZ, Mundodi S, Reoser L, Huala E, Garcia-Hernandez M, et al. (2004) Functional annotation of the Arabidopsis genome using controlled vocabularies. Plant Physiol 135: 745–755.
47. Goodstein DM, Shu S, Howson R, Neupane R, Hayes RD, et al. (2012) Phytozome: a comparative platform for green plant genomics. Nu Acids Res 40: D1178–D1186.
48. Conte MG, Gaillard S, Lanau N, Rouard M, Périn C (2008) GreenPhylDB: a database for plant comparative genomics. Nu Acids Res 36: D991–D998.
49. Hirayama T, Shinozaki K (2010) Research on plant abiotic stress responses in the post-genome era: past, present and future. Plant J 61: 1041–1052.
50. Melotto M, Underwood W, Koczan J, Nomura K, He SY (2006) Plant stomata function in innate immunity against bacterial invasion. Cell 8: 969–980.
51. Sanchez DH, Pieckenstain FL, Szymanski J, Erban A, Bromke M, et al. (2011) Comparative functional genomics of salt stress in related model and cultivated plants identifies and overcomes limitations to translational genomics. Plos One 6: e17094.
52. Krysan PJ, Young JC, Tax F, Sussman MR (1996) Identification of transferred DNA insertions within Arabidopsis genes involved in signal transduction and ion transport. Proc Natl Acad Sci USA 93: 8145–8150.
53. Finkelstein RR (1994) Mutations at two new Arabidopsis ABA response loci are similar to the *abi3* mutations. Plant J 5: 765–771.
54. Bies-Etheve N, da Silva Conceicao A, Giraudat J, Koornneef M, Leon-Kloosterziel K, et al. (1999) Importance of the B2 domain of the Arabidopsis ABI3 protein for Em and 2S albumin gene regulation. Plant Mol Biol 40: 1045–1054.
55. Rohde A, De Rycke R, Beeckman T, Engler G, Van Montagu M, et al. (2000) ABI3 affects plastid differentiation in dark-grown Arabidopsis seedlings. Plant Cell 12: 35–52.
56. Fujii H, Verslues PE, Zhu JK (2007) Identification of two protein kinases required for abscisic acid regulation of seed germination, root growth, and gene expression in Arabidopsis. Plant Cell 19: 485–494.
57. Cheong YH, Kim KN, Pandey GK, Gupta R, Grant JJ, et al. (2003) CBL1, a calcium sensor that differentially regulates salt, drought, and cold responses in Arabidopsis. Plant Cell 15: 1833–1845.
58. Cheong YH, Pandey GK, Grant JJ, Batistic O, Li L, et al. (2007) Two calcineurin B-like calcium sensors, interacting with protein kinase CIPK23, regulate leaf transpiration and root potassium uptake in Arabidopsis. Plant J 52: 223–239.
59. Michard E, Dreyer I, Lacombe B, Sentenac H, Thibaud JB (2005) Inward rectification of the AKT2 channel abolished by voltage-dependent phosphorylation. Plant J 44: 783–797.
60. Ramon M, Rolland F, Thevelein JM, Van Dijck P, Leyman B (2006) ABI4 mediates the effects of exogenous trehalose on Arabidopsis growth and starch breakdown. Plant Mol Biol 63: 195–206.
61. Verslues PE, Bray EA (2006) Role of abscisic acid (ABA) and Arabidopsis thaliana ABA-insensitive loci in low water potential-induced ABA and proline accumulation. J Exp Bot 57: 201–212.
62. Tripathi V, Parasuraman B, Laxmi A, Chattopadhyay D (2009) CIPK6, a CBL-interacting protein kinase is required for development and salt tolerance in plant. Plant J 58: 778–790.
63. Mustilli AC, Merlot S, Vavasseur A, Fenzi F, Giraudat J (2002) Arabidopsis OST1 protein kinase mediates the regulation of stomatal aperture by abscisic acid and acts upstream of reactive oxygen species production. Plant Cell 14: 3089–3099.
64. Juul T, Malolepszy A, Dybkaer K, Kidmose R, Rasmessen JT, et al. (2010) The in vivo toxicity of hydroxyurea depends on its direct target catalase. J Biol Chem 285: 21411–21415.
65. Queval G, Issakidis-Bourguet E, Hoeberichts FA, Vandorpe M, Gakiere B, et al. (2007) Conditional oxidative stress responses in the Arabidopsis photorespiratory mutant cat2 demonstrate that redox state is a key modulator of daylength-dependent gene expression, and define photoperiod as a crucial factor in the regulation of H_2O_2-induced cell death. Plant J 52: 640–657.
66. Boursiac Y, Chen S, Luu DT, Sorieul M, van den Dries N, et al. (2005) Early effects of salinity on water transport in Arabidopsis roots. Molecular and cellular features of aquaporin expression. Plant Physiol 139: 790–805.

The Complete Chloroplast Genome Sequence of *Podocarpus lambertii*: Genome Structure, Evolutionary Aspects, Gene Content and SSR Detection

Leila do Nascimento Vieira[1], Helisson Faoro[2], Marcelo Rogalski[3], Hugo Pacheco de Freitas Fraga[1], Rodrigo Luis Alves Cardoso[2], Emanuel Maltempi de Souza[2], Fábio de Oliveira Pedrosa[2], Rubens Onofre Nodari[1], Miguel Pedro Guerra[1]*

1 Laboratório de Fisiologia do Desenvolvimento e Genética Vegetal, Programa de Pós-graduação em Recursos Genéticos Vegetais, Universidade Federal de Santa Catarina, Florianópolis, Santa Catarina, Brazil, **2** Departamento de Bioquímica e Biologia Molecular, Núcleo de Fixação Biológica de Nitrogênio, Universidade Federal do Paraná, Curitiba, Paraná, Brazil, **3** Departamento de Biologia Vegetal, Universidade Federal de Viçosa, Viçosa, Minas Gerais, Brazil

Abstract

Background: *Podocarpus lambertii* (Podocarpaceae) is a native conifer from the Brazilian Atlantic Forest Biome, which is considered one of the 25 biodiversity hotspots in the world. The advancement of next-generation sequencing technologies has enabled the rapid acquisition of whole chloroplast (cp) genome sequences at low cost. Several studies have proven the potential of cp genomes as tools to understand enigmatic and basal phylogenetic relationships at different taxonomic levels, as well as further probe the structural and functional evolution of plants. In this work, we present the complete cp genome sequence of *P. lambertii*.

Methodology/Principal Findings: The *P. lambertii* cp genome is 133,734 bp in length, and similar to other sequenced cupressophytes, it lacks one of the large inverted repeat regions (IR). It contains 118 unique genes and one duplicated tRNA (*trnN*-GUU), which occurs as an inverted repeat sequence. The *rps16* gene was not found, which was previously reported for the plastid genome of another Podocarpaceae (*Nageia nagi*) and Araucariaceae (*Agathis dammara*). Structurally, *P. lambertii* shows 4 inversions of a large DNA fragment ~20,000 bp compared to the *Podocarpus totara* cp genome. These unexpected characteristics may be attributed to geographical distance and different adaptive needs. The *P. lambertii* cp genome presents a total of 28 tandem repeats and 156 SSRs, with homo- and dipolymers being the most common and tri-, tetra-, penta-, and hexapolymers occurring with less frequency.

Conclusion: The complete cp genome sequence of *P. lambertii* revealed significant structural changes, even in species from the same genus. These results reinforce the apparently loss of rps16 gene in Podocarpaceae cp genome. In addition, several SSRs in the *P. lambertii* cp genome are likely intraspecific polymorphism sites, which may allow highly sensitive phylogeographic and population structure studies, as well as phylogenetic studies of species of this genus.

Editor: Ulrike Gertrud Munderloh, University of Minnesota, United States of America

Funding: This work was supported by Coordenação de Aperfeiçoamento de Pessoal de Nível Superior (CAPES), Conselho Nacional de Desenvolvimento Científico e Tecnológico (CNPq), and Fundação de Amparo à Pesquisa e Inovação do Estado de Santa Catarina (FAPESC 14848/2011-2, 3770/2012, and 2780/2012-4). The funders had no role in study design, data collection and analysis, decision to publish, or preparation of the manuscript.

Competing Interests: The authors have declared that no competing interests exist.

* E-mail: miguel.guerra@ufsc.br

Introduction

Extant gymnosperms are considered the most ancient group of seed-bearing plants that first appeared approximately 300 million years ago [1]. They consist of four major groups, including Gnetophytes, Conifers, Cycads and Ginkgo. Podocarpaceae are considered the most diverse family of Conifers, and much of this diversity has taken place within the *Podocarpus* and *Dacrydium* genera [2]. The Podocarpaceae family comprises 18 genera and 173 species distributed mainly in the Southern Hemisphere, but extending to the north in subtropical China, Japan, Mexico and the Caribbean [3,4].

The *Podocarpus sensu lato* (*s.l.*) genus comprises nearly 100 species, widely spread throughout the Southern Hemisphere and northward to the West Indies, Mexico, southern China and southern Japan [5]. Ledru et al. [6] described that *Podocarpus* populations in Brazil are widely dispersed in eastern Brazil, from north to south, and three endemic species have been reported: *Podocarpus sellowii* Klotzch ex Endl, *Podocarpus lambertii* Klotzch ex Endl, and *Podocarpus brasiliensis* de Laubenfels [7]. *P. lambertii* is a native species from the Araucaria Forest, a subtropical moist forest ecoregion of the Atlantic Forest Biome, which is considered one of the 25 biodiversity hotspots of the world [8]. It is a dioecious evergreen tree of variable height, measuring 1–10 m, shade-tolerant, adapted to high frequency and density of undergrowth [9].

Phylogeny analyses by maximum parsimony of Podocarpaceae family using 18S rDNA gene sequencing and morphological characteristics indicated Podocarpaceae as monophyletic and *Podocarpus s.l.* and *Dacrydium s.l.* genera as unnatural [2]. This author concluded that single-gene studies rarely result in perfect phylogenies, but they could provide a basis for choosing between competing hypotheses. Parks et al. [10] suggested chloroplast (cp) genome sequencing as an efficient option for increasing phylogenetic resolution at lower taxonomic levels in plant phylogenetic and genetic population analyses.

The advancement of next-generation sequencing technologies has enabled the rapid acquisition of whole cp genome sequences at low cost when compared with traditional sequencing approaches. Chloroplast sequences are available for all families of Conifers: Cephalotaxaceae [11], Cupressaceae [12], Pinaceae [13–15], Podocarpaceae (NC_020361.1) and [16], Taxaceae (NC_020321.1), and Araucariaceae [16]. For *Podocarpus* genus, the cp sequence of only one species has recently been obtained: the endemic New Zealand *Podocarpus totara* G. Benn. ex Don (NC_020361.1).

Several studies have proven the potential of cp genomes as tools to understand enigmatic and basal phylogenetic relationships at different taxonomic levels, as well as probe the structural and functional evolution of plants [11,17–20]. Hirao et al. [12] sequenced the cp genome of the first species in the Cupressaceae family, *Cryptomeria japonica.* They reported the deletion of one large inverted repeat (IR), numerous genomic rearrangements, and many differences in genomic structure between *C. japonica* and other land plants, thus supporting the theory that a pair of large IR can stabilize the cp genome against major structural rearrangements and, in turn, providing new insights into both the evolutionary lineage of coniferous species and the evolution of the cp genome [12,21,22].

Chloroplast genome sequencing in gymnosperms also brought insights into evolutionary aspects in Gnetophytes. Wu et al. [23] considered that the reduced cp genome size in Gnetophyte was based on a selection toward a lower-cost strategy by deletions of genes and noncoding sequences, leading to genomic compactness and accelerated substitution rates. More recently, comparative analysis of the cp genomes in cupressophytes and Pinaceae provided inferences about the loss of large IR [11,20]. On one hand, Wu et al. [20] and Wu and Chaw [16] argue that each Pinaceae and cupressophyte lost a different copy of IR. On the other hand, Yi et al. [11] showed that distinct isomers are considered as alternative structures for the ancestral cp genome of cupressophyte and Pinaceae lineages. Therefore, it is not possible to distinguish between hypotheses favoring retention or independent loss of the same IR region in cupressophyte and Pinaceae cp genomes.

The present study focuses on establishing the complete cp genome sequence of a further member of the Podocarpaceae family, the Brazilian endemic species *P. lambertii.* Here, we characterize the cp genome organization of *P. lambertii* and compare its cp genome structure with other conifer species.

Materials and Methods

Plant material and cp DNA purification

Chloroplast isolation of *P. lambertii* was performed from young plants collected at a private area located at Lages, Santa Catarina, Brazil (27° 48′ 57" S, 50° 19′ 33" W), where the species is abundant, with previous permission from the owner (José Antônio Ribas Ribeiro). This species is not considered threatened. Afterwards, the young plants were transplanted to the greenhouse until the collection of needles. The cpDNA isolation was performed according to Vieira et al. [24].

Chloroplast genome sequencing, assembling and annotation

Approximately 50 ng of cp DNA were used to prepare sequencing libraries with Nextera DNA Sample Prep Kit (Illumina Inc., San Diego, CA) according to the manufacturer's instructions. Chloroplast DNA was sequenced using Illumina MiSeq (Illumina Inc., San Diego, CA) at the Federal University of Paraná, Brazil. In total, 495,071 paired-end reads (2×250 bp) were obtained, and *de novo* assembly was performed using Newbler 2.6 v. The obtained paired-end reads were mapped on *P. lambertii* cp genome and the genome coverage estimated using the CLC Genomics Workbench 5.5 software. By using this approach, a total of 377,437 paired-end reads (76.23%) was obtained from cpDNA, resulting in 1,200-fold genome coverage. Initial annotation of the *P. lambertii* cp genome was performed using Dual Organellar GenoMe Annotator (DOGMA) [25]. From this initial annotation, putative starts, stops, and intron positions were determined based on comparisons to homologous genes in other cp genomes. The tRNA genes were further verified by using tRNAscan-SE [26]. A physical map of the cp circular genome was drawn using OrganellarGenomeDRAW (OGDRAW) [27]. The complete nucleotide sequence of *P. lambertii* cp genome was deposited in the GenBank database under accession number KJ010812.

Comparative analysis of genome structure

We used the PROtein MUMmer (PROmer) Perl script in MUMmer 3.0 [28], available at http://mummer.sourceforge.net/ , to visualize gene order conservation (dot-plot analyses) between *P. lambertii* and the non-Pinaceae conifer representatives *P. totara* (Podocarpaceae), *Cephalotaxus oliveri, Cephalotaxus wilsoniana* (Cephalotaxaceae), *Taxus mairei* (Taxaceae), *Taiwania cryptomerioides, T. flousiana* (Cupressaceae), *C. japonica* (Cupressaceae), as well as *Pinus thunbergii*, a Pinaceae representative.

Repeat sequence analysis and IR identification

Simple sequence repeats (SSRs) were detected using MISA perl script, available at (http://pgrc.ipk-gatersleben.de/misa/), with thresholds of eight repeat units for mononucleotide SSRs, four repeat units for di- and trinucleotide SSRs, and three repeat units for tetra-, penta- and hexanucleotide SSRs. Tandem repeats were analyzed using Tandem Repeats Finder (TRF) [29] with parameter settings of 2, 7 and 7 for match, mismatch, and indel, respectively. The minimum alignment score and maximum period size were set as 50 and 500, respectively. All of the repeats found were manually verified, and the nested or redundant results were removed. REPuter [30] was used to visualize the remaining IRs in *P. lambertii* by forward *vs.* reverse complement (palindromic) alignment. The minimal repeat size was set to 30 bp and the identity of repeats ≥90%.

Results and Discussion

Chloroplast genome sequencing, assembling and annotation

P. lambertii cp genome size was determined to be 133,734 bp, which is very similar to *P. totara* (133,259 bp) (NC_020361.1) and larger than the sequenced cp genomes of Pinaceae species, which range from 116,479 bp in *Pinus monophylla* [14] to 124,168 bp in *Picea morrisonicola* [31]. *P. lambertii* cp genome size is smaller than the cp sequences in the cycads *Cycas taitungensis* (163,403 bp) [32]

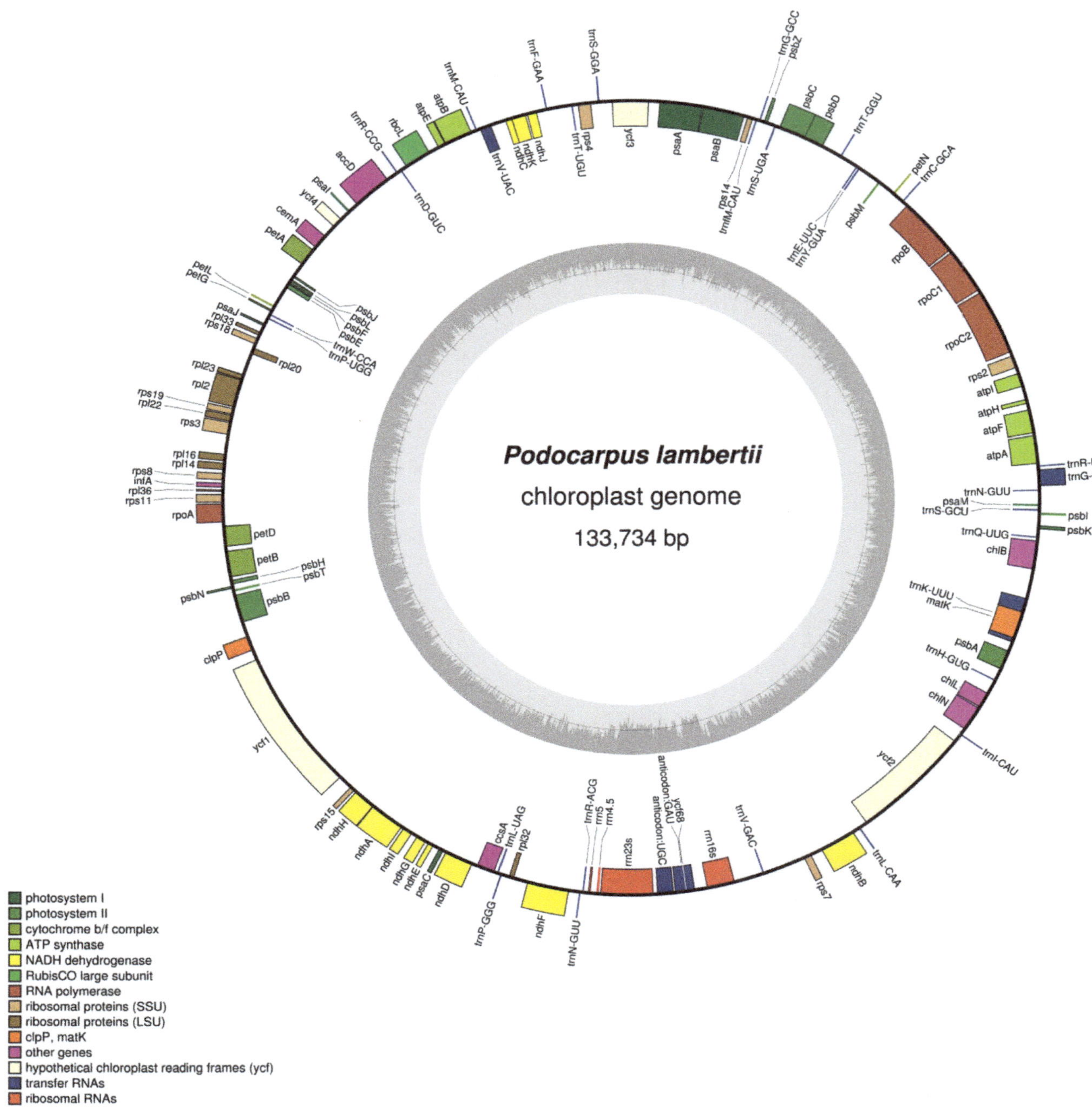

Figure 1. Gene map of *Podocarpus lambertii* chloroplast genome. Genes drawn inside the circle are transcribed clockwise, and genes drawn outside are counterclockwise. Genes belonging to different functional groups are color-coded. The darker gray in the inner circle corresponds to GC content, and the lighter gray corresponds to AT content.

and *Cycas Revoluta* (162,489 bp) (NC_020319.1). The genome size of *P. lambertii* cp is consistent with the size of non-Pinaceae conifer species, which ranges from 127,665 bp in *T. mairei* (NC_020321.1) to 136,196 bp in *C. wilsoniana* [20]. A total of 119 genes were identified in the *P. lambertii* cp genome, of which 118 genes were single copy and one gene, *trnN*-GUU, was duplicated and occurred as an inverted repeat sequence. The following genes were identified and are listed in Figure 1 and Table 1: 4 ribosomal RNA genes, 31 unique transfer RNA genes, 20 genes encoding large and small ribosomal subunits, 1 translational initiation factor, 4 genes encoding DNA-dependent RNA polymerases, 50 genes encoding photosynthesis-related proteins, 8 genes encoding other proteins, including the unknown function gene *ycf2*, and 1 pseudogene, *ycf68*. Among these 118 single copy genes, 14 were genes containing introns (Table 1). The GC content determined for *P. lambertii* cp genome is 37.1%, which is higher than *C. oliveri* (35.2%), *C. wilsoniana* (35.1%), *T. cryptomerioides* (34.6%), and *C. japonica* (35.4%), but lower than *C. taitungensis* (39.5%) and *P. thunbergii* (38.8%).

Table 1. List of genes identified in *Podocarpus lambertii* chloroplast genome.

Category of Genes	Group of gene			Name of gene			
Self-replication	Ribosomal RNA genes	*rrn16*	*rrn23*	*rrn5*	*rrn4.5*		
	Transfer RNA genes	*trnA*-UGC*	*trnC*-GCA	*trnD*-GUC	*trnE*-UUC	*trnF*-GAA	*trnfM*-CAU
		trnG-UCC*	*trnG*-GCC	*trnH*-GUG	*trnI*-CAU	*trnI*-GAU*	*trnK*-UUU*
		trnL-CAA	*trnL*-UAG	*trnM*-CAU	*trnN*-GUU**	*trnP*-GGG	*trnP*-UGG
		trnQ-UUG	*trnR*-ACG	*trnR*-UCU	*trnR*-CCG	*trnS*-GCU	*trnS*-UGA
		trnS-GGA	*trnT*-UGU	*trnT*-GGU	*trnV*-GAC	*trnV*-UAC*	*trnW*-CCA
		trnY-GUA					
	Small subunit of ribosome	*rps2*	*rps3*	*rps4*	*rps7*	*rps8*	*rps11*
		*rps12**	*rps14*	*rps15*	*rps18*	*rps19*	
	Large subunit of ribosome	*rpl2**	*rpl14*	*rpl16*	*rpl20*	*rpl22*	*rpl23*
		rpl32	*rpl33*	*rpl36*			
	DNA-dependent RNA polymerase	*rpoA*	*rpoB*	*rpoC1**	*rpoC2*		
	Translational initiation factor	*infA*					
Genes for photosynthesis	Subunits of photosystem I	*psaA*	*psaB*	*psaC*	*psaI*	*psaJ*	*psaM*
		*ycf3**	*ycf4*				
	Subunits of photosystem II	*psbA*	*psbB*	*psbC*	*psbD*	*psbE*	*psbF*
		psbH	*psbI*	*psbJ*	*psbK*	*psbL*	*psbM*
		psbN	*psbT*	*psbZ*			
	Subunits of cytochrome	*petA*	*petB**	*petD**	*petG*	*petL*	*petN*
	Subunits of ATP synthase	*atpA*	*atpB*	*atpE*	*atpF**	*atpH*	*atpI*
	Large subunit of Rubisco	*rbcL*					
	Chlorophyll biosynthesis	*chlB*	*chlL*	*chlN*			
	Subunits of NADH dehydrogenase	*ndhA**	*ndhB**	*ndhC*	*ndhD*	*ndhE*	*ndhF*
		ndhG	*ndhH*	*ndhI*	*ndhJ*	*ndhK*	
Other genes	Maturase	*matK*					
	Envelope membrane protein	*cemA*					
	Subunit of acetyl-CoA	*accD*					
	C-type cytochrome synthesis gene	*ccsA*					
	Protease	*clpP*					
	Component of TIC complex	*ycf1*					
Genes of unknown function	Conserved open reading frames	*ycf2*					
Pseudogenes		*ycf68*					

*Genes containing introns.
**Duplicated gene.

Gene content differences

The gene content of *P. lambertii* cp genome and that of other conifer cp genomes sequenced to date show high similarity. However, some differences are observed when we compare *P. lambertii* cpDNA with other non-Pinaceae and Pinaceae conifers. One exception is the *rps16* gene, which is absent from the *P. lambertii* cp genome. This result reinforce the apparently loss of rps16 gene in Podocarpaceae and Araucariaceae families. Wu and Chaw [16] reported the *rps16* gene loss in *Nageia nagi* (Podocarpaceae) and *Agathis dammara* (Araucariaceae). This gene is present in other non-Pinaceae conifer cp genomes published so far [11,12,20,32]. The *rps16* gene loss has already been reported in other gymnosperms, such as Pinaceae and Gnetophyte species [23,32,33]. Wu et al. [20] considered *rps16* gene loss as a structural mutation unique to the cpDNAs of gnetophytes and Pinaceae, but since the loss of this gene has been identified in Podocarpaceae and Araucariaceae families, we can consider that some cupressophytes may also present this mutation. This gene is also absent, or nonfunctional, in some angiosperm species of the Fabaceae family, such as *Medicago truncatula*, in which it is completely absent, and in *Phaseolus vulgaris* and *Vigna radiata*, in which it is nonfunctional. In this angiosperm family, the coding sequence contains many internal stop codons and a modified initial stop codon [34,35]. Since this gene was shown to be essential for cell survival in tobacco [36], it was probably transferred to the nucleus, as observed for different species of the Fabaceae family [34,35], and has since become a functional nuclear gene required for normal plastid translation.

The *trnP*-GGG and *trnR*-CCG genes are considered to be relics of plastid genome evolution in gymnosperms, pteridophytes and bryophytes [37]. The *trnP*-GGG gene is present in the *P. lambertii* cp genome, as well as such conifer species as *C. japonica*, *P.*

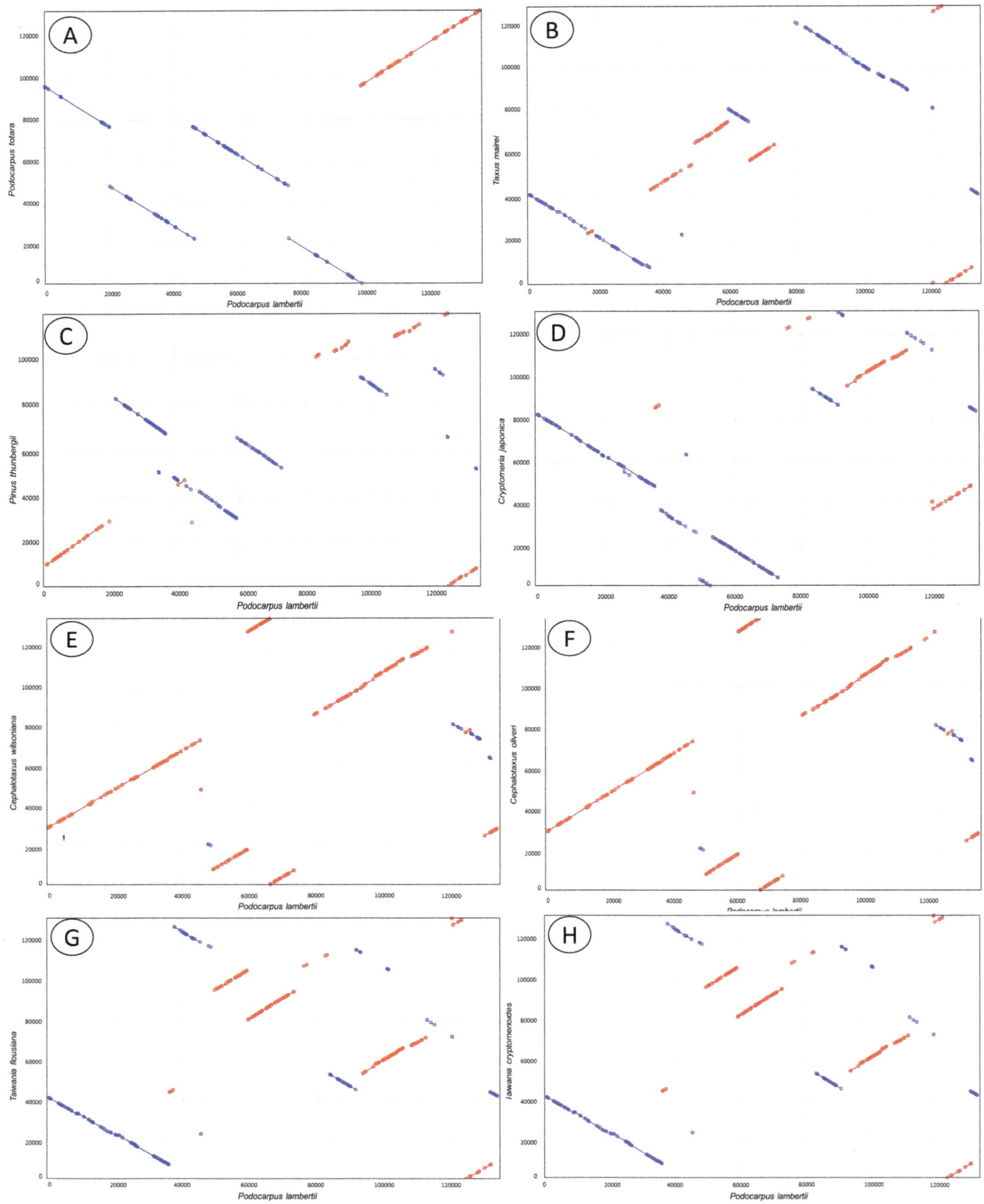

Figure 2. Dot-plot analyses of eight sampled conifer chloroplast DNAs against *Podocarpus lambertii*. A positive slope denotes that the two compared sequences are in the same orientation, whereas a negative slope indicates that the compared sequences can be aligned, but their orientations are opposite. Graphs represents comparisons between *Podocarpus lambertii* (axis X) and *Podocarpus totara* (A), *Taxus mairei* (B), *Pinus*

thunbergii (C), *Cryptomeria japonica* (D), *Cephalotaxus wilsoniana* (E), *Cephalotaxus oliveri* (F), *Taiwania flousiana* (G), and *Taiwania cryptomerioides* (H) in axis Y.

thunbergii, *C. oliveri* and *C. wilsoniana* and other gymnosperm species, such as *C. taitungensis*, *Gnetum* and *Ginkgo*. The *trnR*-CCG gene is present as complete and functional tRNA in *P. lambertii* (Podocarpaceae), as well as the cp genomes of *P. thunbergii* (Pinaceae), *C. taitungensis* (Cycadaceae) [32], whereas it is absent from *C. japonica* (Cupressaceae), *C. oliveri* and *C. wilsoniana* (Cephalotaxaceae), and *T. mairei* (Taxaceae) [11,12]. Hirao et al. [12] suggested that *trnR*-CCG might have been completely lost in the Cupressaceae *s.l.*, which has only relatively recently diverged during the long evolutionary history of plants. These data corroborate the hypothesis based on phytochrome phylogenetic trees, in which the most ancient branch of the conifers seems to be the Pinaceae, and the next split appears to have separated Araucariaceae plus Podocarpaceae from the Taxaceae/Taxodiaceae/Cupressaceae group [38]. This trnR-CCG gene may have been lost during the second split separating Araucariaceae and Podocarpaceae taxa. In addition, *trnT*-GGU occurs as a pseudogene in the *C. japonica* cp genome, with only 43 bp, while it is present and completely functional in *P. lambertii* and *C. oliveri*, *C. wilsoniana*, duplicated in *P. thunbergii*, and totally absent from the *C. taitungensis* cp genome. Interestingly, the *trnT*-GGU gene is highly conserved in angiosperms, and knockout of this gene in tobacco plants produced viable plants, whereas the growth of these plants was strongly affected, suggesting an important role during plastid translation [39]. The loss of the *trnT*-GGU gene in several gymnosperm species suggests that a uridine modification in the anticodon position of the *trnT*-UGU gene occurred during evolution, which would facilitate the reading of threonine codons and makes the *trnT*-GGU gene dispensable in these species [39-42]. Evolutionarily, the loss of this tRNA gene could be used as a tool, or marker gene, to study the possible ways that the conifers diverged during evolution. However, it remains to be determined whether structural differences in the cp ribosome or modification in the structure of this tRNA, between angiosperms and gymnosperms, would facilitate the decoding.

Comparative analysis of genome structure

Chloroplast genome organization is much conserved in angiosperms, as well as the presence of IRs, with very few exceptions. As reported by Terakami et al. [43] in *Pyrus*, *Malus* and *Nicotiana*, neither translocation nor inversion was detected in the three species. In addition, considering the many dicot and monocot species, only one large inversion was reported [43].

In addition to the loss of the large IR in conifers, many genome rearrangements were observed in the cp genome, and such rearrangements appear to play an important role in their evolution. Dot-plot analyses indicate that the structure of the *P. lambertii* cp genome differs significantly from cp genomes of other conifer species, and, surprisingly, it has significant differences when compared to *P. totara* (Figure 2A-H).

For the genus *Cephalotaxus s.l.*, specifically *C. wilsoniana* and *C. Oliveri*, it was shown that the genome structures were almost the same [11]. Similar results were observed in the present study, as revealed by the high similarity in the dot-plot analyses between *Podocarpus* and *Cephalotaxus* genera, as represented by *P. lambertii* × *C. wilsoniana* (Figure 2E) and *P. Lambertii* × *C. oliveri* (Figure 2F), and between the *Podocarpus* and *Taiwania* genera, as represented by *P. lambertii* × *T. flousiana* (Figure 2G) and *P. lambertii* × *T.*

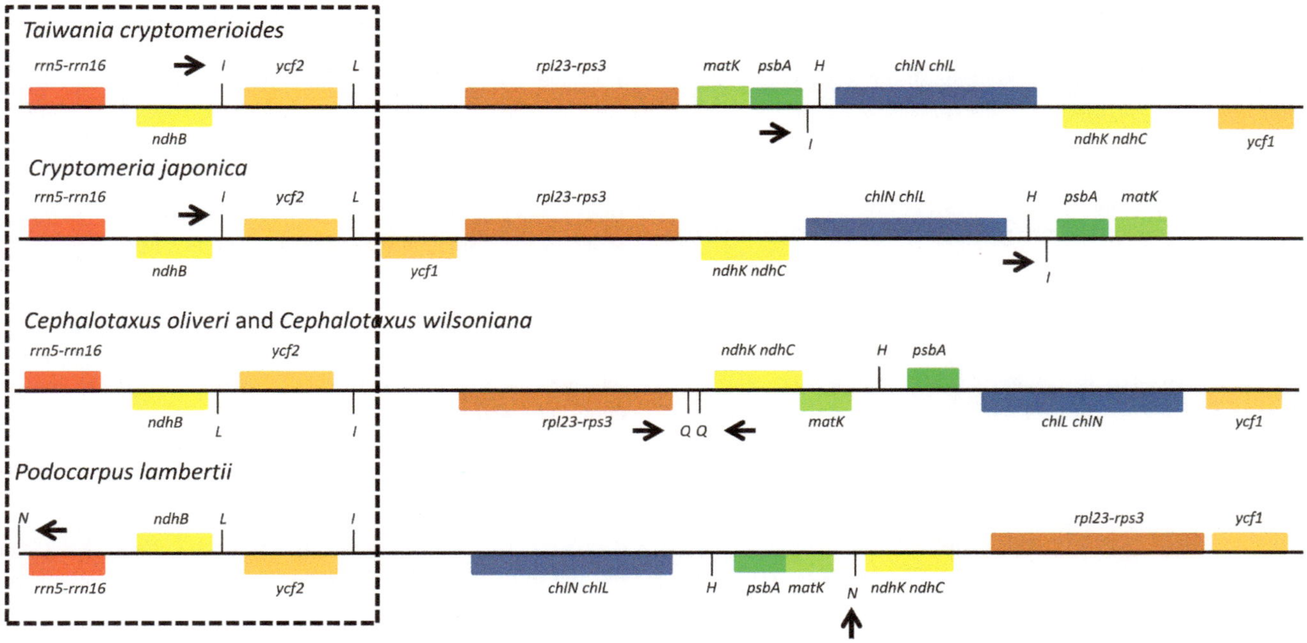

Figure 3. Comparison of IR and genome structure in 5 cupressophytes. Five cupressophyte species from top to bottom are *Taiwania cryptomerioides*, *Cryptomeria japonica*, *Cephalotaxus oliveri*, *Cephalotaxus wilsoniana* and *Podocarpus lambertii*. Genes are represented by boxes extending above or below the baseline, according to the direction of transcription; genes with the same function have the same color. Transfer RNA genes are abbreviated as the type of one letter. Dashed boxes represent the retained IR region, and arrows indicate the short IR on each species. Adapted from Yi et al. (2013).

Table 2. List of simple sequence repeats identified in *Podocarpus lambertii* chloroplast genome.

SSR sequence	Number of repeats													TOTAL
	3	4	5	6	7	8	9	10	11	12	13	14	15	
A/T	–	–	–	–	–	39	14	6	4	6	–	–	1	70
C/G	–	–	–	–	–	–	3	3	–	1	1	–	2	10
AC/GT	–	1	–	1	–	–	–	–	–	–	–	–	–	2
AG/CT	–	21	1	–	–	–	–	–	–	–	–	–	–	22
AT/AT	–	24	7	2	2	–	3	1	–	–	–	–	–	39
AAG/CTT	–	1	–	–	–	–	–	–	–	–	–	–	–	1
AAT/ATT	–	3	–	–	–	–	–	–	–	–	–	–	–	3
AATC/ATTG	2	–	–	–	–	–	–	–	–	–	–	–	–	2
AATG/ATTC	1	–	–	–	–	–	–	–	–	–	–	–	–	1
AATT/AATT	1	–	–	–	–	–	–	–	–	–	–	–	–	1
ACAT/ATGT	1	–	–	–	–	–	–	–	–	–	–	–	–	1
ACCT/AGGT	1	–	–	–	–	–	–	–	–	–	–	–	–	1
AGAT/ATCT	3	–	–	–	–	–	–	–	–	–	–	–	–	3
AAATG/ATTTC	1	–	–	–	–	–	–	–	–	–	–	–	–	1
AGATAT/ATATCT	1	–	–	–	–	–	–	–	–	–	–	–	–	1
TOTAL														**158**

cryptomerioides (Figure 2H). This high similarity in dot-plot analysis indicates the occurrence of exactly the same structural modifications between *P. lambertii* and these two *Cephalotaxus* and *Taiwania species*.

Differently, for *P. lambertii* and *P. totara* (Figure 2A), we observed four large inversions of about 20,000 bp in length each. In both *Cephalotaxus* and *Taiwania* genera, the two sequenced species share the same region of natural occurrence, which is not true for either *Podocarpus* species sequenced. Thus, these large inversions can be explained by, and probably result from, the large distance between the natural occurrence of these two species in that *P. lambertii* occurs in Brazil, while *P. totara* occurs in New Zealand. Moreover, podocarps have a rich fossil record that suggests an origin in the Triassic period (about 220 million years) and a distribution in both the Northern and Southern Hemispheres through the Cretaceous and earliest Tertiary periods, about 100 million years ago [44–46]. Thus, geographic distance and different adaptive traits could explain the structural differences found between these two species of the same genera.

In addition, the loss of one large IR copy already reported in other conifer species were also observed in the *P. lambertii* cp genome [11,12,20]. However, short remaining IR sequences of 326 bp can be found in *P. lambertii*, 544 bp in *C. oliveri*, 530 bp in *C. wilsoniana*, 277 bp in *T. cryptomerioides* and 284 bp in *C. japonica* [11]. These short remaining IR sequences also differ in the nucleic acid sequences and gene content between different conifer species. In *P. lambertii*, *trnN*-GUU remain from the lost IR copy region, while in *T. cryptomerioides* and *C. japonica*, *trnI*-CAU remained after the rearrangements that determined the loss of one IR copy [11]. In *C. oliveri* and *C. wilsoniana*, the *trnQ*-UUG is duplicated; however, this gene is not normally present in the IR region, and its duplication was probably produced by other rearrangements not involved with the IR regions [20]. After much evidence provided by different conifer plastid genomes, it can be concluded that the loss of one IR copy occurred after a reduction in sequence and gene content and that such loss was most likely caused by this reduction [11,12,14,20,23,32,33]. However, this speculation remains to be established. To date, it is not entirely clear whether cupressophytes and Pinaceae species have lost different IR regions [11]. However, we can observe in *P. lambertii* an inversion in the direction of transcription of ribosomal RNA genes spanning *rrn5*-*rrn16* and protein-coding genes, *ndhB* and *ycf2*, when compared to *C. oliveri*, *C. wilsoniana*, *T. cryptomerioides* and *C. japonica* (Figure 3).

Table 3. Distribution of tri-, tetra-, penta-, and hexapolymer simple sequence repeats (SSRs) loci in *Podocarpus lambertii* chloroplast genome.

SSR type	SSR sequence	Size	Start	End	Location
penta	(AATGA)3	15	21884	21898	*trnE-UUC/trnT*-GGU (IGS)
hexa	(AGATAT)3	18	37894	37911	*trnF*-GAA/*ndhJ* (IGS)
tetra	(ATCA)3	12	44346	44357	*atpE/rbcL* (IGS)
tri	(AAG)4	12	75761	75772	*Ycf1* (CDS)
tetra	(AATG)3	12	86350	86361	*ndhA* (intron)
tetra	(TGAT)3	12	97140	97151	*ndhF/trnN*-GUU (IGS)
tetra	(CTAC)3	12	99809	99820	*rrn23* (CDS)
tri	(ATT)4	12	103664	103675	*trnI*-GAU/*rrn16* (IGS)
tri	(ATA)4	12	120539	120550	*rps7/ndhB* (IGS)
tri	(TTA)4	12	122046	122057	*chlL* (CDS)
tetra	(AATT)3	12	122977	122988	*chlL/trnH*-GUG (IGS)
tetra	(CATA)3	12	125437	125448	*psbA/trnK*-UUU (IGS)
tetra	(ATAG)3	12	125570	125581	*psbA/trnK*-UUU (IGS)

CDS, coding sequences; IGS, intergenic spacers.

Table 4. Distribution of tandem repeats in *Podocarpus lambertii* chloroplast genome.

Serial Number	Repeat Length (bp)	Consensus size × Copy number	Start-End	Location
1	32	16×2	3450-3482	*atpA/atpF* (IGS)
2	284	142×2	13170-13454	*rpoC1* (Intron)
3	60	30×2	13496-13557	*rpoC1* (Intron)
4	30	15×2	46625-46653	*trnR*-CCG/*accD* (IGS)
5	90	30×3	47533-47619	*accD* (CDS)
6	42	21×2	48149-48192	*accD* (CDS)
7	52	26×2	57988-58043	*rps18* (CDS)
8	32	16×2	61875-61905	*rpl2/rps19* (IGS)
9	54	18×3	62177-62237	*rps19* (CDS)
10	63	21×3	66568-66630	*rps11* (CDS)
11	32	16×2	75172-75203	*clpP/ycf1* (IGS)
12	104	52×2	75412-75529	*clpP/ycf1* (IGS)
13	36	18×2	79255-79292	*ycf1* (CDS)
14	162	52×3	79351-79504	*ycf1* (CDS)
15	162	81×2	79362-79519	*ycf1* (CDS)
16	108	27×4	79401-79519	*ycf1* (CDS)
17	132	33×4	80478-80619	*ycf1* (CDS)
18	96	24×4	80732-80820	*ycf1* (CDS)
19	273	21×13	81305-81571	*ycf1* (CDS)
20	96	48×2	82408-82528	*ycf1* (CDS)
21	30	15×2	89787-89817	*ndhE/psaC* (IGS)
22	126	42×3	93843-93963	*rpl32* (CDS)
23	64	32×2	97838-97902	*trnR*-ACG/*rrn5* (IGS)
24	300	60×5	109209-109531	*rps12/rps7* (IGS)
25	36	12×3	116515-116547	*ycf2* (CDS)
26	60	20×3	119998-120055	*ycf2/trnI*-CAU (IGS)
27	128	64×2	131733-131853	*trnQ*-UUG/*psbK* (IGS)
28	26	13×2	132530-132556	*psbK/psbI* (IGS)

CDS, coding sequences; IGS, intergenic spacers.

Repeat sequence analysis

The cp genome mode of inheritance, paternal in most gymnosperms, allows us to elucidate the relative contributions of seed and pollen flow to the genetic structure of natural populations by comparison of nuclear and cp markers [47]. The cp microsatellites, or SSRs, may be identified in completely sequenced plant cp genomes by simple database searches, followed by primers designed to screen for polymorphism. To date, studies of cp microsatellites have revealed much higher levels of diversity than have those of cp restriction fragment length polymorphisms (RFLP) [47–49].

We have analyzed the occurrence, type, and distribution of SRRs in the *P. lambertii* cp genome. In total, 156 SSRs were identified. Among them, homo- and dipolymers were the most common with, respectively, 80 and 63 occurrences, whereas tri- (4), tetra- (7), penta- (1), and hexapolymers (1) occur with lower frequency (Table 2). Most homopolymers are constituted by A/T sequences (87.5%), and of the dipolymers, 61.1% were also constituted by multiple A and T bases.In this study, we identified 78 repeats with more than one nucleotide repeat, totaling almost 50% of all SSRs identified. The 13 tri-, tetra-, penta-, and hexapolymers are shown in Table 3, as well as their size and location. From these 13 polymers identified, 9 are localized in intergenic spacers, 3 in coding sequences, and only 1 inside an intron. These results reveal the presence of several SSR sites in *P. lambertii*. Hereafter, these sites can be assessed for the intraspecific level of polymorphism, leading to highly sensitive phylogeographic and population structure studies for this species.

Tandem repeats with more than 30 bp and with a sequence identity of more than 90% have also been examined. Twenty-eight tandem repeats were identified in the *P. lambertii* cp genome (Table 4), of which 15 are located in coding regions of *accD* (2), *rps18* (1), *rps19* (1), *rps11* (1), *ycf1* (8), *rpl32* (1), *ycf2* (1); 11 are distributed in the intergenic spacers of *atpA*/*atpF* (1), *trnR*-CCG/*accD* (1), *rpl2*/*rps19* (1), *clpP*/*ycf1* (2), *ndhE*/*psaC* (1), *trnR*-ACG/*rrn5* (1), *rps12*/*rps7* (1), *ycf2*/*trnI*-CAU (1), *trnQ*-UUG/*psbK* (1), *psbK*/*psbI* (1); and 2 are located in the intron sequence of *rpoC1*. The cp genome of *P. lambertii* has 11 tandem repeats, more than the cp genome of *C. oliveri*, as well as a higher number of repeats in the *ycf1* (6) gene coding sequence [11]. The *ycf1* gene, previously considered as an enigmatic function in the cp genome, has recently been identified as encoding an essential protein component of the cp translocon at the inner envelope membrane (TIC) [50]. In *Salvia miltiorrhiza* and *Cocos nucifera*, two angiosperms, only 7 and 8 tandem repeats, respectively, of about 20 bp were identified, none of them located at the *ycf1* coding sequence [51,52], corroborating the theory that the IR influences the stability of the plastid genome.

Yi et al. [11] attributed the expansion of the *accD* ORF to the presence of tandemly repeated sequences. In the *P. lambertii* cp genome, we identified 2 tandem repeats in *accD* CDS, totaling 132 bp, or 44 codons. The *accD* reading frame length of the *P. lambertii* cp genome is 864 codons, similar to other cupressophyte species, such as *C. oliveri* (936 codons), *C. wilsoniana* (1,056 codons), *C. japonica* (700 codons) and *T. cryptomerioides* (800 codons). In contrast, the reading frame lengths of cycads, Ginkgo and Pinaceae, range from 320 to 359 codons, less than half the size found in cupressophytes. These results support the hypothesis of Hirao et al. [12] and Yi et al. [11] which holds that the *accD* reading frame has displayed a tendency toward enlarging sizes in cupressophytes.

The complete cp genome sequence of *P. lambertii* revealed significant structural changes occurring in the cp genome, even in species from the same genus. These results reinforce the apparently loss of rps16 gene in Podocarpaceae cp genome. In addition, several SSRs in the *P. lambertii* cp genome are likely intraspecific polymorphism sites which may allow highly sensitive phylogeographic and population structure studies, as well as phylogenetic studies, of species of this genus.

Author Contributions

Conceived and designed the experiments: LNV MR MPG RON EMS FOP. Performed the experiments: LNV HPFF HF RLAC. Analyzed the data: LNV HF HPFF RLAC. Contributed reagents/materials/analysis tools: EMS FOP RON MPG. Wrote the paper: LNV MR MPG.

References

1. Murray BG (2013) Karyotype variation and evolution in gymnosperms. In Leitch IJ, Greilhuber J, Dolezel J, Wendel JF, editors. Plant genome diversity 2: physical structure, behaviour and evolution of plant genomes. Springer231243.
2. Kelch DG (1998) Phylogeny of Podocarpaceae: comparison of evidence from morphology and 18S rDNA. Am J Bot 85: 986–996.
3. Farjon A (1998) World checklist and bibliography of conifers. Kew The Royal Botanical Gardens1316
4. Biffin E, Conran J, Lowe A (2011) Podocarp Evolution: A Molecular Phylogenetic Perspective. In: Turner BL, Cernusak LA, editors. Ecology of the Podocarpaceae in Tropical Forests. Washington: Smithsonian Institution Scholarly Press. pp 1–20.
5. Page CN (1990) Coniferophytina. In: Kramer KU, Green PS, editors. The families and genera of vascular plants, Pteridophytous and Gymnosperms: Springer. pp 332–346.
6. Ledru M, Salatino MLF, Ceccantini G, Salatino A, Pinheiro F, et al. (2007) Regional assessment of the impact of climatic change on the distribution of a tropical conifer in the lowlands of South America. Diversity Distrib 13:761–771.
7. de Laubenfels DJD (1985) A taxonomic revision of the genus *Podocarpus*. Blumea 30:251–278.
8. Myers N, Mittermeier RA, Mittermeier CG, Fonseca GAB, Kent J (2000) Biodiversity hotspots for conservation priorities. Nature 403:853–858.
9. Longhi SJ, Brena DA, Ribeiro SB, Gracioli CR, Longhi RV, et al. (2010) Fatores ecológicos deteminantes na ocorrência de *Araucaria angustifolia* e *Podocarpus lambertii*, na floresta Ombrófila mista da FLONA de São Francisco de Paula, RS, Brasil. Cienc Rural 40:57–63.
10. Parks M, Cronn R, Liston A (2009) Increasing phylogenetic resolution at low taxonomic levels using massively parallel sequencing of chloroplast genomes. BMC Biol 7:84.
11. Yi X, Gao L, Wang B, Su Y, Wang T (2013) The complete chloroplast genome sequence of *Cephalotaxus oliveri* (Cephalotaxaceae): evolutionary comparison of cephalotaxus chloroplast DNAs and insights into the loss of inverted repeat copies in gymnosperms. Genome Biol Evol 5:688–698.
12. Hirao T, Watanabe A, Kurita M, Kondo T, Takata K (2008) Complete nucleotide sequence of the Cryptomeria japonica D. Don. chloroplast genome and comparative chloroplast genomics: diversified genomic structure of coniferous species. BMC Plant Biol 8:70.
13. Wakasugi T, Tsudzuki J, Ito S, Nakashima K, Tsudzuki T et al. (1994) Loss of all ndh genes as determined by sequencing the entire chloroplast genome of the black pine *Pinus thunbergii*. Proc Natl Acad Sci USA 91:9794–9798.
14. Cronn R, Liston A, Parks M, Gernandt DS, Shen R et al. (2008) Multiplex sequencing of plant chloroplast genomes using Solexa sequencing-by-synthesis technology. Nucleic Acids Res 36(19):e122.
15. Lin C, Huang J, Wu C, Hsu C, Chaw S (2010) Comparative chloroplast genomics reveals the evolution of Pinaceae genera and subfamilies. Genome Biol Evol 2: 504–517.
16. Wu CS, Chaw SM (2013) Highly rearranged and size-variable chloroplast genomes in conifers II clade (cupressophytes): evolution towards shorter intergenic spacers. Plant Biotechnol J doi: 10.1111/pbi.12141
17. Moore MJ, Bell CD, Soltis PS, Soltis DE (2007) Using plastid genomic-scale data to resolve enigmatic relationships among basal angiosperms. Proc Natl Acad Sci USA 104:19363–19368.
18. Moore MJ, Soltis PS, Bell CD, Burleigh JG, Soltis DE (2010) Phylogenetic analysis of 83 plastid genes further resolves the early diversification of eudicots. Proc Natl Acad Sci USA 107:4623–4628.
19. Jansen RK, Cai Z, Raubeson LA, Daniell H, dePamphilis CW, et al. (2007) Analysis of 81 genes from 64 plastid genomes resolves relationships in angiosperms and identifies genome-scale evolutionary patterns. Proc Natl Acad Sci USA 104: 19369–19374.
20. Wu CS, Wang YN, Hsu CY, Lin CP, Chaw SM (2011) Loss of different inverted repeat copies from the chloroplast genomes of Pinaceae and Cupressophytes and influence of heterotachy on the evaluation of gymnosperm phylogeny. Genome Biol Evol 3:1284–1295.
21. Palmer JD, Thompson WF (1982) Chloroplast DNA rearrangements are more frequent when a large inverted repeat sequence is lost. Cell 29:537–550.
22. Strauss SH, Palmer JD, Howe GT, Doerksen AH (1988) Chloroplast genomes of two conifers lack a large inverted repeat and are extensively rearranged. Proc Natl Acad Sci 85:3898–3902.
23. Wu CS, Lai YT, Lin CP, Wang YN, Chaw SM (2009) Evolution of reduced and compact chloroplast genomes (cpDNAs) in gnetophytes: selection towards a lower cost strategy. Mol Phylogent Evol 52:115–124.
24. Vieira LN, Faoro H, Fraga HPF, Rogalski M, Souza EM et al. (2014) An improved protocol for intact chloroplasts and cpDNA isolation in conifers. PLoS ONE 9(1): e84792. doi:10.1371/journal.pone.0084792
25. Wyman SK, Jansen RK, Boore JL (2004) Automatic annotation of organellar genomes with DOGMA. Bioinformatics 20:3252–3255.
26. Schattner P, Brooks AN, Lowe TM (2005) The tRNAscan-SE, snoscan and snoGPS web servers for the detection of tRNAs and snoRNAs. Nucleic Acids Res 33: W686–689.
27. Lohse M, Drechsel O, Kahlau S, Bock R (2013) OrganellarGenomeDRAW: a suite of tools for generating physical maps of plastid and mitochondrial genomes and visualizing expression data sets. Nucl Acids Res doi: 10.1093/nar/gkt289
28. Kurtz S, Phillippy A, Delcher AL, Smoot M, Shumway M, et al. (2004) Versatile and open software for comparing large genomes. Genome Biol 5:R12.
29. Benson G (1999) Tandem repeats finder: a program to analyze DNA sequences. Nucl Acids Res 27:573–580.
30. Kurtz S, Choudhuri JV, Ohlebusch E, Schleiermacher C, Stoye J et al. (2001) REPuter: The manifold applications of repeat analysis on a genomic scale. Nucl Acids Res 29:4633–4642.
31. Wu CS, Lin CP, Hsu CY, Wang RJ, Chaw SM (2011) Comparative chloroplast genomes of Pinaceae: insights into the mechanism of diversified genomic organizations. Genome Biol Evol 3:309–319.
32. Wu CS, Wang YN, Liu SM, Chaw SM (2007) Chloroplast genome (cpDNA) of *Cycas taitungensis* and 56 cp protein-coding genes of *Gnetum parvifolium*: insights into cpDNA evolution and phylogeny of extant seed plants. Mol Biol Evol 24:1366–1379.
33. Tsudzuki J, Nakashima K, Tsudzuki T, Hiratsuka J, Shibata M, et al. (1992) Chloroplast DNA of black pine retains a residual inverted repeat lacking rRNA genes: nucleotide sequences of *trnQ*, *trnK*, *psbA*, *trnI* and *trnH* and the absence of *rps16*. Mol Gen Genet 232:206–214.
34. Guo X, Castillo-Ramírez S, González V, Bustos P, Fernández-Vázquez JL, et al. (2007) Rapid evolutionary change of common bean (*Phaseolus vulgaris* L.) plastome, and the genomic diversification of legume chloroplasts. BMC Genomics 8: 228.
35. Tangphatsornruang S, Sangsrakru D, Chanprasert J, Uthaipaisanwong P, Yoocha T, et al. (2009) The chloroplast genome sequence of mungbean (*Vigna radiata*) determined by high-throughput pyrosequencing: structural organization and phylogenetic relationships. DNA Res 17:11–22.
36. Fleischmann TT, Scharff LB, Alkatib S, Hasdorf S, Schottler MA, et al. (2011) Nonessential plastid-encoded ribosomal proteins in tobacco: a developmental role for plastid translation and implications for reductive genome evolution. Plant Cell 23:3137–3155.
37. Sugiura C, Sugita M (2004) Plastid transformation reveals that moss *tRNA*(Arg)-CCG is not essential for plastid function. The Plant J 40:314–321.
38. Schmidt M, Schneider-Poetsch HA (2002) The evolution of gymnosperms redrawn by phytochrome genes: the Gnetatae appear at the base of the gymnosperms. J Mol Evol 54:715–724.

39. Alkatib S, Scharff LB, Rogalski M, Fleischmann TT, Matthes A, et al. (2012) The contributions of wobbling and superwobbling to the reading of the genetic code. PLoS Genet. 8(11):e1003076.
40. Ambrogelly A, Palioura S, Soll D (2007) Natural expansion of the genetic code. Nat Chem Biol 3:29–35.
41. Weixlbaumer A, Murphy FV, Dziergowska A, Malkiewicz A, Vendeix FA, et al. (2007) Mechanism for expanding the decoding capacity of transfer RNAs by modification of uridines. Nat Struct Mol Biol 14:498–502.
42. Rogalski M, Karcher D, Bock R (2008) Superwobbling facilitates translation with reduced tRNA sets. Nat Struct Mol Biol 15:192–198.
43. Terakami S, Matsumura Y, Kurita K, Kanamori H, Katayose Y, et al. (2012) Complete sequence of the chloroplast genome from pear (*Pyrus pyrifolia*): genome structure and comparative analysis. Tree Genet Genomes 8:841–854.
44. Hill RS, Brodribb TJ (1999) Southern Conifers in Time and Space. Aust J Bot 47: 639–696.
45. Farjon A (2008) A natural history of conifers. Portland: . Timber Press1304
46. Morley RJ (2011) Dispersal and paleoecology of tropical podocarps. In: Turner BL, Cernusak LA, editors. Ecology of the Podocarpaceae in tropical forests. Washington: Smithsonian Institute Scholarly Press. pp 21–42.
47. Provan J, Powell W, Hollingsworth PM (2001) Chloroplast microsatellites: new tools for studies in plant ecology and evolution. Trends Ecol Evol 16:142–147.
48. Provan J, Corbett G, McNicol JW, Powell W (1997) Chloroplast variability in wild and cultivated rice (*Oryza* spp.) revealed by polymorphic chloroplast simple sequence repeats. Genome 40:104–110.
49. Provan J, Russell JR, Booth A, Powell W (1999) Polymorphic chloroplast simple-sequence repeat primers for systematic and population studies in the genus *Hordeum*. Mol Ecol 8:505–511.
50. Kikuchi S, Bédard J, Hirano M, Hirabayashi Y, Oishi M, et al. (2013) Uncovering the protein translocon at the chloroplast inner envelope membrane. Science 339: 571–574.
51. Huang Y-Y, Matzke AJM, Matzke M (2013) Complete sequence and comparative analysis of the chloroplast genome of coconut palm (*Cocos nucifera*). PLoS ONE 8(8): e74736.
52. Qian J, Song J, Gao H, Zhu Y, Xu J, et al. (2013) The complete chloroplast genome sequence of the medicinal plant *Salvia miltiorrhiza*. PLoS ONE 8(2):e57607.

Quantitative Analysis of Food and Feed Samples with Droplet Digital PCR

Dany Morisset*, Dejan Štebih, Mojca Milavec, Kristina Gruden, Jana Žel

Department of Biotechnology and Systems Biology, National Institute of Biology, Ljubljana, Slovenia

Abstract

In this study, the applicability of droplet digital PCR (ddPCR) for routine analysis in food and feed samples was demonstrated with the quantification of genetically modified organisms (GMOs). Real-time quantitative polymerase chain reaction (qPCR) is currently used for quantitative molecular analysis of the presence of GMOs in products. However, its use is limited for detecting and quantifying very small numbers of DNA targets, as in some complex food and feed matrices. Using ddPCR duplex assay, we have measured the absolute numbers of MON810 transgene and *hmg* maize reference gene copies in DNA samples. Key performance parameters of the assay were determined. The ddPCR system is shown to offer precise absolute and relative quantification of targets, without the need for calibration curves. The sensitivity (five target DNA copies) of the ddPCR assay compares well with those of individual qPCR assays and of the chamber digital PCR (cdPCR) approach. It offers a dynamic range over four orders of magnitude, greater than that of cdPCR. Moreover, when compared to qPCR, the ddPCR assay showed better repeatability at low target concentrations and a greater tolerance to inhibitors. Finally, ddPCR throughput and cost are advantageous relative to those of qPCR for routine GMO quantification. It is thus concluded that ddPCR technology can be applied for routine quantification of GMOs, or any other domain where quantitative analysis of food and feed samples is needed.

Editor: Joshua L. Heazlewood, Lawrence Berkeley National Laboratory, United States of America

Funding: The work was supported by the Slovenian Research Agency (grant Nos.: P4 0165). The funders had no role in study design, data collection and analysis, decision to publish, or preparation of the manuscript.

Competing Interests: The authors have declared that no competing interests exist.

* E-mail: dany.morisset@nib.si

Introduction

In many aspects of basic research, diagnostic tests, and commercial processes, the advent of modern analytical technologies has provided the ability to detect and quantify nucleic acid targets with unprecedented sensitivity and specificity. Currently, the most common technique for analyzing the presence of nucleic acids in food and feed samples is the polymerase chain reaction (PCR) [1–3]. When quantitative analysis is required, the use of real-time quantitative PCR (qPCR) is preferred because of its accuracy and precision [1]. However, its use for target quantification can be seriously limited by a significant bias when the target is present at low concentrations in a background of high numbers of non-target nucleic acids in the sample [4–7]. Another important limitation is its sensitivity to the frequent presence of inhibitors co-extracted with nucleic acid from complex matrices [8].

One example of the need for quantitative nucleic acid analysis in food and feed is the testing for genetically modified organisms (GMOs). Numerous countries have implemented regulations requiring the labeling of products containing GMOs, or materials derived from GMOs, above certain thresholds, therefore emphasizing the requirement for quantification of GMO content [9].

GMO content in food and feed samples is expressed in relative terms as the ratio of the quantity of the transgene (GM target, *i.e.* the nucleic acid fragment introduced in the host genome) to that of the endogene (reference gene in the host genome) [10]. Using qPCR–the common technique for GMO quantification-, standard curves for the quantities of endogene and transgene are prepared separately, using serial dilutions of DNA extracted from reference material [8]. qPCR efficiency and hence quantification by qPCR of endogene and transgene, can be influenced by many factors, including inhibitors, present in food and feed samples, leading to significant under- or over-estimation of GMO content [8,11]. Much effort has been put into improving the performance of qPCR quantification with respect to the inhibition and matrix effects [8] and the low concentration levels of targets in routine samples [12,13]. The lack of certified reference material has also been noted [13]. However, most of the proposed solutions are not practical and reliable quantification of GMOs in food and feed samples still remains a major challenge.

The basis of digital PCR (dPCR) is to quantify the absolute number of targets present in a sample, using limiting dilutions, PCR and Poisson statistics [14]. To do so, the PCR mix is distributed across a large number of partitions containing zero, one or more copies of the target nucleic acid. After end-point PCR amplification, each partition is scrutinized and defined as positive ("1", presence of PCR product) or negative ("0", absence of PCR product) hence the term "digital". The absolute number of target nucleic acid molecules contained in the original sample before partitioning can be calculated directly from the ratio of positive to total partitions, using binomial Poisson statistics [15].

Currently, two approaches are used in commercially available dPCR systems [15,16]. One approach, termed chamber digital PCR (cdPCR), relies on the partitioning of up to a few thousand individual reactions in microfluidic chambers. The second approach, called droplet digital PCR (ddPCR), combines partitioning of the PCR test into several thousands or millions of individual droplets in a water-oil emulsion, with the use of flow cytometry to count positive PCR tests.

dPCR has been adopted for a number of applications, including studies of copy number variation involving allelic discrimination or imbalance, single cell gene expression, hyper-methylation, detection of low copy number nucleic acid targets (reviewed [6,17,18]) and of point mutations. Recently, a cdPCR commercial system has been demonstrated that enables suitable metrological performance for the certification of the copy number ratio of reference materials used in GMO testing [11,19]. Several advantages are proposed for the use of dPCR instead of qPCR in routine GMO testing: 1) it enables absolute target copy number to be detected and avoids the amplification efficiency bias observed with qPCR [11,19], 2) it overcomes dependence on the availability of references or standards [18], 3) it provides data with the high precision and confidence necessary for metrological use [11,19], 4) it provides more accurate data at low target copy numbers than qPCR [20], allowing quantification of low GMO content, and 5), because of its tolerance to inhibitors as an end-point measurement, it can reduce the biases linked to matrix type often observed with qPCR [18]. However, the application of cdPCR is limited by two important factors: the small dynamic range it offers (2–3 logs) and its relatively high price.

Given the larger number of replicates allowed by ddPCR than by cdPCR and the lower price per sample of the former, it has been envisaged that ddPCR could allow better precision [16], confidence and easier adoption of digital PCR technology in laboratories for daily analysis, all at lower cost per sample [15]. The aim of this study was therefore to evaluate the application of ddPCR for quantitative analysis in food and feed samples. Taking GMO testing as a concrete example, ddPCR key performance parameters, using the QX100 droplet system (Bio-Rad, Pleasanton, CA), were compared with current qPCR performance and with the recently studied performance of cdPCR [11,18]. Linearity of response, absolute limits of detection and quantification, repeatability over the dynamic range of the ddPCR endogene and transgene assays were assessed. The applicability of ddPCR with different sample matrices and the practicability of use for routine GMO testing were also evaluated. The results obtained should be applicable to other fields where quantitative testing in food and feed samples is required.

Materials and Methods

Test material

Several MON-ØØ81Ø-6 (MON810) maize seed powder based, certified reference materials (CRM) were purchased from the EU Joint Research Centre, IRMM (Institute for Reference Materials and Measurements, Geel, Belgium). All these CRMs have certified mass/mass (m/m) GM maize/wild-type maize material ratios. Some of them are also certified for the copy/copy (cp/cp) transgene/endogene ratio (see Appendix S1 and Table S1).

Other samples containing the MON810 maize event, and previously assayed by qPCR, were also used in this study. Finally, a limited specificity study was conducted on two samples containing either wild-type maize, or DNA from a milk sample without maize (see Appendix S1 and Table S1).

DNA extraction

DNA was extracted and purified from 200 mg of starting material for all samples using the NucleoSpin Food kit (Macherey-Nagel GmbH & Co. KG, Düren, Germany). In parallel, DNA from sample G147/08 was also extracted and purified according to a modified CTAB method (see Appendix S1) [21].

Dilutions of the stock extracted DNA solutions were made with nuclease- and protease-free water (Sigma-Aldrich Chemie Gmbh, Munich, Germany), using pipettes calibrated with a SAG285 precision weighing module (Mettler-Toledo d.o.o., Ljubljana, Slovenia). All samples were stored at −20°C.

Enzymatic restriction of genomic DNA

Enzymatic digestion of MON810 genomic DNA (gDNA) with *Taq*I (New England Biolabs GmbH, Frankfurt am Main, Germany) was performed as described (see Appendix S1) [19]. 6 µL of digested gDNA were analyzed on a 1% agarose gel to confirm complete digestion.

qPCR reactions and data analysis

The *hmg* gene was used as the endogenous control gene for maize. A unique, single copy DNA integration-border region of the genomic sequence and the inserted sequence element originating from CaMV (35S promoter) were used for specific detection and quantification of the MON810 event. Probe and primer nucleotide sequences were the same as in the inter-laboratory validated protocol [22] but the TAMRA quencher in the probes was replaced by the Black Hole Quencher 1 (BHQ-1). The same primers and probes were used for both qPCR and ddPCR experiments (see Appendix S1 and Table S2).

MON810 content was determined by qPCR, using relative quantification according to the standard curve approach. Standard curves were prepared from five serial dilutions of the copy/copy ratio certified reference material ERM-BF413gk (starting from approximately 100 ng to 1 ng DNA per reaction) and used in two replicates. For each sample, the quantification was done based on two replicates of three dilutions. Results of quantification performed with CRM certified for transgene/endogene copy ratio were expressed as percentages of the copy/copy ratio.

Droplet Digital PCR reactions and data analysis

Duplex ddPCR reaction mixes were prepared as follows. 10 µL of 2× ddPCR Master Mix (Bio-Rad, Pleasanton, CA) and 1 µL of each primer (final concentration of 300 nM) and probe (final concentration of 180 nM) were mixed, and 4 µL of DNA template added. For singleplex reactions, 3 µL of nuclease- and protease-free water (Sigma-Aldrich Chemie Gmbh, Munich, Germany) were added to complete a 20 µL reaction volume. Final primer and probe concentrations (purchased at Eurofins MWG Operon, Ebersberg, Germany) in ddPCR mixes were identical to the qPCR conditions used in this study, and to those used in the previously described chamber digital PCR (cdPCR) conditions [11] (see Appendix S1 and Table S2).

ddPCR workflow and data analysis were performed as described (see Appendix S1) [15].

Determination of ddPCR key performance parameters

Comparison of singleplex and duplex reactions. The ddPCR duplex assay was evaluated using three 8-well cartridges containing the singleplex *hmg*, the singleplex MON810 and the duplex ddPCR assay respectively. In each cartridge, one well was filled with a non template control (NTC) ddPCR mix, while the seven other wells contained ddPCR mixes with DNA extracted from the ERM-BF413ek CRM (average of 46,571 *hmg* copies and 324 MON810 copies). Droplets were generated for each individual cartridge, and those droplets containing the PCR mixes of the three cartridges were transferred onto a single PCR plate for amplification followed by droplet count.

Dynamic range, repeatability, limits of detection and quantification. A dilution series was prepared with MON810 maize DNA extracted from the ERM-BF413gk CRM. DNA quantification in the initial MON810 maize DNA solution was estimated by qPCR as described [23]. The dilution series consisted of 14 solutions containing from approximately 118000 to 0.02 copies of *hmg*, and from 4300 to 0.0008 copies of MON810 per 20 µL, respectively). Five replicates of the dilution series and of a non template control (NTC) were measured by ddPCR. For qPCR, measurements were made in duplicate. Linearity over the dynamic range was determined by the coefficient of correlation R^2, calculated on the average of the target copy numbers measured in the replicated dilution series for both qPCR and ddPCR. Repeatability over the dynamic range was determined by the coefficient of variation (cv) of the measured target copy number or of the MON810 content between the replicates of the dilution series. The absolute limit of quantification (aLOQ) and absolute limit of detection (aLOD) for qPCR and ddPCR were determined on these experimental results.

An additional set of experiments was performed to establish repeatability between different emulsification runs. Intra- and inter-cartridge (ddPCR 8 well chips) repeatability was determined on five independent series consisting of seven replicates of the ERM-BF413ek (approximately 100 ng and an average of 46571 *hmg* copies and 324 copies of MON810 per 20 µl reaction) and one NTC. Four series were prepared by two operators and droplet amplification reactions were performed simultaneously on the same 96-well plate. The fifth series was prepared by one operator on a second day and ddPCR reactions were performed on another 96-well plate. A total of 35 replicate positive ddPCR results were then analyzed.

Specificity. The DNA extracts from samples G053/12 (approximately 6,200 copies of *hmg*) and G031/12 (approximately 100 ng DNA) were tested with the duplex MON810/*hmg* ddPCR assay. A total of eight replicate reactions per sample were performed (see Appendix S1 for more details).

Applicability. Samples representative of four different maize-containing matrices from routine GMO testing were used to test the applicability: seed-powder flour, corn flakes, wheat seed-powder flour with maize contamination, and feed containing maize (see Appendix S1 and Table S1).

Results and Discussion

Given the limitations of qPCR for the quantification of GMO in food and feed samples, especially at low target levels and in some complex matrices, the use of ddPCR in routine GMO quantification was evaluated, following the generally accepted minimum performance requirements for analytical methods [24,25].

In order to avoid biases as much as possible when comparing qPCR and ddPCR quantification, we transferred the inter-laboratory validated qPCR *hmg*- and MON810-specific assays to ddPCR technology with the minimum of adaptation. Therefore, apart from the mastermix and settings that are specific to the QX100 droplet system, primers and probe nucleotide sequences and concentrations, DNA concentration, and PCR thermo-profile were kept identical to those in the qPCR assays.

ddPCR can readily be set as a duplex application

Because GMO content is calculated based on the ratio of transgene/endogene quantities, it would be more practical to perform endogene–transgene duplex reactions to reduce costs. For this reason, evaluation of duplex qPCR and ddPCR assays was performed and compared to the performance of singleplex assays. The primers and probes of both the *hmg* and MON810 systems were mixed in the qPCR or ddPCR volumes to final concentrations equal to those in the original singleplex assays.

Evaluation of qPCR duplex systems has shown that the *hmg* system performed identically in duplex and singleplex qPCR reactions, while MON810 amplification was significantly affected in the duplex reactions, showing signal values approximately 5.5 Cq higher than in singleplex reactions (data not shown). An attempt to optimize the duplex qPCR assay was made by varying primers and probe concentration of both systems. However, the *hmg* and MON810 were affected differently under all the tested conditions, resulting in under-estimations of the MON810 content and/or loss of sensitivity (data not shown). These results are not surprising: the difficulty of multiplexing qPCR assays is well documented, including its application to GMO quantification [26–28]. One of the limitations is the need to choose target sequences with similar, short lengths. Another difficulty linked to GMO detection is that the event-specific targets needed for reliable and specific GMO quantification are usually the junction regions between the transgene and the host plant genome, leaving a very narrow window for design, and decreasing further the flexibility for a multiplex design. Also, the need is usually to quantify low concentrations of transgene (down to 0.1%) in a background of high endogene quantities. This asymmetry in concentration renders establishment of a qPCR duplex assay targeting both the MON810 transgene and the *hmg* endogene even more difficult. In the following experiments, results of only the singleplex qPCR assays were used for comparison with ddPCR assay results.

For ddPCR, no significant variation of the measured target copy number was observed between the singleplex and duplex ddPCR assays for both *hmg* (bias = −1.8%) and MON810 systems (bias = 3.7%). Similarly, no significant variation of the measured MON810 content was observed between the singleplex and the duplex ddPCR assays (bias = 5.8%) (Table S3). Additionally, the repeatability of the duplex assay in measurements of the MON810 content appears slightly better that of the singleplex assays. From these results, it was concluded that the duplex ddPCR assay performs as well as the singleplex ddPCR assays without any additional optimization. This was further confirmed by the successful establishment of several additional duplex ddPCR assays from singleplex qPCR assays, targeting GM events or screening elements, without optimization (data not shown).

Enzymatic restriction of genomic DNA

It was recently suggested that it is preferable to expose gDNA to endonuclease restriction in order to improve amplification efficiency and to increase the accuracy of GMO target copy number measurement with cdPCR [19]. gDNA endonuclease restriction for ddPCR testing was also evaluated and found unnecessary (Table S4).

Digestion of DNA may be necessary for digital PCR analysis in the case of a possible linkage between targets, such as multiple copies of target physically linked on the same chromosome, or if different targets present on a same plasmid need to be quantified. In the case of GMO quantification, the target of the event-specific assay and the target chosen for endogene quantification are present as unique copies in the host plant genome and are not linked. In the case of MON810 maize, results of the two dimensional analyses of the droplet signals in digested and undigested DNA samples (Figure S1 in Supporting Information) suggests the absence of linkage between MON810 positive droplets and *hmg* positive droplets, thus confirming the independence of the two targets. Further, ddPCR performance was evaluated using non-digested gDNA.

Dynamic Range, precision and limits of quantification

A recent study has estimated that the theoretical ddPCR dynamic range is 10^5 target copies, and it has been established experimentally that the dynamic range covers more than 4 orders of magnitude [15].

The ddPCR duplex assay response was investigated over target concentrations ranging from approximately 0.02 to 118,000 *hmg* copies and from approximately 0.0008 to 4300 MON810 copies per 20 µL of ddPCR reaction. Due to pipetting errors, that were noted after loading the 8-well cartridges with ddPCR mixes, data from one reaction (1.4% of the total dataset) were excluded from the analysis. The average number of droplets read for each ddPCR reaction included in the data analysis was 13,606 with a standard deviation of 931 droplets (coefficient of variation cv = 6.8%).

The ddPCR response was linear over concentrations ranging from an average of 5 to 118,000 *hmg* copies (0.02% to 99.5% positive droplets) with a coefficient of correlation (R^2) of 0.9990. Similarly, the ddPCR response for MON810 was, on average, linear from 6 to 4,340 MON810 copies ($R^2 = 0.9993$; 0.03% to 17.9% positive droplets in average) (Figure 1). This performance was similar to those of the two singleplex qPCR assays, which was linear over the same dynamic ranges (R^2 *hmg* = 0.9939 and R^2 MON810 = 0.9958) (data not shown). The ddPCR linear response for the MON810/*hmg* duplex assays covered a broader range than the same assays tested in cdPCR which was limited to 2–3 orders of magnitude [18,19]. This wider range of concentrations can be attributed to the large number of partitions available for reactions in ddPCR (13,606 droplets on average in this work) compared to the number (765) available for cdPCR test. It has already been asserted that qPCR offers a much broader dynamic range than digital PCR [16]. The dynamic range observed for ddPCR covers the whole range of target concentrations usually needed by a laboratory for routine GMO testing (0.1% to 100% transgene/endogene ratio cp/cp).

For individual targets and for GMO content, the coefficient of correlation R^2 obtained with ddPCR met the requirements ($R^2 > 0.98$) set by the European Union Reference Laboratory for GM Food and Feed [25] for acceptance of a quantitative PCR-based detection method for GMO.

All samples used for determining the dynamic range came from serial dilution of a single stock MON810 maize DNA sample. At a higher concentration (118,000 *hmg* copies per 20 µl ddPCR mix), each droplet contained, on average, 5.9 *hmg* molecules, which is the upper recommended concentration for use of the droplet system (Bio-Rad, personal communication). This finding supports the fact that the duplex MON810/*hmg* ddPCR assay can be used over a wide range of target concentrations to determine the MON810 content in a given sample, and that values around 115,000 copies constitute the upper range of quantification with ddPCR.

The absolute limit of quantification (aLOQ) is the lowest target copy number in a sample that can be reliably quantified with an acceptable level of precision and accuracy [25]. The aLOQ of the *hmg* or MON810 ddPCR systems was estimated as the lowest copy number within the dynamic range with a coefficient of variation (cv) of the measured copy number≤25% [25]. Based on this criterion, aLOQ was estimated to be around 5 copies for the *hmg* system, and 18 copies for the MON810 system and for the duplex ddPCR assay (Table S5). As a comparison, it is usually agreed that aLOQ of qPCR assays range from 30 to 100 target copies per reaction [12,29]. The aLOQ of the qPCR MON810 specific method used in this study was initially estimated at 10 copies of the target MON810 sequence [22], and at a higher range of 31–63 copies in another recent study [18]. In our laboratory, the aLOQ estimated for the qPCR assay on the same DNA is 18 copies. The aLOQ of the duplex ddPCR assay was therefore equal to or better than that estimated for the qPCR performance. Similarly, the aLOQ of the duplex ddPCR assay was also in a range similar to the aLOQ measured in cdPCR (15–56 transgene copies), assessed only on the MON810 assay [18].

Sensitivity

The absolute limit of detection (aLOD) is the lowest target copy number in a sample that can be reliably detected, but not necessarily quantified [25]. For this study, aLOD was calculated based on experimental data obtained to determine the dynamic range. aLOD was determined as the lowest concentration level for which all five ddPCR replicates resulted in at least two positive droplets per reaction. aLOD was estimated to be five copies for the *hmg* system, and six copies for the MON810 system, which is suitable for routine GMO testing. The absolute sensitivity is, according to our own assessment (data not shown), lower than the one observed for the MON810 singleplex qPCR assay at around 6–18 copies, and comparable to the one observed in cdPCR [18].

Repeatability

Intra- and inter-cartridge repeatability of the ddPCR was assessed by two different operators and over two different days for both *hmg* and MON810 target copy number determinations and for MON810 content determination. Less than 10% variability was observed within each of the five cartridges for the determination of *hmg* copies, MON810 copies, and MON810 content. Similarly, comparison of the values between the five cartridges showed low variability (cv<10%) for all the three measured parameters (Figure 2 and Table S6).

The overall repeatability could also be estimated by analyzing the results of the experiment performed for aLOQ and dynamic range determination. All along the dynamic range, the cv of the determined *hmg* copies, MON 810 copies, and MON810 content remained below the threshold for acceptance of quantitative methods (cv<25%) (Table S5).

It has already been observed with both ddPCR [15] and cdPCR [19] that the relative uncertainty in target copy number varies across the dynamic range, with higher uncertainty, and consequently higher measurement variability, as the target copy number decreases. In this study, similar variability of the measured target copy numbers (Table S5 and Figure 1) and of the MON810 content was observed at lower target concentrations (Table S5 and Figure 3).

In all experiments and for all three parameters (hmg and MON 810 copy number, MON810 content), the coefficient of variability measured at each point of the dynamic range was far below the 25% threshold set in international guidance documents for validation of GMO testing methods [24,25]. These experiments demonstrate that, using ddPCR, one can obtain repeatable and comparable quantitative estimates of GMO target number or content.

Trueness

Trueness is defined as the closeness of agreement between the average value obtained from a series of test results and an accepted reference value [25]. Acceptance criterion for trueness is that the

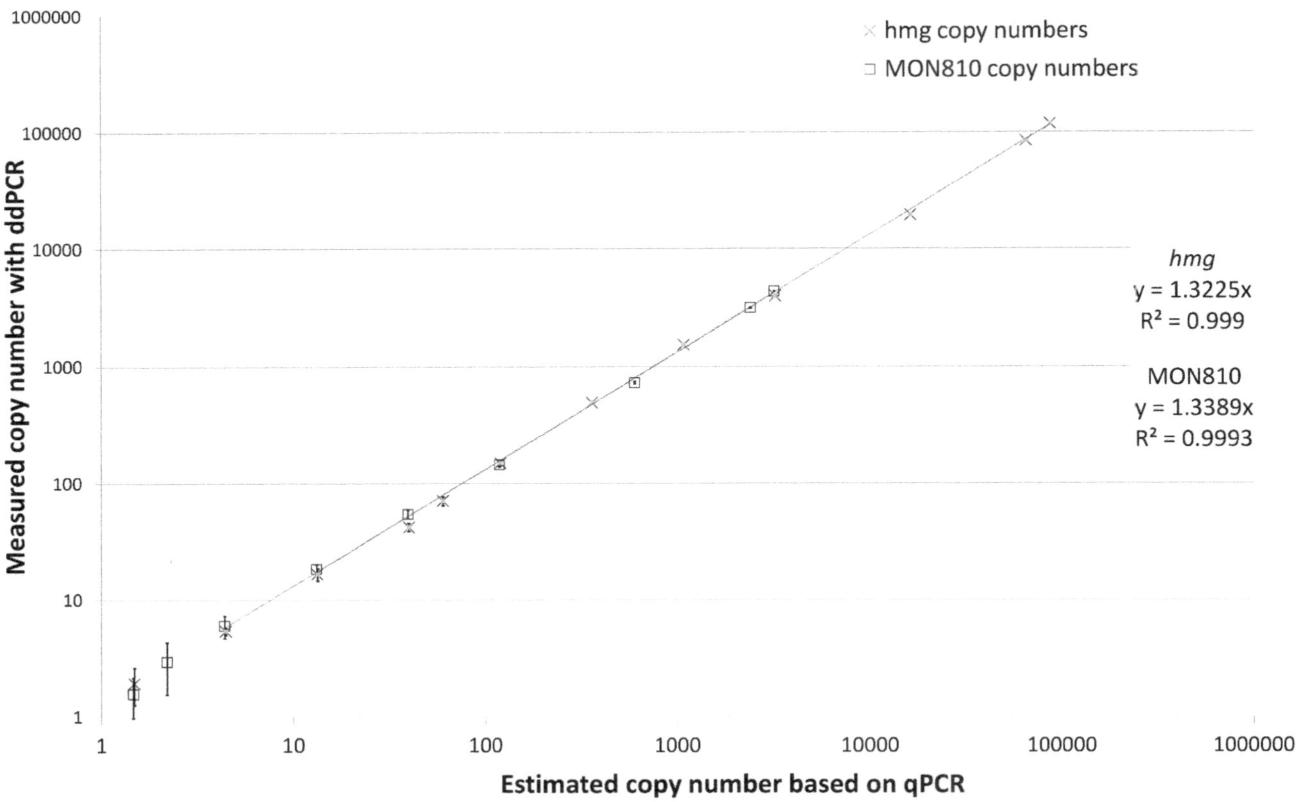

Figure 1. Dynamic range of the ddPCR duplex assay. Five replicates for each data point. Error bars represent the standard deviation between the five replicates at each target concentration.

measured value has to be within ±25% of the accepted reference value over the whole dynamic range [24,25].

To assess trueness, data generated by the experiments for dynamic range determination and for intra- and inter-cartridge repeatability performance were used. In the absence of a DNA reference material certified for absolute copy number concentration, trueness could only be assessed for the MON810 content.

In the experiment performed to determine the intra- and inter-cartridge repeatability (on CRM ERM-BF413ek), the average value of the pooled ddPCR data showed good agreement with the certified value (Table 1) and, in every case, the MON810 content measured by ddPCR (within the dynamic range) was within ±25% of the certified value (Figure 2). By comparison, the MON810 content measured by qPCR was close to the limit of acceptance (Table 1).

The average value of the pooled ddPCR duplex assay data at each dilution level of the dynamic range showed good agreement with the reference value of 3.85% cp/cp (ERM-BF413gk) (Table 1). The MON810 content measured by qPCR was similar to that of the ddPCR value, the latter being slightly closer to the target value (Table 1). Throughout the dynamic range, each individual ddPCR measurement of the MON810 content fell within ±25% of the certified value (Figure 3). It is noteworthy that the deviation between the MON810 content measured by ddPCR and the reference value tended to increase with lower target copy number. Nevertheless, the MON810/hmg ddPCR duplex assay met trueness acceptance criteria throughout the whole dynamic range. Trueness was also evaluated on two additional CRMs and on samples from the ILC-EURL-GMFF proficiency program: ddPCR results showed better agreement with the target values than qPCR and were in accordance with the trueness acceptance criterion (Table 1).

Specificity

The ddPCR amplicons used for this study are the same as for the qPCR singleplex assays that were subjected to inter-laboratory validation and for which specificity was thoroughly checked at this stage. Moreover, the key assay parameters (primer and probe concentration, thermal profiles) were not modified. Therefore, the specificity of the ddPCR amplicons was not thoroughly verified as it is inferred that specificity should be the same as for the qPCR singleplex assays. A limited specificity study was however conducted on two samples containing either wild-type maize, or DNA from milk sample without maize. In all cases, a false-positive rate far below 5%, which is the generally used acceptance criterion, was observed (Table S7). In addition, for all experiments performed in this study, NTC resulted in no signal (no positive droplet).

Applicability

Another important factor when introducing new methods or technologies for testing GMOs in food and feed is their applicability. More specifically, their ability to perform well with different sample matrices and within a range of concentrations relevant for GMO testing has to be demonstrated [24].

MON810 contents measured by the ddPCR duplex MON810/*hmg* assay in maize seed-powder flour samples and corn flakes samples are in good accordance with the values measured with the qPCR singleplex MON810 and *hmg* assays (Table 1).

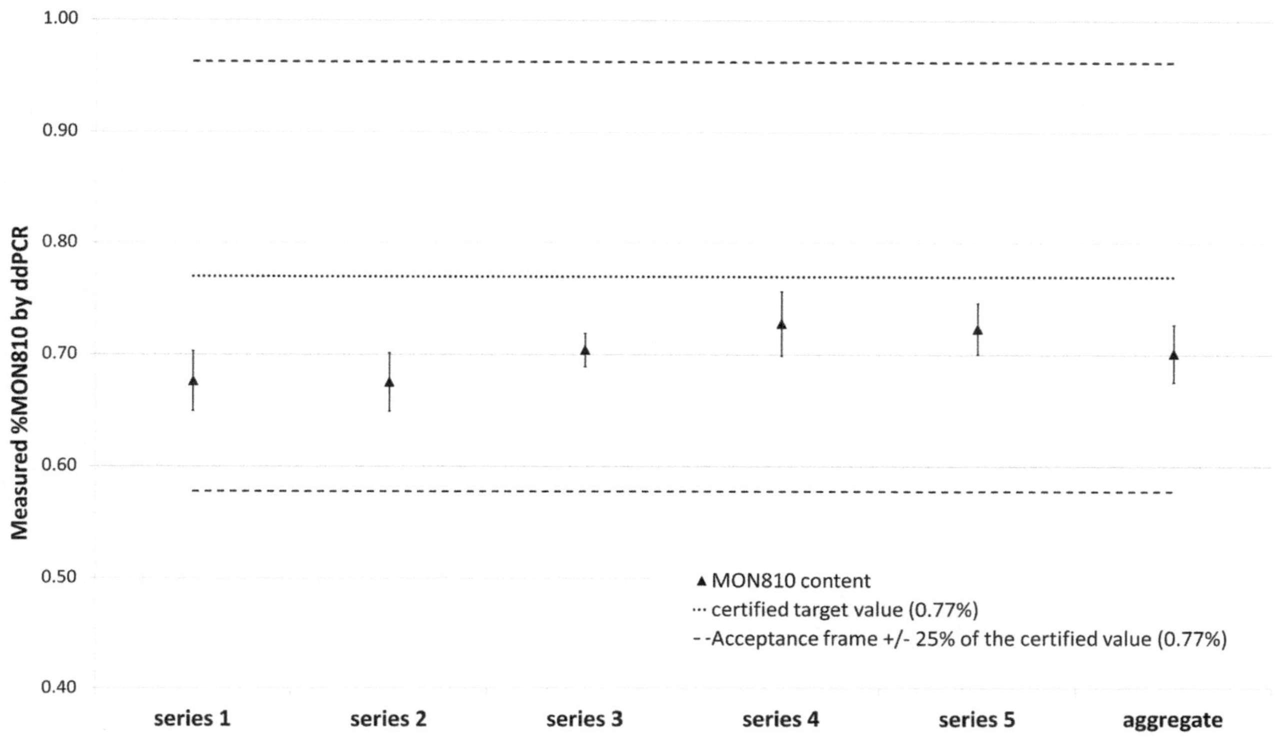

Figure 2. Repeatability results of the ddPCR duplex assay. MON810 content measured by ddPCR in five series of seven replicates. The aggregate represents the sum of the five series. The target certified MON810 content (0.77%) is indicated by a dotted line. Acceptance criterion for repeatability is ±25% of the target content (from 0.58% to 0.96%) represented by the dashed lines. Error bars represent the standard deviation between the replicates for each series or in the aggregate.

During qPCR tests, we detected the presence of inhibition in the stock DNA solution of two samples (wheat seed-powder flour with maize contamination and maize feed), as indicated by the large differences in MON810 content (cv>25%) calculated from different sample dilutions. Consequently, diluted DNA samples were used to determine the MON810 content with qPCR. In the wheat sample, the generally used NucleoSpin Food kit (NSF) extraction protocol resulted in strong inhibition of the qPCR amplifications, and new DNA extraction following the CTAB protocol was needed. Both stock and diluted DNA solutions from the CTAB extract could be used for MON810 content determination with qPCR. The ddPCR measurements of MON810 content were in agreement with the accepted values obtained with qPCR but with a bias slightly above the acceptance limit for the wheat seed-powder flour sample (Tables 1 and S6). It has to be noted that there is substantial empirical knowledge about the efficacy of DNA extraction methods related to sample matrix. GMO laboratories use adapted DNA extraction methods that in some cases may reduce inhibition during the qPCR analysis. Furthermore, inhibition is already assessed during the screening and identification phases of GMO testing, so there is no risk that substances in a DNA extract would totally inhibit the qPCR or ddPCR amplification reaction, resulting in false negatives.

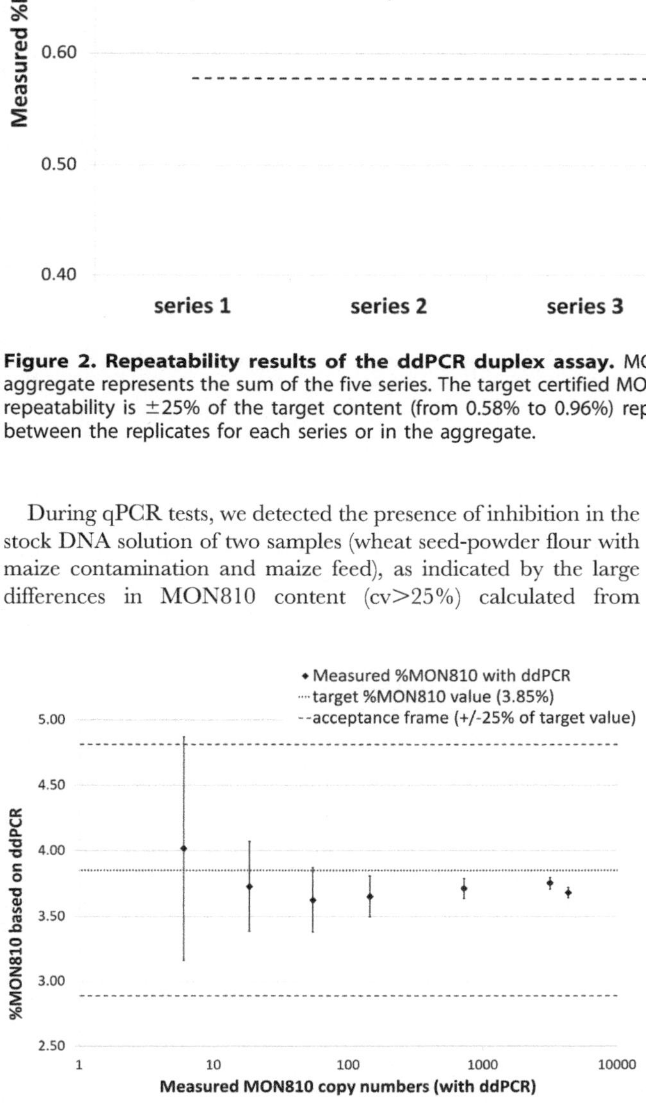

Figure 3. Precision of the duplex ddPCR assay as a function of the target concentration. MON810 content measured by ddPCR in five series of seven target concentrations. The target MON810 content (3.85%) is indicated by a dotted line. Acceptance criterion for precision is ±25% of the target content (from 2.89% to 4.81%) represented by the dashed lines. Error bars represent the standard deviation of the measured MON810% by ddPCR at each target concentration (five replicates per target concentration).

Interestingly, for the DNA extracts obtained with NSF, very low differences were observed between the MON810 content determined by ddPCR in the stock (presenting inhibition with qPCR) and the diluted DNA solutions for both matrices (Table S8 in Appendix S2). This result suggests that the ddPCR duplex assay is more tolerant to inhibitors found in some complex food and feed matrices than the qPCR assays, as suggested earlier [18].

In summary, the ddPCR MON810/*hmg* duplex assay can be applied for routine quantification of the MON810 maize, as demonstrated on a large range of transgene content (experimentally from 0.04% to 29.6%) usually found in samples. Moreover, its

Table 1. Application of the ddPCR duplex assay on different sample matrices.

Sample	Target value	Average MON810% (ddPCR)	Bias MON810% (ddPCR)	Average MON810% (qPCR)	Bias MON810% (qPCR)
ERM-BF413d	0.57%[a]	0.62%	8.0%	0.46%	−19.3%
ERM-BF413f	2.85%[a]	2.92%	2.5%	2.29%	−19.6%
ERM-BF413ek	0.77%[a]	0.70%	−9.0%	0.58%	−24.7%
ERM-BF413gk	3.85%[a]	3.68%	−4.1%	3.66%	−4.9%
G0009/04	0.29%[c]	0.26%	−11.7%	0.29%	n.a.
G0180/07	0.04%[c]	0.04%	2.9%	0.04%	n.a.
G211/10	0.45%[b]	0.46%	−1.8%	0.50%	11.1%
G212/10	2.10%[b]	2.32%	10.4%	2.30%	9.5%
G231/11	2.64%[c]	2.31%	−12.4%	2.64%	n.a.
G0147/08	29.6%[c]	21.7%	−26.7%	29.6%	n.a.
G254/11	3.82%[c]	3.47%	−9.2%	3.82%	n.a.

Target value: MON810 content expressed as a percentage in cp/cp ratio.
[a]Target value certified by the CRM provider or evaluated from the certified value of another CRM in the same series. [b]Target value in cp/cp ratio attributed in the ILC-CRL-GMFF proficiency program. [c]Target value estimated by qPCR, using a CRM certified in cp/cp ratio (same as "Average MON810% (qPCR)").
Average MON810% (ddPCR): Average of the MON810 content measured by ddPCR.
Bias MON810% (ddPCR): Bias of the average MON810 content measured by ddPCR with the target value.
Average MON810% (qPCR): Average of the MON810 content measured by qPCR.
Bias MON810% (qPCR): Bias of the MON810 content measured by qPCR with the target certified value.
n.a.: not applicable (value determined by qPCR is the target value).

use for several types of food, feed and seed matrices commonly found in routine samples has been verified.

Practicability

Before introducing a new technology in a laboratory, one has to verify its practicability for daily use [24].

To do so, calculations were made based on the simultaneous quantification of four samples-the average number run in parallel in middle-size GMO laboratories. The quantitative analysis of four samples using *hmg* and MON810 singleplex qPCR assays requires a total of 96 reactions (see set up A, Table S9 in Appendix S2). This number of reactions is high, due mainly to the use of inter-laboratory validated singleplex methods, the need for monitor inhibition that requires additional dilutions, and the need for a standard curve. With ddPCR, the standard curve is not needed and a duplex MON810/*hmg* assay with a wide dynamic range is used. Since inhibition is assessed during the screening and identification phases of GMO testing and given the demonstrated tolerance of the assay for amplification inhibition, it is not necessary to control the inhibition at the ddPCR quantification stage. Thus, a simple ddPCR testing set-up is proposed (set up B, Table S9 in Appendix S2) that included two replicate reactions for each test portion of the sample, in accordance with the ISO 21570:2005 standard [30]. Including NTC and quantification control reactions in the set-up, a total of 20 reactions would be necessary to reliably quantify MON810 in four samples.

Based on the experience acquired during this study and assuming the samples and mixes are already prepared, the simultaneous analysis of these four samples with ddPCR would require approximately 190 minutes. For comparison, qPCR would take 160 minutes to generate results (see Table S10 in Appendix S2). In terms of hands-on-time, ddPCR would require approximately 15 min less than qPCR. The main difference between the two approaches can be attributed to the time needed by the droplet reader to analyze individual droplets.

Considering the cost of reagents, consumables and labor at NIB and the above proposed set-ups for GMO testing, the quantification of a given transgene in four samples with ddPCR would cost approximately US$20.9 per sample and US$22.3 using qPCR. If more samples must be handled simultaneously, ddPCR shows even better throughput and smaller cost than qPCR (see Tables S11 and S12 in Appendix S2). The use of cdPCR, taking into account prices per chip or plate (from US$150 to US$400 each) [16] and using the proposed set-up for ddPCR (four runs, i.e. chips or plate per sample), would lead to a cost per sample much greater than that required for current routine detection of GMOs in most laboratories.

Quantification of routine samples using ddPCR is therefore practical and has the potential to provide better throughput and cost-effectiveness than qPCR for GMO laboratories. A MIQE checklist is made available in Appendix S3.

Conclusion

The intention of this study was to demonstrate the usability of ddPCR in real-life routine diagnostics, rather than to re-investigate the recently reported metrological characteristics of this technology [15]. The applicability of ddPCR was investigated for the quantification of GMO in food, feed and seed samples. The ddPCR MON810/*hmg* duplex assay presented here and implemented without optimization from the inter-laboratory validated singleplex qPCR assays, achieves a wide dynamic range close to five orders of magnitude with an upper limit of quantification of about 118,000 target copies. It also shows very good sensitivity, suitable for GMO testing. The excellent performance of the tested parameters enables the quantification of samples from different matrices, using DNA extracted with common methods without up-front DNA quantity estimation. The limits of quantification, trueness and repeatability of the duplex assay comply with international recommendations [24,25] and are comparable or superior to those of the inter-laboratory validated qPCR singleplex assays (Table 2).

Table 2. Summary table of qPCR, ddPCR and cdPCR performance for MON810 detection and quantification.

Parameter tested	qPCR	ddPCR	cdPCR
"plexing" of the assay	Singleplex	Duplex	Duplex
Need for endonuclease treatment	No	No	Yes
Dynamic range	5 logs	5 logs	2–3 logs [18,19]
Absolute limit of quantification	18 copies (MON810, this study), 30–60 copies [18]	5 copies (*hmg*), 18 copies (MON810)	15–56 copies (MON810) [18]
Absolute limit of detection	1–10 copies (this study and [18])	5 copies (*hmg*), 6 copies (MON810)	1–10 copies (MON810) [11,18,19]
Repeatability through the dynamic range	Cv<35% (%Mon810) [22]	Cv<25% (*hmg* cp nb), Cv<20% (Mon810 cp nb), Cv<19% (%Mon810)	Cv<11% (*hmg* cp nb), Cv<22% (Mon810 cp nb), Cv<23% (%Mon810) [11]
Trueness	From −16.7% to 2.3% [22]. From −24.7% to 11.1% (this study)*	From −9.0% to 10.4% (this study)*	From −21.20 to 41.4% [18]
Applicability			
- *Matrices tested in this study*	Food, feed and seeds	Food, feed and seeds	Seeds, plasmid
- *GMO content range tested in this study*	0.04% to 26.4%	0.04% to 26.4%	0.07% to 57.1% [11]
- *Sensitivity to inhibitors*	Sensitivity to inhibition (use dilutions, alternative DNA extraction)	Less (not?) sensitive to inhibition	N.A.
Practicability			
- *Number samples/96-well plate*	4 samples	23 samples	N.A.
- *Time for results/96-well plate*	3 hours	6 hours	N.A.
- *Price/sample if 96-well plate*	US$23.3	US$16.1	N.A.

qPCR: data produced in this study, or obtained from the literature, when indicated [18,22]
ddPCR: data produced in this study.
cdPCR: data produced on a BioMark System (Fluidigm, South San Francisco) using the 12.765 digital arrays (Fluidigm) and obtained from the literature [11,18,19].
Repeatability through the dynamic range: assessed through the coefficient of variation (Cv) of the target copy numbers or the MON810 content between repeats.
Trueness: assessed through the calculation of the bias between the MON810 content measured and the target MON810 content. * For our study, trueness is indicated only when qPCR and ddPCR results could be compared to a third, independent value (obtained from the CRM provider or proficiency test organizer).
Time for results/96 well plate: Total time needed from DNA pipetting to the analysis of the results; reaction mixes are already prepared.
Price/sample if 96-well plate: Price based on material and reagent costs available at NIB, including labor cost.
N.A.: not evaluated.

Applicability of the technology has been verified on representative matrices found in routine samples, and on the range of GMO content usually found in routine samples and relevant to different international labeling requests. Unlike qPCR, quantification by ddPCR has been found to be insensitive to the amplification inhibition present in some DNA extracts. It is also very precise at very low levels of target content. The use of the ddPCR duplex assay in routine GMO analysis was shown to be practical, following the new test set-up proposed in this study.

It was recently discussed as to whether price is a limiting factor to the adoption of dPCR in the laboratory [16]. The data provided here show that, in the context of GMO quantification, ddPCR running costs are lower than those of the standard qPCR technology, given the superior throughput, and especially when numerous samples are handled simultaneously (Table 2). Increasing the multiplexing will certainly give further, additional advantage to ddPCR in terms of cost and throughput, and could allow its use already at the screening and/or identification steps. The establishment of duplex reactions is straightforward and does not need optimization, which is also encouraging. This characteristic has the advantages of reducing the cost of analysis, and of decreasing the uncertainty linked to droplet volume variation and dilution pipetting errors [19].

To be employed in routine testing, methods based on ddPCR shall be properly validated through ring-trials and verified during their introduction in laboratories to demonstrate their fitness for the purpose. However, the ddPCR performance demonstrated in this study on real routine samples should lead to greater confidence and easier adoption of digital PCR technology, to generating more precise data on everyday tests, and at overall better cost per sample. The demonstrated application of ddPCR for routine quantification of GMO content in food and feed samples should act as an inducement to introduce this technology in other areas where precise analytical testing is required in food and feed samples.

Supporting Information

Figure S1 VIC *vs.* FAM channel clustering plot of droplets for non-digested and *Taq*I digested MON810 DNA. Upper frame: Non digested DNA. Lower frame: *Taq*I digested DNA. Upper left quadrant: FAM (*hmg*) positive-VIC (MON810) negative droplet cluster. Upper right quadrant: FAM (*hmg*) positive-VIC (MON810) positive droplet cluster. Lower left quadrant: FAM (*hmg*) negative-VIC (MON810) negative droplet cluster. Lower right quadrant: FAM (*hmg*) negative-VIC (MON810) positive droplet cluster.

Table S1 Test material used in this study.

Table S2 Primers and probes used in this study.

Table S3 Comparison of quantification using singleplex and duplex ddPCR assays.

Table S4 Comparison of duplex ddPCR quantification on digested and non-digested genomic DNA.

Table S5 Results from the dilution series used for the dynamic range, the aLOD and aLOQ determination, and the overall repeatability.

Table S6 Inter- and intra-cartridge repeatability.

Table S7 False-positive rates observed with ddPCR.

Appendix S1 Supporting material and methods.

Appendix S2 Table S8. Inhibition effect on MON810 quantification in qPCR and ddPCR. Table S9. Set-up and number of reactions needed for simultaneous quantification of four samples with qPCR and ddPCR. Table S10. Time needed for quantification with qPCR and ddPCR. Table S11. Set-up and number of reactions needed for simultaneous quantification of 23 samples with qPCR and ddPCR. Table S12. Set-up and number of reactions needed for quantification with qPCR and ddPCR (full 96 well-plate).

Acknowledgments

We thank Prof. Dr. Roger Pain for reviewing the manuscript. This work was co-financed by the Slovenian Ministry of Economic Development and Technology, Metrology Institute (MIRS) in the frame of the contract no. 640118/2008/67 on performing activities as holder of the national standard MIRS/NIB/FITO for the amount of substance in food of plant origin.

Author Contributions

Conceived and designed the experiments: DM DS. Performed the experiments: DM DS. Wrote the paper: DM MM KG JZ.

References

1. Holst-Jensen A (2009) Testing for genetically modified organisms (GMOs): Past, present and future perspectives. Biotechnol Adv 27: 1071–1082.
2. Hoorfar J, Wolffs P, Radstrom P (2004) Diagnostic PCR: validation and sample preparation are two sides of the same coin. APMIS 112: 808–814.
3. Settanni L, Corsetti A (2007) The use of multiplex PCR to detect and differentiate food- and beverage-associated microorganisms: A review. Journal of Microbiological Methods 69: 1–22.
4. Hindson BJ, Ness KD, Masquelier DA, Belgrader P, Heredia NJ, et al. (2011) High-Throughput Droplet Digital PCR System for Absolute Quantitation of DNA Copy Number. Anal Chem 83: 8604–8610.
5. Pekin D, Skhiri Y, Baret JC, Le Corre D, Mazutis L, et al. (2011) Quantitative and sensitive detection of rare mutations using droplet-based microfluidics. Lab Chip 11: 2156–2166.
6. Pohl G, Shih IM (2004) Principle and applications of digital PCR. Expert Rev Mol Diagn 4: 41–47.
7. Sanders R, Huggett JF, Bushell CA, Cowen S, Scott DJ, et al. (2011) Evaluation of Digital PCR for Absolute DNA Quantification. Anal Chem 83: 6474–6484.
8. Cankar K, Stebih D, Dreo T, Zel J, Gruden K (2006) Critical points of DNA quantification by real-time PCR–effects of DNA extraction method and sample matrix on quantification of genetically modified organisms. BMC Biotechnol 6: 37.
9. Gruere GP, Rao SR (2007) A review of international labeling policies of genetically modified food to evaluate India's proposed rule. AgBioForum 10: 51–64.
10. Weighardt F (2006) Quantitative PCR for the detection of GMOs. In: Querci M, Jermini.M, Van den Eede G, editors. Training course on the Analysis of Food Samples for the Presence of Genetically Modified Organisms. Luxembourg, Luxembourg: Luxembourg: Office for Official Publications of the European Communities.pp. 1–19.
11. Corbisier P, Bhat S, Partis L, Rui Dan Xie V, Emslie K (2010) Absolute quantification of genetically modified MON810 maize (*Zea mays* L.) by digital polymerase chain reaction. Anal Bioanal Chem 396: 2143–2150.
12. Berdal KG, Boydler C, Tengs T, Holst-Jensen A (2008) A statistical approach for evaluation of PCR results to improve the practical limit of quantification (LOQ) of GMO analyses (SIMQUANT). Eur Food Res Technol 227: 1149–1157.
13. Lee D, La Mura M, Greenland A, Mackay I (2010) Quantitation using informative zeros (QUIZ): application for GMO detection and quantification without recourse to certified reference material. Food Chem 118: 974–978.
14. Sykes PJ, Neoh SH, Brisco MJ, Hugues E, Condon J, et al. (1992) Quantitation of targets for PCR by use of limiting dilution. Biotechniques 13: 444–449.
15. Pinheiro LB, Coleman VA, Hindson CM, Herrmann J, Hindson BJ, et al. (2012) Evaluation of a Droplet Digital Polymerase Chain Reaction Format for DNA Copy Number Quantification. Anal Chem 83: 1003–1011.
16. Baker M (2012) Digital PCR hits its stride. Nat Meth 9: 541–544.
17. Devonshire AS, Sanders R, Wilkes TM, Taylor MS, Foy CA et al. (2012) Application of next generation qPCR and sequencing platforms to mRNA biomarker analysis. Methods. In press.
18. Burns MJ, Burrell AM, Foy C (2010) The applicability of digital PCR for the assessment of detection limits in GMO analysis. Eur Food Res Technol 231: 353–362.
19. Bhat S, Herrmann J, Armishaw P, Corbisier P, Emslie KR (2009) Single molecule detection in nanofluidic digital array enables accurate measurement of DNA copy number. Anal Bioanal Chem 394: 457–467.
20. Whale AS, Huggett JF, Cowen S, Speirs V, Shaw J, et al. (2012) Comparison of microfluidic digital PCR and conventional quantitative PCR for measuring copy number variation. Nucleic Acids Res 40: e82.
21. Somma M (2006) Extraction and purification of DNA. In: Querci M, Jermini.M, Van den Eede G, editors. Training Course on the Analysis of Food Samples for the Presence of Genetically Modified Organisms-User Manual.Luxembourg: Office for Official Publications of the European Communities.
22. European Union Reference Laboratory for GM Food and Feed (2006) CRL assessment on the validation of an event-specific method for the relative quantitation of maize line MON 810 DNA using real-time PCR as carried out by Federal Institute for Risk Assessment (BfR). Available: http://gmo-crl.jrc.ec.europa.eu/summaries/Mon810_validation_report.pdf. Accessed 29-1-2013.
23. Cankar K, Chauvensy-Ancel V, Fortabat MN, Gruden K, Kobilinsky A, et al. (2008) Detection of nonauthorized genetically modified organisms using differential quantitative polymerase chain reaction: application to 35S in maize. Anal Biochem 376: 189–199.
24. Codex Committee On Methods Of Analysis And Sampling (2010) Guidelines On Performance Criteria And Validation Of Methods For Detection, Identification And Quantification Of Specific DNA Sequences And Specific Proteins In Foods. Available: http://www.codexalimentarius.org/download/standards/11667/CXG_074e.pdf. Accessed 29-1-2013.
25. European Union Reference Laboratory for GM Food and Feed (2008 October) Definition of minimum Performance requirements for analytical methods of GMO testing. Available: http://gmo-crl.jrc.ec.europa.eu/doc/Min_Perf_Requirements_Analytical_methods.pdf. Accessed 29-1-2 2013.
26. Nadal A, Coll A, La Paz JL, Esteve T, Pla M (2006) A new PCR-CGE (size and color) method for simultaneous detection of genetically modified maize events. Electrophoresis 27: 3879–3888.
27. Chaouachi M, Chupeau G, Berard A, McKhann H, Romaniuk M, et al. (2008) A High-Throughput Multiplex Method Adapted for GMO Detection. J Agric Food Chem 56: 11596–11606.
28. Heide B, Dromtorp S, Rudi K, Heir E, Holck A (2008) Determination of eight genetically modified maize events by quantitative, multiplex PCR and fluorescence capillary gel electrophoresis. Eur Food Res Technol 227: 1125–1137.
29. Peccoud J, Jacob C (1996) Theoretical uncertainty of measurements using quantitative polymerase chain reaction. Biophys J 71: 101–108.
30. International Organization for Standardization (2005) ISO 21570:2005Foodstuffs–Methods of analysis for the detection of genetically modified organisms and derived products–Quantitative nucleic acid based methods. Geneva, Switzerland: International Organization for Standardization.

12

True-Breeding Targeted Gene Knock-Out in Barley Using Designer TALE-Nuclease in Haploid Cells

Maia Gurushidze[1]*, Goetz Hensel[1], Stefan Hiekel[1], Sindy Schedel[1], Vladimir Valkov[1,2], Jochen Kumlehn[1]

1 Leibniz Institute of Plant Genetics and Crop Plant Research (IPK), Plant Reproductive Biology, Gatersleben, Germany, **2** Institute of Genetics and Biophysics, Naples, Italy

Abstract

Transcription activator-like effector nucleases (TALENs) are customizable fusion proteins able to cleave virtually any genomic DNA sequence of choice, and thereby to generate site-directed genetic modifications in a wide range of cells and organisms. In the present study, we expressed TALENs in pollen-derived, regenerable cells to establish the generation of instantly true-breeding mutant plants. A *gfp*-specific TALEN pair was expressed via *Agrobacterium*-mediated transformation in embryogenic pollen of transgenic barley harboring a functional copy of *gfp*. Thanks to the haploid nature of the target cells, knock-out mutations were readily detected, and homozygous primary mutant plants obtained following genome duplication. In all, 22% of the TALEN transgenics proved knocked out with respect to *gfp*, and the loss of function could be ascribed to the deletions of between four and 36 nucleotides in length. The altered *gfp* alleles were transmitted normally through meiosis, and the knock-out phenotype was consistently shown by the offspring of two independent mutants. Thus, here we describe the efficient production of TALEN-mediated gene knock-outs in barley that are instantaneously homozygous and non-chimeric in regard to the site-directed mutations induced. This TALEN approach has broad applicability for both elucidating gene function and tailoring the phenotype of barley and other crop species.

Editor: Hector Candela, Universidad Miguel Hernández de Elche, Spain

Funding: This study was made financially possible through a grant number FKZ 0315965 from the German Federal Ministry of Education and Research. The funders had no role in study design, data collection and analysis, decision to publish, or preparation of the manuscript.

Competing Interests: The authors have declared that no competing interests exist.

* E-mail: maiko@ipk-gatersleben.de

Introduction

The targeted manipulation of genes, involving either their knock-out, replacement or activation, is an attractive technology in the context of both basic and applied genetic research. In a plant breeding context, its major relevance would be to induce mutations specifying novel phenotypes not readily achievable using conventional breeding. A key requirement for such genome engineering is the introduction of double strand breaks (DSBs) in the DNA sequence present at a pre-determined genomic site. Engineered zinc-finger nucleases (ZFNs), transcription activator-like effector nucleases (TALENs), homing endonucleases and RNA-guided nucleases (CRISPRs) have all been reported to be capable of inducing DSBs [1,2]; of these, the ZFNs have been the most frequently employed. Despite considerable progress made in employing ZFNs for DNA targeting, however, their routine utilization is hampered owing to limited reliability, off-target effects and comparatively high costs. TALENs are customizable fusion proteins between a specific DNA binding domain and FokI endonuclease, the former directing the nuclease to the user-defined genomic site. The binding domain is derived from transcription activator-like (TAL) effectors of pathogenic *Xanthomonas* spp., while the FokI nuclease domain introduces DSBs when present in a dimeric form, achieved by introducing a pair of TALEN units. The resulting DSBs are processed by the host cell's DNA repair machinery, which can leave behind alterations at the breakage sites. As each repeat of the DNA binding domain of the TAL effector recognizes nucleotides in a largely context-independent manner [3,4], the selection of a target DNA motif and the subsequent customization of the TALEN is made relatively straightforward. Reliable platforms for the modular assembly of TALENs are provided by Golden Gate cloning [5–8], high-throughput solid phase assembly [9], and ligation-independent cloning techniques [10].

TALENs have been proven to be highly efficient for targeted genetic manipulation in yeast, human cell lines and animal species. However, their use has been quite limited in plants, including the generation of targeted knock-out mutations in *Nicotiana benthamiana* leaf cells [11], in *Arabidopsis thaliana* [6,12], in *Brachypodium distachyon* calli and in rice plants [13]. Li et al. [7] have successfully targeted the promoter of a bacterial blight susceptibility gene in rice to generate mutations which converted the host's susceptibility into resistance. A recent demonstration that TALEN-induced DSBs can generate deletions in the promoter of a barley phytase gene has been provided by Wendt et al. [14], but neither the resulting phenotypes nor the sexual transmission of the induced mutations were analysed. Finally, Zhang et al. [15], working with *N. tabacum* protoplasts, achieved TALEN-mediated targeted gene insertion and gene replacement *via* homologous recombination.

A broad spectrum of genomic resources has been developed in barley, along with various enabling technologies such as the large scale production of doubled haploids [16] and effective transformation protocols [17,18]. The sequence-enriched physical map of barley [19] has provided an invaluable platform for both genome-assisted research and crop improvement. Nevertheless, the targeted modification of specific barley genes remains quite intractable. Establishing usable and efficient protocols to achieve gene targeting would open a number of both basic and applied research areas in this leading temperate cereal species, which is also regarded as a viable model for the genetically more complex, but more important crop species wheat. Here, we describe progress in generating targeted gene knock-outs in barley using TALEN technology. The target chosen to demonstrate the success of the experiments was the reporter gene encoding green fluorescent protein, since this provided a simple bioassay for knock-out events (Fig. S1 in File S1). The TALEN constructs were introduced into embryogenic pollen cultures consisting primarily of haploid cells able to be converted *via* pollen embryogenesis and genome duplication into fertile doubled haploid plants [17]. We show that customized TALENs can induce targeted mutations in barley, and that these are faithfully transmitted to the progeny of primary mutant plants.

Materials and Methods

Plant Materials

Grains of non-transformed winter barley (*Hordeum vulgare* L.) cv. 'Igri' and of the two transgenic lines PV 89 and BPI 09, which each carry a single copy of *gfp*, were germinated (growth chamber, 14/12°C day/night, 16 h photoperiod 20,000 lux), vernalized for eight weeks (growth chamber, 4°C, 9 h photoperiod) and raised in a glasshouse (18/14°C, 16 h photoperiod at approximately 25,000 lux) to flowering. Spikes were harvested when the awns had just emerged from the sheath of the flag leaf, as described by Kumlehn et al. [17].

Plasmids and Barley Transformation

A pair of *gfp*-specific TALEN sequences was designed and assembled by Cellectis Bioresearch (Paris, France) using a modified version of *X. campestris* pv. *vesicatoria* AvrBs3 as the backbone for both TALEN units. To evaluate potential off-target cleavage sites in the barley genome (The International Barley Genome Sequencing Consortium, 2012), the kmasker tool (http://webblast.ipk-gatersleben.de/kmasker/) was applied, using standard parameters. The alignments were manually refined, and forty best matching genomic regions were identified.

The left (pTAL.pLess.009677) and right (pTAL.pLess.009678) TALEN units were each introduced into the *Spe*I/*Hin*dIII cloning sites of pUbi-AB-M (DNA Cloning Service, Hamburg, Germany) to form an expression cassette under the control of the maize *UBIQUITIN-1* promoter [20]; the *Agrobacterium tumefaciens NOPALINE SYNTHASE* gene provided the termination sequence (*NOST*, Fig. 1). The two intermediate vectors (pUbi-TALEN-Left and pUbi-TALEN-Right) and synthetic sequences encoding the simian virus S40 (SV40) NLS, and NLS-HA (SV40 NLS with an added hemagglutinin tag) were digested with *Asc*I/*Eco*47III, and ligated such that the NLS became attached to the N-terminus of pUbi-TALEN-Left and NLS-HA to the N-terminus of pUbi-TALEN-Right. Each TALEN expression cassette was introduced into the *Sfi*I cloning site of the binary vector p7d35S, which harbours *BAR*, a gene for bialaphos resistance, driven by the cauliflower mosaic virus doubled enhanced *35S* (*CaMV d35S*) promoter [21] (Fig. 1).

Constructs were introduced *via* electroporation into *A. tumefaciens* strain LBA4404pSB1, which carries the disarmed Ti plasmid pAL4404 and the hypervirulence-conferring vector pSB1 [22]. Microspore isolation and the *Agrobacterium*-mediated transformation of embryogenic pollen of cv. 'Igri' with the *gfp*-harbouring plasmid pYF133 [23] was carried out as described by Kumlehn et al. [17]. The two *gfp*-specific TALEN sequences were introduced into embryogenic pollen cultures made from lines PV 89 and BPI 09 *via* the same method, except that bialaphos (rather than hygromycin B) was used as the selective agent, since these lines already harboured hygromycin resistance. The density of the transformed LBA4404pSB1 cells was determined spectrophotometrically, and immediately prior to the inoculation the two strains were mixed in a 1:1 ratio to obtain a final cell density of 2.5×10^7 per mL.

Genomic DNA Isolation, PCR, Sequencing and DNA Gel Blot Analysis

A 200–400 mg sample of fresh leaf tissue was snap-frozen in liquid nitrogen, and total genomic DNA was isolated following Palotta et al. [24]. Each 20 µL PCR assay of this DNA was based on 50–100 ng genomic DNA as template, using the sets of primer pairs detailed in Table S1 in File S1. The *gfp* amplicons were purified using a QIAquick PCR purification kit (QIAGEN, Hilden, Germany), and were either directly sequenced, or sequenced following cloning into pGEM-T (PROMEGA, Madison, WI). DNA gel blot analysis was employed to derive the number of integrated T-DNA copies present: for this purpose, a 25 µg aliquot of genomic DNA was digested with either *Hin*dIII or *Ssp*I, electrophoresed through a 0.8% agarose gel and transferred onto a positively charged nylon membrane (Hybond N+, ROCHE, Mannheim, Germany), which was then hybridized with a DIG-labelled probe (*gfp*, *Fok*I or *BAR*) generated from a PCR based on the relevant primer pairs (Table S1 in File S1). Signal was visualized using a DIG System for Filter Hybridisation according to the manufacturer's instructions (ROCHE, Mannheim, Germany).

RNA Isolation and Reverse Transcriptase Reaction

Total RNA was isolated using the TRIzol reagent (INVITROGEN, Life Technologies, Darmstadt, Germany), following the manufacturer's protocol. After a DNase I treatment (DNA-free kit, AMBION, Life Technologies, Darmstadt, Germany), the cDNA first strand was synthesized using an iScript Select cDNA synthesis kit in the presence of oligo $(dT)_{20}$ primers (BIO-RAD, Munich, Germany). The resulting cDNA served as template for subsequent PCRs based on the *Fok*I domain primer pair (Table S1 in File S1).

Flow Cytometry and Induced Genome Duplication

The ploidy level of the primary transgenic plants was determined by flow cytometry (Ploidy Analyser 1, PARTEC, Muenster, Germany) according to the manufacturer's instructions. To induce genome duplication (and thereby achieve instant homozygosity of the transgene), the haploid plantlets were treated with 0.1% w/v colchicine, based on a well-established procedure [25].

Results

GFP Reporter Lines

To facilitate the detection of site-specific mutagenesis, *gfp* was used as the TALEN target. To this end, single copy transgenic reporter lines were generated *via Agrobacterium*-mediated transformation of cv. 'Igri' embryogenic pollen cultures. Instant homozygosity of *gfp* was achieved by treating haploid primary transgenics with colchicine. The number of *gfp* copies present in the progeny of the resulting genome-doubled lines was checked by DNA gel blot hybridization. Two single-copy lines (BPI 09 and PV 89) in which there was no segregation with respect to either the presence (Fig. S2a in File S1) or the expression (Fig. S2b in File S1) of *gfp* were used for the subsequent TALEN experiments.

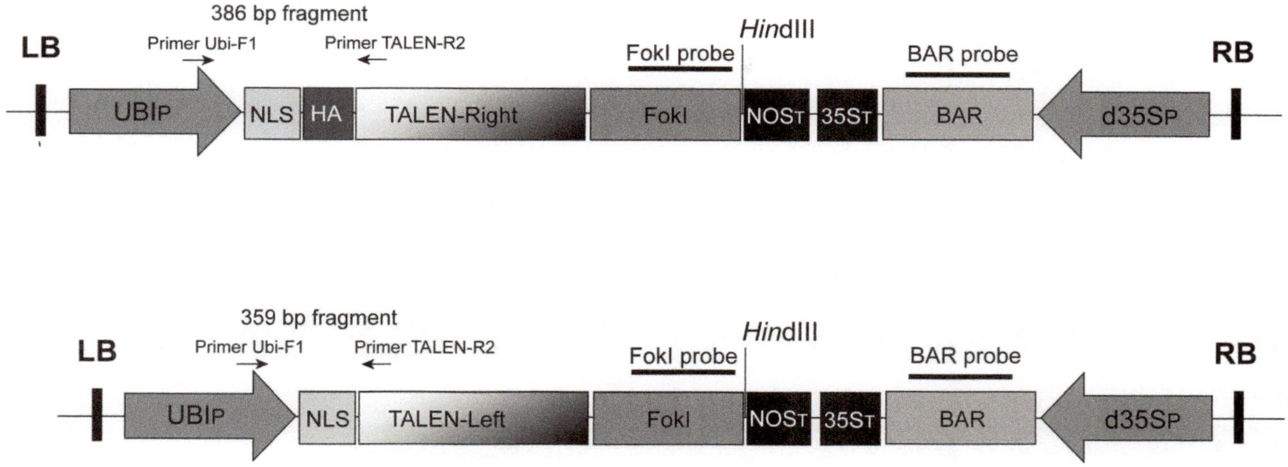

Figure 1. Schematic representation of the T-DNA TALEN constructs. The right-TALEN unit is shown above the left one (not drawn to scale). Primer binding sites are indicated by arrows and the *FokI* and *BAR* probe regions used for DNA gel blot analysis by bars. LB: T-DNA left border, UBIP: maize *UBIQUITIN-1* promoter, NLS: SV40 nuclear localization signal, HA: hemagglutinin tag, TALEN-Right: customized DNA binding domain of right TALEN unit, TALEN-Left: customized DNA binding domain of left TALEN unit, FokI: DNA cleavage domain of *FokI* endonuclease, NOST: *A. tumefaciens NOPALINE SYNTHASE* terminator, 35ST: *CaMV 35S* terminator, BAR: bialaphos resistance gene, d35SP: *CaMV 35S* promoter harbouring a doubled enhancer sequence, RB: T-DNA right border, *HindIII*: restriction site targeted in the DNA gel blot analysis.

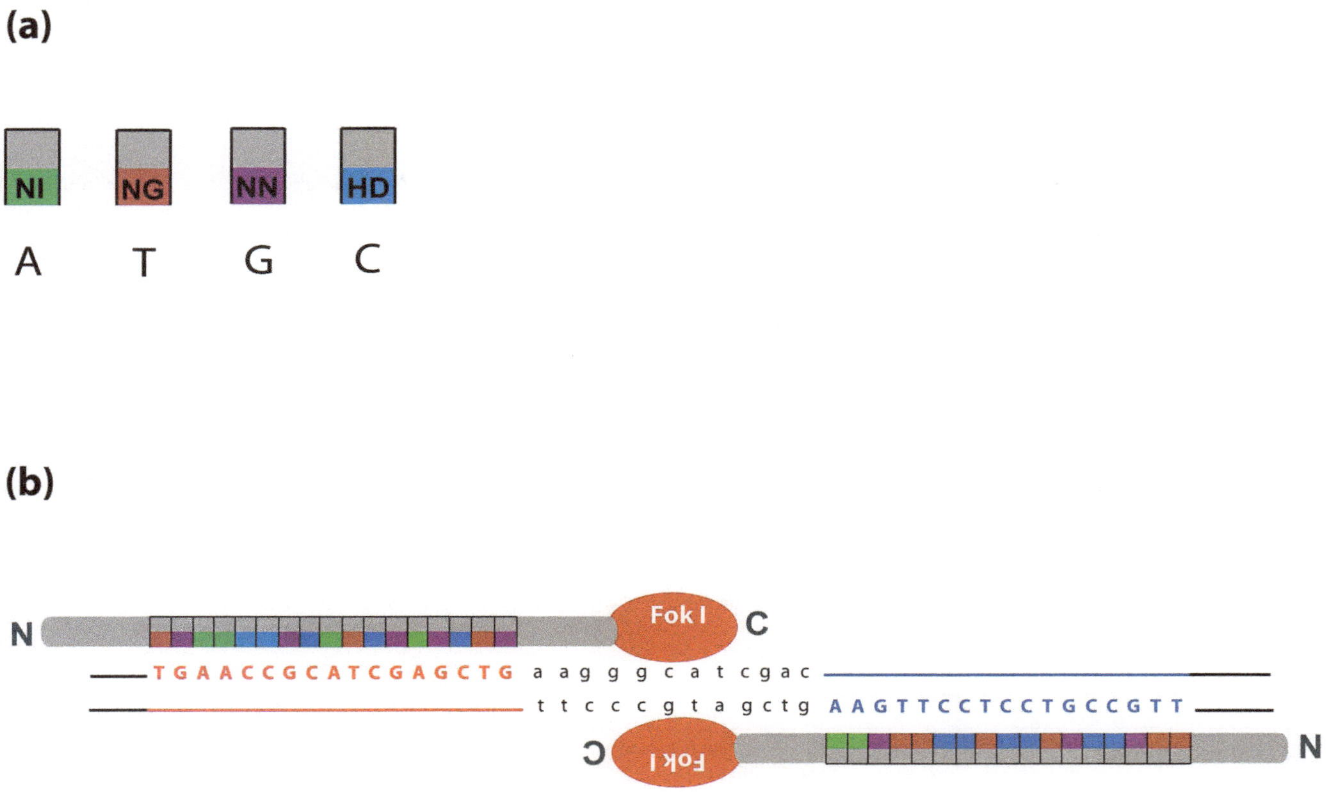

Figure 2. Customized TALENs targeting *gfp*. (a) Four basic modular amino acid repeats with their repeat variable diresidues indicated by the standard single letter codes shown in the colored boxes. The preferred nucleotide recognized by the individual repeat is shown below the repeat. (b) TALENs and their target sequences in the *gfp*. DNA binding specificity was achieved by the arrays of repeats assembled according to the target sequence. The nucleotide sequence shown in red represents the left and the one shown in blue the right-hand TALEN binding site. Lower case letters represent the region where FokI domains are expected to dimerize and introduce DSBs.

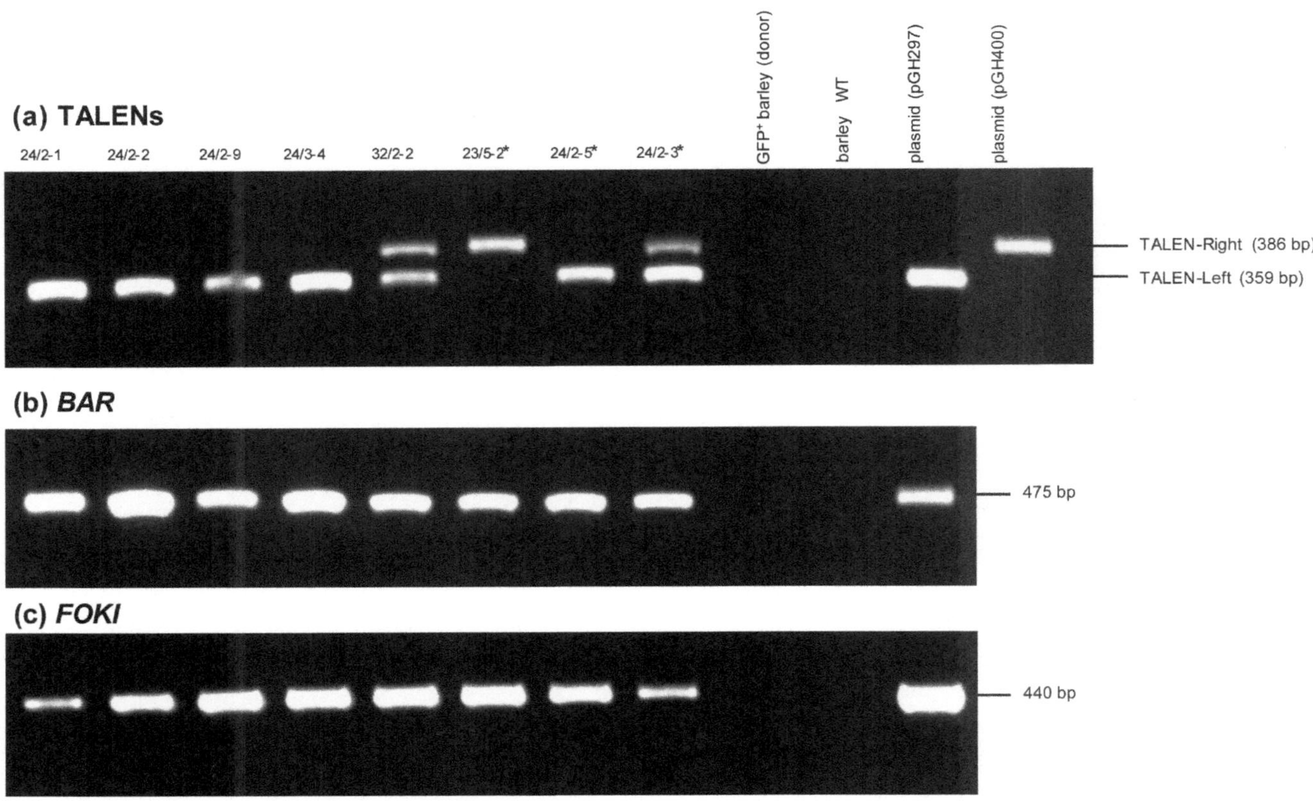

Figure 3. PCR amplification from TALEN transgenics and the parental *gfp* reporter line. Non-mutated transformants marked by asterisks beside their identifier. Amplicons derived from (a) the left (359 bp) and right (386 bp) TALEN units produced using the primer pair Ubi-F1/TALEN-R2 (b) *BAR* (primer pair: GH-Bar-F1/GH-Bar-R1) (c) the *FokI* endonuclease domain (primer pair: FokI-F1/FokI-R). WT: non-transformed cv. 'Igri'.

Primary TALEN Transgenics and Mutants

Each TALEN unit comprised 17 modular amino acid repeats specifically targeting selected DNA motifs within *gfp*, along with a C-terminally fused FokI endonuclease domain (Fig. 2). BPI 09 and PV 89 were retransformed with the TALEN constructs, followed by bialaphos selection and the screening of regenerants for the presence of GFP fluorescence, *Fok*I, *BAR* and the TALENs, as well as for *Fok*I expression and ploidy level. Since the *Xanthomonas AvrBs*3 endogenous nuclear localization signals (NLS) had been removed from the standard scaffold to increase TALEN nuclease efficiency [26], the NLS of simian virus S40 was attached to the N-terminus of each TALEN unit. In case of the right TALEN unit, the NLS was combined with a hemagglutinin (HA) tag to facilitate the discrimination between the left and right TALEN units in subsequent molecular analyses (Fig. 1). PCR based on one primer placed within the *UBIQUITIN-1* promoter and the other at the N-terminus of the TALEN sequence yielded a 359 bp amplicon when the left TALEN unit was present, a 386 bp amplicon when the right unit was present and both fragments when both TALEN T-DNAs were successfully incorporated into the host genome (Figs 1, 3a). Of the 18 retransformed regenerants, fourteen proved to carry just the left TALEN unit, two just the right TALEN unit and two harboured both TALEN units (Table S2 in File S1). Four out of the 18 transformants (plants 24/2-1, 24/2-2, 24/2-9 and 24/3-4) did not accumulate GFP (Table 1, Fig. 4). Sequence analysis indicated that each of the *gfp* knock-outs was caused by a genetic modification in the target sequence (Fig. 5), whereas the target sequence in plants still expressing *gfp*, with the exception of 32/2-2, was not modified. The *gfp* amplicon of 32/2-2 was heterogeneous (two different sequences), suggesting that the plant was either heterozygous or chimeric; one sequence was the intact *gfp*, and the other had a 4 nt deletion. PCR analysis showed that

Table 1. Summary of *gfp*-specific TALEN transformation outputs.

gfp-reporter line	Experiment Number	Number of processed spikes	Cultivated pollen [million]	Bialaphos-resistant regenerants	*gfp* mutant alleles	Mutant plant identifier
BPI 09	1	11	1.2	9	3 (2 identical)	24/2-1 24/2-2 24/2-9
BPI 09	2	15	1.6	5	1	24/3-4
PV 89	3	43	4.8	4	1	32/2-2
total	3	69	7.6	18	4 unique	5

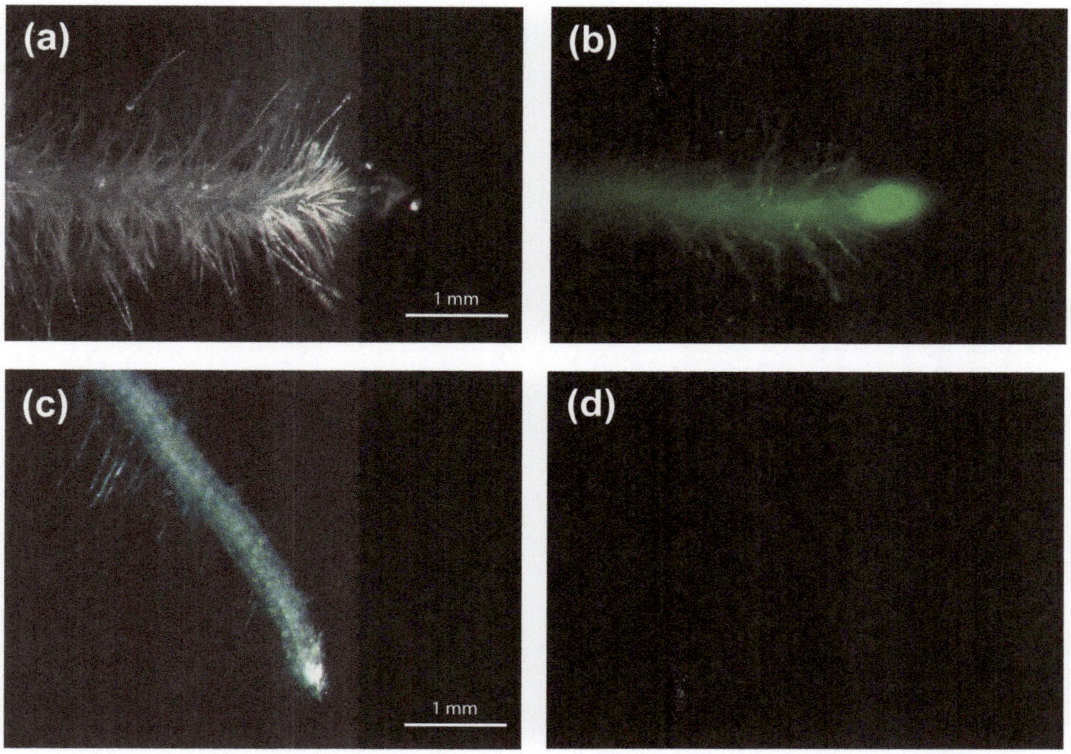

Figure 4. Root tips of reporter and mutated lines visualized. A root tip of a reporter line (a) under white light and (b) when excited with far blue light. A root tip of a TALEN *gfp* knock-out individual (24/2-1) visualized (c) under white light and (d) when excited by far blue light.

(a)

```
GFP-WT   AGGTGAAGTTCGAGGGCGACACCCTGGTGAACCGCATCGAGCTGAAGGGCATCGACTTCAAGGAGGACGGCAACATCCTGGGGCACAAGCTGGAG
Donor    AGGTGAAGTTCGAGGGCGACACCCTGGTGAACCGCATCGAGCTGAAGGGCATCGACTTCAAGGAGGACGGCAACATCCTGGGGCACAAGCTGGAG
24/2-1   AGGTGAAGTTCGAGGGCGAC------------------------------------TTCAAGGAGGACGGCAACATCCTGGGGCACAAGCTGGAG   -36
24/2-9   AGGTGAAGTTCGAGGGCGAC------------------------------------TTCAAGGAGGACGGCAACATCCTGGGGCACAAGCTGGAG   -36
24/2-2   AGGTGAAGTTCGAGGGCGACACCCTGGTG---------------CCCCTGGTGTACTTCAAGGAGGACGGCAACATCCTGGGGCACAAGCTGGAG   -25 +10
24/3-4   AGGTGAAGTTCGAGGGCGACACCCTGGTGAACCGCATCGAG-----------CGACTTCAAGGAGGACGGCAACATCCTGGGGCACAAGCTGGAG   -11
32/2-2   AGGTGAAGTTCGAGGGCGACACCCTGGTGAACCGCATCGAGCTGAAGG---- CGACTTCAAGGAGGACGGCAACATCCTGGGGCACAAGCTGGAG   -4
```

(b)

```
                 10        20        30        40        50        60        70        80        90       100       110       120       130       140
         ....|....|....|....|....|....|....|....|....|....|....|....|....|....|....|....|....|....|....|....|....|....|....|....|....|....|....|....|
GFP_WT   MVSKGEELFTGVVPILVELDGDVNGHKFSVSGEGEGDATYGKLTLKFICTTGKLPVPWPTLVTTFTYGVQCFSRYPDHMKQHDFFKSAMPEGYVQERTIFFKDDGNYKTRAEVKFEGDTLVNRIELKGIDFKEDGNILGH
donor    MVSKGEELFTGVVPILVELDGDVNGHKFSVSGEGEGDATYGKLTLKFICTTGKLPVPWPTLVTTFTYGVQCFSRYPDHMKQHDFFKSAMPEGYVQERTIFFKDDGNYKTRAEVKFEGDTLVNRIELKGIDFKEDGNILGH
24/2-1   MVSKGEELFTGVVPILVELDGDVNGHKFSVSGEGEGDATYGKLTLKFICTTGKLPVPWPTLVTTFTYGVQCFSRYPDHMKQHDFFKSAMPEGYVQERTIFFKDDGNYKTRAEVKFEGD------------FKEDGNILGH
24/2-9   MVSKGEELFTGVVPILVELDGDVNGHKFSVSGEGEGDATYGKLTLKFICTTGKLPVPWPTLVTTFTYGVQCFSRYPDHMKQHDFFKSAMPEGYVQERTIFFKDDGNYKTRAEVKFEGD------------FKEDGNILGH
24/2-2   MVSKGEELFTGVVPILVELDGDVNGHKFSVSGEGEGDATYGKLTLKFICTTGKLPVPWPTLVTTFTYGVQCFSRYPDHMKQHDFFKSAMPEGYVQERTIFFKDDGNYKTRAEVKFEGDTLV-----PLVYFKEDGNILGH
24/3-4   MVSKGEELFTGVVPILVELDGDVNGHKFSVSGEGEGDATYGKLTLKFICTTGKLPVPWPTLVTTFTYGVQCFSRYPDHMKQHDFFKSAMPEGYVQERTIFFKDDGNYKTRAEVKFEGDTLVNRIERLQGGRQHPGAQAGV
32/2-2   MVSKGEELFTGVVPILVELDGDVNGHKFSVSGEGEGDATYGKLTLKFICTTGKLPVPWPTLVTTFTYGVQCFSRYPDHMKQHDFFKSAMPEGYVQERTIFFKDDGNYKTRAEVKFEGDTLVNRIELKATSRRTATSWGTS

                150       160       170       180       190       200       210       220       230       240       250       260
         ....|....|....|....|....|....|....|....|....|....|....|....|....|....|....|....|....|....|....|....|....|....|....|....|...
GFP_WT   KLEYNYNSHNVYIMADKQKNGIKVNFKIRHNIEDGSVQLADHYQQNTPIGDGPVLLPDNHYLSTQSALSKDPNEKRDHMVLLEFVTAAGITHGMDELYKSEKDELYAAARLQIVQTFGNKVS*
donor    KLEYNYNSHNVYIMADKQKNGIKVNFKIRHNIEDGSVQLADHYQQNTPIGDGPVLLPDNHYLSTQSALSKDPNEKRDHMVLLEFVTAAGITHGMDELYKSEKDELYAAARLQIVQTFGNKVS*
24/2-1   KLEYNYNSHNVYIMADKQKNGIKVNFKIRHNIEDGSVQLADHYQQNTPIGDGPVLLPDNHYLSTQSALSKDPNEKRDHMVLLEFVTAAGITHGMDELYKSEKDELYAAARLQIVQTFGNKVS*
24/2-9   KLEYNYNSHNVYIMADKQKNGIKVNFKIRHNIEDGSVQLADHYQQNTPIGDGPVLLPDNHYLSTQSALSKDPNEKRDHMVLLEFVTAAGITHGMDELYKSEKDELYAAARLQIVQTFGNKVS*
24/2-2   KLEYNYNSHNVYIMADKQKNGIKVNFKIRHNIEDGSVQLADHYQQNTPIGXGPVLLPDNHYLSTQSALSKDPNEKRDHMVLLEFVTAAGITHGMDELYKSEKDELYAAARLQIVQTFGNKVS*
24/3-4   QLQQPQRLYHGRQAEERHQGELQDPPQHRGRQRAARRPLPAEHPHRRRPRAAARQPLPEHPVRPEQRPQREARSHGPAGVRDRRRDHSRHGRAVQV*
32/2-2   WSTTTTATTSISWPTSRRTASR*
```

Figure 5. Modifications of the *gfp* sequence induced by TALENs. (a) Alignment of the *gfp* sequences of the various mutants and the donor line. The underlined sequence in red represents the left and that in blue the right TALEN binding site; the sequence shown in green is a duplication of the underlined motif. The number of nucleotides deleted (dashes) or inserted (upper case letters in green) is shown to the right of each sequence. (b) Intact *gfp* (WT) and *gfp* knock-out peptide sequences. Stop codons indicated by asterisks.

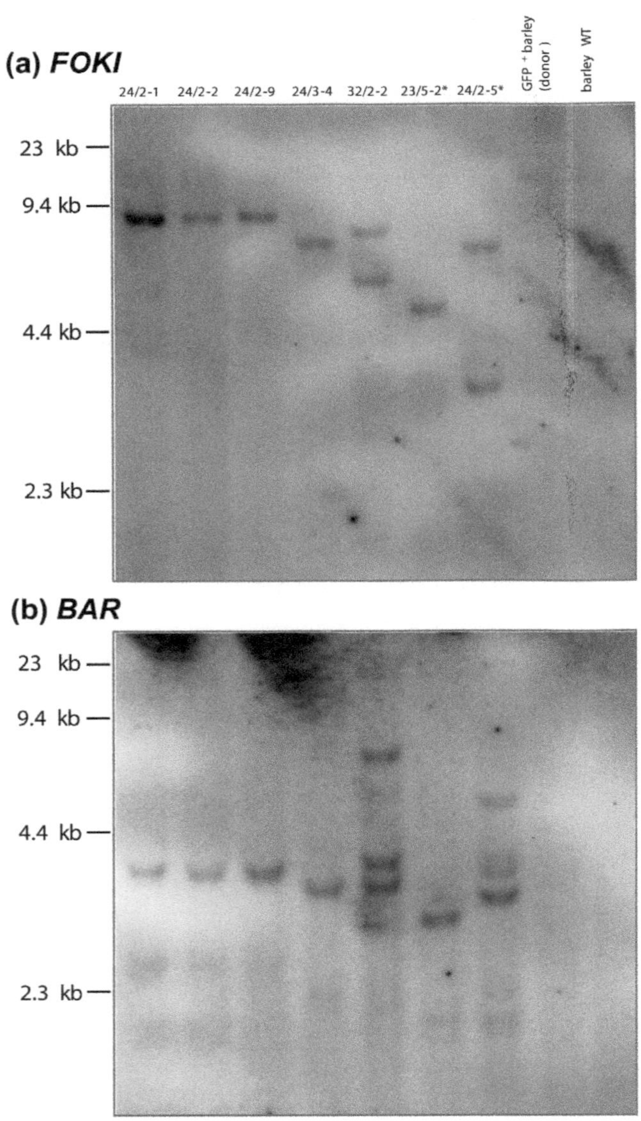

Figure 6. DNA gel blot analysis. Non-mutated transformants marked by asterisks beside their identifier. (a) *Hin*dIII digested DNA hybridized with a *Fok*I probe. (b) *Ssp*I digested DNA hybridized with a *BAR* probe. Note that two copies of *Fok*I and three copies of *BAR* are present in the mutant 32/2-2 suggesting that one of integrated three T-DNA copies is incomplete.

24/2-1, 24/2-2, 24/2-9 and 24/3-4 each possessed stably integrated only the left TALEN T-DNA, while 32/2-2 harboured both the left and right TALEN T-DNA (Fig. 3a), in agreement with the DNA gel blot analysis, which revealed that 24/2-1, 24/2-2, 24/2-9 and 24/3-4 each harboured a single copy of the *Fok*I sequence, while 32/2-2 harboured two copies of the nuclease domain (Fig. 6a). All mutants expressed *Fok*I as was shown using reverse transcription polymerase chain reaction (Fig. S3 in File S1).

The loss of GFP function was caused by deletions ranging in size from 4 to 36 nt. The mutations induced frame shifts and/or the formation of a premature stop codon. Plant 24/2-2 carried a 25 nt deletion plus a 10 nt insertion, where 8 nt of the insertion was

Table 2. Characterization of primary *gfp* knock-out plants.

Mutant identifier	TALEN units stably integrated	*Fok*I copy number[1]	*BAR* copy number[2]	DNA modification (bps)	*Fok*I expression (RT-PCR)	Ploidy of regenerant	Grain set[3]	Zygosity of induced mutation
24/2-1	Left	1	1	−36	+	1n	+	homozygous
24/2-9	Left	1	1	−36	+	1n	+	homozygous
24/2-2	Left	1	1	−25 + 10	+	2n	+	homozygous
24/3-4	Left	1	1	−11	+	1n/2n chimeric	-	not determined
32/2-2	Left + Right	2	3	−4	+	2n	+	heterozygous or chimeric

[1]according to DNA gel blot analysis of *Hin*dIII-digested DNA.
[2]according to DNA gel blot analysis of *Ssp*I-digested DNA.
[3]haploid plantlets treated with colchicine to induce chromosome doubling.

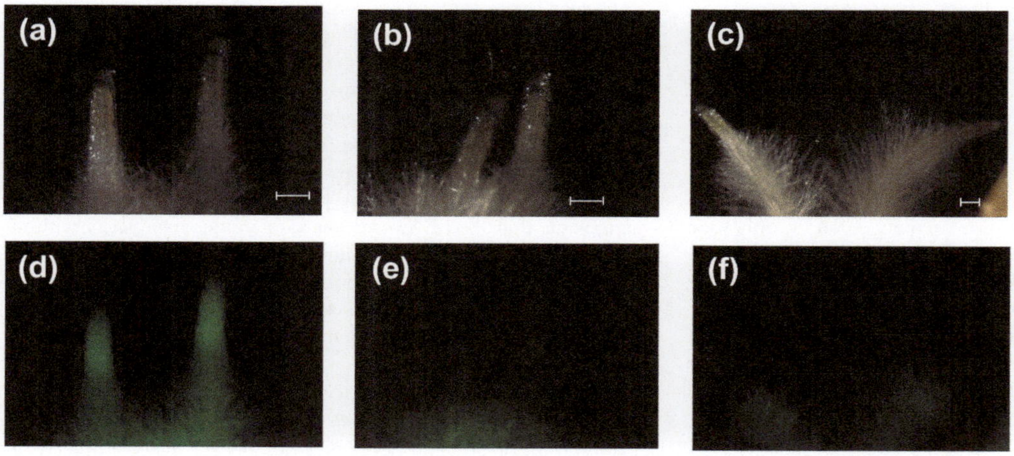

Figure 7. The inheritance of mutated *gfp* in the T_1 generation bred from plants 24/2-1 and 24/2-2 all of which carried a knock-out copy of *gfp*. Scale bar: 0.2 mm. Root tips imaged (a–c): under bright field, (d–f): after excitation with far blue light. (a, d) *gfp* donor, (b, e) 24/2-1, (c, f) 24/2-2. Some auto-fluorescence was detectable in the root hairs of the mutants (see e, f) and also in those of non-transformed plants (not shown).

result of a duplication of an adjacent sequence (Fig. 5). Apart from plant 32/2-2, at least one of the TALEN binding motifs was altered in the target sequence, thereby abolishing the capability of the TALEN to bind. Overall, five independent targeted mutagenesis events were observed (Table 2). Two events (24/2-1 and 24/2-9) derived from the same transformation experiment, and both carried the same 36 nt deletion.

The potential off-targets that might allow forming TALEN hetero- or homodimers were ranked according to the number of mismatches. Forty barley genomic regions that are best matching with the pre-defined TALEN target sequence are shown in Fig. S4 in File S1. The analysis revealed that there is no sequence with less than seven mismatches to the target (Fig. S4 in File S1). Consequently, the presence of an effective off-target in the barley genome is anticipated to be extremely unlikely.

Heritability of the *gfp* Mutations

Except for plant 24/3-4, each of the TALEN transgenics grew normally and were self-fertile following either colchicine treatment-induced or spontaneous chromosome doubling. The mutant 24/3-4 was a ploidy chimera involving both 1n and 2n sectors. Apparently only haploid rather than the diploid tissue contributed to the germ line causing sterility of this mutant. This phenomenon is not unusual for regenerants derived from embryogenic pollen cultures [27,28]. Respectively, the transmission of the mutation was determined for the colchicine-treated doubled haploid 24/2-1 and the spontaneously diploidized mutants 24/2-2 and 32/2-2. The zygosity of the primary transgenic plants were determined by scoring for GFP fluorescence in the pollen (Fig. S5 in File S1) and in T_1 seedlings (Fig. 7). In addition, the *gfp* sequence was acquired from a sample of T_1 individuals to identify whether the T_0 mutants carried more than one independent sequence alteration. At least two grains per spike were germinated, resulting in sets of 21, 46 and 32 T_1 individuals derived from 24/2-1, 24/2-2 and 32/2-2, respectively. In the cases of 24/2-1 and 24/2-2, every T_1 individual carried the same mutated *gfp* sequence as did their primary transformant parent, and none expressed functional *gfp*. Whereas in the progeny of 32/2-2, segregation with regard of GFP fluorescence was detected. Furthermore, sequencing of the *gfp* in the progeny of the primary mutant 32/2-2 showed that various new *gfp* alleles were present in the T_1 that had not been detected in the parental plant. In addition, several T_1 plants carried the intact (wild-type) *gfp* allele. The new mutant *gfp* alleles observed in the T_1 possessed indels at the TALEN target site (Fig. 8). Notably, the mutant allele recovered in the T_0 primary mutant 32/2-2 (4 nt deletion) was no longer identified among the sequenced T_1 individuals.

Discussion

TALEN-Mediated Gene Modification in Barley

Thanks to the combination of targeting *gfp* and exploiting haploid technology, it was possible to provide unequivocal evidence for targeted gene knock-out in barley. We have demonstrated for the first time that customizable designer nucleases can be expressed and function in haploid cells, which in turn can be instantly regenerated into homozygous mutants. The sample of mutants showed that the altered alleles were faithfully transmitted through meiosis and so were passed on to the sexual progeny of the primary transgenics.

The recent acquisition of the barley genome sequence (The International Barley Genome Sequencing Consortium, 2012)

```
32/2-2_A  GCCGAGGTGAAGTTCGAGGGCGACACCCTGGTGAACCGCATCGAGCTGAAGGGCATCGACTTCAAGGAGGACGGCAACATCCTGGGGCACAAGCTGGAGTACAACTACAA
32/2-2_C  GCCGAGGTGAAGTTCGAGGGCGACACCCTGGTGAACCGCATCGAGCTGAAGGG--------TCAAGGAGGACGGCAACATCCTGGGGCACAAGCTGGAGTACAACTACAA  -8
32/2-2_D  GCCGAGGTGA-------------------------------------------------CTTCAAGGAGGACGGCAACATCCTGGGGCACAAGCTGGAGTACAACTACAA  -49
32/2-2_E  GCCGAGGTGAAGTTCGAGGGCGACACCCTGGTGAACCGCATCGAG---------GACGACTTCAAGGAGGACGGCAACATCCTGGGGCACAAGCTGGAGTACAACTACAA  -11 +2
32/2-2_F  GCCGAGGTGAAGTTCGAGGGCGACACCCTGGTGAACCGCATCGA---------------CTTGAAGGASGACKGCAACATCCTGGGGCACAAGCTGGAGTACAACTACAA  -15
```

Figure 8. Various *gfp* alleles (A–F) present in T_1 individuals derived from the primary transformant 32/2-2. The underlined sequence in red represents the left and that in blue the right TALEN binding site. Note that allele 32/2-2_B (not shown in this figure) that was identified in the primary mutant could not be detected in the progeny.

largely facilitates the identification of appropriate target sites within genes known to determine traits of interest to barley breeders, and offers the possibility of excluding (or at least minimizing) off-target effects and achieving a high specificity of the designed nucleases. We screened the barley genome for potential off-target cleavage sites. However, even the best matches we found contained at least seven mismatches compared to the TALEN target nucleotide sequence (Fig. S4 in File S1). Recent studies showed an extremely low off-target activity even if only one nucleotide mismatch was present [15,29]. Considering this high specificity of TALENs, we did not find any barley genomic sequences that could be seriously considered as potential off-targets. In conclusion, we assume that the unintended induction of mutations in the barley genome by the *gfp*-specific TALENs is at least indistinguishable from the occurrence of background mutations.

The frequency with which mutation was induced (~20%) and the characteristics of the induced mutations were comparable to what has been documented in other examples of the TALEN approach [12–14]. Two cases of rather larger deletions were observed, one involving the loss of 36 nt, including the entire left TALEN binding site, and the other of 25 nt which was accompanied by a 10 nt insertion. This sort of modifications has been associated with the repair of DSBs *via* non-homologous end joining. Most modifications induced were deletions, which is in accord with the observations made by Kim et al. [30] who reported that in contrast to the modifications induced by ZFNs, most TALEN events are deletions, and speculated that the reason for this may be associated with the comparatively large spacer region provided between the two TALEN binding sites.

Four of the five primary mutant plants showed no evidence of heterozygosity for the *gfp* sequence, while the remaining one (32/2-2) harboured two distinct *gfp* sequences. The presence of two different *gfp* alleles in the latter primary transgenic implies that the plant was either heterozygous or chimeric. TALEN-induced mosaicism involving as many as nine variant sequences of the *HvPAPhy_a* promoter region in a single barley primary transgenic plant has recently been recorded [14]. No such extreme cases of multiple sequences were observed here, suggesting that the use of embryogenic pollen as the explant encourages the formation of single, rather than multiple targeted modification events and that the TALENs preferentially induce mutations at the early stage of embryogenic development. Our hypothesis is that the early disruption of the TALEN binding sites avoids repeated binding of the TALENs, thereby ensuring that no further diversification of mutations will be induced in subsequent daughter cells.

The two primary mutants 24/2-1 and 24/2-9 carried an identical 36 nt deletion. The repeated recovery of specific designer endonuclease-derived modifications has been reported in maize [31], zebrafish [32], and rat [33]. The latter authors have suggested the possibility that microhomology-mediated reclosure of DSBs could explain the phenomenon. In the present case, however, the same deletion was recovered from the same experiment, which would suggest that the two individuals were in fact clones derived from a single mutagenesis event.

More notable was the finding that four of the five induced mutants possessed only a single stably integrated TALEN unit, as confirmed by DNA gel blot analysis. The creation of DSBs requires FokI dimerization [34,35], so the assumption is that the non-integrated TALEN unit must have only been transiently expressed in the cells of embryogenic pollen from which these mutants were derived. If so, the implication is that the transient expression of TALENs can induce DSBs at a very early stage of pollen embryogenesis. A similar non-integration of a TALEN construct has recently been observed in experiments conducted using tobacco protoplast-derived calli [15]. The possibility that whole plants harbouring targeted mutations can be generated from cells modified by the transient expression of TALENs offers the prospect of inducing targeted mutations and transgene-free mutants *via* non-integrative expression. In other words, it may be feasible to avoid the laborious procedure of removing the TALEN-coding expression cassettes by segregation. In this context, it is significant that TALENs can act as a mutagen *in planta* without necessarily requiring the maintenance of recombinant DNA. In this sense, the use and regulation of mutant crop plants generated *via* TALENs may not be as troublesome as those which are legislatively classed as being genetically engineered.

The Transmission and Zygosity of the Mutations

The transmission of the *gfp* mutations through meiosis was explored in T_1 progenies derived from 24/2-1, 24/2-2 and 32/2-2. The reason for the choice of the primary mutant 24/2-1 was that it was a haploid mutant regenerant treated with colchicine to induce chromosome doubling, which ensured that the plant was homozygous for the *gfp* mutation. In contrast, 24/2-2 and 32/2-2 were spontaneously chromosome doubled plants, so could have been either homozygous or heterozygous with regard to the *gfp* allele disruption, depending on whether the mutation event occurred prior to or after diploidization. The progenies 24/2-1 and 24/2-2 were not showing any segregation for GFP accumulation, either among their pollen or among their selfed offspring. The conclusion is that the *gfp* mutation must have occurred in a haploid cell in both cases. The ability to mutagenize haploid cells permits the generation of instantly homozygous mutants, which is highly advantageous compared to previously published methods of gene targeting. It has also been possible to show that the altered *gfp* alleles were transmitted through meiosis to the gametophyte and hence to the progeny of primary mutants, just as has been demonstrated in rice in the case of an altered *Os11N3* promoter sequence [7]. Among the progeny of 32/2-2, multiple *gfp* alleles were observed that had not been detected in the T_0 primary mutant. On the other hand, the allele present in the primary mutant was not recovered among 32 sequenced T_1 individuals. The reason might be that the T_0 primary mutant was chimeric concerning the mutant alleles present, and the cells harbouring the observed allele (4 nt deletion) did not contribute to the germline of the primary mutant. Alternatively, as the primary transgenic mutant 32/2-2 carried both TALEN units stably integrated, the TALENs might have kept binding the target sites repeatedly producing new mutant alleles until at least one of the target sites had been destroyed, that rendered the target region inaccessible to the TALE nucleases.

In conclusion, we have demonstrated a practical method for inducing heritable targeted gene disruption in barley using TALENs. This success should encourage the application of this technology to elucidate gene function and develop improved or new traits not only in barley, but also in other cereal species, in particular wheat and maize. The results obtained may also pave the way for the establishment of ways to use TALENs to edit cereal and other crop plant sequences based on DSB repair *via* homologous recombination.

Supporting Information

File S1 Figures S1–S5 and Tables S1–S2. Figure S1: Embryogenic pollen cultures developed from donor *gfp* transgenic barley. Figure S2: DNA gel blot and *gfp* expression analysis of the donor line PV89. Pollen culture of independent hemizygous transgenic plant is shown as a control. Figure S3: PCR amplification of the *Fok*I domain from a template of cDNA prepared from each of the primary *gfp* mutants, the *gfp* donor and non-transformed cv. 'Igri'. *ACTIN* served as a positive control. Figure S4: Best matches found in the barley genome as compared to the target sequence. Figure S5: Immature pollen produced by primary transgenic *gfp* knock-out plant 24/2-1 and *gfp* donor. Table S1: Primer sequences used for the identification of T-DNA elements. Table S2: Transgenic plants characterized for ploidy level and the presence and expression of TALEN T-DNA. Sequences of the used binary vectors.

Acknowledgments

We particularly wish to thank Ingrid Otto, Andrea Müller, Cornelia Marthe, Sibylle Freist and Petra Hoffmeister for their technical assistance, and Waltraut Wackermann, Kerstin Jacobs and Enk Geyer for their careful maintenance of plant materials. We acknowledge the constructive input of Drs. Raheleh Karimi-Ashtiyani and Andreas Houben. We also thank Dr. Hector Candela and three anonymous reviewers for their comments and suggestions to improve the manuscript.

Author Contributions

Conceived and designed the experiments: JK MG GH. Performed the experiments: MG GH SS VV. Analyzed the data: MG SH. Wrote the paper: MG JK GH.

References

1. Voytas DF (2013) Plant genome engineering with sequence-specific nucleases. Annu Rev Plant Biol 64: 327–350.
2. Puchta H, Fauser F (2013) Synthetic nucleases for genome engineering in plants: prospects for a bright future. Plant J DOI: 101111/tpj12338.
3. Boch J, Scholze H, Schornack S, Landgraf A, Hahn S, et al. (2009) Breaking the code of DNA binding specificity of TAL-Type III effectors. Science 326: 1509–1512.
4. Moscou MJ, Bogdanove AJ (2009) A simple cipher governs DNA recognition by TAL effectors. Science 326: 1501.
5. Weber E, Gruetzner R, Werner S, Engler C, Marillonnet S (2011) Assembly of designer TAL effectors by golden gate cloning. PLoS ONE 6 e19722. doi:10.1371/journal.pone.0019722.
6. Cermak T, Doyle EL, Christian M, Wang L, Zhang Y, et al. (2011) Efficient design and assembly of custom TALEN and other TAL effector-based constructs for DNA targeting. Nucleic Acids Res 39: e82.
7. Li T, Liu B, Spalding MH, Weeks DP, Yang B (2012) High-efficiency TALEN-based gene editing produces disease-resistant rice. Nat Biotech 30: 390–392.
8. Sanjana NE, Cong L, Zhou Y, Cunniff MM, Feng G, et al. (2012) A transcription activator-like effector toolbox for genome engineering. Nat Protoc 7: 171–192.
9. Reyon D, Tsai SQ, Khayter C, Foden JA, Sander JD, et al. (2012) FLASH assembly of TALENs for high-throughput genome editing. Nat Biotech 30: 460–465.
10. Schmid-Burgk JL, Schmidt T, Kaiser V, Höning K, Hornung V (2013) A ligation- independent cloning technique for high-throughput assembly of transcription activator–like effector genes. Nat Biotech 31: 76–81.
11. Mahfouz MM, Li L, Shamimuzzaman M, Wibowo A, Fang X, et al. (2011) De novo-engineered transcription activator-like effector (TALE) hybrid nuclease with novel DNA binding specificity creates double-strand breaks. Proc Natl Acad Sci USA 108: 2623–2628.
12. Christian M, Qi Y, Zhang Y, Voytas DF (2013) Targeted mutagenesis of *Arabidopsis thaliana* using engineered TAL effector nucleases (TALENs). G3 doi:10.1534/g3.113.007104.
13. Shan Q, Wang Y, Chen K, Liang Z, Li J, et al. (2013) Rapid and efficient gene modification in rice and Brachypodium using TALENs. Mol Plant doi:10.1093/mp/sss162.
14. Wendt T, Holm PB, Starker CG, Christian M, Voytas DF, et al. (2013) TAL effector nucleases induce mutations at a pre-selected location in the genome of primary barley transformants. Plant Mol Biol doi:10.1007/s11103-013-0078-4.
15. Zhang Y, Zhang F, Li X, Baller JA, Qi Y, et al. (2013) Transcription Activator-Like Effector Nucleases enable efficient plant genome engineering. Plant Physiol 161: 20–27.
16. Maluszynski M, Kasha KJ, Forster BP, Szarejko I (2003) Published doubled haploid protocols in plant species. In: Maluszynski M, Kasha KJ, Forster BP, Szarejko I, editors. Doubled Haploid Production in Crop Plants: A Manual. Dordrecht: Kluwer Academic Publishers. pp 309–335.
17. Kumlehn J, Serazetdinova L, Hensel G, Becker D, Loerz H (2006) Genetic transformation of barley (*Hordeum vulgare* L.) via infection of androgenetic pollen cultures with *Agrobacterium tumefaciens*. Plant Biotechol J 4: 251–261.
18. Hensel G, Kastner C, Oleszczuk S, Riechen J, Kumlehn J (2009) *Agrobacterium*-mediated gene transfer to cereal crop plants: current protocols for barley wheat triticale and maize. Int J Plant Genomics 2009 835608. doi:10.1155/2009/835608.
19. The International Barley Genome Sequencing Consortium (2012) A physical genetic and functional sequence assembly of the barley genome. Nature 491: 711–716.
20. Christensen AH, Quail PH (1996) *Ubiquitin* promoter-based vectors for high-level expression of selectable and/or screenable marker genes in monocotyledonous plants. Transgen Res 5: 213–218.
21. O'dell JT, Nagy F, Chua NH (1985) Identification of DNA sequences required for activity of the cauliflower mosaic virus 35S promoter. Nature 313: 810–812.
22. Komari T, Hiei Y, Saito Y, Murai N, Kumashiro T (1996) Vectors carrying two separate TDNAs for co-transformation of higher plants mediated by *Agrobacterium tumefaciens* and segregation of transformants free from selection markers. Plant J 10: 165–174.
23. Fang YD, Akula C, Altpeter F (2002) *Agrobacterium*-mediated barley (*Hordeum vulgare* L.) transformation using green fluorescent protein as a visual marker and sequence analysis of the T-DNA: barley genomic DNA junctions. J Plant Physiol 159: 1131–1138.
24. Palotta M, Graham R, Langridge P, Sparrow D, Barker S (2000) RFLP mapping of manganese efficiency in barley. Theor Appl Genet 101: 1100–1108.
25. Thiebaut J, Kasha KJ (1978). Modification of the colchicine technique for chromosome doubling of barley haploids. Can J Genet Cytol 20: 513–521.
26. Miller JC, Tan S, Qiao G, Barlow KA, Wang J, et al. (2011) A TALE nuclease architecture for efficient genome editing. Nat Biotech 29: 143–148.
27. Kasha KJ, Hu TC, Oro R, Simion E, Shim YS (2001) Nuclear fusion leads to chromosome doubling during mannitol pretreatment of barley (*Hordeum vulgare* L.) microspores. J Exp Bot 52: 1227–1238.
28. González-Melendi P, Ramrez C, Testillano PS, Kumlehn J, Risueño MC (2005) Three dimensional confocal and electron microscopy imaging define the dynamics and mechanisms of diploidisation at early stages of barley microspore-derived embryogenesis. Planta 222: 47–57.
29. Mussolino C, Morbitzer R, Lütge F, Dannemann N, Lahaye T, et al. (2011) A novel TALE nuclease scaffold enables high genome editing activity in combination with low toxicity. Nucleic Acids Res, 39: 9283–9293.
30. Kim Y, Kweon J, Kim J-S (2013) TALENs and ZFNs are associated with different mutation signatures. Nat Methods 10: 185.
31. Gao H, Smith J, Yang M, Jones S, Djukanovic V, et al. (2010) Heritable targeted mutagenesis in maize using a designed endonuclease. Plant J 61: 176–187.
32. Sander JD, Cade L, Khayter C, Reyon D, Peterson RT, et al. (2011) Targeted gene disruption in somatic zebrafish cells using engineered TALENs. Nat Biotech 29: 697–698.
33. Tesson L, Usal C, Ménoret S, Leung E, Niles BJ, et al. (2011) Knockout rats generated by embryo microinjection of TALENs. Nat Biotech 29: 695–696.
34. Bitinaite J, Wah DA, Aggarwal AK, Schildkraut I (1998) FokI dimerization is required for DNA cleavage. Proc Natl Acad Sci USA 95: 10570–10575.
35. Wah DA, Hirsch JA, Dorner LF, Schildkraut I, Aggarwal AK (1997) Structure of the multimodular endonuclease FokI bound to DNA. Nature 388: 97–100.

13

Scanning of Transposable Elements and Analyzing Expression of Transposase Genes of Sweet Potato [*Ipomoea batatas*]

Lang Yan[1], Ying-Hong Gu[1], Xiang Tao[2], Xian-Jun Lai[1], Yi-Zheng Zhang[1], Xue-Mei Tan[1]*, Haiyan Wang[1]*

1 Key Laboratory of Bio-resources and Eco-environment, Ministry of Education, Sichuan Key Laboratory of Molecular Biology and Biotechnology, Center for Functional Genomics and Bioinformatics, College of Life Sciences, Sichuan University, Chengdu, Sichuan, People's Republic of China, **2** Chengdu Institute of Biology, Chinese Academy of Sciences, Chengdu, Sichuan, People's Republic of China

Abstract

Background: Transposable elements (TEs) are the most abundant genomic components in eukaryotes and affect the genome by their replications and movements to generate genetic plasticity. Sweet potato performs asexual reproduction generally and the TEs may be an important genetic factor for genome reorganization. Complete identification of TEs is essential for the study of genome evolution. However, the TEs of sweet potato are still poorly understood because of its complex hexaploid genome and difficulty in genome sequencing. The recent availability of the sweet potato transcriptome databases provides an opportunity for discovering and characterizing the expressed TEs.

Methodology/Principal Findings: We first established the integrated-transcriptome database by *de novo* assembling four published sweet potato transcriptome databases from three cultivars in China. Using sequence-similarity search and analysis, a total of 1,405 TEs including 883 retrotransposons and 522 DNA transposons were predicted and categorized. Depending on mapping sets of RNA-Seq raw short reads to the predicted TEs, we compared the quantities, classifications and expression activities of TEs inter- and intra-cultivars. Moreover, the differential expressions of TEs in seven tissues of *Xushu* 18 cultivar were analyzed by using Illumina digital gene expression (DGE) tag profiling. It was found that 417 TEs were expressed in one or more tissues and 107 in all seven tissues. Furthermore, the copy number of 11 transposase genes was determined to be 1–3 copies in the genome of sweet potato by Real-time PCR-based absolute quantification.

Conclusions/Significance: Our result provides a new method for TE searching on species with transcriptome sequences while lacking genome information. The searching, identification and expression analysis of TEs will provide useful TE information in sweet potato, which are valuable for the further studies of TE-mediated gene mutation and optimization in asexual reproduction. It contributes to elucidating the roles of TEs in genome evolution.

Editor: Turgay Unver, Cankiri Karatekin University, Turkey

Funding: This work was supported by funds from the National Science and Technology Pillar Program of China (No. 2007BAD78B03, http://kjzc.jhgl.org/) and the "Eleventh-Five" Key Project of Sichuan Province, China (No. 07SG111-003-1, http://www.scst.gov.cn/). The funders had no role in study design, data collection and analysis, decision to publish, or preparation of the manuscript.

Competing Interests: The authors have declared that no competing interests exist.

* E-mail: hayawang@scu.edu.cn (HYW); txmyyf@scu.edu.cn (XMT)

These authors contributed equally to this work.

Introduction

Sweet potato [*Ipomoea batatas*] is the world's seventh largest food crop cultivated worldwide due to its high yield, wide adaptability and strong resistance. It is grown on about 9 million hectares in the world, yielding 140 million tons per year, and over 97% of the world output of sweet potato is produced from developing countries. The cultivated area and yield of sweet potato in China, about 6.6 million hectares and 100 million tons, account for 70% and 85% of total area and yield of the world, respectively [1,2]. Sweet potato belongs to the *Convolvulaceae* family, *Ipomoea* genus, *Batatas* section. It is the only hexaploid ($2n = 6x = 90$) plant in this section with a huge genome (2,200 to 3,000 Mbp) [3–5] and complicated genetic structure. Many questions like genome sequencing and genetic evolution mechanism are still unresolved so far. In general, the organisms adapt to the changing environment through favorable mutation and chromosome recombination in the process of sexual reproduction. Since sweet potato mostly performs asexual reproduction, how does it reorganize its genetic substance in the process of evolution, while lacking of gametic recombination?

Transposable elements (TEs) are one of the important genetic factors for genome reorganization. They are the most abundant genomic components in eukaryotes, even accounting for more than 50% of the entire genome [6–9], especially in some large cereal genomes such as maize (85%), wheat (80%) and barley (84%) [10–13]. The TEs affect the genome as mutagenic agents by replicating and translocating to generate plasticity, producing structural changes in single gene or overall genome followed by altered spatial and temporal patterns of gene expression and, ultimately, gene function [14]. Although mutations may be harmful, and could lead to different diseases and even death of the individuals, they are the basis of biological evolution for the species. Thus, mutations generate diversity that may provide adaptive advantages to the changing environments, being further selected as a result of the natural selection [15]. The role of TEs in evolution was proposed by Barbara McClintock in the 1980s, since then progress had been made to understand the significance of TEs in genome evolution through the comprehensive study of the structure and function of TEs in different organisms [16]. For example, it has been reported that Helitrons, a kind of DNA-TEs in *Zea mays*, could capture and move gene fragments to an extent that around 20% genes in maize genome were found located differently between two maize lines [17]. And in *Arabidopsis thaliana*, Helitrons could proliferate themselves in the genome after acquiring the exon fragments [18]. Since many of the exon fragments captured by TEs are expressed, the TE-mediated exon shuffling might lead to the appearance of novel genes. A series of studies have been carried on the mutational capacity of TEs, their ability to regulate genetic systems, and their sensitivity to environmental stress. These studies have demonstrated that the TEs could not only shape the structure and function of the genomes, but also generate genetic polymorphisms favoring population adaptation, which plays a major role in genome evolution [16]. However, the TEs are still poorly understood in sweet potato as its hexaploidy genome makes genetic manipulation and genome sequencing very challenging.

In this study, we searched and identified TEs in sweet potato on the basis of the integrated-transcriptome database, which provides a large amount of expressed TE homologues. Such a large number of TEs, which were scarcely identified in sweet potato before, represent the collection of TEs with the largest number and the most complex classification in sweet potato by far. The searching, identification and expression analysis of TEs provides useful resources and information of TEs in sweet potato, which may be valuable for the study of TE-mediated gene mutation and optimization in asexual reproduction. Our result provides a new method for TE searching on species with transcriptome sequences while lacking genome information. Such amounts of TEs found in sweet potato are important data resources and material bases for studying the TE functions further. It also contributes to the elucidation of the roles of TEs in genome evolution.

Results

Integrated-transcriptome Database of Sweet Potato

There were totally four transcriptome databases of sweet potato which had been established from primary sequencing data of three cultivars in China. Two of them were established by our laboratory from the vegetative organ [19] and flowers [20] of *Xushu* 18 cultivar (XS 18) in 2012. The other two came from the fibrous and tuberous root of *Guangshu* 87 cultivar (GS 87) in 2010 [21] and the tuberous root of a purple sweet potato *Jingshu* 6 cultivar (JS 6) in 2012 [22], respectively. The characteristics and details of the four transcriptome databases are listed in Table 1.

Sweet potato integrated-transcriptome database was established by integrating the above four databases. All the raw reads from these databases were combined and *de novo* assembled [23–26]. The resulted integrated-transcriptome database comprised of totally 279,473 transcripts, 118,309 of which were >200 nt in length, and the longest one was 13,067 nt. For annotation of the assembled transcripts, sequence-similarity search was conducted against the NCBI non-redundant (Nr) protein database through Basic Local Alignment Search Tool (BLAST) alignment [27]. The transcripts longer than 200 nt were submitted for annotation through Blast2GO, and 60,976 transcripts were annotated. The integrated and the other four transcriptome databases were used for searching TEs and for analyzing the quantities, distributions and expression levels of TEs inter- and intra- cultivars, respectively.

Prediction of TEs in the Integrated-transcriptome Database of Sweet Potato

Functional annotation and homologous sequence alignment were used to search TEs in the integrated-transcriptome database. There were totally 3,677 TEs found in 60,976 annotated transcripts using keyword searching, including 2,626 retro-TEs (class I) and 1,051 TEs (class II). Different keywords got respective results: 151 "transposon", 659 "retrotransposon", 432 "transposase", 783 "reverse transcriptase", 28 "transposable element", 333 "retroelement", 131 "hAT", 92 "En/spm", 122 "Mutator", 71 "MULE", 719 "Non-LTR", 16 "PIF", 192 "Copia", 140 "Gypsy", 8 "Mariner". In addition, 1,284 TE sequences from ten kinds of other higher plants were downloaded for homologous sequence alignment with all transcripts in the integrated-transcriptome database. Among the 1284 TEs, 434 TEs got hits with similarity >70%, in which 93 got hits with similarity >90% through BLASTn with a stringent cut-off value of e-10. Among the 434 hit sequences in sweet potato integrated-transcriptome, 402 were annotated as TEs, including 106 retro-TEs and 296 TEs. As to the third searching method, 3 TEs were obtained in the pair-wise alignment between assembled sequences and the sub-terminal conserved sequence of TEs in leguminous plants with similarity > 70% [28]. All above TEs were combined, the redundant TEs and the non-TE-like reverse-transcription virus were excluded. We finally identified 1,405 TEs, including 883 retro-TEs (Class I) and 522 TEs (Class II), and also found that there were 257 TEs with full length of ORF longer than 1000 nt from the Galaxy ORF prediction [29–30].

Among the 883 retro-TEs, there were 247 with LTR sequences which could be classified into Ty1/Copia and Ty3/Gypsy superfamily, 507 without LTR and 129 unclassified. In terms of the 522 TEs belonging to class II, 501 of them could be classified into 6 superfamilies, including Tc1/Mariner, hAT, Mutator, PIF-harbinger, CACTA and Helitron, and 21 were unclassified. The unclassified TEs showed sequence similarity lower than the threshold value set in BLAST alignment with the known TEs and thus couldn't be classified. However, they can still be further classified if the TE-specific conserved sequences analysis was carried out. For example, the conserved sequence CTAG and its preceding 18 bp palindromic sequence was suggested to be able to produce a hairpin loop to capture gene fragments by an unknown mechanism possibly associated with their rolling circle (RC) replication process [31]. Depending on this characteristic, the initially unclassified *Arabidopsis thaliana* TE Basho was lately grouped into Helitron superfamily. This kind of TE-specific conserved sequence was also existed in the 3′ terminus of 4 unclassified sweet potato TEs. Therefore, the detailed analysis of the complete sequences and some experiment validations will be helpful to identified unclassified TEs. It is noteworthy that few of the 1,405 TEs were collected with full length in plant repeat databases like PlantTribes or Repbase [32,33]. All the TEs identified in sweet potato are summarized in Table 2.

The 1,405 TEs were submitted to GenBank for similarity alignment using BLASTn program and the result showed that the

Table 1. Characteristics and details of four primary transcriptome databases.

	XS18-v	XS18-f	GS87-r	JS6-r
Sequencing depth (folds)	49.6	Not measured	48.36	137.1
Number of reads	48,716,884	Not measured	59,233,468	25,888,890
Length of reads (bp)	75	75	75	75
Identified genes	51,763	45,698	35,051	40,280
Number of SSRs	4,249	Not measured	4,114	851
Number of Contigs	128,052	70,412	208,127	473,238
Average length of contigs (bp)	321	628	202	138
N50 (bp)	509	895	252	118

XS18-v: transcriptome from a mixed sample of roots, stems and leaves in cultivar *Xushu 18.* **XS18-f:** transcriptome from flowers in cultivar *Xushu 18.* **GS87-r:** transcriptome from roots in cultivar *Guangshu 87.* **JS6-r:** transcriptome from roots in cultivar *Jjingshu 6.*

TEs in sweet potato shared deep homologies with those from at least 28 species of higher plants. The plant which got the most BLAST hits in number was *Vitis vinifera* (20.6%), followed by *Arabidopsis thaliana* (15.0%), *Oryza sativa* (13.7%), *Ipomopea trifida* (10.2%), etc (Figure 1). In addition, the plant which got the highest identity to sweet potato TEs was *Ipomopea trifida*, with sequence-similarity distribution ranging from 58% to 100%, followed by *Vitis vinifera* (50%~98%), *Glycine max* (46%~95%), *Populus trichocarpa* (43%~90%), *Arabidopsis thaliana* (39%~89%), *Zea mays* (47%~87%), *Oryza sativa* (43%~84%), etc. The high homologies of TEs between sweet potato and other plants revealed that these TEs were widely distributed in the vast majority of the genomes of higher plants, and thus provided further evidence for the evolutionary relationships between sweet potato and other dicotyledonous plants.

Differences in the Number of TEs Inter- and Intra-cultivars

To compare the number of the expressed TEs in three different sweet potato cultivars and different tissues of XS 18, the reads of the four primary databases were mapped to the 1,405 TEs. Firstly, the two databases from the same XS 18 cultivar were combined, then the number of TEs among three different sweet potato cultivars (XS 18, GS 87 and JS 6) can be determined. Secondly, the two databases established from the same XS 18 cultivar, corresponding to the vegetative and reproductive organs of XS 18, were used to analyze TE expression difference and specificity between these two kinds of tissues in XS 18. Finally, the vegetative transcriptome database of XS 18 was used to analyze TE expression difference and specificity among seven different vegetative organs and tissues, including YL (young leaves), ML (mature leaves), stem, FR (fibrous roots), ITR (initial tuberous roots), ETR (expanding tuberous roots) and HTR (harvest tuberous roots).

Table 2. Transposable elements identified and collected in the integrated-transcriptome database of sweet potato.

Classification		TE numbers	Length distribution (bp)	Similarity distribution
Order	Superfamily			
Retrotransposon				
LTR	Ty1/Copia	137	201–4207	51–100%
	Ty3/Gypsy	95	209–1840	46–95%
Non-LTR		507	201–6019	43–100%
Unclassified		144	201–10813	39–100%
DNA transposon				
TIR	Tc1/Mariner	6	207–926	67–77%
	hAT	136	208–3297	47–97%
	Mutator	93	204–3959	44–98%
	PIF-harbinger	23	625–2861	43–87%
	MITEs	3	416–1117	49–76%
	CACTA	80	205–3719	43–99%
Non-TIR	Helitron	159	201–3300	48–99%
IS family		4	201–1911	67–100%
Unclassified		18	201–3718	39–100%
Total		1,405	201–10813	39–100%

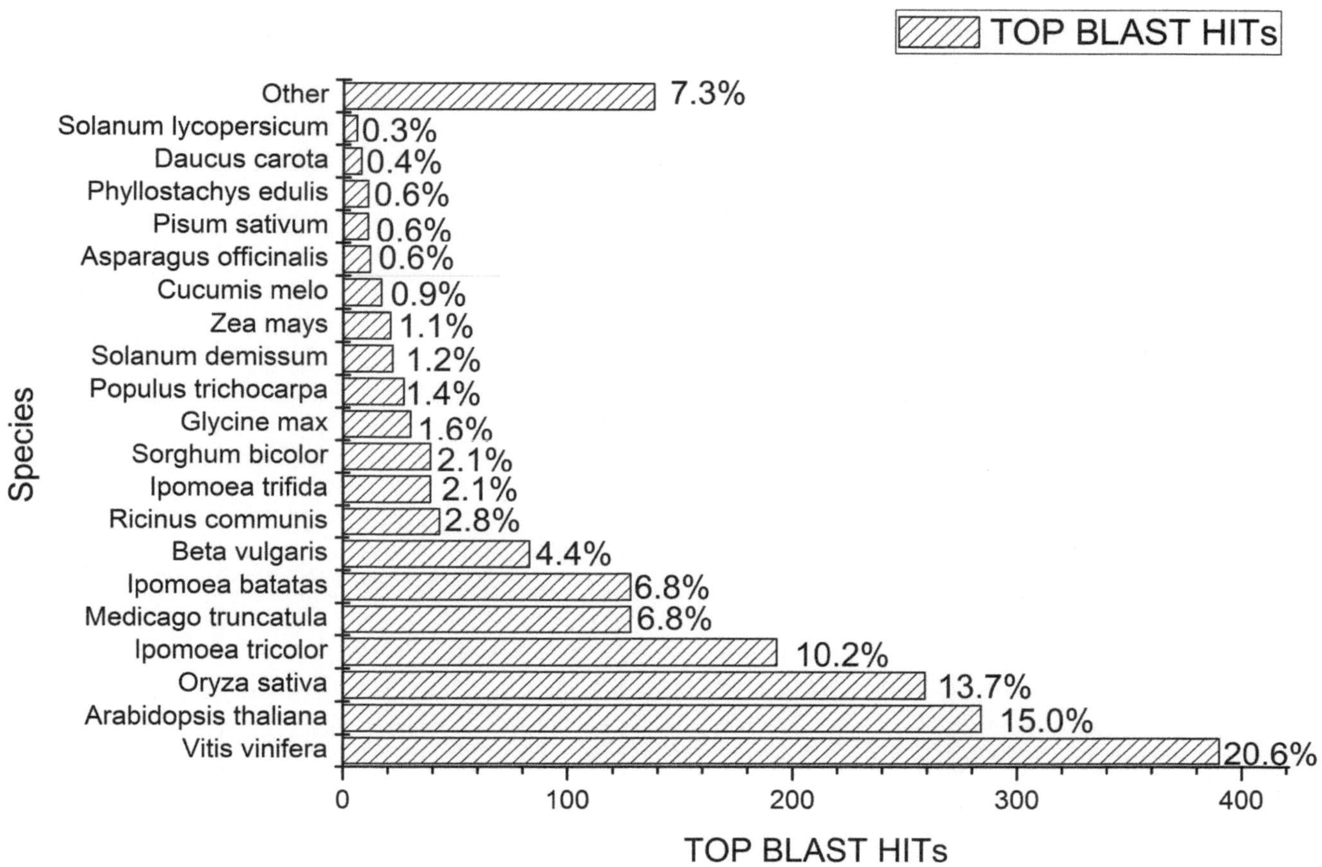

Figure 1. Top-Hit species distribution. 1,405 BLASTX-hit TE sequences were calculated. More than 50% of the identified TEs have the highest homology with *Vitis vinifera, Arabidopsis thaliana, Oryza sativa* and *Ipomoea tricolor*. Less than 5% of the top matches hit sweet potato itself due to the limited number of the sweet potato protein sequences available in the NCBI database.

Among 1,405 TEs expressed in sweet potato, there were 1,209 TEs identified in XS 18, 994 TEs in GS 87 roots and only 5 TEs in JS 6 roots. However, except for the five TEs found in JS6 which were also expressed in other two cultivars, we found that the expressions of some TEs were restricted to one cultivar. The pair-wise alignments of TEs showed that 412 TEs were specifically expressed in XS 18 (called XS18-specifically expressed TEs, XS18_SETEs) and 197 TEs in GS 87 (called GS87-specifically expressed TEs, GS87_SETEs). In the XS 18 cultivar, there were 1,030 TEs expressed in vegetative organs and 832 in reproductive organs. Similarly, there were 157 TEs expressed specifically in the vegetative organs (called vegetative organ-specifically expressed TEs, VO_SETEs) while only 124 in the reproductive organs (called reproductive organ-specifically expressed TEs, RO_-SETEs) (Figure 2). The TE numbers in each of the seven vegetative organs and tissues were determined by digital gene expression (DGE) tag profiling [19]. Among 328,383 distinct clean tags generated in the DGE tag profiling, only 689 tags could be mapped to 417 TEs expressed in these seven tissues, implying that the expressions of other 988 TEs lacking the recognition bases CATG could not be detected through this method. Figure 3 demonstrated that the average number of expressed TEs in one tissue was 232. The tissue which had the lowest number of expressed TEs was YL (192 TEs), accounting for 46.04% of all the detected TEs. The highest one was FR (270 TEs), accounting for 64.75%. In addition, there were 101 TEs expressed in only one tissue, accounting for 24.22% and 109 TEs expressed in all the seven tissues, accounting for 26.14%. These two kinds of TEs were more than half of all the detected TEs and the rest TEs expressed in 2–6 tissues were relatively less (<50%). Moreover, among the tissue-specifically expressed TEs, the number expressed in FR was 2–3 times more than that in other six tissues.

Differences in the Type of TEs Inter- and Intra-cultivars

In accordance with the above results, the types of the expressed TEs were analyzed further. The TEs expressed in each cultivar were classified into superfamilies and the results are shown in Table 3. The superfamily possessing the most number of expressed TEs in both XS 18 and GS 87 was Helitron, with 145 and 140 TEs, respectively, followed by superfamily hAT (123 and 116 TEs). Even so, there were some differences between them. In XS 18, the most abundant superfamily of retro-TEs was the Ty1/Copia, while in GS 87, Ty3/Gypsy had a slight advantage in numbers of the expressed TEs. As to the DNA-TEs, the superfamilies CACTA and PIF contained more TEs in GS 87 than in XS 18. Taken into consideration the total numbers of TEs in the two cultivars, the importance of these two superfamilies to GS87 was more obvious.

The TE types between the vegetative and reproductive organs of XS 18 were almost the same, and the Helitron was the most abundant superfamily in both organs. The situation was the same when the TE types in seven tissues of vegetative organs in XS 18

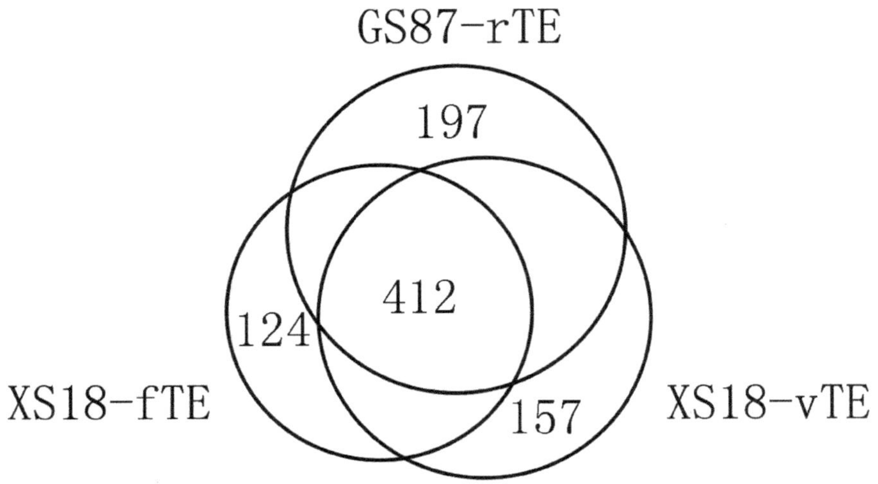

Figure 2. Expressed TE number in different cultivars of sweet potato. TEs were compared pair-wisely to detect the expressed TE number in different cultivars of sweet potato. Each circle represented the TE number expressed in a certain cultivar or tissue, and the cross part meant the co-expressed TE number in both two cultivars. Abbreviations GS87-rTE meant TEs in roots of cultivar *Guangshu*87, XS18-fTE meant TEs in flowers of cultivar *Xushu*18 and XS18-vTE meant TEs in vegetative organs of cultivar *Xushu*18. This overlapping circle diagram was made by auto CAD 2004 software.

were analyzed. The superfamily with the most expressed TEs was Helitron and the least was Ty3/Gypsy. The expressed TEs in superfamily hAT and Mutator were relatively less in the vegetative organs of XS 18 (Table 4).

Differences in the Expressing Activities of TEs Inter- and Intra-cultivars

By mapping all the reads of each cultivar transcriptome database to the identified TEs in the integrated-transcriptome database, we could calculate the relative expression level of every TE in each database and analyze the differences in the expressing activities at three levels. RPKM (reads per kilobases per million reads) was used as the standardized unit, which normalized the gene length and sequencing depth to make the expression levels of genes in different transcriptomes comparable. The comparisons among cultivars demonstrated that some TEs had stable expression levels but others showed significant differences. The major characteristics were as follow: All of the 5 TEs in JS 6 cultivar showed high expression levels, ranging from 30,000~190,000 RPKM, much higher than their expression levels in the other two cultivars (about 2000 RPKM). As to the TEs expressed in XS 18 and GS 87, the expression levels of TEs ranged from 0.42~70,000 RPKM and 5.59~30,000 RPKM, respectively, indicating that the expression levels of various TEs varied greatly in different cultivars.

Meanwhile, a large number of TEs were differentially expressed in XS 18 and GS 87. For example, Ib_DTC_34571 belonging to CACTA superfamily, expressed highly in GS 87 (2759.25 RPKM) but lowly in XS 18 (0.42 RPKM). Oppositely, Ib_RN_13038, non-LTR retro-TE, had highly expression level in XS 18 (9280.42 RPKM) but lowly expressed in GS 87 (32.82 RPKM). Except for that, we also found some TEs have relatively stable expression levels in two cultivars. For example, Ib_DTM_2831 belonging to Mutator superfamily, had high and similar expression level in XS 18 (4479.14 RPKM) and GS 87 (4181.12 RPKM).

In addition, the TE expression levels between the vegetative and reproductive organs of XS 18 were compared and some TEs expressed specifically in flowers were found. For example, Ib_RU_704, unclassified retro-TE, had a high expression in flower (3243.17 RPKM) but a low expression in vegetative organs (40.09 RPKM). And Ib_RLC_116580, retro-TE belonging to LTR-Copia superfamily, was expressed highly in flower (3837.77 RPKM), but not expressed in vegetative organs. Therefore, Ib_RLC_116580 could be defined as the flower-specially expressed TE.

DGE Analysis of TEs in Vegetative Tissues of Xushu18

The DGE tag profiling was used to analyze TE expression levels among vegetative tissues of XS 18. We found lots of typical TEs differentially expressed in different vegetative tissues. For example, Ib_DU_31235, a unclassified TE, was expressed highly in ML (260.71 TPM, tags per million reads), but its expressions in other tissues were around 30 TPM; Ib_RN_25697, a non-LTR retro-TE, was expressed highly in stem with 63.41 TPM but lowly in other tissues (around 2 TPM), even not expressed in YL and HTR; Ib_RN_6012, a non-LTR retro-TE, was expressed lowly in YL (13.13 TPM), but highly in other 6 tissues (60.66–128.63 TPM) (Figure 4). In addition, some TEs were stably expressed in 7 tissues. For example, Ib_DTH_65331, a TE belonging to hAT superfamily, was expressed highly in seven tissues with TPM value higher than 100; Ib_DTC_11049, a TE belonging to CACTA superfamily, was expressed lowly in all tissues with TPM value around 1.

To analyze differential expression patterns of TEs among seven tissues, we pair-wisely compared them and obtained 21 pairs of comparisons. There were numerous TEs showing differential expression (DETEs) and specific expression (SETEs). Among 417 TEs expressed in vegetative organs, the number of DETEs between each two tissues, including the up-regulated and the down-regulated TE, ranged from 7 to 46 and the average number was 27 (Figure 5). The largest difference was observed between ETR and HTR, and there were 25 TE transcripts up-regulated and 21 down-regulated. The smallest difference occurred between FR and ITR, in which only 7 DETEs were identified. In addition, a large number of SETEs between each two tissues were also detected. In the 21 pair-wisely comparisons, the number of SETEs

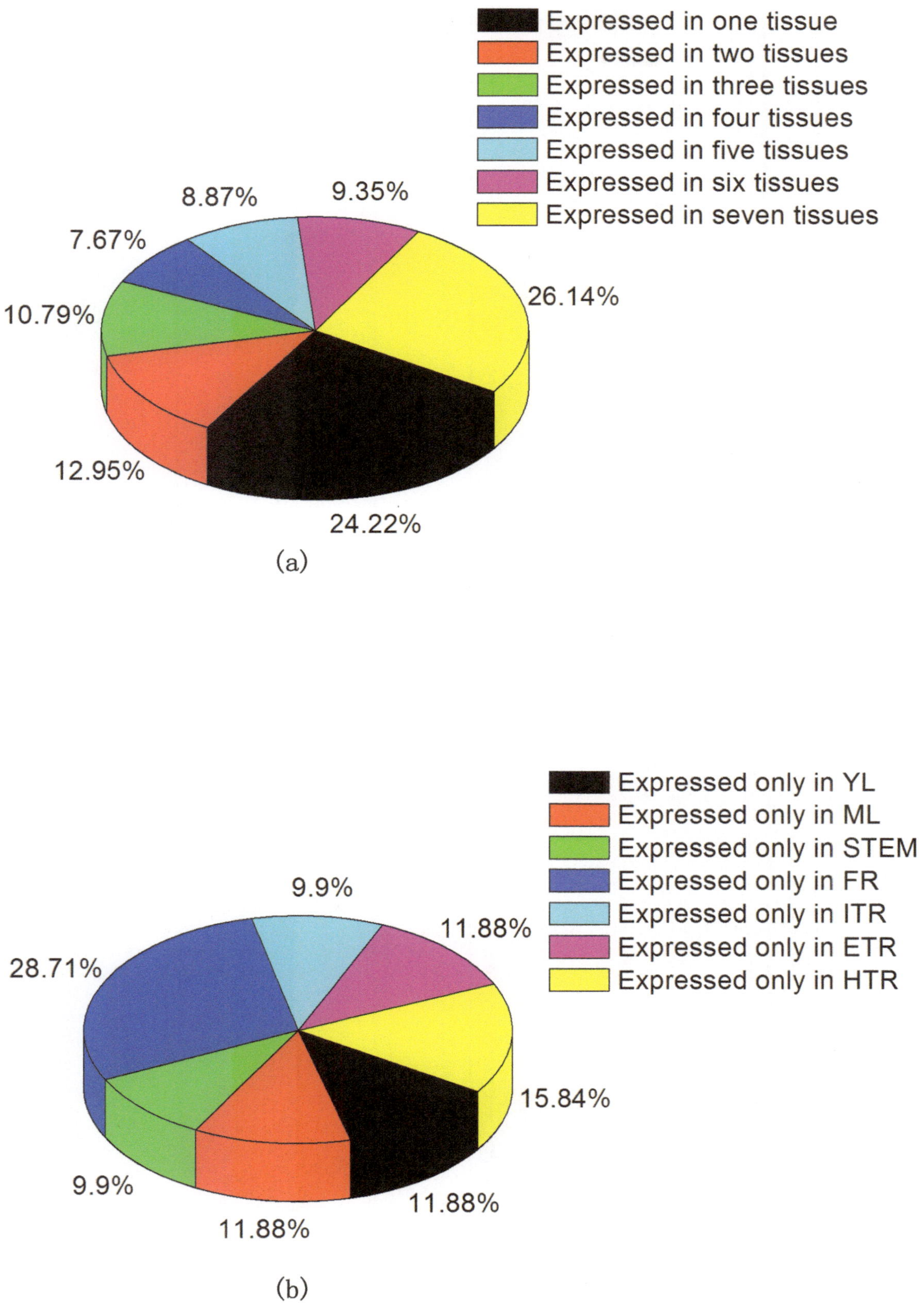

Figure 3. Analysis of digital gene expression (DGE) tag profiling. The analysis included (a) the statistics of the TEs expressed in 1–7 tissues. (b) the expressed TE number only in one tissues. The explanations for abbreviations were in bracket: YL (young leaves), ML (mature leaves), FR (fibrous roots), ITR (initial tuberous roots), ETR (expanding tuberous roots) and HTR (harvest tuberous roots).

Table 3. Classifications of the TEs in different sweet potato cultivars.

	Retro-TEs				DNA TEs						
	Ty1/Copia	Ty3/Gypsy	Non-LTR	Unclassify	Mariner	hAT	Mutator	PIF	Helitron	CACTA	unclassify
XS18-vTE	99	68	351	53	5	115	76	19	139	60	45
XS18-fTE	76	43	319	44	3	85	59	13	88	53	50
GS87-rTE	68	76	327	48	6	116	72	20	140	70	51
JS6-rTE	0	0	2	1	0	1	1	0	0	0	0

XS18-vTE: TEs in vegetative organs of cultivar *Xushu*18. XS18-fTE: TEs in flowers of cultivar *Xushu*18. GS87-rTE: TEs in roots of cultivar *Guangshu*87. JS6-rTE: TEs in roots of cultivar *Jingshu*6.

Table 4. Statistics of the digital gene expression (DGE) tag profiling of TEs in the main superfamily.

Family name	TE members						
	YL	ML	Stem	FR	ITR	ETR	HTR
LTR-retrotransposon	17	24	17	27	14	18	21
Non-LTR retrotransposon	35	38	44	48	44	36	43
hAT transposon	28	34	32	44	37	33	39
Mutator transposon	23	29	26	25	30	27	27
PIF-harbinger transposon	7	8	7	7	6	5	9
CACTA transposon	12	18	16	16	15	12	14
Helitron transposon	37	52	54	58	35	27	59

YL: young leaves, **ML:** mature leaves, **FR:** fibrous roots, **ITR:** initial tuberous roots, **ETR:** expanding tuberous roots, **HTR:** harvest tuberous roots.

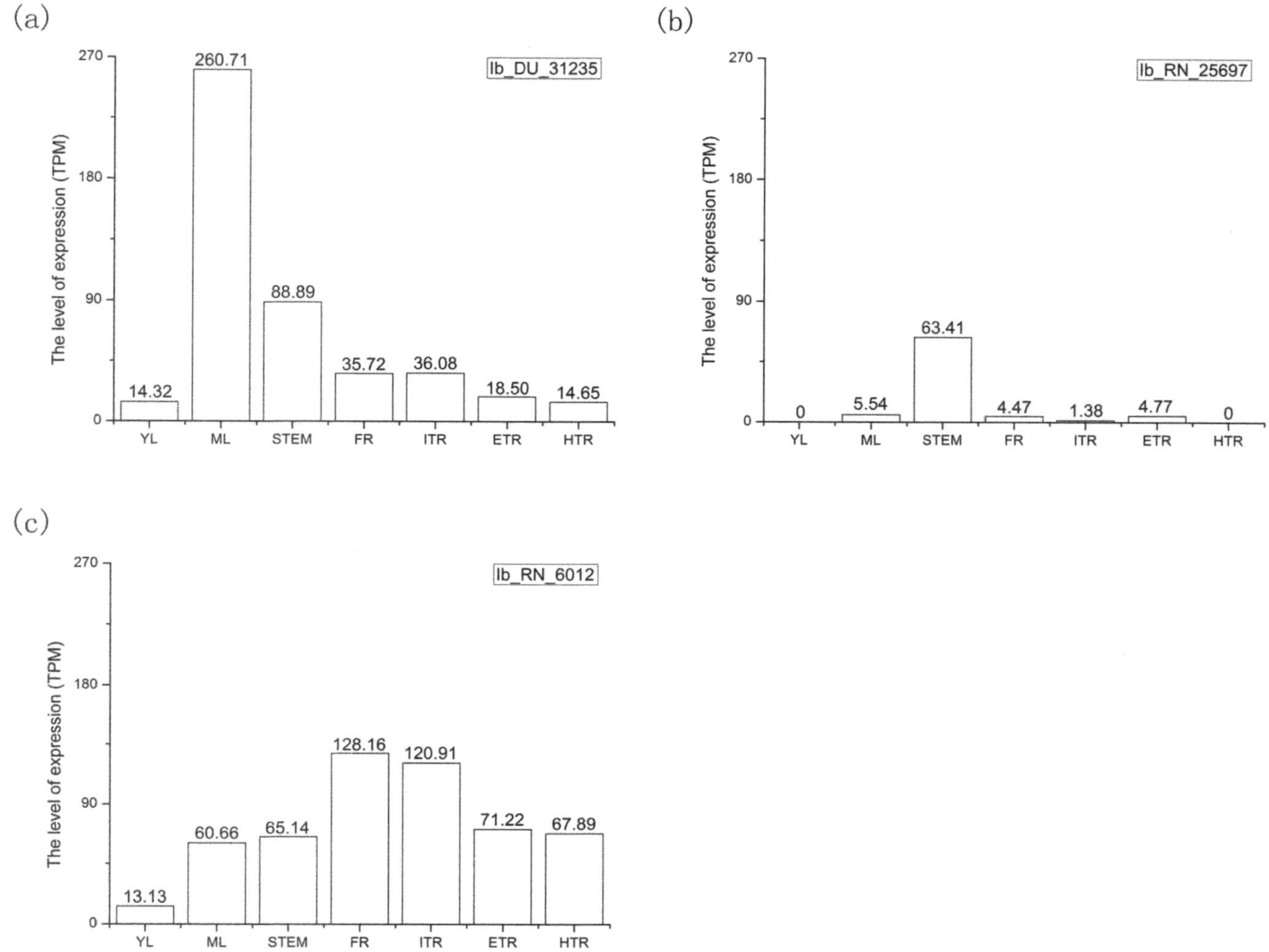

Figure 4. Representatives of differentially expressed TEs in different tissues and organs of *Xushu*18. (a) Differentially expressed TE Ib_DU_31235, which expressed extremely higher in ML than in other tissues. (b) Specially expressed TE Ib_RN_25697, which could not express in all tissues. (c) Differentially expressed TE Ib_RN_6012, which expressed extremely lower in YL than in other tissues. Abbreviations TPM means tags per million reads, the unit of gene expression level.

which expressed in only one of the two compared tissues ranged from 108 to 166, with an average number of 137 (Figure 6). The SETE patterns among tissues revealed that the largest difference was shown between YL and FR. There were 45 TEs expressed in YL but not in FR and 121 TEs expressed in FR but not in YL, oppositely. The smallest difference occurred between YL and ETR, in which 63 and 45 TEs were specifically expressed in YL and ETR, respectively.

Evolutionary Analysis of Transposase Genes

Sixteen TEs in sweet potato belonging to superfamily hAT, Mutator and PIF, respectively, and showing high homologies with transposase genes in other higher plants, were chosen for further studies of transposase genes. Bioinformatics analyses on these 16 TEs, involving gene length, predicted protein molecular weight, isoelectric points, etc are shown in Table S2. The coding sequences of 16 transposase genes were cloned and sequenced. The sequence similarity between the gene-clone sequences using Sanger sequencing and the assembled transcripts using Illumina sequencing were all higher than 95%, indicating that the predicted TE sequences were reliable. The comparisons between predicted and sequenced TEs are shown in Table 5.

The 16 sequenced transposase genes were then used for evolutionary analyses. The three phylogenetic trees of TEs in three superfamilies (Figure 7, a-c) showed that there were close relationships between TEs in sweet potato and dicotyledons like *Vitis vinifera*, *Glycine max* and *Populus trichocarpa*. The TE superfamily showing the highest sequence similarity with that in *Vitis vinifera* was Mutator, such as Ib_DTM_1770, Ib_DTM_4635, etc. On the other hand, we analyzed the evolutionary relationships of 16 TEs with each other. The phylogenetic tree diagram (Figure 7d) demonstrated that the evolutionary relationships of TE in Mutator superfamily were very close, with the bootstrap of 100 between certain TEs. However, there appeared to be a clear differentiation within this family, since bootstrap 84 presented at the boundary of the two subfamilies of Mutator (Mutator-FAR and Mutator-PB1). Therefore, these Mutator-TEs could be classified in more detail. Noteworthily, Ib_DU_2831 was originally not classified, but since it was highly homologous with Ib_DTM_3260, it could be classified into Mutator superfamily. Similarly, the unclassified TE Ib_DU_1235 could be classified into hAT superfamily because of

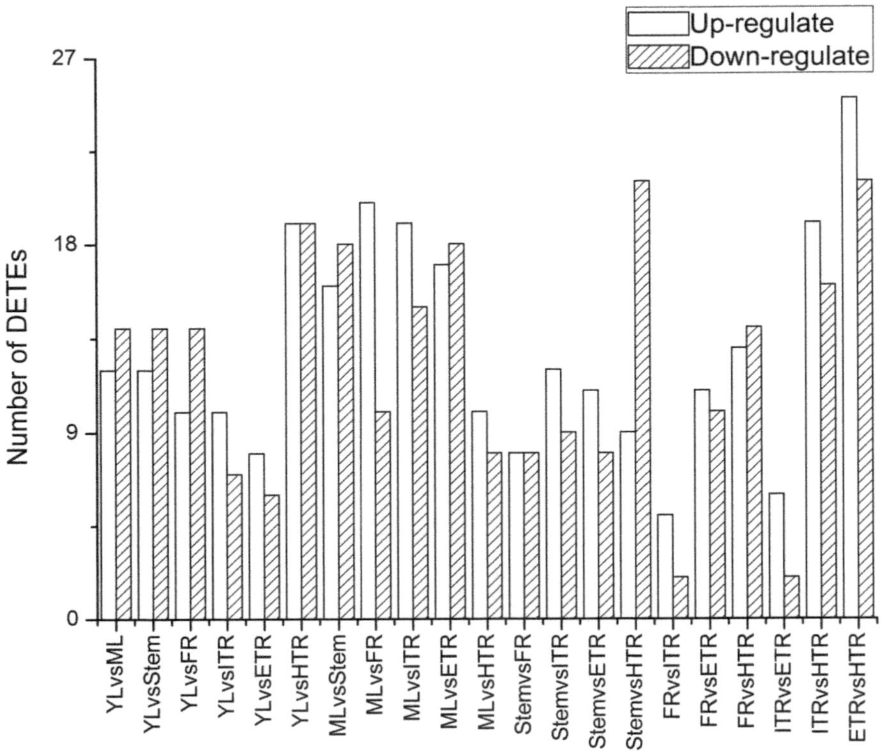

Figure 5. Differentially expressed TEs (DETEs) in seven tissues. Seven libraries were compared pair-wisely to detect the differentially expressed TEs between each two libraries respectively by edgeR. 21 pairs of comparison were implemented and Up-(black) and down-regulated (red) DETEs were quantified.

its closest genetic relationships with Ib_DTH_1962. However, from the phylogenetic tree diagram we could see that the PIF superfamily member Ib_DTP_11286 showed low homology with the same superfamily member Ib_DTP_9943, even lower than the hAT superfamily member Ib_DTH_1962. This might indicate that these two superfamilies were relatively close in sequence similarity and even in component structure. The alignment of 1,405 TEs with each other by using BLASTn was employed to analyze the genetic relationships among TEs. The results showed that 78 of 162 unclassified TEs had their relatives with similarity more than 78%, so that the evolutionary relationship analysis could be used to identify some unclassified TEs.

Identification of Transposase-gene Copy Number

Real-time polymerase chain reaction (PCR)-based methodology for the determination of transposase-gene copy number was introduced and demonstrated [34–36]. Absolute quantification determines the exact copy number of transposase gene by relating the *Ct* value to a standard curve. According to the standard curves of 11 transposase genes and one single-copy S8e gene [37], the copy number of each transposase gene in genomic DNA per microliter was tested, as well as that of S8e. The transposase-gene copy number in sweet potato genome can be determined as the copy ratio of transposase gene to S8e in one sample.

The standard curves for Ib_DTM_FAR14362 and S8e gene, each ranging from 10^6 to 10^9 copies per microliter are shown in Figure 8 (the standard curves for other ten transposase genes are shown in Figure S1). The *Ct* values of Ib_DTM _FAR14362 in each dilution ranged from 16 to 21, while the ranges of *Ct* values in S8e gene were 20–28. Figure 8 shows that both curves were highly linear (R2>0.99) in the range tested by the duplicate reactions and the slopes of the standard curves were −3.84 and −3.52, respectively. From the slope of each curve, PCR amplification efficiency (E) with 0.82 and 0.92 were calculated in the investigated range, respectively. The results of absolute quantification and the calculated transposase gene copy number are shown in Table 6. It was found that the copy number was around 1 to 3. These 11 transposase genes belonged to low copy number genes, in which there were one gene with 3 copies and 6 genes with 2 copies in genome, and there were 4 genes belonging to single-copy genes. In addition, the expression levels of these 11 transposase genes in XS 18 were also calculated. They all had relatively high expression ranging from 939.33–14372.48 RPKM. As listed in Table 6, there were not any proportional relationships between gene copy number in genome and their expression levels in transcriptome.

Discussion

A New Perspective for TEs Scanning

Transposable elements may be important motors of genetic variability, they account for the majority of genome (as in maize), and have the ability to generate genetic polymorphisms favoring population adaptation [38]. Researchers have been working to identify different TE types, in order to elucidate the molecular mechanism of TE transposition. The most widely used method to search for TE in the genome is to design element-specific primers for gene cloning by recognizing the sub-terminal conserved sequences of TEs in each family [28]. As for distinguishing different types of the TEs in model organisms, it was in accordance

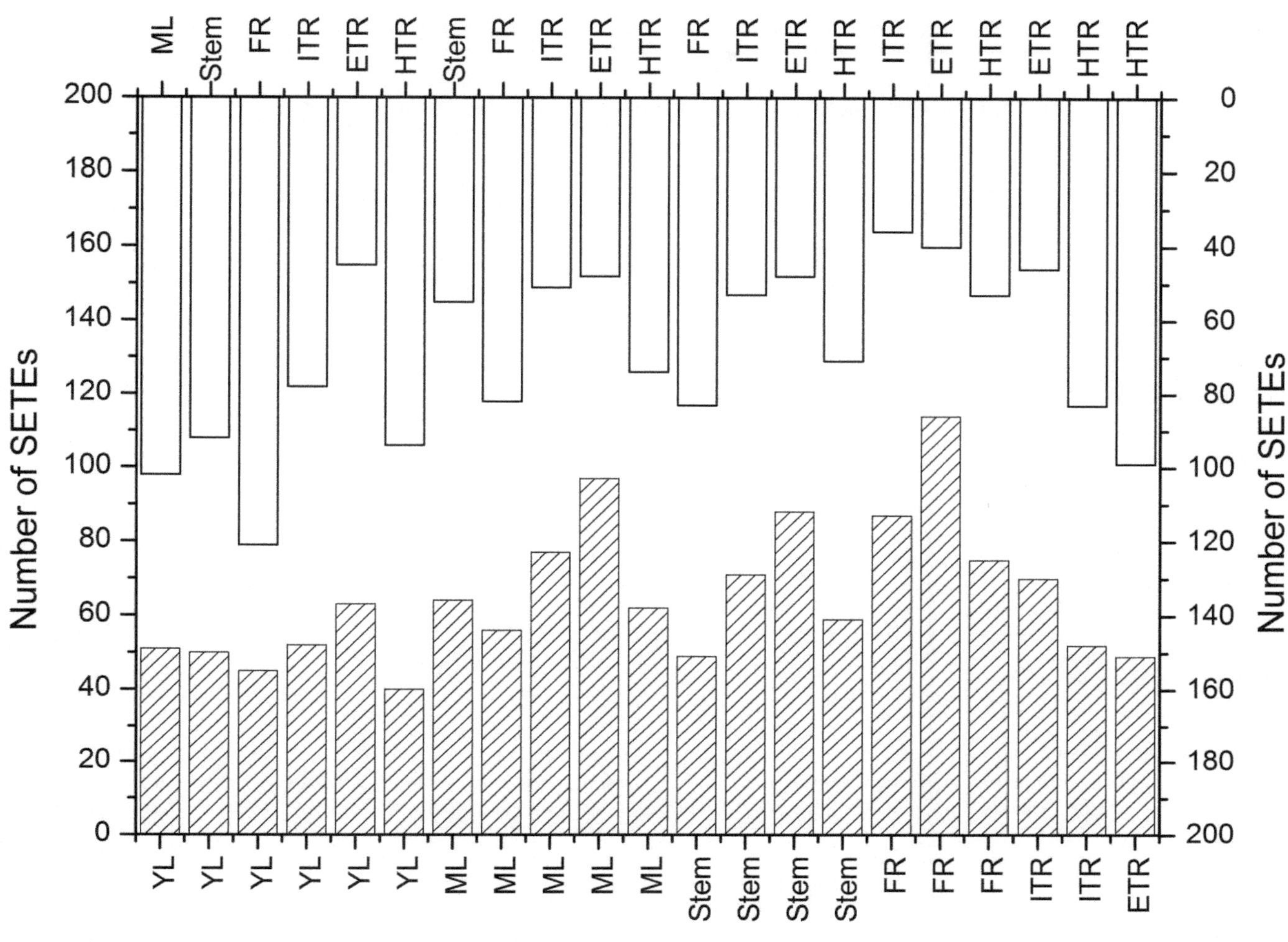

Figure 6. Specifically expressed TEs (SETEs) in seven tissues. Seven libraries were compared pair-wisely to detect the specifically expressed TEs between each two libraries respectively. 21 pairs of comparison were implemented and the SETEs between each two samples were quantified in red and black histograms (top vs. bottom).

with their transposition chemistry, that is, by the enzymes that catalyze the DNA-strand cleavage and transfer steps necessary for their movement, such as DDE transposases in IS sequences and members of Tc/Mariner superfamily [39]. However, these methods, which detect TEs one-by-one, not only require significant labor, being costly and time-consuming, but also may miss some TE or generate false-positive clones because of genetic variations or nonspecific cloning. The availability of genome sequencing results of some model or non-model plants made genome-widely TEs prediction practical and feasible. Therefore, a few databases of plant TEs like SoyTEdb in *Glycine max* [14], as well as some integrated repetitive-element databases of eukaryotes like Repbase Update and TIGR [33,40], were established. Therefore, the methods of TE identification in known genomes were based on the sequence alignment against the fully characterized elements in some TE databases. Along with more and more transcriptomes being sequenced, a method depending on transcriptome sequencing and gene annotation was developed, to search and identify TEs in plants without prior genome information. The advantage of this novel approach is that a global overview of potentially active TEs can be obtained. In contrast, the classic strategies required the TE to be analyzed on a one-by-one basis, being firstly identified at the DNA level, which allowed further transcription and transposition studies. For example, the 276 expressed TEs were identified in *Saccharum officinarum*, by searching from 260,781 transcriptome sequences in the sugarcane expressed-sequence tag (SUCEST) database [41]. In this paper, an improved method was used in TE searching in sweet potato, and 1,405 expressed TEs were identified after redundancy reduction. Noteworthy, the quantities and abundance of TEs searched from sweet potato were more than those searched from sugarcane. This may come from the more sensitive searching method and more abundant integrated-transcriptome database. Three different ways were used for TE searching in this study. The integrated-transcriptome database was built by re-assembling all the raw sequencing reads from four independent transcriptome databases of sweet potato representing three cultivars. It comprised plenty of transcripts with longer length and higher integrity and provided the possibilities to compare the differences of expressed TEs intra- or inter- cultivars. Some TEs were randomly selected for cloning and sequencing and the results showed that the similarities of the cloned and assembled sequences were higher than 95%, implying that the assembled TEs are considerably reliable.

Since there were few TEs in sweet potato reported before, the large numbers of TEs obtained in this study will provide an abundant resource for further studies like cloning and functional

Table 5. Comparison of predicted and measured transposase genes from the genome of Sweet potato.

Gene name	Predicted/measured length (bp)	Intron numbers	Intron length (bp)	BDP	ADP
Ib_DTM_FAR11812	2,475/2,475	0	\	0.28%(7/2475)	0.24%(2/824)
Ib_DTM_FAR14118	2,082/2,207	1	112	0.58%(12/2082)	0.29%(2/693)
Ib_DTM_FAR14362	2,076/2,587	2	104 and493	0.77(16/2076)	0.43%(3/692)
Ib_DTM_PB12217	2,238/2,238	0	\	0.31%(7/2238)	0.27(2/745)
Ib_DTM_PB13260	2,313/2,313	0	\	0.39%(9/2313)	0.39%(3/770)
Ib_DTM_PB14635	1,752/1,752	0	\	0.86%(15/1752)	0.86%(5/583)
Ib_DTP_9943	1,332/1,332	0	\	0.75%(10/1332)	1.12%(5/443)
Ib_DTH_1962	2,043/2,468	2	98and132	0.59%(12/2043)	0.44%(3/680)
Ib_DTM_5847	1,662/1,662	0	\	0.48%(8/1,662)	0.36% (2/553)
Ib_DTM_1664	2,538/2,754	1	209	0.55% (14/2,538)	0.59% (5/845)
Ib_DTM_2890	2,001/2,001	0	\	0.45% (9/2,001)	0.45% (3/666)
Ib_DTM_1770	2,601/2,863	1	247	0.38% (10/2,601)	0.69% (6/866)
Ib_DTM_3282	2,361/2,361	0	\	0.38% (9/2,361)	0.63% (5/786)
Ib_DTH_1235	2,712/2,889	1	168	0.59% (16/2,712)	0.66% (6/903)
Ib_DTP_11286	1,176/1,176	0	\	0.68% (8/1,176)	0.36% (3/824)
Ib_DU_2831	2,367/2,367	0	\	0.46% (11/2,367)	0.38% (3/788)

BDP: bases differences percentage.
ADP: Amino acids differences percentage.

identification of interested TEs. The intensive study of their transpositions will contribute to elucidating the role of TEs in genetic variation in the process of asexual reproduction and the evolution of sweet potato.

Prediction of the Size of TEs in the Sweet Potato Genome

Depending on transcriptome sequencing and gene annotation, 1,405 TEs were predicted from the sweet potato integrated-transcriptome through different searching methods. Even though these transcribed TE sequences could not completely reflect all the TEs existed in sweet potato genome, they represented the active part in these cultivars and tissues. Owing to the availability of the genome and transcriptome sequences of *Oryza sativa*, the TEs of rice can be predicted from the annotation information of genome and transcriptome, respectively. TEs in *Oryza sativa* genome could be acquired from the annotation information of the *Oryza sativa* genome from the web of the Rice Genome Annotation Project, http://rice.plantbiology.msu.edu/index.shtml [42]. TEs in the *Oryza sativa* transcriptome in same cultivar were obtained by employing the same method used in sweet potato (the *Oryza sativa* transcriptome were assembled ourselves and are not published). As a result, there are 17,272 TEs found in the genome and 2,021 TEs in the transcriptome,indicating that the former is about 8.6-fold more than the later and most of the TEs in the genome were silent. As to the 17,272 TEs found in rice genome, most of them could be classified into retro-TEs (12,143) and DNA-TEs (3,968). In terms of the 2,021 TEs predicted from transcriptome, there are 1,290 retro-TEs and 747 DNA-TEs. The statistics above implied that there were about 10% genomic retro-TEs and 19% genomic DNA-TEs expressed. There also existed difference of the expressed members among TE superfamilies. For example, the TE members in Ty3/Mariner superfamily identified in rice genome was twice more than that in transcriptome, meaning that less than half members were expressed. While for Non-LTR subfamily, the difference was more than one hundred times higher and the expressed TEs in this superfamily were few. Conceivably, some subfamilies, such as TNP and SNF2, were not detected to express in transcriptome. Noteworthy, for some superfamilies such as hAT and Mutator, the TE members identified in transcriptome were more than those identified in genome (85 versus 27 for hAT, and 167 versus 65 for Mutator). The TE number of these two certain superfamilies in transcriptome was 2.5–3 fold higher than that in genome, the reasons may come from the transcriptional regulation or alternative splicing, and may be associated with the discontinuity in transcriptome splicing. Even so, it is certain that the numbers and types of TEs in the genome are more than the expressed TEs in transcriptome in *Oryza sativa*. Although the ratios of TE numbers and types predicted from the genomes to those from transcriptomes showed differences among various model plant species, to a certain extent, they could be used to predict the TE numbers and types in the genome of non-model plant species on the basis of transcriptome database. At this point we can speculate that the number of TEs in sweet potato genome might be 15,000 or more.

Diversity of Expressed TEs among Sweet Potato and other Plant Species

As for the integrated-transcriptome database of sweet potato, 3 superfamlies of retro-TEs and 7 superfamilies of DNA-TEs were identified and characterized. The distributions of TE superfamilies in transcriptomes of sweet potato and other species were compared, and some similarities among them were found. For example, Gypsy-TEs, a superfamily of LTR-retro-TEs, account for minority of retro-TEs in *Oryza sativa* (285 Gypsy-TEs in 1283 retro-TEs, about 22.2%) and the sugarcane expressed sequence tag (SUCEST) database (19 gypsy-TEs in 128 retro-TEs, about 14.8%) [41]. This is consistent with what we found in sweet potato (95 expressed Gypsy-TEs in 883 retro-TEs,10.8%). It was reported that these long terminal repeat (LTR) retro-TEs have the ability to trigger TE expression by their cis-regulatory elements in 5′ LTR. These regulatory sequences are similar to the well-characterized motifs required for the activation of stress-responsive gene

Figure 7. The evolutionary tree diagram of 16 transposase genes in the sweet potato. The evolutionary tree diagrams were drawn depended on the gene sequences similarity between the sweet potato and other higher plants. Each species name with an alphanumeric number means the homologous sequences in different species of each TE in sweet potato with the similarity above 80% (labeled with same digital label). The 16 sweet potato TEs were marked in red boxes. (a–c) the phylogenetic relationships of TEs in three superfamilies named Mutator, PIF-harbinger and hAT. (d) the evolutionary relationships of the 16 TEs with each other.

expression [43]. Therefore, these TE activations caused by environmental changes could eventually result in mutagenesis in the genome, which may help the organism adapt to new environmental conditions. These TEs also played a key role in translating changes in the external environment into changes at the genomic level. Indeed, Gypsy-TEs were found to respond directly to some specific stress situations [44], but the members of this superfamily in sweet potato and other plant species was few. This is probably due to the fine transcriptional control which makes difficulty for the expression of Gypsy-TEs under normal conditions [45].

On the other hand, the expressed DNA-TEs belonging to superfamily hAT (115 TEs) and Mutator (113 TEs) accounted for 43.7% of all the DNA-TEs in sweet potato, which were similar

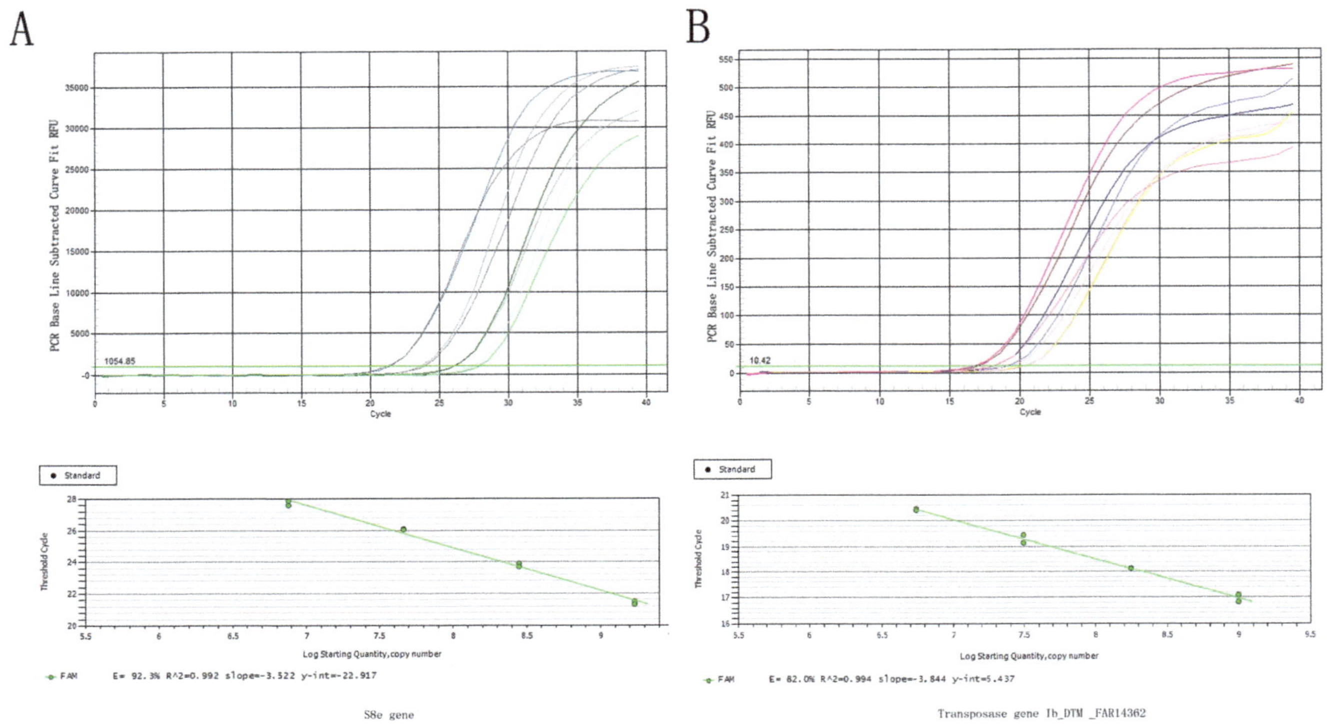

Figure 8. The SYBR Green I based real-time PCR for absolute quantification. Amplification graph and standard curve constructed using serially diluted DNA plasmid standards of (A) S8e gene and (B) transposase gene Ib_DTM _FAR14362, each ranging from 10^6 to 10^9 copies per microliter. The baseline and amplification parameters as *Ct* values in each dilution were analyzed through fluorescence data automatically.

with the distribution in other higher plants. For example, in *Arabidopsis thaliana*, the most abundant DNA-TEs in its genome were Mutator-like element (MULE), which reached 108 TEs, accounting for 17.33% among total 623 DNA-TEs [46]. It suggested that these TEs may play an important role in genome restructure [47,48]. A process called transduplication is a potentially rich source of novel coding sequences, reflecting that the activities of these TEs have a substantial impact on the evolution of new genes in plants, by their capacity to capture and mobilize genes or fragments [49]. Just as three thousand Pack-MULEs in rice, they have mobilized fragments of more than 1,000 genes. Many of these gene fragments are likely to be non-functional pseudogenes. However, 42% of these retro-genes have recruited new exons to become chimeric genes and show some degree of function through expression [45].

However, the Non-LTR elements in retro-TEs were generally abundant in sweet potato, which is inconsistent with that in *Oryza sativa*. The reason why the Non-LTR elements were the most expressed retro-TEs in sweet potato will only be determined after a detailed analysis of their complete sequences.

TE Distribution Differences and Environmental Adaptation

The major TE superfamilies reported in other plants are also present in three cultivars of sweet potato, but the family members and their expression levels differ enormously among cultivars. The TE number identified in *Jingshu*6 cultivar, a purple sweet potato, was extremely low, probably reflecting this cultivar may has distant genetic relationships with other two cultivars. The reasons for such big differences among cultivars need further analysis. Although the total TE numbers identified in *Xushu*18 and *Guangshu*87 were almost equal, the types and expression levels of TEs in the two cultivars were obviously different (Table 3). Specifically expressed TEs were existed in both cultivars, and differentially expressed TEs showed varied expression levels between these cultivars. To some TEs expressed in two cultivars, their expression level varied, even to the extent of hundred folds. In addition, the differentially and specifically expressed TEs were also found between the vegetative organs and reproductive organs of XS 18. There were almost 500 TEs expressed specifically within the vegetative organs while ~200 TEs in the reproductive organs. This suggested that the expression of a TE could be changed due to the cultivars, tissues and organs.

The reason for the differential TEs expression level among cultivars may come from the environmental influences on cultivars, such as temperature and rainfall [43]. JS 6 is grown up in north china with dry climate and low rainfall, while XS 18 and GS 87 are cultivated in the south of Yangtze River with humid and rainy climate. We can speculate that the growth environment conditions of sweet potato cause the difference in the number of expressed TEs and their expression level. So the TE-induced adaptive mutations suggested a widespread role of TEs in environmental adaptation. As reported, TEs played an important role in the responsive capacity of their hosts in the face of environmental challenges [44]. For example, TEs might directly regulate the function of individual genes, provide a mechanism for rapidly acquiring new genetic material and disseminate regulatory elements that can lead to the creation of stress-inducible regulatory networks [50]. And stress-activated TEs might generate the raw diversity that species require to survive among different stressful environments. Rather than being redundant, the presence of many TEs and different expression pattern among cultivars are required

Table 6. Identification of transposase gene copy number of Sweet potato using real-time PCR-based absolute quantification.

Gene name	Standards starting quantity over a dilution series (copy number)				Ct value of standards in duplicate				Ct value of sample	Absolute copies/ul genomic DNA	Copies	Expression in SX18 (RPKM)
S8e	7.6×10^6	4.7×10^7	2.8×10^8	1.7×10^9	27.6/27.9	26.1/26.1	23.7/23.9	21.3/21.5	26.4	2.57×10^7	1	\
Ib_DTP_9943	5.6×10^6	3.2×10^7	1.8×10^8	1.0×10^9	20.7/21.0	19.7/19.6	18.7/18.6	17.6/17.5	19.7	2.63×10^7	1	5,070.19
Ib_DTM_1664	5.4×10^6	3.2×10^7	1.7×10^8	1.0×10^9	20.6/19.8	19.4/19.6	18.2/18.3	16.9/18.9	18.9	5.25×10^7	2	8,415.1
Ib_DTM _FAR11812	5.6×10^6	3.2×10^7	1.8×10^8	1.0×10^9	20.1/20.8	19.5/19.6	18.0/18.3	16.9/17.1	18.8	7.59×10^7	3	5,439.92
Ib_DTM _FAR14362	5.4×10^6	3.2×10^7	1.8×10^8	1.0×10^9	20.4/20.5	19.1/19.5	18.1/18.1	16.8/17.1	19.0	5.25×10^7	2	1,008.72
Ib_DTM _PB12217	5.4×10^6	3.2×10^7	1.8×10^8	1.0×10^9	23.4/23.4	22.2/22.4	21.0/21.2	19.7/20.0	22.4	2.63×10^7	1	14,372.48
Ib_DTM _PB1 4635	5.3×10^6	3.2×10^7	1.7×10^8	1.0×10^9	23.0/22.8	22.1/21.6	20.7/20.8	19.2/19.3	21.9	2.69×10^7	1	3,265.49
Ib_DU_2831	5.4×10^6	3.2×10^7	1.7×10^8	1.0×10^9	20.3/20.3	19.1/19.2	17.9/17.9	16.6/16.8	18.7	5.62×10^7	2	8,816.56
Ib_DTM_2890	5.5×10^6	3.2×10^7	1.8×10^8	1.0×10^9	19.8/19.6	18.4/18.5	17.3/17.5	16.0/16.5	18.1	5.37×10^7	2	939.33
Ib_DTM _FAR14118	5.6×10^6	3.2×10^7	1.8×10^8	1.0×10^9	20.8/20.5	19.2/19.7	17.8/17.9	16.6/16.6	18.8	5.62×10^7	2	1,368.12
Ib_DTH_1962	5.6×10^6	3.2×10^7	1.8×10^8	1.0×10^9	20.3/19.6	18.4/18.5	17.8/17.6	16.2/16.6	18.3	5.37×10^7	2	3,787.79
Ib_DTM_3260	5.3×10^6	3.2×10^7	1.7×10^8	1.0×10^9	21.4/21.4	20.3/20.4	19.2/19.3	17.3/17.7	20.4	2.63×10^7	1	11,441.81

to overcome the challenges imposed by different environmental conditions [43].

Highly Expressed TE Superfamily and Genome Evolution

The result from TEs identification revealed a surprising amount of expressed TE homologues. It provided an enormous source of variability that can be used to create novel genes or modify genetic functions for genome evolution. The DGE profiling analyses demonstrated that the hAT superfamily had obvious advantages in TE expression levels, because in some highly expressed superfamilies, the expression level of TEs in Mutator was 3.40 TPM in average (the maximum value was 17.27 and the minimum value was 0.07 with 38 members), in CACTA was 1.79 TPM (the maximum was 17.16 and the minimum was 0.08 with 27 members), but in hAT it reached 42.2 TPM (the maximum was 211 and the minimum was 0.32 with 56 members). Particularly, a member (Ib_DTH_29406) of hAT superfamily had the highest average expression level in all 7 tissues of *Xushu*18. The high transcription activities of TEs in hAT superfamily in sweet potato may be related to their functional specificity. It was reported before that the hAT TEs were a diverse and ancient transposon superfamily which had insertion specificities, suggesting that they may be the most frequent contributors to genome evolution [51]. From the earliest discovered TEs like Ac in *Zea mays*, Tam in *Antirrhinum majus*, hobo in *Drosophila melanogaster* to the domesticated TEs like DAYSLEEPER in *Arabidopsis thaliana* which has been exapted for new function rather than transposition [52–53], these hATs has been verified to have high activity in the process of plant adaptative evolution. In addition to simply altering the gene structure, hATs insertions can lead to some positive or negative regulatory functions. For example, Ac in *Zea mays* tends to transpose into the 5′ ends of plant genes and the promoter and enhancer elements within these TEs could potentially alter gene expression [54]. No matter what the gross effects on the overall architecture of genomes caused by hAT are, or the broad range of changes in gene expression and function, from subtle quantitative effects to the rewiring of regulatory networks and the evolution of entirely new genes, it is suggesting that hAT-induced mutations has played a key part in adaptive evolution over longer periods of time in plants [55]. Therefore, whether the high expression of hAT TEs is related to the genome evolution in sweet potato should be identified in the future studies.

High Expression Level and Low Genome Copy Number of TEs

TEs have played an important role in determining the size and structure of a complex plant genome. Every aspect of TE life cycle has the potential for genome alteration and somaclonal variation, such as increasing gene copy number and genome size, mobilization and rearrangement of gene fragments and epigenetic silencing of genes, horizontal gene transferring and chromosomal rearrangements [56]. It is crucial that TE functional characteristics are essential for explaining the dynamics and evolution of plant genomes. How the TEs acted on the various aspects of chromosome structure and evolution depended on the numbers and predominant types of TEs that expressed [57]. According to the results described above, TEs belonging to hAT and mutator superfamilies were dominant at expression level in sweet potato, but their copy number was low in the genome. The reasons for the low copy of individual transposase gene were numerous, one of them is that the total amount of TEs in Mutator and hAT is large, the copy number of each certain TE in genome should be restricted. Thus, it is not surprising that plants devote considerable resources to TE control [45].

However, the TEs in these two superfamilies generally displayed high expression and this is possibly associated with the high transposition effectiveness of the individual TE. The highly expressed TEs were predominantly active with highly mutagenic ability. And the high activity of these TEs may be useful not only for their own transposition effect, but also for providing the active components like transposase for other defective TEs with incomplete structure, such as the SINEs with huge amount and high copy number in genome. Mutation is the ultimate source of genetic variation and TEs are also likely to play a relevant role in adaptation because of their ability to generate mutations of great variety and magnitude, and their capacity to be responsive and susceptible to environmental changes [43]. From this perspective, these highly expressed but low copy number TEs could be more important than those TEs which have higher copy number and ordinary expression in the genome alteration and somatic mutation. As reported before, Mutator TEs in *Zea mays* is by far the most mutagenic plant transposon, causing new mutations at up to a hundred times the spontaneous rate [47]. The high transposition frequency and the tendency to insert into low copy sequences for such transposon have made it the primary means by which genes are mutagenized in maize (*Zea mays L.*) [48].

Materials and Methods

Plant Material

The sweet potato transcriptome databases used in this study were established from three cultivars in China. These cultivars were different in phenotype, planted area, main uses and so on. XS 18 is the leading cultivar in the Yangtze River Basin of China in terms of annual hectareage and total production with widely growth habit. It has green cordate and slightly toothed leaves, and elliptic roots with red skin and white flesh with purple rings in some places. It is mainly used for starch processing and the production of ethanol [58]. The second cultivar is GS 87, a local cultivar planted mostly in South China. Its vines are short, the lobed leaves are green at all stages of growth, the tuberous roots have red skin and orange flesh with good eating quality [59]. The third one is JS 6, an important food cultivar which is mainly planted in the north of China. It has spreading growth habit, green and long vines, triangular slightly lobed green leaves, and the roots are spinning and purple outside and inside [60].

Stem cuts of XS 18 were planted in May, 2011, and grown under natural light and temperature in experimental field of college of life sicences in Sichuan University, Chengdu, Sichuan Province of China. The college gave permission to conduct the study on this site and the field studies did not involve endangered or protected species. Samples from young leaves were used for PCR, RT-PCR and real-time quantitative PCR in this study. All tissue samples collected were snap-frozen immediately in nitrogen and stored at −80°C until further processing.

Sweet Potato Transcriptome Reads Data and Databases

The reads used for assembling the sweet potato integrated transcriptome were from four transcriptome databases. The first one was the transcriptome database of vegetative organs of cultivar XS 18, including roots, stems and leaves [19]. The reads sequences were obtained in NCBI Short Read Archive (SRA, http://www.ncbi.nlm.nih.gov/Traces/sra) under accession number SRA043582, and the transcriptome sequences were in http://cfgbi.scu.edu.cn/SweetPotato/index.php. The second one was the transcriptome database of flowers of cultivar XS 18 and the reads sequences were in SRA under accession number SRA043584 [20]. The other two transcriptome databases were established from fibrous and tuberous roots of GS 87 [21] and tuberous roots of JS 6 [22], and the reads sequences were obtained all in SRA with accession number SRA022988 and SRX090758, respectively.

Extraction of RNA and Genomic DNA

Total RNAs were extracted using the Trizol Reagent (Invitrogen, USA), and treated with DNase I (Fermentas, USA) according to the manufacturer's instructions. RNA quality and purity were assessed with $OD_{230/260}$ ratio. Total cDNAs were synthesized from RNA with Moloney murine leukemia virus (M-MLV) reverse transcriptase (Invitrogen, CA, USA) using oligo (dT) as primer following the manufacturer's instructions. Genomic DNA extraction was performed using the CTAB method [61]. Agarose gel electrophoresis was used to show the integrity of the DNA, while spectrophotometry was employed to display the concentration and cleanliness.

De novo Integrated Transcriptome Assembly and Annotation

All the assemblies were run on a 64-bit Linux system (Ubuntu 10.10) with 32G physical memory. Reads from four databases above were qualitatively assessed and assembled with *de novo* assemblers of Trinity (http://trinityrnaseq.sourceforge.net) [23], SOAP de novo v1.04 (http://soap.genomics.org.cn) [24] and Velvet v1.0.12 (http://www.ebi.ac.uk/~zerbino/velvet) [25] using different parameters, respectively. All of the assemblies from each assembler with optimized parameters were combined and treated with CD-HITEST to reduce redundancy (http://www.bioinformatics.org/cd-hit), and then the remains were reassembled with CAP3 (http://pbil.univ-lyon1.fr/cap3.php) [26].

Searching for the TEs

The TEs searching was conducted in three independent screenings. Firstly, keyword searching from the annotated transcripts was used. The transcripts (>200 nt) in the sweet potato integrated-transcriptome database were submitted for annotation through BLAST using Blast2GO software v2.4.4 (http://www.blast2go.com/b2ghome) [27]. For BLASTX against the NR database, the threshold was set to E-value$\leq 10^{-5}$. Based on the annotation information, TEs were retrieved from the database by keyword searching. One kind of keywords included transposon-related words, such as "transposon", "retrotransposon", "transposase", "retrotransposase", "transposable element" etc. The other kind of keywords was based on the order names and superfamily names of eukaryotic TEs, like "Non-LTR", "Copia", "Gypsy", "Mutator", "Harbinger" etc [6]. It is noteworthy that Blast2GO may be failed to give accurate annotated information when the firstly hit gene was annotated as hypothetical protein even the other hits were annotated as transposase gene. Therefore, the second method based on sequence similarity was used as a supplementary. The CDS and full-length sequences of fully characterized transposase genes of the higher plants were downloaded from GenBank (mainly from *Arabidopsis thaliana, Vitis vinifera, Ricinus communis, Glycine max, Sorghum bicolor, Oryza sativa, Populus trichocarpa, Zea mays,* etc.), and compared with transcripts in sweet potato integrated-transcriptome database using BLASTn program. Due to the enormous data and to avoid spurious matches, a very stringent expectation cut-off value (e-10 or better) was used. Thirdly, TEs were searched depending on the homology of the sub-terminal conserved sequence of TE families as reported in the literatures before [28]. All the searching results were compared pair-wisely to remove redundant, and detailed manual evaluation was further conducted to exclude the non-TEs like reverse transcription virus etc. The ORF predictions of TEs were carried out by using Galaxy (http://main.g2.bx.psu.edu/) [29,30].

Cloning and Sequencing of Transposase Genes

In PCR and RT-PCR reactions, primers for amplification were designed according to assembled transcripts using Primer Premier 5.0 (PREMIER Biosoft International, CA, USA) and synthesized by GENEWIZ, Inc. (http://www.genewiz.com.cn). Primer sequences are shown in Table S1. The transposase genes were amplified using KOD FX DNA polymerase (TOYOBO, Japan), under the cycling conditions as 94°C for 5 min, followed by 35 cycles consisting of 94°C for 45 s, 60°C for 45 s, and 72°C for 1 min, and a final extension cycle of 72°C for 15 min. The PCR products were fractionated and recovered in a 1% agarose gel, then ligated to 50ng vector pMD-18T (TIANGEN BIOTECH, Beijing, China) using T4 DNA ligase (TAKARA BIO, Japan). Recombinant plasmids were transformed into *Escherichia coli* JM109 competent cells [62] and clones were picked for validation through colony PCR, plasmid electrophoresis and restriction enzyme digestion (Fermentas, USA). The positive plasmids were sequenced at BGI-Shenzhen, Shenzhen, China (http://www.genomics.cn).

Evolutionary Analysis of Transposase Gene Sequence

Transposase gene sequences of sweet potato were aligned with those of other higher plants in NCBI NR database on the bases of sequence similarity. The transposase genes from various higher plants showing similarity above 80% were screened and their sequences were downloaded from NCBI (mainly from *Vitis vinifera, Populus trichocarpa, Glycine max, Zea mays, Sorghum bicolor, Arabidopsis thaliana, Ricinus communis, Oryza sativa,* etc.). Using software CluxtalX (2.0) and MEGA4, we determined the evolutionary distance depending on the homologous differences and drew the evolutionary diagram.

Analysis of TEs Expression in Inter- and Intra-cultivars

Expression analyses of TEs among sweet potato cultivars were carried out using the util/alignReads.pl script in Trinity software. In order to obtain the expression level of TEs in different cultivars, we aligned the reads in each of four databases to the Trinity-assembled transcripts of the integrated database. As for the expression analyses of TEs in *Xushu*18 vegetative organ, we used the DGE tag profiling [19]. According to all the tags generated from seven sequenced DGE libraries of sweet potato, including YL (young leaves), ML (mature leaves), stem, FR (fibrous roots), ITR (initial tuberous roots), ETR (expanding tuberous roots) and HTR (harvest tuberous roots), we searched CATG with the downstream 17 base pairs in the assembled TEs and the resulted 21 base pairs tags became the new expression tags related to TEs. These tags were mapped to the distinct clean tags in DGE tag profiling of transcriptome and the resulted TEs expression tags were aligned to the TE sequences using Bowtie available at the Galaxy website to detect the expression level of TEs. The edgeR package (Empirical analysis of DGE in R) was used for differential and specifical expression analysis of TEs in DGE tag profiling [63]. We normalized tag distribution for gene expression level in each library to make an effective library size and extracted differentially expressed genes (DEGs) with p value $\leqq 0.05$ and log2 fold-change $\geqq 1$. And we compared libraries pair-wise and used hypergeometric test to identify differentially expressed TEs (DETEs) and specifically expressed TEs (SETEs).

Identification of TE Gene Copy Number

Real-time PCR-based absolute quantification was used to identify the copy number of transposase genes in sweet potato genome. The PCR standards of transposase genes and S8e gene was amplified by conventional PCR from sweet potato genomic DNA, respectively, purified using the E.Z.N.A^{TM} Gel Extraction Kit (OMEGA, USA). The purified fragment was cloned into the TA-cloning site of a vector pMD-18T (TIANGEN BIOTECH, Beijing, China). The positive recombinant plasmids were purified using the QIAprep Spin Miniprep Kit (Qiagen) and linearized by restriction enzyme digestion. A dilution series is prepared from the cloned plasmids to provide a measure of absolute standard abundance to generate a standard curve. Linearized plasmid was quantified using a spectrophotometer and copy number was calculated for all standards by the following formula [34–36]:

$$\text{Number of copies}/\mu l = \frac{6.02 \times 10^{23} \times \text{DNA concentrations g}/\mu l}{\text{Number of bases pairs} \times 660 \text{ daltons}}$$

The SYBR Green based real-time PCR primer sets were designed using Beacon Designer 3.0 (Premier Biosoft International, CA). Primers stocks were prepared at 100 μM in TE (10mM Tris, pH 8.0, 1mM EDTA), and working solutions were diluted to 10 μM. All real-time PCR runs were performed in duplicate, and each reaction mixture was prepared using SYBR Premix Ex Taq kit (Takara). PCR amplifications were carried out in a total volume of 20μl, containing 6.4 μl PCR-grade water, 0.8 μl of each primer, 10 μl 2×SYBR Premix Ex Taq, and 2.0 μl appropriately diluted template DNA. It is necessary for the unknown target genes to be diluted to a point where the resulting PCR signal follows within the linear range of the standard curve, which must be determined empirically. At least 5-fold dilution may often minimizes potentially interfering substances that inhibit PCR amplification. To minimize pipetting errors and achieve better reproducibility, a master mix of the common components should be prepared and aliquoted into each well of the sample plate.

The thermal cycling protocol was as follows: initial denaturation for 10 min at 95°C followed by 40 cycles of 5 s at 95°C, 5 s at 60°C, and 5 s at 72°C. The fluorescence signal was measured at the end of each extension step at 72°C. After the amplification, a melting peak analysis with a temperature gradient of 0.1°C per second from 60 to 95°C was performed to confirm that only the specific products were amplified. Finally, the samples were cooled down to 40°C for 30s. The baseline and amplification parameters as *Ct* values in each dilution were analyzed through fluorescence data automatically. The *Ct* values were plotted against the logarithm of their initial standard copy number. Each standard curve was generated by a linear regression of the plotted points. Real-time PCR amplifications were carried on 11 transposase genes and one single copy gene S8e simultaneously from sweet potato genomic DNA with the standard dilutions in a run. Based on each standard curve, the absolute copy number of 11 unknown transposase genes from genomic DNA per μl had been derived. And then, the transposase gene copy number in sweet potato genome was calculated by dividing the copy number of transposase gene by that of S8e gene. The copy ratio of transposase gene to single-copy S8e gene equals to the transposase gene copy number. These procedures were optimized for 96-well format using a Bio-Rad IQ detection system which use fluorescein as an internal passive reference dye for normalization of well-to-well optical variation.

Acknowledgments

We thank Mr. Chuan-Qi Liu for computer procuring and configuring, and Mr. Chuan-Wu Zhao for writing some scripts.

Author Contributions

Conceived and designed the experiments: LY YZZ HYW. Performed the experiments: LY YHG XT. Analyzed the data: LY. Contributed reagents/materials/analysis tools: YHG XT XJL XMT. Wrote the paper: LY HYW. Revised the manuscript: LY HYW.

References

1. Thottappilly G (2009) Introductory Remarks. In: Loebenstein G, Thottappilly G, eds. The sweetpotato: Springer. pp 3–8.
2. Loebenstein G (2009) Origin, Distribution and Economic Importance. In: Loebenstein G, Thottappilly G, eds. The sweetpotato: Springer. pp 9–12.
3. Firon N, LaBonte D, Villordon A, McGregor C, Kfir Y, et al. (2009) Botany and Physiology: Storage Root Formation and Development. In: Loebenstein G, Thottappilly G, eds. The sweetpotato: Springer. pp 13–26.
4. Gaba V, Singer S (2009) Propagation of Sweetpotatoes, In Situ Germplasm Conservation and Conservation by Tissue Culture. In: Loebenstein G, Thottappilly G, eds. The sweetpotato: Springer. pp 65–80.
5. Srinivas T (2009) Economics of sweetpotato production and marketing. In: Loebenstein G, Thottappilly G, eds. The sweetpotato: Springer. pp 235–267.
6. Wicker T, Sabot F, Hua-Van A, Bennetzen JL, Capy P, et al. (2007) A unified classification system for eukaryotic transposable elements. Nat Rev Genet 8: 973–982.
7. International Rice Genome Sequencing Project (2005) The map-based sequence of the rice genome. Nature 436: 793–800.
8. Adams MD, Celniker SE, Holt RA, Evans CA, Gocayne JD, et al. (2000) The genome sequence of Drosophila melanogaster. Science 287: 2185–2195.
9. International Human Genome Sequencing Consortium (2001) Initial sequencing and analysis of the human genome. Nature 409: 860–921.
10. Morgante M (2006) Plant genome organisation and diversity: the year of the junk! Curr Opin Biotechnol 17: 168–173.
11. Dolgin E (2009) Maize genome mapped. Nature News 1098.
12. The International Brachypodium Initiative (2010) Genome sequencing and analysis of the model grass Brachypodium distachyon. Nature 463: 763–768.
13. The International Barley Genome Sequencing Consortium (2012) A physical, genetic and functional sequence assembly of the barley genome. Nature 491: 711–716.
14. Du J, Grant D, Tian Z, Nelson RT, Zhu L, et al. (2010) SoyTEdb: a comprehensive database of transposable elements in the soybean genome. BMC Genomics 11: 113.
15. Yurchenkoa NN, Kovalenkoa LV, Zakharova IK (2011) Transposable Elements: Instability of Genes and Genomes. Russian Journal of Genetics: Applied Research 1: 489–496.
16. Biemont C (2010) A brief history of the status of transposable elements: from junk DNA to major players in evolution. Genetics 186: 1085–1093.
17. Morgante M (2005) Massive changes of the maize genome are caused by Helitrons. Heredity 95: 421–422.
18. Hollister JD, Gaut BS (2007) Population and evolutionary dynamics of Helitron transposable elements in Arabidopsis thaliana. Mol Biol Evol 24: 2515–2524.
19. Tao X, Gu YH, Wang HY, Zheng W, Li X, et al. (2012) Digital gene expression analysis based on integrated de novo transcriptome assembly of sweet potato [Ipomoea batatas (L.) Lam]. PLoS One 7: e36234.
20. Tao X, Gu YH, Jiang YS, Zhang YZ, Wang HY (2013) Transcriptome analysis to identify putative floral-specific genes and flowering regulatory-related genes of sweet potato. Biosci Biotechnol Biochem 77: 2169–2174.
21. Wang Z, Fang B, Chen J, Zhang X, Luo Z, et al. (2010) De novo assembly and characterization of root transcriptome using Illumina paired-end sequencing and development of cSSR markers in sweetpotato (Ipomoea batatas). BMC Genomics 11: 726.
22. Xie F, Burklew CE, Yang Y, Liu M, Xiao P, et al. (2012) De novo sequencing and a comprehensive analysis of purple sweet potato (Impomoea batatas L.) transcriptome. Planta 236: 101–113.
23. Grabherr MG, Haas BJ, Yassour M, Levin JZ, Thompson DA, et al. (2011) Full-length transcriptome assembly from RNA-Seq data without a reference genome. Nat Biotechnol 29: 644–652.
24. Li R, Zhu H, Ruan J, Qian W, Fang X, et al. (2010) De novo assembly of human genomes with massively parallel short read sequencing. Genome Res 20: 265–272.
25. Zerbino DR, Birney E (2008) Velvet: Algorithms for de novo short read assembly using de Bruijn graphs. Genome Res 18: 821–829.
26. Huang X, Madan A (1999) CAP3: a DNA sequence assembly program. Genome Res 9: 868–877.
27. Conesa A, Gotz S, Garcia-Gomez JM, Terol J, Talon M, et al. (2005) Blast2GO: a universal tool for annotation, visualization and analysis in functional genomics research. Bioinformatics 21: 3674–3676.
28. Holligan D, Zhang X, Jiang N, Pritham EJ, Wessler SR (2006) The transposable element landscape of the model legume Lotus japonicus. Genetics 174: 2215–2228.
29. Goecks J, Nekrutenko A, Taylor J (2010) Galaxy: a comprehensive approach for supporting accessible, reproducible, and transparent computational research in the life sciences. Genome Biol 11: R86.
30. Hillman-Jackson J, Clements D, Blankenberg D, Taylor J, Nekrutenko A (2012) Using Galaxy to perform large-scale interactive data analyses. Curr Protoc Bioinformatics Chapter 10: Unit10 15.
31. Hollister JD, Gaut BS (2007) Population and evolutionary dynamics of Helitron transposable elements in Arabidopsis thaliana. Mol Biol Evol 24: 2515–2524.
32. Wall PK, Leebens-Mack J, Muller KF, Field D, Altman NS, et al. (2008) PlantTribes: a gene and gene family resource for comparative genomics in plants. Nucleic Acids Res 36: D970–976.
33. Jurka J, Kapitonov VV, Pavlicek A, Klonowski P, Kohany O, et al. (2005) Repbase Update, a database of eukaryotic repetitive elements. Cytogenet Genome Res 110: 462–467.
34. Beltran J, Jaimes H, Echeverry M, Ladino Y, Lopez D, et al (2009) Quantitative analysis of transgenes in cassava plants using real-time PCR technology. In Vitro Cellular and Development Biology- Plant 45: 48–56.
35. Yi CX, Zhang J, Chan KM, Liu XK, Hong Y (2008) Quantitative real-time PCR assay to detect transgene copy number in cotton (Gossypium hirsutum). Anal Biochem 375: 150–152.
36. Kihara T, Zhao CR, Kobayashi Y, Takita E, Kawazu T, et al. (2006) Simple identification of transgenic Arabidopsis plants carrying a single copy of the integrated gene. Biosci Biotechnol Biochem 70: 1780–1783.
37. Duarte JM, Wall PK, Edger PP, Landherr LL, Ma H, et al. (2010) Identification of shared single copy nuclear genes in Arabidopsis, Populus, Vitis and Oryza and their phylogenetic utility across various taxonomic levels. BMC Evol Biol 10: 61.
38. Schnable PS, Ware D, Fulton RS, Stein JC, Wei F, et al. (2009) The B73 maize genome: complexity, diversity, and dynamics. Science 326: 1112–1115.
39. Nagy Z, Chandler M (2004) Regulation of transposition in bacteria. Res Microbiol 155: 387–398.
40. Ouyang S, Buell CR (2004) The TIGR Plant Repeat Databases: a collective resource for the identification of repetitive sequences in plants. Nucleic Acids Res 32: D360–363.
41. Rossi M, Araujo PG, Van Sluys M-A (2001) Survey of transposable elements in sugarcane expressed sequence tags (ESTs). Genetics and Molecular Biology 24: 147–154.
42. Ouyang S, Zhu W, Hamilton J, Lin H, Campbell M, et al. (2007) The TIGR Rice Genome Annotation Resource: improvements and new features. Nucleic Acids Res 35: D883–887.
43. Casacuberta E, Gonzalez J (2013) The impact of transposable elements in environmental adaptation. Mol Ecol 22: 1503–1517.
44. Bennetzen JL (2000) Transposable element contributions to plant gene and genome evolution. Plant Mol Biol 42: 251–269.
45. Lisch D (2013) How important are transposons for plant evolution? Nat Rev Genet 14: 49–61.
46. Le QH, Wright S, Yu Z, Bureau T (2000) Transposon diversity in Arabidopsis thaliana. Proc Natl Acad Sci U S A 97: 7376–7381.
47. Hershberger RJ, Benito MI, Hardeman KJ, Warren C, Chandler VL, et al. (1995) Characterization of the major transcripts encoded by the regulatory MuDR transposable element of maize. Genetics 140: 1087–1098.
48. Diao XM, Lisch D (2006) Mutator transposon in maize and MULEs in the plant genome. Yi Chuan Xue Bao 33: 477–487.
49. Hoen DR, Park KC, Elrouby N, Yu Z, Mohabir N, et al. (2006) Transposon-mediated expansion and diversification of a family of ULP-like genes. Mol Biol Evol 23: 1254–1268.
50. de Souza FS, Franchini LF, Rubinstein M (2013) Exaptation of transposable elements into novel cis-regulatory elements: is the evidence always strong? Mol Biol Evol 30: 1239–1251.
51. Xu Z, Dooner HK (2005) Mx-rMx, a family of interacting transposons in the growing hAT superfamily of maize. Plant Cell 17: 375–388.
52. Knip M, de Pater S, Hooykaas PJ (2012) The SLEEPER genes: a transposase-derived angiosperm-specific gene family. BMC Plant Biol 12: 192.
53. Bundock P, Hooykaas P (2005) An Arabidopsis hAT-like transposase is essential for plant development. Nature 436: 282–284.

54. Shimatani Z, Takagi K, Eun CH, Maekawa M, Takahara H, et al. (2009) Characterization of autonomous Dart1 transposons belonging to the hAT superfamily in rice. Mol Genet Genomics 281: 329–344.
55. Cowley M, Oakey RJ (2013) Transposable elements re-wire and fine-tune the transcriptome. PLoS Genet 9: e1003234.
56. Shapiro JA (2010) Mobile DNA and evolution in the 21st century. Mob DNA 1: 4.
57. Tollis M, Boissinot S (2012) The evolutionary dynamics of transposable elements in eukaryote genomes. Genome Dyn 7: 68–91.
58. Carpena AL (2009) Important cultivars, varieties, and hybrids. In: Loebenstein G, Thottappilly G, eds. The sweetpotato: Springer. pp 27–40.
59. Luo WL (2012) Characteristic and Measures for High Yield Cultivation of Sweet Potato Cultivar Guangshu 87 with Good Quality. Fujian Journal of Agricultural Sciences 2: 135–140.
60. Wu QY, Qin YZ, Liu MY, Xiong XY (2011) Comparative Experiment for High-quality Sweet-potato Varieties. Hunan Agricultural Sciences 5: 4–5.
61. He XJ, Zheng T, Su JQ, ChenZD (2011) DNA Extraction of 7 Species Plant of Melastomaceae Using Modified CTAB Method. Chinese Journal of Tropical Agriculture 31:10.
62. Inoue H, Nojima H, Okayama H (1990) High efficiency transformation of Escherichia coli with plasmids. Gene 96: 23–28.
63. Robinson MD, McCarthy DJ, Smyth GK (2010) edgeR: a Bioconductor package for differential expression analysis of digital gene expression data. Bioinformatics 26: 139–140.

14

Nested Insertions and Accumulation of Indels Are Negatively Correlated with Abundance of *Mutator*-Like Transposable Elements in Maize and Rice

Dongyan Zhao, Ning Jiang*

Department of Horticulture, Michigan State University, East Lansing, Michigan, United States of America

Abstract

Mutator-like transposable elements (MULEs) are widespread in plants and were first discovered in maize where there are a total of 12,900 MULEs. In comparison, rice, with a much smaller genome, harbors over 30,000 MULEs. Since maize and rice are close relatives, the differential amplification of MULEs raised an inquiry into the underlying mechanism. We hypothesize this is partly attributed to the differential copy number of autonomous MULEs with the potential to generate the transposase that is required for transposition. To this end, we mined the two genomes and detected 530 and 476 MULEs containing transposase sequences (candidate coding-MULEs) in maize and rice, respectively. Over 1/3 of the candidate coding-MULEs harbor nested insertions and the ratios are similar in the two genomes. Among the maize elements with nested insertions, 24% have insertions in coding regions and over half of them harbor two or more insertions. In contrast, only 12% of the rice elements have insertions in coding regions and 19% have multiple insertions, suggesting that nested insertions in maize are more disruptive. This is because most nested insertions in maize are from LTR retrotransposons, which are large in size and are prevalent in the maize genome. Our results suggest that the amplification of retrotransposons may limit the amplification of DNA transposons but not vice versa. In addition, more indels are detected among maize elements than rice elements whereas defects caused by point mutations are comparable between the two species. Taken together, more disruptive nested insertions combined with higher frequency of indels resulted in few (6%) coding-MULEs that may encode functional transposases in maize. In contrast, 35% of the coding-MULEs in rice retain putative intact transposase. This is in addition to the higher expression frequency of rice coding-MULEs, which may explain the higher occurrence of MULEs in rice than that in maize.

Editor: I. King Jordan, Georgia Institute of Technology, United States of America

Funding: This study is supported by grants DBI-0607123 and MCB-1121650 from National Science Foundation to NJ. The funders had no role in study design, data collection and analysis, decision to publish, or preparation of the manuscript.

Competing Interests: The authors have declared that no competing interests exist.

* E-mail: jiangn@msu.edu

Introduction

Transposable elements (TEs) are genomic sequences that are capable of moving from one position to another. Based on the transposition intermediate, TEs can be divided into two classes. Class I, also called RNA or retrotransposons, transpose via an RNA intermediate. Class II, DNA transposons, transpose via a DNA intermediate. TEs can also be grouped into autonomous or non-autonomous elements, where the former encode proteins (transposase for class II elements) responsible for the transposition of themselves as well as their corresponding non-autonomous counterparts.

TEs constitute large fractions of most plant genomes sequenced to date, *i.e.*, ~85% of maize [1], ~62% of soybean [2] and sorghum [3], ~63% of tomato [4], ~43% of papaya [5], and ~35% of rice genomes [6]. Comparative analysis reveals that the abundance of different classes of TEs (*e.g.*, RNA and DNA TEs) varies dramatically in different plants. In rice, the genomic coverage of RNA TEs (20%) is 1.5-fold of that of DNA TEs (13%) whereas in maize the difference is 8-fold between RNA (76%) and DNA (9%) TEs [1,6]. This difference is the greatest in the papaya genome, which contains ~43% RNA TEs and very few DNA TEs (0.2%). In general, the amount of retrotransposons is correlated with plant genome size whereas such correlation is not found for DNA elements [7].

Mutator-like transposable elements (MULEs), first discovered in maize [8,9], are widespread in plants [10,11], fungi [12,13], and animals [14]. MULEs are one of the most complex TE families, with dramatic variation in structure, sequence, size, and abundance within a genome and among different plant genomes [1,10,15]. Based on the similarity of their terminal inverted repeats (TIRs), MULEs can be grouped into TIR-MULEs or non-TIR-MULEs, where TIR-MULEs are characterized by long TIRs (100–500 bp) with high sequence similarity and non-TIR-MULEs have relatively short TIRs with low sequence similarity between their terminal sequences [10]. Within the TIR-MULE group, a special subgroup was discovered in several plants, which contains tandem TIRs (two consecutive TIRs) flanking the internal sequence [15]. In addition, some MULEs harbor gene and/or gene fragments, referred to as Pack-MULEs [16]. Their high transposition frequency combined with capability to duplicate gene fragments suggest that MULEs play important roles in genome evolution [17,18]. Individual genomes can contain all forms of MULEs, including TIR- and tandem TIR-MULEs, non-TIR-MULEs, and Pack-MULEs, albeit with differential abundance [10,15]. Rice by far contains the largest amount of known MULEs (n = 32,000, 5.5% of the genome) and Pack-MULEs (n = 2,924) [15,19]. As a close relative, maize contains 12,900 MULEs (1% of the genome) and 276 Pack-MULEs, which are relatively less abundant given that its genome size is over 5 times larger than rice (2,066 Mb vs. 370 Mb) [1,6]. The papaya genome represents an extreme case, which is almost void of DNA TEs including MULEs. Despite the prevalence and importance of MULEs, the mechanism underlying the differential amplification of these elements or DNA elements in general remains largely unknown.

The first autonomous *Mutator* element discovered is *MuDR* in maize. It encodes two proteins, *MURA*, the major protein responsible for its transposition, and *MURB*, a helper protein found only in the *Zea* genus and for which the function is still unclear [17,20,21]. MULE transposases belong to the DDE transposase family, which commonly consists of a helix-turn-helix (HTH) DNA-binding domain at the amino terminus and a DDE catalytic domain at the carboxyl terminus [22,23]. *MURA*-like transposase sequences have been discovered in many other organisms, including plants, fungi, and animals [24–26]. In addition to the differences in TIR length, *MURA*-like transposase sequences are also divergent among MULEs as demonstrated by distinct subfamilies [11,25]. Furthermore, some MULE transposase sequences, such as *FAR1* and *MUSTANG*, have been domesticated as cellular genes and are no longer associated with any mobility [27,28].

The abundance of TEs is a result of the interplay between the amplification through transposition, duplication, horizontal transfer and loss via excision, sequence erosion, deletion *etc.* [29]. Since autonomous elements are responsible for the transposition of both themselves and their corresponding non-autonomous counterparts, their abundance and activity influence the speed of amplification, and therefore contribute to the abundance of TEs in a genome. Maize and rice belong to the same family (*Poaceae*) with a common ancestor occurring 50–70 million years ago [30]. As the most important crops worldwide, the genome of both maize and rice were sequenced through a hierarchical method using bacterial artificial chromosome clones (BACs) accompanied by high-density genetic maps [1,6]. The availability of high quality genomic sequence and well-annotated TEs allow a comparative study of TEs in these two organisms. In this study, we report the detection and analysis of all MULEs containing transposase sequences (candidate coding-MULEs) in the maize and rice genomes. In addition, we dissected the possible factors involved in the loss of coding capacity of those elements, which facilitate the understanding about the underlying mechanisms for differential abundance of MULEs in the two genomes.

Materials and Methods

Genomic sequences and TE libraries

The B73 maize genomic sequence RefGen_v2 was downloaded from the MaizesSquence.org (http://www.maizesequence.org/) and the Nipponbare rice genomic sequence Release 7 was downloaded from the Rice Genome Annotation Project at Michigan State University (http://rice.plantbiology.msu.edu/index.shtml). TE library for rice and MULE TIR libraries (MULE_TIR) for both maize and rice were constructed and curated by the Jiang Lab. The TE library for maize was downloaded from the Maize Transposable Element Database (http://maizetedb.org/~maize/) in August 2011.

Identification of candidate coding-MULEs

To maximize the possibility of detecting coding-MULEs, all the transpositionally active *MURA*-related transposases (*e.g. MURA*, *Hop*, *Jittery*, *Os3378*, *AtMu1*) [8,9,12,24,26,31] and some *MURA*-related transposases with conceptual translations were collected and used in the NCBI TBLASTN search (e $<10^{-5}$, http//www.ncbi.nlm.nih.vov/blast/) against the maize and rice genomes, respectively. Detailed information about the MULEs used as queries is provided in Table S1. Sequences producing significant alignments (e $<10^{-5}$) were retained and their 50 kb flanking sequences were retrieved. The resulting sequences were masked by the MULE_TIR libraries of maize and rice, respectively, using the RepeatMasker program (www.repeatmasker.org). A candidate coding-MULE must satisfy the following criteria. First, the pairwise identity between its two TIRs should be higher than 75% and in an inverted orientation with the TIR-ends facing outwards. Second, sequence homologous to transposases (NCBI BLASTX, e $<10^{-5}$) must be located within the two TIRs. Lastly, a TSD (8–11 bp) immediately flanking the TIRs must be present. For TSDs of 8 bp, one mismatch or indel (one nucleotide) is allowed and for those equal or larger than 9 bp, a maximum of 2 mismatches (or one mismatch plus one indel) is allowed.

Detection of nested TE insertions in the candidate coding-MULEs and estimation of ages of nested LTR retroelements

The candidate coding-MULEs were masked using the maize and rice TE libraries, respectively, using the RepeatMasker program (www.repeatmasker.org) and the resulting output file was used to determine the number and type of nested TEs in the candidate coding-MULEs. The coordinates of the coding regions (as defined in the following section) and the inserted TEs were compared in order to determine whether the inserted TEs interrupt the open reading frame of the candidate coding-MULEs. If the coordinates of the inserted TE are within the coordinates of the coding region, the TE is considered to be inserted in the coding region of the candidate coding-MULE.

Estimation of the insertion time of intact long terminal repeat (LTR) retrotransposons was based on the divergence of the two LTRs. Sequences of the two LTRs of one element were aligned using the MUSCLE program [32] and the number of substitutions per site between the two LTRs was obtained using MEGA 5.10. Nucleotide substitution rate 1.3×10^{-8} per site per year generated by Ma *et al.* [33] was used to calculate the approximate age of the LTR elements.

Determination of the coding capacity of the candidate coding-MULEs

To determine the coding region of the candidate coding-MULEs, nested insertions were first masked using the maize and rice TE libraries (without MULE sequences), respectively, using the RepeatMasker program (www.repeatmasker.org). The masked sequences were used to search (NCBI BLASTX with e $<10^{-5}$) against the total protein datasets of maize and rice, which contain annotated TE proteins and were downloaded from the MaizeSequence.org (http://www.maizesequence.org/) and the Rice Genome Annotation Project (http://rice.plantbiology.msu.edu/index.shtml), respectively. To ensure that the annotated proteins producing significant alignments (smallest e-value) with the candidate coding-MULEs were indeed MULE transposase, these protein sequences were used to search against the above mentioned *MURA* and *MURA*-related transposase sequences using the NCBI BLASTP (e $<10^{-5}$) program. If the annotated proteins had significant alignments (e $<10^{-5}$) with *MURA* or *MURA*-related sequences, the coding region was defined according to the annotated protein. For the annotated proteins without any significant alignment with known transposases, it suggested that the transposase sequence was not annotated in an open reading frame (ORF) because of some defects (*e.g.*, frameshift, premature stop codon, or large deletions). In this case, the nucleotide sequences of the elements were used to search (NCBI BLASTX with e $<10^{-5}$) against the *MURA* and *MURA*-related transposase sequences directly and the regions with significant alignments with those transposase sequences were defined as the coding region. For each candidate coding-MULE, the alignment profiles were manually examined and custom python scripts were used to retrieve the sequences encoding transposases of the candidate coding-MULEs. After obtaining the transposase coding sequences of the candidate coding-MULEs, their coding capacity was evaluated for the presence of DNA binding and catalytic domains. The HTH DNA binding domain was determined using the Jpred3 program (http://www.compbio.dundee.ac.uk/www-jpred/), which was capable of detecting all the experimentally defined HTH domains of some well-annotated transposases (*e.g.*, *Hermes*, *Tc3*, *Mos1*, *Phage Mu*) [34]. To locate the DDE domain, a multiple sequence alignment was conducted using the transposase coding sequences of the candidate coding-MULEs in maize and rice separately, with the *MURA* protein sequence as a reference. The conserved catalytic domain was established by comparing candidate coding-MULEs with that of *MURA* and positions of the DDE amino acids in the transposases were recorded. Specifically, if the three amino acids (DDE) are all present, and the length from the first "D" to the "E" is longer than 100 amino acids, without premature stop codon(s) and frameshift(s), the element was considered to contain a catalytic domain.

Determination of indels in the candidate coding-MULEs

The number and length of indels between two homologous coding-MULEs were determined using the following procedure. First, sequences of the candidate coding-MULEs (after removing nested TEs) were used to conduct an all vs. all search using the NCBI BLASTN program (e $<10^{-10}$). Second, after the self-match was excluded, the element that produced the most significant alignment (the two sequences with the longest alignment) was retained to form a pair with the query coding-MULE, for which pairwise alignment was conducted using the "gap" program available from the GCG package (version 11.0, Accelrys Inc., San Diego, CA). Lastly, the number and length of indels were normalized based on the total length of the two elements to make them comparable between maize and rice. That is, the number of indels was divided by the total length of the two elements, giving a value of the average number of indels per kb (NIK). Similarly, the normalized length of indels is the average length of indels per kb (LIK). Grouping of different coding-MULE pairs was based on two parameters, *i.e.*, pairwise nucleotide identities and synonymous substitution rates (Ks), respectively. To calculate the Ks values, coding sequences (as defined in the previous section) of these element pairs were aligned based on amino acid codons, premature stop codons and frameshifts were removed to achieve correct coding frame. The resulting aligned sequences in fasta format were changed to.axt format using a custom Python script and Ks values were determined using KaKs_Calculator program [35]. Comparison of NIK and LIK of coding-MULE pairs with similar identity or Ks between maize and rice was conducted using the SAS9.3 program at the High Performance Computing Centre, Michigan State University. Nested insertions and one copy of the TSD were excluded upon alignment. The pairs of candidate coding-MULEs used in the alignments were provided in Table S3 and their original sequences were provided in fasta format in Data S1.

In addition, the indel numbers in the LTR sequences of the *Copia*-like LTR retrotransposon in the two genomes were calculated. The sequences of LTRs in rice and maize were from the rice and maize TE libraries mentioned above. The LTR sequences were used to search against the maize and rice genomes using the RepeatMasker program (www.repeatmasker.org). Custom Python scripts were used to extract the coordinates of intact LTRs and retrieve their sequences. Pairwise sequence identity and determination of average indel numbers were similar to that conducted for coding-MULEs.

Phylogenetic analysis

Amino acid sequences of the catalytic region (corresponding to 331–489 amino acids of *MURA*) of the candidate coding-MULEs were used to conduct the phylogenetic analysis in MEGA 5.2.2 [36]. Frameshifts were corrected and premature stop-codons were excluded to ensure appropriate alignment. Figure 1A contains known MULEs and representative elements with catalytic domains in the two genomes. The representatives were chosen as following: if two elements shared 80% identity in 80% of the element region, only one element was retained. In this way, 211 rice MULEs and 195 maize MULEs with catalytic regions were excluded. Figure 1B contains known MULEs in Figure 1A and all MULEs with putative intact transposase, with the exception of one element family in rice. This element family contains 108 members and 59 of them contain putative intact transposases, and only four representative elements were chosen to be included in Figure 1B (clade III) in order to attain a manageable size. The Maximum Likelihood method was used to infer the evolutionary history and the tree with the highest log likelihood value is shown. A discrete Gamma distribution was used to model evolutionary rate differences among sites (5 categories (+G, parameter = 4.0540)). All positions with higher than 95% site coverage were used in the analysis and only bootstrap (500 replicates) values equal or higher than 50% were shown.

Determination of the expression status of the candidate coding-MULEs

Expression Sequence Tags (ESTs) for maize (516,425 ESTs) and rice (1,253,557 ESTs) were downloaded from the NCBI dbEST database (http://www.ncbi.nlm.nih.gov/dbEST/index.html) on January 4th, 2013. The full-length cDNA (flcDNA) library for maize (27,455) was downloaded from The Maize Full Length cDNA Project (http://www.maizecdna.org/) on February 2nd, 2013 and that for rice (37,139) was from the Knowledge-based Oryza Molecular Biological Encyclopedia (http://cdna01.dna.affrc.go.jp/cDNA/) on Oct 1st, 2008. The EST and flcDNA libraries were concatenated into one file (EST-flcDNA) for both maize and rice, respectively. The candidate coding-MULE sequences were searched against the EST-flcDNA library using the NCBI BLASTN program (e $<10^{-5}$). A coding-MULE is considered to have expression evidence if the identity of matched sequence between the coding-MULE and EST/flcDNA is higher than 99.5% over the entire length of the EST/flcDNA.

Results

Candidate MULEs containing transposase sequences (coding-MULEs) in maize and rice

In this study, we focused on TIR-MULEs (see Introduction) containing *MURA*-related transposase sequences (candidate cod-

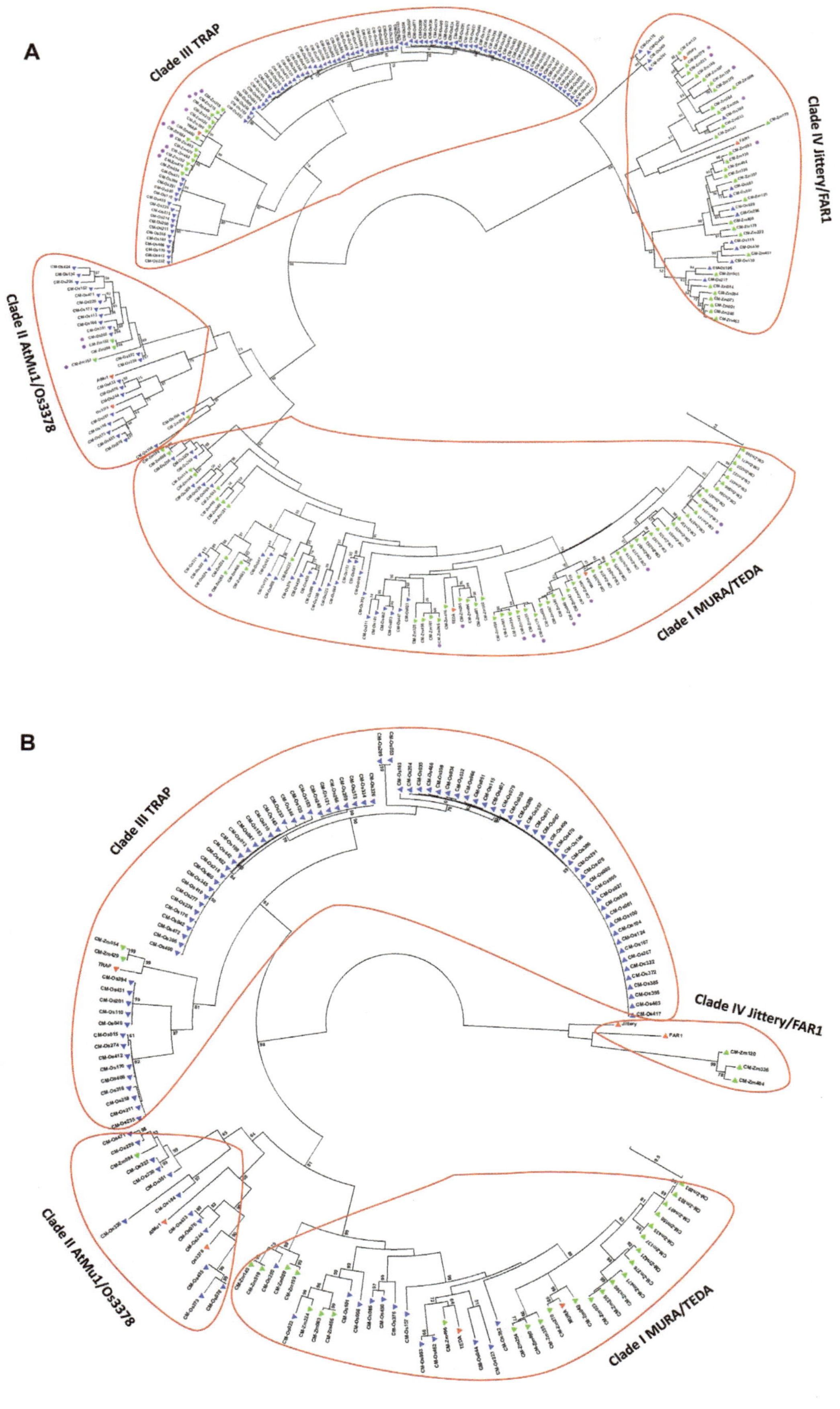
A
Clade III TRAP
Clade IV Jittery/FAR1
Clade II AtMu1/Os3378
Clade I MURA/TEDA
B
Clade III TRAP
Clade IV Jittery/FAR1
Clade II AtMu1/Os3378
Clade I MURA/TEDA

Figure 1. Phylogenetic analysis of the candidate coding-MULEs with DDE domain. A). All representative candidate coding-MULEs containing DDE domain; B). Candidate coding-MULEs with putative intact transposases. Candidate coding-MULEs in maize are denoted by green triangles; those in rice are denoted by blue triangles; known MULE transposases are denoted by red triangles, maize coding-MULEs with nested LTR insertions are denoted with purple round dots.

ing-MULEs) in the maize and rice genomes. Non-TIR MULEs (see Introduction) were not considered because of the difficulty in defining the boundary of their termini and target site duplication with high confidence. To retrieve candidate coding-MULEs, a collection of *MURA*-related transposase sequences from several organisms, especially plants (Table S1) was used to search against (NCBI TBLASTN, $e<10^{-5}$) the maize and rice genomic sequences. The following criteria (see Materials and Methods for more details) were employed for defining a candidate coding-MULE. First, the element should contain a pair of TIRs and the distance between the two TIRs should be 2 to 30 kb (including nested TE insertions inside the element). Second, the sequence located within the TIRs should have homology (NCBI TBLASTN, $e<10^{-5}$) to known transposase. Third, a TSD should be immediately flanking the TIRs of a single element. With these criteria, we detected 530 and 476 candidate coding-MULEs in maize and rice, respectively (Table S2A, S2B). Due to the presence of nested TE insertions in maize and rice [37–40], the candidate coding-MULEs were first masked using TE libraries and nested insertions were identified based on the output of RepeatMasker. After removing nested TE insertions, most candidate coding-MULEs (>95%) are between 2 to 10 kb in size in both maize and rice. However, the number of elements within different size ranges exhibits different patterns in the two species. In maize, a considerable portion of candidate coding-MULEs ranges from 2 to 3.5 kb (~20%) while only 9% of the rice elements fall into the same category ($\chi^2 = 25.7360$, $p<0.0001$; Figure 2). The minimum length of known autonomous MULEs in plants is around 3.5 kb (Table S1), suggesting that elements smaller than 3.5 kb are unlikely to encode an intact transposase (transposase containing both DNA-binding and catalytic domains and no other coding defects). Most (71%) candidate coding-MULEs in rice are between 3.5 to 8 kb while only 58% of the elements in maize are within this range. As a result, candidate coding-MULEs with relatively small (2–3.5 kb) and large size (> 8 kb) are more prevalent in maize (42%) than that in rice (29%) ($\chi^2 = 20.3473$, $p<0.0001$)

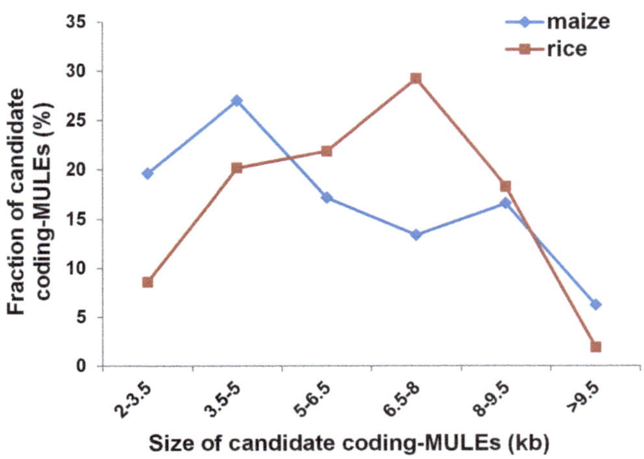

Figure 2. Fraction of candidate coding-MULEs within different size ranges after removing nested TE insertions.

To determine the phylogenetic relationship of these candidates, protein sequences containing the catalytic (DDE) domain within the elements were used to construct a phylogenetic tree. The phylogenetic analysis revealed that all elements belong to four distinct phylogenetic clades (Figure 1A). Clade I is represented by *MURA*, the first known MULE transposase [8]; and *TEDA*, a recently discovered MULE in maize [41]. Clade II is represented by *AtMU1*, an active element in *Arabidopsis* [24] and *Os3378*, an active MULE element in rice [31]. The representative element for clade III is *TRAP*, a MULE in maize [42]. The remainder of elements comprises clade IV, which is represented by *Jittery* and *FAR1*. *Jittery* is another active MULE from maize [26] and *FAR1* is a gene domesticated from a MULE transposase in *Arabidopsis* [27]. As shown in Figure 1A, clade I and IV contain more maize elements (green triangles) while rice elements (blue triangles) are more expanded in clade II and III.

Nested insertions within the candidate coding-MULEs

As mentioned above, nested insertions of some TE families are very common in maize and rice [37–40]. If a TE inserts into a coding-MULE, it may interrupt the transposase ORF and abolish its function. To test whether this is the case, all TEs located within the candidate coding-MULEs were examined. The numbers of candidate coding-MULEs containing nested insertions are largely comparable between maize (n = 191, 36%) and rice (n = 195, 41%) ($\chi^2 = 2.5761$, $p = 0.1085$). However, among the 195 elements with nested insertions in rice, 107 of them are highly similar to each other, where 40 contain an intact Os0548 (a *Tourist*-like miniature inverted repeat transposable element, MITE, which is 274 bp in length) and 67 contain a partial Os0548 at the same site. As a result, the 107 elements are likely copies derived from a single insertion event where 40 of them are amplifications of the element with the intact Os0548 and meanwhile one of them may have experienced partial deletion of Os0548 followed by proliferation to 67 copies. Apparently, the insertion of this element did not abolish its capability for further transposition. In contrast, the 191 elements in maize all harbor independent insertions. If we exclude the 107 elements containing Os0548 from rice, the fraction of candidate coding MULEs with nested insertions is much lower in rice ([195–107] / [476–107] * 100 ≈ 24%) than that in maize (36%) ($\chi^2 = 15.1021$, $p<0.0001$). Moreover, the number and pattern of inserted TEs in individual candidate coding-MULE is different between maize and rice. Among the elements with nested insertions, most (~81%) contain only a single TE insertion in rice whereas that in maize is only less than 47% ($\chi^2 = 49.6349$, $p<0.0001$) (Figure 3A). The fact that candidate coding-MULEs in maize contain more nested TEs than that in rice is also obvious when comparing the average number of nested insertions per coding-MULE with nested TEs (2 nested insertions in maize vs. 1 in rice).

Based on the class (see Introduction) of the inserted TEs, coding-MULEs with nested insertions were divided into three categories, *i.e.*, coding-MULEs with RNA-TE insertions (all inserted TEs were RNA TEs), coding-MULEs with DNA-TE insertions (all inserted TEs were DNA TEs), and those with RNA-DNA-TE insertions (with insertions from both RNA and DNA TEs). It turned out that over 72% of the candidate coding-MULEs in maize contained RNA-TE insertions compared with 29% of

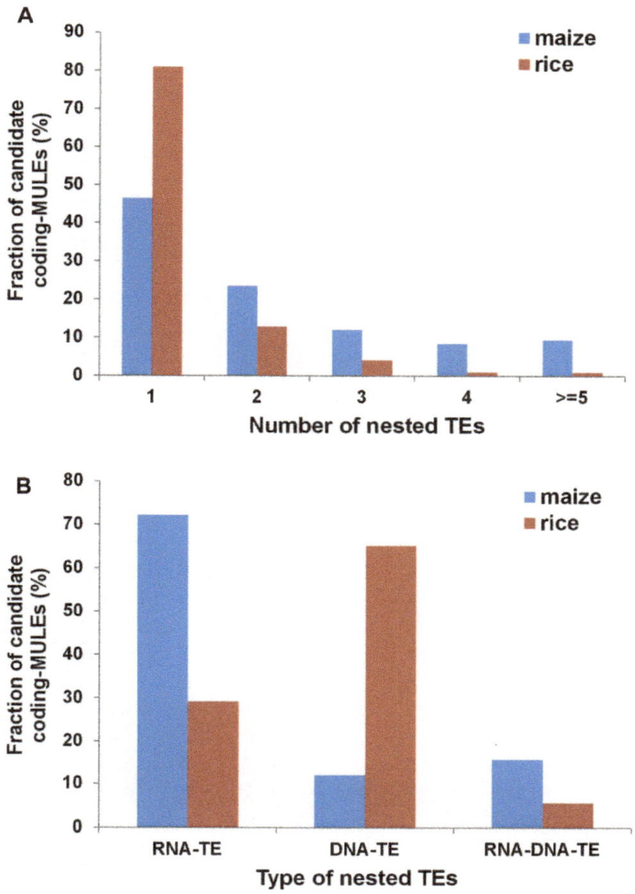

Figure 3. Fractions of candidate coding-MULEs containing different numbers (A) and types (B) of nested TE insertions.

that in rice ($\chi^2 = 71.4396$, $p<0.0001$, Figure 3B). In contrast, there are more candidate coding-MULEs contain DNA-TE insertions in rice (65%) than that in maize (12%) ($\chi^2 = 112.5264$, $p<0.0001$). Furthermore, the average size of the inserted TEs in maize is about 3-fold of that in rice for DNA-TE (708 bp vs. 245 bp, $p = 0.0016$, t-test) and 1.3-fold for RNA-TE (6818 bp vs. 5081 bp, $p = 0.0006$, t-test) (Table 1). Accordingly, maize candidate coding-MULEs contain more independent and larger TE insertions, with most insertions from LTR retrotransposons. Interestingly, most maize MULEs (10 out of 14 vs. 26% of genome average, elements with purple dots in Figure 1A) grouping with *TRAP* (clade III) are associated with nested insertions of LTR retrotransposons while they are not particularly older or distribute differently than other MULEs. This suggests there might be some structural features of these elements that attract LTR elements.

To date the insertion time of the nested TEs, we analyzed all the intact LTR retrotransposons within the candidate coding-MULEs, which are characterized with long terminal repeat (LTR) at both ends of the element. The LTRs of one element are identical when inserted into a new location and become divergent over time. As a result, the identity between the LTRs has been used to estimate the age of the insertion [43]. In light of this, we calculated the approximate ages of 133 and 25 intact LTR elements within the candidate coding-MULEs in maize and rice, respectively (Figure 4). The largest fractions were estimated to be inserted within 1 million years in both maize (~65%) and rice (~69%), followed by

Table 1. Number and average length of nested TE insertions based on TE classes in the candidate coding-MULEs.

	RNA-TE*	DNA-TE*	Total
Maize	329 (6818a†)	74 (708a)	403
Rice	99 (5081b)	152 (245b)	251

*Numbers in parenthesis represent the average length (bp) of nested TEs.
†Means in parenthesis in each column followed by different letters are significantly different (p<0.005, t-test).

insertions which have occurred 1 to 2 million years ago. Few LTR insertions were older than 3 million years in both genomes. Therefore, the insertion of LTR elements into the candidate coding-MULEs occurred within the same evolutionary time frame in both genomes despite the fewer insertions observed in rice.

As a comparison, we also examined how many coding-MULEs inserted into other TEs. It turned out that only 9% of the maize candidate elements and 7% of the rice elements inserted into other TEs (Table 2). This is in contrast to the fact that over 1/3 of the elements harbor insertions from other TEs, suggesting the coding-MULEs are more likely serving as targets for other elements rather than targeting other TEs. In addition, more maize candidate coding-MULEs (7%) inserted into RNA TEs than the rice elements (3%) (Table 2), which is consistent with the fact that there are many more RNA TEs in the maize genome than that in rice [1,6].

Coding capacity of the candidate coding-MULEs

Previous studies indicate that class II transposases, including MULE transposase, consist of an N-terminal helix-turn-helix (HTH) DNA-binding domain and a C-terminal DDE catalytic domain [22,23,44,45]. The DNA binding domain is responsible for binding to the transposon DNA especially the TIR sequences and the catalytic domain is responsible for excision and integration of the element [34,46]. To determine whether an individual candidate coding-MULE contains both domains, the coding regions of these elements were extracted from the relevant gene annotation or annotated manually. Subsequently, the presence of a DNA-binding domain was determined using the Jpred3 program and that of the catalytic domain by examination of a multiple

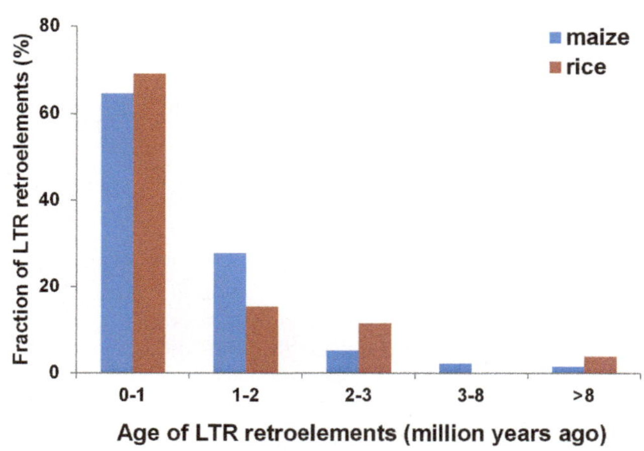

Figure 4. Ages of intact LTR retroelements in the candidate coding-MULEs in maize and rice.

Table 2. Insertion preference of candidate coding-MULEs (CMs).

Types of TEs	Maize			Rice		
	RNA-TE (% of total CMs*)	DNA-TE (% of total CMs*)	RNA/DNA	RNA-TE (% of total CMs*)	DNA-TE (% of total CMs*)	RNA/DNA
Nested TE insertion events in coding-MULEs	329 (32%)	74 (10%)	4.45	99 (19%†)	46 (9%†)	2.15
Coding-MULE insertion events in other TEs	35 (7%)	9 (2%)	3.89	15 (3%)	19 (4%)	0.79

*Note that individual coding-MULE may contain different number or types of nested TEs.
†In rice, 107 candidate coding-MULEs are copies of two elements resulting from one TE insertion event (see results), which was corrected in both the number of DNA-TE insertion events and the total candidate coding-MULEs (476–107+1 = 370).

sequence alignment containing *MURA* protein as the reference (see Materials and Methods for details). A transposase containing both DNA-binding and catalytic domains and without other defects (*e.g.*, frameshift, premature stop codon) in its ORF is considered a potentially intact transposase.

To obtain a better view, the candidate coding-MULEs were divided into six groups based on their coding capacity. Group 1 includes coding-MULEs with putative intact transposase, *i.e.*, at least one HTH domain (some elements have more than one HTH domain, *e.g.*, *Mos1* [34]) close to the amino terminus and a DDE catalytic domain close to the carboxyl terminus of the transposase protein. Group 2 is comprised of coding-MULEs with mutation of either translation start codon (ATG) or key amino acids (D, D, E) in the catalytic domain. Group 3 and 4 include coding-MULE transposases containing frameshift and premature stop codon, respectively. Those containing both frameshift and premature stop codon were assigned to Group 5. Group 6 consists of coding-MULEs with various forms of deletions in their transposases, including deletions of the DNA-binding domain and/or the catalytic domain or other regions. Since many candidate coding-MULEs contain several types of coding defects, one element may be assigned to more than one group (Table 3A). Meanwhile, a more specific classification where each element is only assigned to one sub-type is available in Table 3B.

Overall, rice contains many more candidate coding-MULEs with putative intact transposase than maize (35.29% vs. 5.85%) ($\chi^2 = 137.0185$, $p<0.0001$; Table 3A), and deletion is likely the most important factor contributing to the difference. Most of the elements in maize (>89%) are associated with various deletions within the transposase compared to only 41% of the elements with deletions in rice. The presence of premature stop codon is the second most frequent defect, but the frequency is similar between maize (32.64%) and rice (33.40%) ($\chi^2 = 0.0658$, $p = 0.7975$; Table 3A). For elements with other defects (mutation of DDE motif or start codon, frameshift, and presence of both frameshift and premature stop codon), no dramatic difference was observed between the two species (Table 3A). In addition, many more maize elements (~40%) contain more than one type of defect than that in rice (< 25%) ($\chi^2 = 25.1084$, $p<0.0001$).

When classifying these elements based on whether they contain nested TE insertions or not, it is clear that in maize there are more candidate coding-MULEs containing putative intact transposase if there are no nested TEs (Figure 5). For example, only 2 out of 191 (1.05%) elements with nested insertions seem to have putative intact transposase, which is in contrast to 29 out of 339 (8.55%) elements without nested insertions in maize ($\chi^2 = 12.5035$, $p = 0.0004$; Table 3A). In contrast, ratios of candidate coding-MULEs harboring putative intact transposases for those with (35.90%) and without (34.88%) nested TE insertions ($\chi^2 = 0.0526$, $p = 0.8185$; Table 3A) are comparable in rice. Nevertheless, if the high copy elements (those containing intact and partial Os0548 element) were excluded, only 12.5% candidate coding-MULEs with nested insertions harbor putative intact transposase in rice, suggesting the destructive effect of nested TE insertions in both species, or elements with nested insertions are older than these without nested insertions so more mutations have accumulated. A close examination indicated that more nested TEs (n = 46) in maize directly disrupted transposases than that in rice (n = 23) ($\chi^2 = 23.0278$, $p<0.0001$), and most of them are LTR retrotransposons. Again, this indicates that nested TEs are more deleterious to coding-MULEs in maize than that in rice.

A second phylogenetic tree was constructed based on the catalytic domain of the elements with putative intact transposase (Figure 1B). Compared with Figure 1A, the number of elements in all the four clades declines, which is not surprising. This decrease is more obvious in maize, as revealed by the existence of only one clade (clade I) containing more than 10 elements and the other clades having four or less elements left. In contrast, rice (Nipponbare) still has more than 10 coding-MULEs in three clades (clade I, II, and III) and clade III seems to be active or recently active as evidenced by the short branch lengths (Figure 1B). In contrast, transposition activity in clades II to IV might be limited currently or in the future in maize (B73). This is consistent with the fact that there are either few maize elements (clades II and III) or many maize elements bear relatively long branches (clade IV) in these clades (Figure 1A), which are the signatures for loss of transposition activity.

Other insertions and deletions within the candidate coding-MULEs

The analysis above indicated that deletions in coding regions are prevalent in maize coding-MULEs, and this is possibly the most dominant factor for loss of coding capacity of these elements. The abundance of elements with deletions in maize could be attributed to a high deletion frequency in maize. Alternatively, the element with deletions may have an advantage in transposition (for instance, because of small size), so they have achieved relatively high copy numbers. To determine whether the frequency of insertion/deletion within the candidate coding-MULEs is similar between maize and rice, pairwise comparison was conducted using individual coding-MULEs and their most similar homologs. For accuracy, only pairs bearing higher than 95% identity in both maize and rice were considered. This portion of coding-MULE pairs were analyzed and the number and length of indels in those pairs were calculated and normalized to the average number of indels per kb (NIK) and the average length of indels per kb (LIK) (see Materials and Methods). The results show that the number of indels is significantly higher ($p<0.05$, t-test) in maize than that in rice in all the identity ranges except 98–99% (Figure 6A). The

Table 3. Candidate coding-MULEs (CMs) with or without distinct defects in their coding regions (redundant grouping), and candidate coding-MULEs (CMs) with or without distinct defects in their coding regions (non-redundant grouping).

A. Candidate coding-MULEs (CMs) with or without distinct defects in their coding regions (redundant grouping).								
Group	ORF* status		Maize			Rice		
			With TE (% of total CMs with nested TEs)	Without TE (% of total CMs without nested TEs)	Total	With TE (% of total CMs with nested TEs)	Without TE (% of total CMs without nested TEs)	Total
Group I	ORF with putative intact transposase		2 (1.05%)	29 (8.55%)	31 (5.85%)	70 (35.90%)	98 (34.88%)	168 (35.29%)
Group II	ORF with DDE or start codon mutation		19 (9.95%)	20 (5.90%)	39 (7.36%)	13 (6.67%)	14 (4.98%)	27 (5.67%)
Group III	ORF with frameshift		36 (18.85%)	37 (10.91%)	73 (13.77%)	34 (17.44%)	45 (16.01%)	79 (16.60%)
Group IV	ORF with premature stop codon		82 (42.93%)	91 (26.84%)	173 (32.64%)	64 (32.82%)	95 (33.81%)	159 (33.40%)
Group V	ORF with both frameshift and premature stop codon		19 (9.95%)	12 (3.54%)	31 (5.85%)	18 (9.23%)	33 (11.74%)	51 (10.71%)
Group VI	ORF with deletions		187 (97.91%)	288 (84.96%)	475 (89.62%)	66 (32.85%)	129 (45.91%)	195 (40.97%)
B. Candidate coding-MULEs (CMs) with or without distinct defects in their coding regions (non-redundant grouping).								
Group	ORF status		Maize			Rice		
			With TE (% of total CMs with nested TEs)	Without TE (% of total CMs without nested TEs)	Total	With TE (% of total CMs with nested TEs)	Without TE (% of total CMs without nested TEs)	Total
Group I	ORF with putative intact transposase		2 (1.05%)	29 (8.55%)	31 (5.85%)	70 (35.90%)	98 (34.88%)	168 (35.29%)
Group II	ORF with DDE or start codon mutation only		1 (0.52%)	4 (1.18%)	5 (0.94%)	3 (1.54%)	7 (2.49%)	10 (2.10%)
Group III	ORF with frameshift		1 (0.52%)	9 (2.65%)	10 (1.89%)	10 (5.13%)	4 (1.42%)	14 (2.94%)
Group IV	ORF with premature stop codon		0 (0%)	8 (2.36%)	8 (1.51%)	37 (18.97%)	28 (9.96%)	65 (13.66%)
Group V	ORF with both frameshift and premature stop codon		0 (0%)	1 (0.29%)	1 (0.19%)	9 (4.62%)	15 (5.34%)	24 (5.04%)
Group VI	ORF with deletions	deletion only	81 (42.41%)	185 (54.57%)	266 (50.19%)	37 (18.97%)	64 (22.78%)	101 (21.22%)
		with DDE or start codon mutation only	8 (4.19%)	5 (1.47%)	13 (2.45%)	5 (2.56%)	5 (1.78%)	10 (2.10%)
		with frameshift	16 (8.38%)	16 (4.72%)	32 (6.04%)	6 (3.08%)	8 (2.85%)	14 (2.94%)
		with premature stop codon	63 (32.98%)	71 (20.94%)	134 (25.28%)	9 (4.62%)	34 (12.10%)	43 (9.03%)
		with both frameshift and premature stop codon	19 (9.95%)	11 (3.24%)	30 (5.66%)	9 (4.62%)	18 (6.41%)	27 (5.67%)
Total			191	339	530	195	281	476

*ORF: open reading frame.

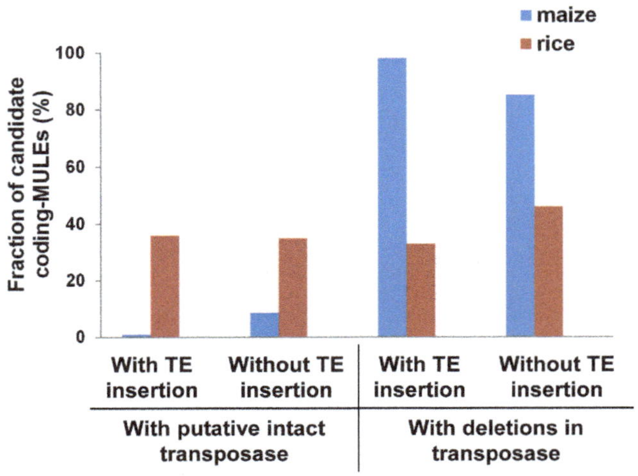

Figure 5. Fractions of candidate coding-MULEs containing putative intact transposases and transposases with deletions. The fractions were calculated within each category, *e.g.*, fraction of candidate coding-MULEs with TE insertions containing putative intact transposases is based on the total number of elements with TE insertions.

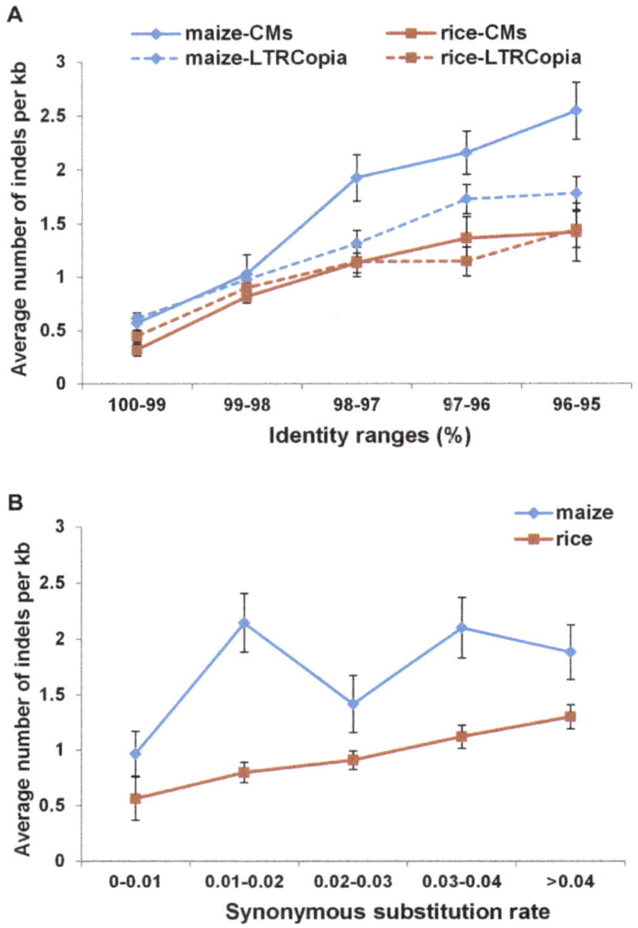

Figure 6. Average number of indels in maize and rice TEs. A). Average number of indels in candidate coding-MULEs (maize-CMs and rice-CMs) and *Copia*-like LTR elements using pairwise nucleotide identity as the grouping criterion; B). Average number of indels in candidate coding-MULEs using synonymous substitution rate as the grouping criterion.

number of indels is increasing when the coding-MULE pair is less similar, with a ~4.5 fold increase when the similarity is 95–96% compared to that of over 99% in both maize and rice. Meanwhile, the most significant difference between maize and rice is observed with the group containing element pairs with 95–96% identity. For this group, the NIK in maize is ~2.48 and that in rice is ~1.40, suggesting that there is at least one more indel per kb per coding-MULE pair generated in maize than those in rice. When comparing the LIK of coding-MULE pairs between maize and rice, no significant difference was observed and the average LIK values are 87.36 bp ± 16.06 for maize and 89.24 bp ± 11.21 for rice ($p = 0.9235$, t-test). To ensure the nucleotide identity reflect the divergence of homologous coding-MULE pairs, we also calculated indel rates based on synonymous substitution rates (Ks) between these pairs of elements in coding regions (Figure 6B), and a similar trend was observed. Taken together, it seems that coding-MULEs in maize experienced more indels than that in rice.

To determine whether the higher frequency of indels in maize is specific to coding-MULEs or a generic feature for the entire genome, we also conducted analysis of indels in LTR sequences of individual *Copia*-like LTR elements in the two genomes. Like that observed for coding-MULEs, maize LTR sequences seem to have experienced more frequent indels than that in rice. However, the difference is only evident when the nucleotide identity is within 96–97% ($p = 0.0227$, t-test). It should be noted that indel frequencies between LTRs and coding-MULEs in rice are not significantly different within the same identity range. In contrast, the average indel numbers in coding-MULEs are about two-fold of those in LTRs when the sequence identity ranges from 95% to 98% (Figure 6A). Collectively, these results demonstrate that maize may be more prone to incidence of indels than rice at the whole genome level. More importantly, within the maize genome, coding-MULEs seem to have experienced more indels than LTRs if we assume point mutation rate is comparable across different families of TEs.

Expression evidence of the candidate coding-MULEs in maize and rice

A functional transposase is only available for transposition if the element is expressed. To determine how many candidate coding-MULEs are expressed, we searched the publically available EST and full-length cDNA (fl-cDNA) databases for evidence of expression. The completeness of these datasets was roughly assessed by determining the proportion of non-TE genes that have at least one match in the database with 99.5% or higher identity over the entire matched sequence. Under this criterion, we found 36% of non-TE genes in maize and 45% of that in rice having expression evidence, suggesting that the dataset for rice is 1.2-fold more comprehensive than that of maize. We did not exploit the data from next generation sequencing such as RNA-seq because most candidate coding-MULEs have highly similar copies in the genome and the short reads of RNA-seq experiments do not allow us to determine which specific element is expressed.

An element was considered to be expressed if it matches an fl-cDNA or EST sequence with at least 99.5% identity over the entire matched sequence (at least 300 bp). The results revealed that there were more coding-MULEs with either EST or fl-cDNA evidence in rice (n = 28) than that in maize (n = 9) ($\chi^2 = 12.3933$, p = 0.0004; Table 4A). Even if we take into account the 1.2-fold enrichment of expression evidence in rice than maize, rice still contains more elements (n = 23) ($\chi^2 = 7.9969$, p = 0.0047) with expression evidence. In both genomes, elements with expression evidence include coding-MULEs with and without nested TE insertions. In rice, expression evidence was detected for 4% (n = 7) of elements with putative intact transposase and 7% (n = 21) of elements with defects in transposases whereas those in maize are only 3% (n = 1) and 1.6% (n = 8), respectively (Table 4B). In either case, it does not appear that elements with intact transposases are more frequently transcribed. Interestingly, among the 7 elements containing putative intact transposases that are expressed in rice, three have a MITE insertion (among the 107 elements containing the Os0548 element) in the C-terminal region of the relevant coding-MULEs (~2.5 kb downstream of the stop codon of the transposase ORF), which does not seem to affect the expression of the elements, at least at the transcription level (Table 4A). Overall, only a small subset of candidate coding-MULEs are expressed, yet rice has many more expressed elements than maize does, which may explain why the rice genome harbor many more MULEs. However, we must be cautious about interpreting this difference because it is known that expression of TEs is often induced under stress and the higher number of coding-MULEs expressed in rice may be due to the overrepresentation of stress conditions in the rice dataset.

Discussion

Differential amplification of MULEs including coding-MULEs in rice and maize

The first active MULE element, *Mutator*, was found in maize due to the high mutation frequency it caused [8]. With the availability of a myriad of genome sequences, analysis of TEs has been carried out at the whole genome level, resulting in the discovery of the prevalence of MULEs in most living organisms [10–14,25]. Being members of the grass family, both maize and rice contain thousands of MULEs [1,15]. However, compared with rice (~370 Mb available sequence), maize (~2.1 Gb available sequence) contains less MULEs per unit genomic sequence (~84 per Mb in rice vs. ~8 per Mb in maize) as well as per unit coding sequence of non-TE genes (~770/Mb in rice vs. ~318/Mb in maize, based on maize genome annotation v2 and MSU rice annotation version 7). The persistence of TEs is an interaction between amplification through transposition, duplication of genomic sequences, and loss of TEs by excision and sequence erosion. Transposase, the protein encoded by autonomous elements, catalyzes transpositions, and is therefore responsible for increasing the copy number of the elements. In this study, we detected comparable numbers (530 vs. 476) of candidate coding-MULEs in maize and rice. Since all the candidate elements contain partial or complete transposases, they should have been derived from putative autonomous elements at a certain evolutionary stage. This implies that the resource for generating transposases was comparable between maize and rice in the traceable past, and the loss of such resource has been accelerated in maize in the recent past, which led to the reduced abundance of MULEs.

Maize and rice share an ancestor about 50–70 million years ago [30] and should have inherited largely the same set of transposable elements including MULEs. From this point of view, it is surprising that the phylogenetic composition of coding-MULEs is different between maize and rice, in addition to the difference in total copy numbers. The maize elements are abundant in clade I and IV, while the rice elements were amplified in clade II and III. This suggests that either the amplification of TEs is a rather fortuitous process or different genome environments may favor the survival of distinct elements.

Factors involved in degeneration of coding-MULEs

In this study, we dissected various factors that may lead to the loss of coding capacity of MULE transposases. Apparently, for both maize and rice, deletion is the most devastating factor (Table 3A), which causes loss of the coding sequence or frameshift. Deletion of sequence is much more prevalent in maize than that in rice (90% vs. 40% of the elements). Point mutation is the second most important factor, which resulted in the mutation of DDE motif, the loss of a start codon, and the formation of premature stop codon. Nevertheless, the frequency of this type of mutation (~40%) is comparable between maize and rice and therefore does not explain the differential amplification of MULEs between the two species. In addition, it is likely that point mutations within regions other than the DDE motif and the start codon may also lead to dysfunctional transposase. As a result, the number of functional transposases might be overestimated in this study. The third important factor is the insertion by other TEs, which may interrupt the ORF of the transposases or interrupt the *cis*-elements that are required for further transposition. Again, the number of elements that harbor nested insertions is comparable between maize and rice. However, the nested insertions in maize are more harmful than that in rice for a variety of reasons (see below). Thus, among the three major factors that demolish the coding capacity of MULEs, two of them are significantly enforced in maize.

Based on pairwise comparison of homologous elements, indels occur more frequently in maize than that in rice. Certainly, this is based on the assumption that the point mutation rate is comparable in the two species. Given the fact that number of elements with defects caused by point mutation is similar in maize and rice (see above), this is likely the case. In addition, our analysis using synonymous substitution rate led to similar results, further suggesting that functional constraint on nucleotide substitution is similar in the two species. Our analyses confirm previous studies of low stability of the maize genome [47]. By comparing an orthologous region of the maize, sorghum, and rice, Ilic *et al.* [48] found that the maize genome experienced more sequence deletions than rice, leading to the conclusion that the maize genome is less stable compared with the rice genome. The instability was partly attributed to the polyploidization event of maize. Several studies suggest that genome rearrangement was more frequent in species that experienced a polyploid event (reviewed in [49]). Unequal homologous recombination and illegitimate recombination are the two most proposed mechanisms responsible for DNA removal [50]. Woodhouse *et al.* [51] suggests that genes were preferentially deleted from one of the two maize homeologs possibly through a similar illegitimate recombination, which is also the primary source for TE removal in the maize genome.

The comparison of indel frequency between *Copia*-like LTR elements and coding-MULEs shed new lights on this issue. If we assume the frequency of point mutation is comparable across the two genomes, it is obvious that both types of elements demonstrate elevated indel frequency in maize, suggesting that in general indels occur more often in maize than that in rice. However, what is somehow unexpected is that coding-MULEs in maize seem to be subject to higher indel frequency than that of LTR elements and therefore contributes to their degeneration. Such difference could be attributed to the fact that they are located in different regions of the genomes and the indel rate is influenced by the recombination

Table 4. Candidate coding-MULEs with expression evidence in maize and rice, and summary of candidate coding-MULEs with expression evidence in maize and rice.

A. Candidate coding-MULEs with expression evidence in maize and rice.

Species	Coding-MULE	With nested TE	EST/fl-cDNA*	ORF status
Maize	CM-Zm373	Yes	BT069140	deletion only
	CM-Zm391	Yes	gi|211046304|gb|FK977962.1|FK977962	deletion only
	CM-Zm138	Yes	gi|211501343|gb|FL022227.1|FL022227	deletion only
	CM-Zm216	Yes	gi|211515319|gb|FL470580.1|FL470580	deletion only
	CM-Zm352	Yes	gi|211161841|gb|FK974923.1|FK974923	deletion & premature_stop_codon & frameshift
	CM-Zm418	No	BT041283	deletion only
	CM-Zm091	No	BT070085	deletion only
	CM-Zm369	No	gi|211378400|gb|FL229430.1|FL229430	deletion only
	CM-Zm137	No	gi|211249173|gb|FL479070.1|FL479070	putative intact
Rice	CM-Os180	Yes	AK066496	deletion only
	CM-Os087	Yes	gi|87004716|gb|CI306800.1|CI306800	deletion only
	CM-Os392	Yes	gi|88668382|gb|CI721944.1|CI721944	deletion only
	CM-Os222	Yes	gi|88692462|gb|CI736037.1|CI736037	deletion only
	CM-Os464	Yes	gi|88846512|gb|CI373280.1|CI373280	deletion only
	CM-Os133	Yes	AK072808	deletion & premature_stop_codon & frameshift
	CM-Os028	Yes	AK288956	deletion & premature_stop_codon & frameshift
	CM-Os046	Yes	AK120202	frameshift & premature_stop_codon
	CM-Os181	Yes	AK067920	premature_stop_codon
	CM-Os252	Yes	gi|88718219|gb|CI741729.1|CI741729	putative intact
	CM-Os388	Yes	gi|88695069|gb|CI739107.1|CI739107	putative intact
	CM-Os364	Yes	gi|88695796|gb|CI737373.1|CI737373	putative intact
	CM-Os446	No	AK073736	deletion only
	CM-Os006	No	gi|58687855|gb|CK076542.1|CK076542	deletion only
	CM-Os083	No	gi|86437298|gb|CI119020.1|CI119020	deletion only
	CM-Os089	No	gi|86827995|gb|CI274907.1|CI274907	deletion only
	CM-Os422	No	gi|87030711|gb|CI361925.1|CI361925	deletion only
	CM-Os209	No	gi|88276743|gb|CI397545.1|CI397545	deletion only
	CM-Os317	No	gi|88727053|gb|CI750708.1|CI750708	deletion only
	CM-Os217	No	AK066465	deletion & frameshift
	CM-Os370	No	gi|88846132|gb|CI372900.1|CI372900	deletion & frameshift
	CM-Os350	No	gi|88297974|gb|CI555989.1|CI555989	frameshift
	CM-Os302	No	gi|29629482|gb|CB634491.1|CB634491	frameshift & premature_stop_codon
	CM-Os018	No	gi|88730292|gb|CI753906.1|CI753906	premature_stop_codon
	CM-Os395	No	gi|88279043|gb|CI400043.1|CI400043	putative intact
	CM-Os200	No	gi|5701669|gb|C28952.2|C28952	putative intact
	CM-Os044	No	AK100632	putative intact
	CM-Os352	No	gi|29632126|gb|CB637135.1|CB637135	putative intact

B. Summary of candidate coding-MULEs with expression evidence in maize and rice.

	Coding-MULEs with putative intact transposase			Coding-MULEs with defects in transposase		
	With expression evidence	Without expression evidence	Total	With expression evidence	Without expression evidence	Total
Maize	1 (3.13%)	31 (96.88%)	32	8 (1.61%)	490 (98.39%)	498
Rice	7 (3.95%)	170 (96.05%)	177	21 (7.02%)	278 (92.98%)	299

*EST/fl-cDNA: expressed sequence tag/full length cDNA.

rate of the regions because it is known that TEs in different chromosomal domains evolve differently [52]. Alternatively, indels occur during transposition, so it could be due to the different transposition mechanism of different families of TEs as well. No matter what the underlying mechanism is, the difference in indel frequency explains, at least to some degree, why LTR elements are more successful than DNA TEs in maize.

In addition to the lower abundance of elements with putative intact transposase, the number of expressed candidate coding-MULEs in maize is only 1/3 of that in rice. The fewer expressed elements in maize is consistent with the low activity of MULEs in the genome, which is in contrast with the recent burst of amplification of LTR retrotransposons. It is well established that enormous variation of TE activities and compositions exist among different organisms (reviewed by [53,54]). Some plant species, such as moss (*Physcomitrella patens*) and maize (*Zea mays*), exhibited relatively high LTR retrotransposon activity and low DNA TE activity while some animal species, *e.g.*, nematode (*Caenorhabditis elegans*) and brown bat (*Myotis lucifugus*), are more active in DNA TEs and less in LTR elements [54]. This suggests that genomes may have distinct mechanisms for silencing different types of elements that led to differential amplification of distinct TE families [53]. In fact, LTR elements, especially *gypsy*-like LTRs, express more frequently than DNA TEs in maize as revealed by more expressed sequence tags (ESTs) mapped to LTR retrotransposons than that to DNA TEs [55]. So far, it is unclear whether the lack of expression may further accelerate the degeneration of coding-MULEs in maize. For normal genes, those that are not expressed evolve more rapidly than genes that are highly expressed [56–59], and there is a correlation between expression level, Ka/Ks, and pseudogenization [60,61]. As a result, it would not be surprising if the low expression frequency of the maize elements contributes to the loss of coding capacity of MULEs.

Interaction between different TEs

Plant genomes harbor many distinct superfamilies of TEs and it is not known whether the amplification of some TEs impacts the amplification/survival of other TEs. Based on the genome-wide analyses of TEs, the rice genome harbors a total of 166,700 TEs [62] while that for maize is 1,283,000 [1]. If this is translated to TE insertions per unit sequence, the density of TEs in maize is 1.4-fold of that in rice (613 TEs/Mb sequence in maize vs. 439 TEs/Mb sequence in rice). From this point of view, the total number of nested TE insertions found in the candidate coding-MULEs (403 in maize vs. 251 in rice) is largely comparable to the genomic average and to each other in rice and maize. Nevertheless, nested TE insertions are more detrimental on the candidate coding-MULEs in maize than that in rice due to the following facts. First, the majority (~88%) of the candidate coding-MULEs with TE insertions in maize contains RNA or retrotransposons (RNA-TE and DNA-RNA-TE), which are usually larger than DNA-TEs and are more disruptive (Figure 3B). Second, more candidate coding-MULEs in maize harbor two or more TE insertions (53%) compared with the majority of elements in rice that contain only one TE insertion (81%) (Figure 3A). This further increases the chance to abolish the functionality of the coding-MULEs, which is consistent with the fact that in maize, the proportion of elements with their transposases interrupted by nested TEs is twice of that in rice. Collectively, these nested TE insertions impaired a large fraction of coding-MULEs in maize, which is reflected by the presence of only two elements (~1%) with potentially intact transposases among all the coding-MULEs containing nested insertions (Table 3A). On the other hand, few candidate coding-MULEs have inserted into other TEs including retrotransposons. This is likely because MULEs preferentially insert into low copy or genic sequences [63]. Such target specificity confers certain evolutionary advantages. For instance, the elements in genic regions are more likely to be expressed. Nevertheless, it may also bring about vulnerability to these elements when other elements such as retrotransposons are actively transposing. This is because retrotransposons tend to insert into other repetitive sequences so their activity is deleterious to other elements while the amplification of MULEs or other DNA transposons rarely interrupt retrotransposons. This is consistent with the results from a previous study where it was shown the incidents of insertions of other elements (including LTR elements) into miniature inverted repeat transposable elements (MITEs) were 65 times more often than that of MITEs into LTR and other DNA elements [40]. This indicates that MITEs are similar to coding-MULEs in terms of serving as targets for other elements rather than targeting other TEs.

Both theoretical modeling and empirical results demonstrated that mating systems play potential roles in shaping the abundance and diversity of transposable element within a genome [64–67]. A recent study showed that mode of reproduction contributed to the different transposon profiles in self-fertilizing *Arabidopsis thaliana* and its outcrossing relative *Arabidopsis lyrata* [65]. This study and other studies led to the conclusion that a reduced efficacy of natural selection against TE insertions in selfing populations [68–70]. Such reduced pressure against TE insertions provides more advantage for DNA TEs than RNA TEs because the insertion of DNA TEs are more likely in the genic regions which are subject to more intensive selection. Our comparison between rice (a selfing plant) and maize (an outcrossing plant) provides additional understanding about the possible influence of mating system on the dynamics of TEs. This is because outcrossing offers continuous opportunity for stochastic introduction of novel autonomous elements that may initialize new transposition activity. If a new retrotransposon is introduced, it is conceivable that the increased amplification of retrotransposons gradually abolishes the activity of MULEs or other DNA transposons through insertion into them. In contrast, the introduction of DNA TEs is not as harmful for retrotransposons. Taken together, the genomes of outcrossing plants are less favorable to DNA TEs due to the elevated efficacy of natural selection as well as the increased chances being attacked by retrotransposons. From this point of view, plants experiencing significant outcrossing are less likely to contain abundant DNA TEs and papaya is an excellent example.

Supporting Information

Table S1 *MURA* and *MURA*-related transposases used in the study.

Table S2 A. Genomic coordinates and target site duplications (TSDs) of the candidate coding-MULEs in maize. B. Genomic coordinates and target site duplications (TSDs) of the candidate coding-MULEs in rice.

Table S3 Pairs of candidate coding-MULEs from maize (CM-Zm) and rice (CM-Os) used in calculating indels.

Acknowledgments

We thank Ann Ferguson and Stefan Cerbin (Michigan State University) for critical reading of the manuscript.

Author Contributions

Conceived and designed the experiments: NJ DZ. Performed the experiments: DZ. Analyzed the data: DZ. Contributed reagents/materials/analysis tools: DZ NJ. Wrote the paper: NJ DZ.

References

1. Schnable PS, Ware D, Fulton RS, Stein JC, Wei FS, et al. (2009) The B73 maize genome: complexity, diversity, and dynamics. Science 326: 1112–1115.
2. Schmutz J, Cannon SB, Schlueter J, Ma JX, Mitros T, et al. (2010) Genome sequence of the palaeopolyploid soybean. Nature 463: 178–183.
3. Paterson AH, Bowers JE, Bruggmann R, Dubchak I, Grimwood J, et al. (2009) The *Sorghum bicolor* genome and the diversification of grasses. Nature 457: 551–556.
4. Sato S, Tabata S, Hirakawa H, Asamizu E, Shirasawa K, et al. (2012) The tomato genome sequence provides insights into fleshy fruit evolution. Nature 485: 635–641.
5. Ming R, Hou SB, Feng Y, Yu QY, Dionne-Laporte A, et al. (2008) The draft genome of the transgenic tropical fruit tree papaya (*Carica papaya Linnaeus*). Nature 452: 991–996.
6. Matsumoto T, Wu JZ, Kanamori H, Katayose Y, Fujisawa M, et al. (2005) The map-based sequence of the rice genome. Nature 436: 793–800.
7. Slotkin RK, Nuthikattu S, Jiang N (2012) The evolutionary impact of transposable elements on gene and genome regulation. In: Greilhuber J, Wendel J, editors. Molecular Biology and Evolution of the Plant Genome. Springer Press. pp. 35–58.
8. Robertson DS (1978) Characterization of a *Mutator* system in maize. Mutation Research 51: 21–28.
9. Bennetzen JL, Swanson J, Taylor WC, Freeling M (1984) DNA insertion in the first intron of maize Adh1 affects message levels: cloning of progenitor and mutant *Adh1* alleles. Proc Natl Acad Sci U S A 81: 4125–8.
10. Yu ZH, Wright SI, Bureau TE (2000) *Mutator*-like elements in *Arabidopsis thaliana*: structure, diversity and evolution. Genetics 156: 2019–2031.
11. Rossi M, Araujo PG, de Jesus EM, Varani AM, Van Sluys MA (2004) Comparative analysis of *Mutator*-like transposases in sugarcane. Molecular Genetics and Genomics 272: 194–203.
12. Chalvet F, Grimaldi C, Kaper F, Langin T, Daboussi MJ (2003) Hop, an active *Mutator*-like element in the genome of the fungus *Fusarium oxysporum*. Molecular Biology and Evolution 20: 1362–1375.
13. Neuveglise C, Chalvet F, Wincker P, Gaillardin C, Casaregola S (2005) *Mutator*-like element in the yeast *Yarrowia lipolytica* displays multiple alternative splicings. Eukaryotic Cell 4: 615–624.
14. Marquez CP, Pritham EJ (2010) *Phantom*, a new subclass of *Mutator* DNA transposons found in insect viruses and widely distributed in animals. Genetics 185: 1507–1517.
15. Ferguson AA. Jiang N (2012) *Mutator*-like elements with multiple long terminal inverted repeats in plants. Comparative and Functional Genomics. DOI: 10.1155/2012/695827.
16. Jiang N, Bao ZR, Zhang XY, Eddy SR, Wessler SR (2004) Pack-MULE transposable elements mediate gene evolution in plants. Nature 431: 569–573.
17. Lisch D (2002) *Mutator* transposons. Trends in Plant Science 7: 498–504.
18. Lisch D (2005) Pack-MULEs: theft on a massive scale. Bioessays 27: 353–5.
19. Ferguson A, Zhao D, Jiang N (2013) Selective acquisition and retention of genomic sequences by Pack-MULEs based on GC content and breadth of expression. Plant Physiology 163: 1419–1432.
20. Lisch D, Girard L, Donlin M, Freeling M (1999) Functional analysis of deletion derivatives of the maize transposon *MuDR* delineates roles for the *MURA* and *MURB* proteins. Genetics 151: 331–341.
21. Kim SH,Walbot V (2003) Deletion derivatives of the *MuDR* regulatory transposon of maize encode antisense transcripts but are not dominant-negative regulators of *Mutator* activities. Plant Cell 15: 2430–2447.
22. Hua-Van A, Capy P (2008) Analysis of the DDE motif in the *Mutator* superfamily. Journal of Molecular Evolution 67: 670–681.
23. Yuan YW, Wessler SR (2011) The catalytic domain of all eukaryotic cut-and-paste transposase superfamilies. Proc Natl Acad Sci U S A 108: 7884–9.
24. Singer T, Yordan C, Martienssen RA (2001) Robertson's *Mutator* transposons in *A. thaliana* are regulated by the chromatin-remodeling gene *Decrease in DNA Methylation* (*DDM1*). Genes & Development 15: 591–602.
25. Lisch DR, Freeling M, Langham RJ, Choy MY (2001) *Mutator* transposase is widespread in the grasses. Plant Physiology 125: 1293–1303.
26. Xu ZN, Yan XH, Maurais S, Fu HH, O'Brien DG, et al. (2004) *Jittery*, a *Mutator* distant relative with a paradoxical mobile behavior: excision without reinsertion. Plant Cell 16: 1105–1114.
27. Hudson ME, Lisch DR, Quail PH (2003) The *FHY3* and *FAR1* genes encode transposase-related proteins involved in regulation of gene expression by the phytochrome A-signaling pathway. Plant Journal 34: 453–471.
28. Cowan RK, Hoen DR, Schoen DJ, Bureau TE (2005) *MUSTANG* is a novel family of domesticated transposase genes found in diverse angiosperms. Molecular Biology and Evolution 22: 2084–2089.
29. Pritham EJ (2009) Transposable elements and factors influencing their success in eukaryotes. Journal of Heredity 100: 648–655.
30. Wolfe KH, Gouy ML, Yang YW, Sharp PM, Li WH (1989) Date of the monocot-dicot divergence estimated from chloroplast DNA sequence data. Proc Natl Acad Sci U S A 86: 6201–6205.
31. Gao DY (2012) Identification of an active *Mutator*-like element (MULE) in rice (*Oryza sativa*). Molecular Genetics and Genomics 287: 261–271.
32. Edgar RC (2004) MUSCLE: multiple sequence alignment with high accuracy and high throughput. Nucleic Acids Research 32: 1792–1797.
33. Ma JX, Devos KM, Bennetzen JL (2004) Analyses of LTR-retrotransposon structures reveal recent and rapid genomic DNA loss in rice. Genome Research 14: 860–869.
34. Nesmelova IV, Hackett PB (2010) DDE transposases: structural similarity and diversity. Advanced Drug Delivery Reviews 62: 1187–1195.
35. Zhang Z, Li J, Zhao XQ, Wang J, Wong GK, et al. (2006) KaKs_Calculator: calculating Ka and Ks through model selection and model averaging. Genomics Proteomics Bioinformatics 4: 259–263.
36. Tamura K, Peterson D, Peterson N, Stecher G, Nei M, et al. (2011) MEGA5: molecular evolutionary genetics analysisusing Maximum Likelihood, Evolutionary Distance, and Maximum Parsimony methods. Molecular Biology and Evolution 28: 2731–2739.
37. Kronmiller BA, Wise RP (2008) TEnest: automated chronological annotation and visualization of nested plant transposable elements. Plant Physiology 146: 45–59.
38. Kronmiller BA, Wise RP (2009) Computational finishing of large sequence contigs reveals interspersed nested repeats and gene islands in the *rf1*-associated region of maize. Plant Physiology 151: 483–495.
39. SanMiguel P, Tikhonov A, Jin YK, Motchoulskaia N, Zakharov D, et al. (1996) Nested retrotransposons in the intergenic regions of the maize genome. Science 274: 765–768.
40. Jiang N, Wessler SR (2001) Insertion preference of maize and rice miniature inverted repeat transposable elements as revealed by the analysis of nested elements. Plant Cell 13: 2553–2564.
41. Li YB, Harris L, Dooner HK (2013) *TED*, an autonomous and rare maize transposon of the *Mutator* superfamily with a high gametophytic excision frequency. Plant Cell 25: 3251–3265.
42. Comelli P, Konig J, Werr W (1999) Alternative splicing of two leading exons partitions promoter activity between the coding regions of the maize homeobox gene *Zmhox1a* and *Trap* (transposon-associated protein). Plant Molecular Biology 41: 615–625.
43. SanMiguel P, Gaut B, Tikhonov A, Nakajima Y, Bennetzen J (1998) The paleontology of intergene retrotransposons of maize. Nature 20: 43–45.
44. Benito MI, Walbot V (1997) Characterization of the maize *Mutator* transposable element *MURA* transposase as a DNA-binding protein. Molecular and Cellular Biology 17: 5165–5175.
45. Babu MM, Iyer LM, Balaji S, Aravind L (2006) The natural history of the WRKY-GCM1 zinc fingers and the relationship between transcription factors and transposons. Nucleic Acids Research 34: 6505–6520.
46. Hennig S, Ziebuhr W (2010) Characterization of the transposase encoded by *IS256*, the prototype of a major family of bacterial insertion sequence elements. Journal of Bacteriology 192: 4153–4163.
47. Vitte C, Bennetzen JL (2006) Analysis of retrotransposon structural diversity uncovers properties and propensities in angiosperm genome evolution. Proc Natl Acad Sci U S A 103: 17638–17643.
48. Ilic K, SanMiguel PJ, Bennetzen JL (2003) A complex history of rearrangement in an orthologous region of the maize, sorghum, and rice genomes. Proc Natl Acad Sci U S A 100: 12265–12270.
49. Wendel JF (2000) Genome evolution in polyploids. Plant Molecular Biology 42: 225–249.
50. Bennetzen JL, Ma JX, Devos K (2005) Mechanisms of recent genome size variation in flowering plants. Annals of Botany 95: 127–132.
51. Woodhouse MR, Schnable JC, Pedersen BS, Lyons E, Lisch D, et al. (2010) Following tetraploidy in maize, a short deletion mechanism removed genes preferentially from one of the two homeologs. Plos Biology 8: e1000409. DOI: 10.1371/journal.pbio.1000409.
52. Tian ZX, Rizzon C, Du JC, Zhu LC, Bennetzen JL, et al. (2009) Do genetic recombination and gene density shape the pattern of DNA elimination in rice long terminal repeat retrotransposons? Genome Research 19: 2221–2230.
53. Feschotte C, Pritham EJ (2007) DNA transposons and the evolution of eukaryotic genomes. Annual Review of Genetics 41: 331–368.
54. Huang CRL, Burns KH, Boeke JD (2012) Active transposition in genomes. Annual Review of Genetics 46: 651–675.
55. Vicient CM (2010) Transcriptional activity of transposable elements in maize. BMC Genomics 11: 601. DOI: 10.1186/1471-2164-11-601.
56. Hastings KEM (1996) Strong evolutionary conservation of broadly expressed protein isoforms in the troponin I gene family and other vertebrate gene families. Journal of Molecular Evolution 42: 631–640.

57. Duret L, Mouchiroud D (1999) Expression pattern and, surprisingly, gene length shape codon usage in *Caenorhabditis*, *Drosophila*, *Arabidopsis*. Proc Natl Acad Sci U S A 96: 4482–4487.
58. Duret L, Mouchiroud D (2000) Determinants of substitution rates in mammalian genes: expression pattern affects selection intensity but not mutation rate. Molecular Biology and Evolution 17: 68–74.
59. Davidson RM, Gowda M, Moghe G, Lin HN, Vaillancourt B, et al. (2012) Comparative transcriptomics of three Poaceae species reveals patterns of gene expression evolution. Plant Journal 71: 492–502.
60. Frith MC, Wilming LG, Forrest A, Kawaji H, Tan SL, et al. (2006) Pseudo-messenger RNA: Phantoms of the transcriptome. Plos Genetics 2: 504–514.
61. Zou C, Lehti-Shiu MD, Thibaud-Nissen F, Prakash T, Buell CR, et al. (2009) Evolutionary and expression signatures of pseudogenes in *Arabidopsis* and rice. Plant Physiology 151: 3–15.
62. Jiang N, Panaud O (2013) Transposable element dynamics in rice and its wild relatives. In: Zhang Q, Wing RA, editors. Genetics and Genomics of Rice, Plant Genetics and Genomics: Crops and Models. Springer Science+Business Media New York. DOI: 10.1007/978-1-4614-7903-1_5.
63. Liu SZ, Yeh CT, Ji TM, Ying K, Wu HY, et al. (2009) *Mu* transposon insertion sites and meiotic recombination events co-localize with epigenetic marks for open chromatin across the maize genome. Plos Genetics 5: e1000733. DOI: 10.1371/journal.pgen.1000733.
64. Boutin TS, Le Rouzic A, Capy P (2012) How does selfing affect the dynamics of selfish transposable elements? Mobile DNA 3.
65. Lockton S, Gaut BS (2010) The evolution of transposable elements in natural populations of self-fertilizing *Arabidopsis thaliana* and its outcrossing relative *Arabidopsis lyrata*. BMC Evolutionary Biology 10.
66. Wright SI, Schoen DJ (1999) Transposon dynamics and the breeding system. Genetica 107: 139–148.
67. Wright S, Finnegan D (2001) Genome evolution: Sex and the transposable element. Current Biology 11: R296–R299.
68. Dolgin ES, Charlesworth B, Cutter AD (2008) Population frequencies of transposable elements in selfing and outcrossing *Caenorhabditis nematodes*. Genet Res 90: 317–329.
69. Tam SM, Causse M, Garchery C, Burck H, Mhiri C, et al. (2007) Thedistribution of *copia*-type retrotransposons and the evolutionary history of tomato and related wild species. J Evol Biol 20: 1056–1072.
70. Wright SI, Quang HL, Schoen DJ, Bureau TE (2001) Population dynamics of an *Ac*-like transposable element in self- and cross-pollinating *Arabidopsis*. Genetics 158: 1279–1288.

Citrus sinensis Annotation Project (CAP): A Comprehensive Database for Sweet Orange Genome

Jia Wang[1,2]♮, Dijun Chen[1]♮, Yang Lei[1]♮, Ji-Wei Chang[1], Bao-Hai Hao[1], Feng Xing[1], Sen Li[1], Qiang Xu[3], Xiu-Xin Deng[3], Ling-Ling Chen[1]*

1 Center for Bioinformatics, College of Life Science and Technology, Huazhong Agricultural University, Wuhan, P.R. China, **2** School of Science, Huazhong Agricultural University, Wuhan, P.R. China, **3** Key Laboratory of Horticultural Plant Biology of Ministry of Education, Huazhong Agricultural University, Wuhan, P.R. China

Abstract

Citrus is one of the most important and widely grown fruit crop with global production ranking firstly among all the fruit crops in the world. Sweet orange accounts for more than half of the *Citrus* production both in fresh fruit and processed juice. We have sequenced the draft genome of a double-haploid sweet orange (*C. sinensis* cv. Valencia), and constructed the *Citrus sinensis* annotation project (CAP) to store and visualize the sequenced genomic and transcriptome data. CAP provides GBrowse-based organization of sweet orange genomic data, which integrates *ab initio* gene prediction, EST, RNA-seq and RNA-paired end tag (RNA-PET) evidence-based gene annotation. Furthermore, we provide a user-friendly web interface to show the predicted protein-protein interactions (PPIs) and metabolic pathways in sweet orange. CAP provides comprehensive information beneficial to the researchers of sweet orange and other woody plants, which is freely available at http://citrus.hzau.edu.cn/.

Editor: Meng-xiang Sun, Wuhan University, China

Funding: The present study was supported by the National Basic Research Program of China (2012AA10A304) and the National Natural Science Foundation of China (31271406 and 31222047). The funders had no role in study design, data collection and analysis, decision to publish, or preparation of the manuscript.

Competing Interests: The authors have declared that no competing interests exist.

* E-mail: llchen@mail.hzau.edu.cn

♮ These authors contributed equally to this work.

Introduction

Citrus is one of the most important and widely grown fruit crop in the world, with global production and total acreage ranking firstly among all the fruit crops. *Citrus* is a large genus with more than ten major species. Among them, sweet orange is responsible for about 60% of production for both fresh fruit and processed juice consumption [1]. Besides their economical and nutritional importance, *Citrus* fruits also have unique botanical characteristics such as nucellar embryony [2]. Normal sweet oranges are diploids with nine pair of chromosomes, and the estimated genome size is about 367 Mb [2]. Recently, we sequenced the draft genome of a double-haploid sweet orange (*C. sinensis* cv. Valencia) by using whole genome shotgun approach combined with long paired-end DNA sequencing technology [3]. The double-haploid genome was assembled to 4,811 scaffolds with N50 equal to 1.7 Mb. The total contig length (320.5 Mb) covers about 87% of the sweet orange genome, and scaffolds were aligned and oriented to the *Citrus* linkage map, about 80% of the assembled genome was anchored to nine pseudo-chromosomes [3]. An integrative strategy combining *ab initio* gene prediction, homology search, and experimental evidence including expressed sequence tags (ESTs), RNA-seq and RNA-paired end tags (RNA-PETs) was employed to annotate protein-coding genes in sweet orange genome, finally we obtained 29,445 protein-coding gene loci with 44,387 transcripts [3]. The availability of the sweet orange genome sequence provides a valuable genomic resource for citrus genetics and breeding improvement. To intuitively provide the sweet orange genome sequence and annotation, we constructed *Citrus sinensis* annotation project (CAP), which is a portal site for various types of sweet orange data.

CAP provides an integrative platform for GBrowse-based organization of sweet orange genomic data and links many public databases, which includes overview of the pseudo-chromosomes and scaffolds, gene annotation containing *ab initio* gene prediction, EST, RNA-seq and RNA-PET evidence. Detailed protein coding gene information is provided in a keyword search system including predicted function, homologs in model plants, RNA and protein fold prediction and transcriptome evidence. In addition, we construct a user-friendly web interface to show the predicted protein-protein interactions (PPIs) in sweet orange, and supply metabolic pathways based on the Kyoto Encyclopedia of Genes and Genomes (KEGG) pathways [4]. CAP can provide comprehensive information beneficial to the researchers of sweet orange and other woody plants.

Results and Discussion

Gene annotation

Precise gene prediction is one of the most important goals in genome annotation. We combined *ab initio* gene finding programs

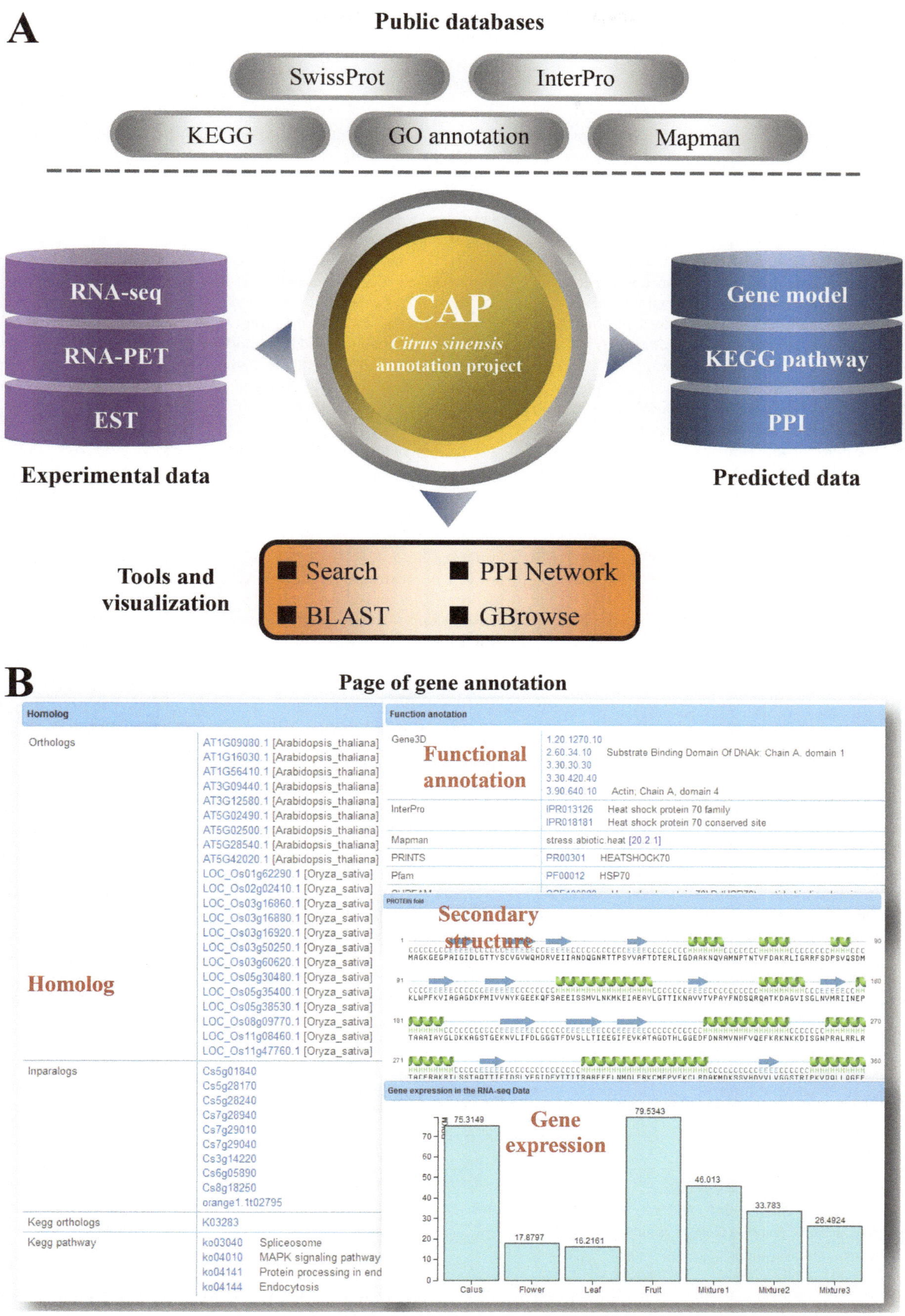
A
Public databases
SwissProt
InterPro
KEGG
GO annotation
Mapman
RNA-seq
RNA-PET
EST
Experimental data
CAP
Citrus sinensis annotation project
Gene model
KEGG pathway
PPI
Predicted data
Tools and visualization
Search
PPI Network
BLAST
GBrowse
B
Page of gene annotation
Homolog
Function anotation
Functional annotation
Secondary structure
Gene expression
Gene expression in the RNA-seq Data
Calus
Flower
Leaf
Fruit
Mixture1
Mixture2
Mixture3

Figure 1. The related public resources of CAP database and its gene annotation. (**A**) The framework and linked public databases in CAP. (**B**) The major gene annotation page in CAP, including homologs, functional information, secondary structure and RNA-seq gene expression values in four tissues (callus, leaf, flower and fruit) and three mixtures of these tissues.

and evidence-based annotation including homology searches, EST, RNA-Seq and RNA-PET experimental evidence to identify protein-coding genes in sweet orange genome. Detailed process is described in [3]. In CAP website, gene annotation page provides convenient searching items for gene information. Users can search the system by gene locus, Gene Ontology (GO) [5], InterPro category [6] or functional information. Comprehensive gene annotation is linked to public resources (Figure 1A). Figure 1B illustrates the detailed gene annotation, including functional information in SwissProt [7], othologs in *Arabidopsis thaliana* and *Oryza sativa*, inparalogs in sweet orange, KEGG orthologs [4], GO [5] and Mapman [8] category, protein fingerprints in PRINTS [9], protein families and domains in Pfam [10] and SUPFAM [11], prediction RNA fold and protein secondary structure, as well as RNA-seq Reads Per Kilo bases per Million (RPKM) values for different sweet orange tissues. For the annotated protein-coding gene models, 93.5% have RNA-seq evidence support, 78.2% is supported with proteins in public non-redundant database, and 66.7% is supported with EST evidence, only a very small fraction of genes are solely predicted with *ab initio* gene-finding programs [3]. Table 1 lists the statistical information for functions of the protein-coding genes. More than 18,000 sweet orange genes have homologs in public databases. Furthermore, more than 26,000 protein-coding genes contain protein family and domain information. Only 4,930 genes have no functional information, which are annotated as "hypothetical proteins or conserved hypothetical proteins" (Table 1).

GBrowse

It is well known that GBrowse is one of the most important genomic viewers for manipulating and displaying annotations on genomes [12], which has been extensively used to construct database for a variety of model organisms, such as Flybase [13], WormBase [14], SGD [15] and SilkDB [16]. CAP provides GBrowse-based integration of sweet orange genome annotation, including *ab initio* gene prediction, EST, RNA-seq and RNA-PET evidence-based gene annotation. Users can easily browse any interested regions in the sweet orange genome. According to the position on a scaffold, users can access a variety of track features, including scaffolds, protein-coding gene models, non-coding RNA, repetitive sequences, *ab initio* gene prediction, general information including GC content, 3-frame or 6-frame translation, RNA-seq and RNA-PET data from four sweet orange tissues (callus, leaf, flower and fruit) and three mixtures of these tissues, and ESTs from sweet orange and other citrus species (Figure 2A). Figure 2B illustrates a protein-coding gene Cs8g01880 in chromosome 8, detailed information includes the final gene model, RNA-seq and RNA-PET data from different tissues, ESTs from sweet orange and other citrus species, and four *ab initio* gene prediction tools, i.e., Genscan [17], GeneID [18], FgeneSH (http://linux1.softberry.com/berry.phtml?topic = fgenesh&group = programs&subgroup = gfind) and GlimmerHMM [19]. Gene mode page in GBrowse is available for each gene, including gene name, position, length, exon and intron position, 5′ and 3′ un-translated region, genomic sequence and transcripts (Figure 2C).

Protein-protein interactions (PPIs)

The sweet orange PPI network is predicted with ortholog-based and domain-combination methods, and then K-nearest neighbors (KNN) method is used to verify and filter the predicted PPIs, the final PPI network contains 124,491 interactions involving 8,195 proteins [20]. The web interface of PPI is constructed with JAVA and hosted on an Apache web server. The gene search page is linked to PPI, users can also submit one or more gene ID numbers to PPI search page, and then the server will return proteins that interact with the query proteins. The query protein and its interaction partners are visualized with Cytoscape software [21]. Figure 3A shows the PPI network of Cs8g02750.1 and Cs4g05680.2, Cs8g02750.1 is a proteasome subunit with 112 interacting partners, and Cs4g05680.2 is a serine/threonine-protein kinase with 236 interacting proteins. The two proteins have common and specific protein interactions. Cs8g02750.1 mainly interacts with other proteasome subunit proteins, while Cs4g05680.2 contains a wide variety of interacting partners, including many protein kinases, ribosomal proteins, 14-3-3 proteins, v-ATPases, tubulin proteins *etc.* In Figure 3A, nodes are colored according to Mapman functional categories [8]. Solid line between two nodes indicates interaction predicted with ortholog-based method, and dash line indicates interaction predicted with domain-combination method. Thickness and color of the solid line denotes different score levels, the higher the orthologous score, the thicker the line is. If a user clicks a node in the PPI network, its

Table 1. Statistics of functional information for protein-coding genes in sweet orange.

Category	Description	Number
I	High similarity to known proteins in SwissProt (identity >90%) (identity >50%)intellectual (identity >90%)	524
II	Medium similarity to known proteins in SwissProt (identity >50%)	10,613
III	Low similarity to known proteins (identity >30%)	18,368
IV	InterPro domain-containing protein	26,916
V	Conserved hypothetical or hypothetical protein	4,930

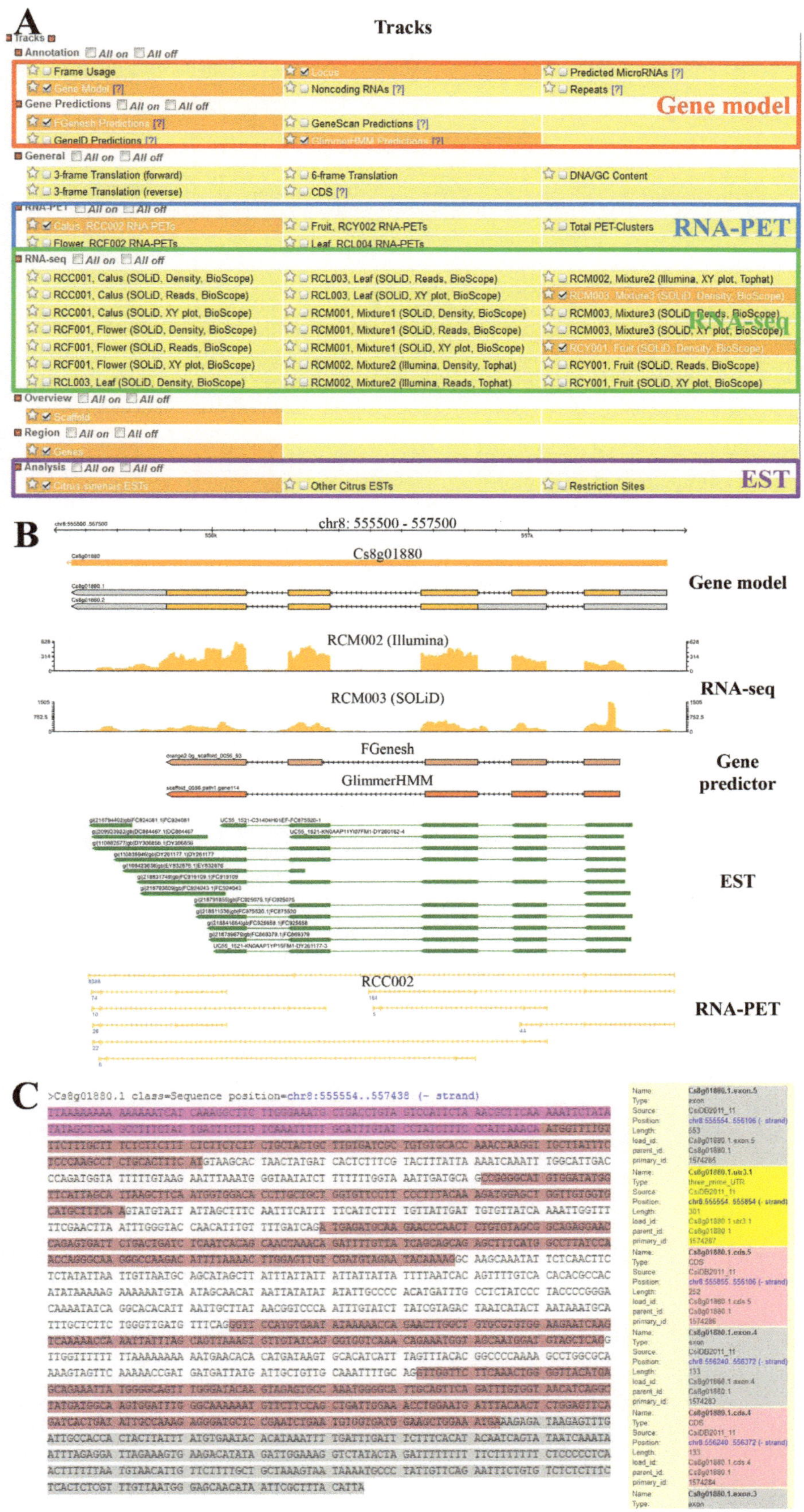
A
Tracks
Annotation All on All off
Frame Usage
Locus
Predicted MicroRNAs [?]
Gene Model [?]
Noncoding RNAs [?]
Repeats [?]
Gene model
Gene Predictions All on All off
FGenesh Predictions [?]
GeneScan Predictions [?]
GeneID Predictions [?]
GlimmerHMM Predictions [?]
General All on All off
3-frame Translation (forward)
6-frame Translation
DNA/GC Content
3-frame Translation (reverse)
CDS [?]
RNA-PET All on All off
Calus, RCC002 RNA-PETs
Fruit, RCY002 RNA-PETs
Total PET-Clusters
RNA-PET
Flower, RCF002 RNA-PETs
Leaf, RCL004 RNA-PETs
RNA-seq All on All off
RCC001, Calus (SOLiD, Density, BioScope)
RCL003, Leaf (SOLiD, Reads, BioScope)
RCM002, Mixture2 (Illumina, XY plot, Tophat)
RCC001, Calus (SOLiD, Reads, BioScope)
RCL003, Leaf (SOLiD, XY plot, BioScope)
RCM003, Mixture3 (SOLiD, Density, BioScope)
RCC001, Calus (SOLiD, XY plot, BioScope)
RCM001, Mixture1 (SOLiD, Density, BioScope)
RCM003, Mixture3 (SOLiD, Reads, BioScope)
RCF001, Flower (SOLiD, Density, BioScope)
RCM001, Mixture1 (SOLiD, Reads, BioScope)
RCM003, Mixture3 (SOLiD, XY plot, BioScope)
RCF001, Flower (SOLiD, Reads, BioScope)
RCM001, Mixture1 (SOLiD, XY plot, BioScope)
RCY001, Fruit (SOLiD, Density, BioScope)
RCF001, Flower (SOLiD, XY plot, BioScope)
RCM002, Mixture2 (Illumina, Density, Tophat)
RCY001, Fruit (SOLiD, Reads, BioScope)
RCL003, Leaf (SOLiD, Density, BioScope)
RCM002, Mixture2 (Illumina, Reads, Tophat)
RCY001, Fruit (SOLiD, XY plot, BioScope)
RNA-seq
Overview All on All off
Scaffold
Region All on All off
Genes
Analysis All on All off
Citrus sinensis ESTs
Other Citrus ESTs
Restriction Sites
EST
B
chr8: 555500 - 557500
Cs8g01880
Gene model
RCM002 (Illumina)
RNA-seq
RCM003 (SOLiD)
FGenesh
Gene predictor
GlimmerHMM
EST
RCC002
RNA-PET
C
>Cs8g01880.1 class=Sequence position=chr8:555554..557438 (- strand)

Figure 2. GBrowse in CAP. (A) GBrowse tracks in CAP. The tracks include general overview, gene model, RNA-seq, RNA-PET and EST evidence. **(B)** Graphic example of Cs8g1880 gene annotation in GBrowse. Gene model shows the exon-intron structure of the gene. Gene predictor shows the prediction results of some *ab initio* gene-finding programs. RNA-seq, RNA-PET and EST are the experimental evidence to support the gene model. **(C)** Text file of Cs8g1880 gene model in GBrowse.

Mapman annotation, functional information and expression value will be shown (Figure 3A).

Metabolic pathways

KEGG pathway maps are graphical diagrams representing knowledge of reaction networks for metabolism, and each map summarizes experimental evidence in published literatures [4]. Based on KEGG Orthology (KO) groups, we obtained the KEGG orthologous genes in sweet orange genome, and generated the sweet orange metabolic pathways. KEGG modules in each pathway map are produced by converting nodes to gene identifier nodes and colored in blue. Sweet orange pathways include four categories, *i.e.*, metabolism, genetic information processing, environmental information processing and cellular processes. Each category contains many pathways. When a user clicks a pathway, the reference KEGG pathway will be shown, and enzymes or proteins which have KEGG orthologs in sweet orange are colored in blue. Figure 3B shows the metabolic pathway of citrate cycle (TCA cycle). When mouse moves to an enzyme with blue color, orthologs in sweet orange and their expression values in different tissues are shown, which are also linked to the corresponding gene annotation.

Search modules

CAP provides various query interface and graphical visualization to facilitate the retrieve and demonstration of sweet orange data. As mentioned above, "gene search" is the principal search system, which allows users to enter keywords such as gene locus, GO [5] or InterPro category [6], and functional information. The retrieving result links to GBrowse and PPI. Users can also submit the gene locus to search its protein interaction in the PPI menu. In addition, users can perform a BLAST sequence search to retrieve homologous sequences in sweet orange genome. BLAST search results include graphical summary of the sequence alignment, briefly and detailed description of the alignment. All the search results performed by the above search modules can be further used for functional investigation.

Conclusions

The present work provides a comprehensive collection of sweet orange genomic and transcriptomic data, which are organized and deposited in an online database CAP. Convenient web interface is designed to show gene annotation, protein interaction and metabolic pathway. CAP serves the plant research community by providing a reference genome and annotation for sweet orange.

In the near future, CAP will collect the experimentally validated data for sweet orange genes. In addition, small RNA and degradome sequencing data will be added to CAP. New high-throughput DNA-sequencing technologies are being developed and it is expected that the number of *Citrus* species sequences will grow rapidly. These new sequences will be incorporated into the CAP by comparison to the *C. sinensis* reference genome in the future. With the update of sweet orange genome annotation, CAP will update to new version.

Methods

Data source and website architecture

The genomic data for sweet orange has been submitted to NCBI GenBank under the accession number AJPS00000000 and BioProject ID PRJNA86123. The raw data for sweet orange genome sequencing, assembling and annotation are available from sweet orange annotation project [3]. All the data are organized and stored in MySQL database (http://www.mysql.com/). Besides, the sequence information and functional annotation for protein-coding genes are provided in CAP. A genome browser is developed on the basis of GBrowse [12]. CAP is implemented in JSP language and deployed on Apache Tomcat web server (http://tomcat.apache.org/). The integrated network browser is created by Cytoscape web program (http://cytoscapeweb.cytoscape.org/) [21]. The architecture and linked public databases are shown in Figure 1A. CAP can be accessed through IE 6.0 or higher, Netscape 7.0 or higher, Safari, Opera, Chrome and Firefox from multiple platforms. JavaScript is required to use all the functions of CAP.

Gene annotation and linked databases

SwissProt homologs are obtained by using BLASTP based on bi-directional best hit (BBH) method to search against UniProtKB/SwissProt [7]. Thresholds for BLASTP search are sequence coverage >0.7, identity >30%, e value <1e-10 and bit-score >60. Pfam category is predicted by using hmmer program [22]. Mapman annotation is obtained using BLASTP based on BBH method between *A. thaliana* and sweet orange genes. Gene3D, InterPro, PRINTS and SUPFAM annotation is predicted with Interproscan program [6]. RNA secondary structure is predicted with RNAfold program in ViennaRNA [23], and protein secondary structure is predicted with Psipred program [24].

Gene annotation in CAP is linked to many public databases. For example, Orthologs in *A. thaliana* and *O. sativa* are linked to gene model in TAIR [25] and MSU rice gene models (http://rice.plantbiology.msu.edu/), respectively. GO annotation links to gene ontology in EMBL database (http://www.embl.org/), Gene3D links to the corresponding CATH Superfamily (http://www.cathdb.info/), InterPro links to EMBL database (http://www.cathdb.info/), PRINTS links to SPRINT database (http://www.bioinf.manchester.ac.uk/dbbrowser/sprint/), Pfam links to corresponding Pfam category (http://pfam.sanger.ac.uk/), and SUPFAM links to superfamily database (http://supfam.org/).

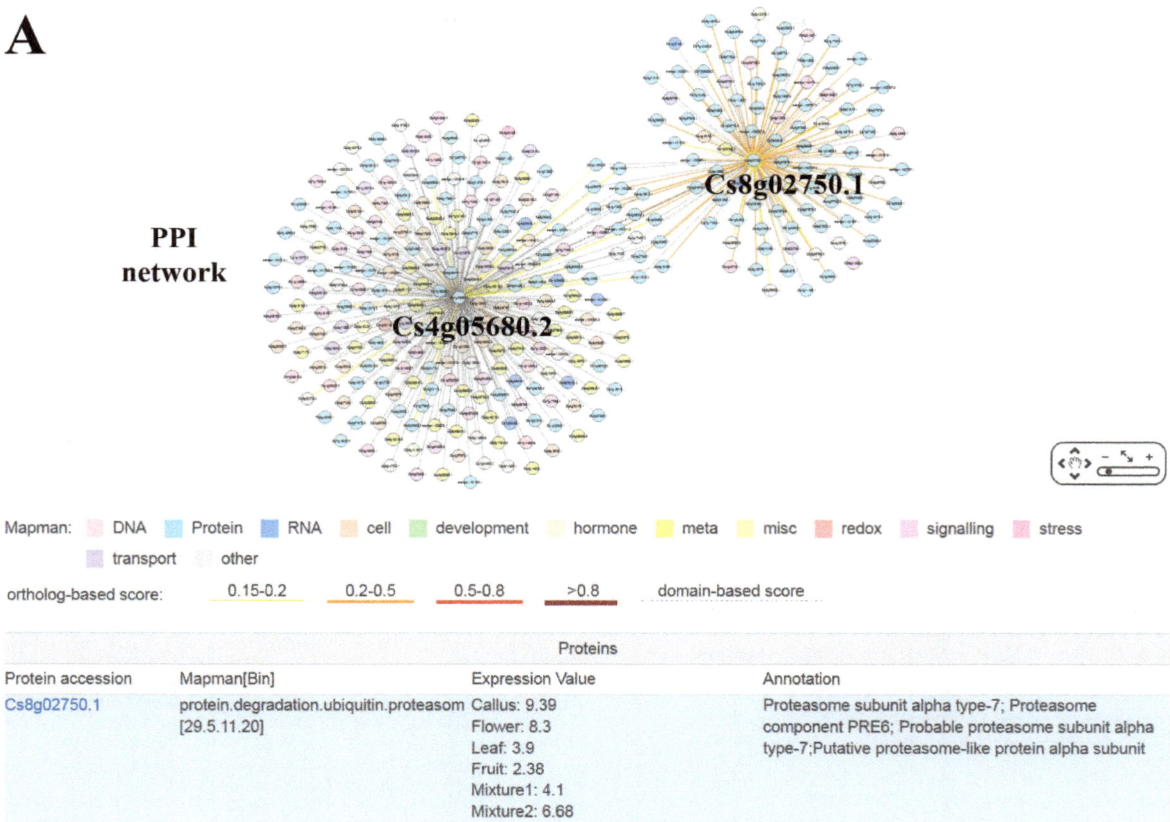
A
PPI network
Cs8g02750.1
Cs4g05680.2
Mapman: DNA Protein RNA cell development hormone meta misc redox signalling stress transport other
ortholog-based score: 0.15-0.2 0.2-0.5 0.5-0.8 >0.8 domain-based score
Proteins
Protein accession | Mapman[Bin] | Expression Value | Annotation
Cs8g02750.1 | protein.degradation.ubiquitin.proteasom [29.5.11.20] | Callus: 9.39 Flower: 8.3 Leaf: 3.9 Fruit: 2.38 Mixture1: 4.1 Mixture2: 6.68 | Proteasome subunit alpha type-7; Proteasome component PRE6; Probable proteasome subunit alpha type-7;Putative proteasome-like protein alpha subunit

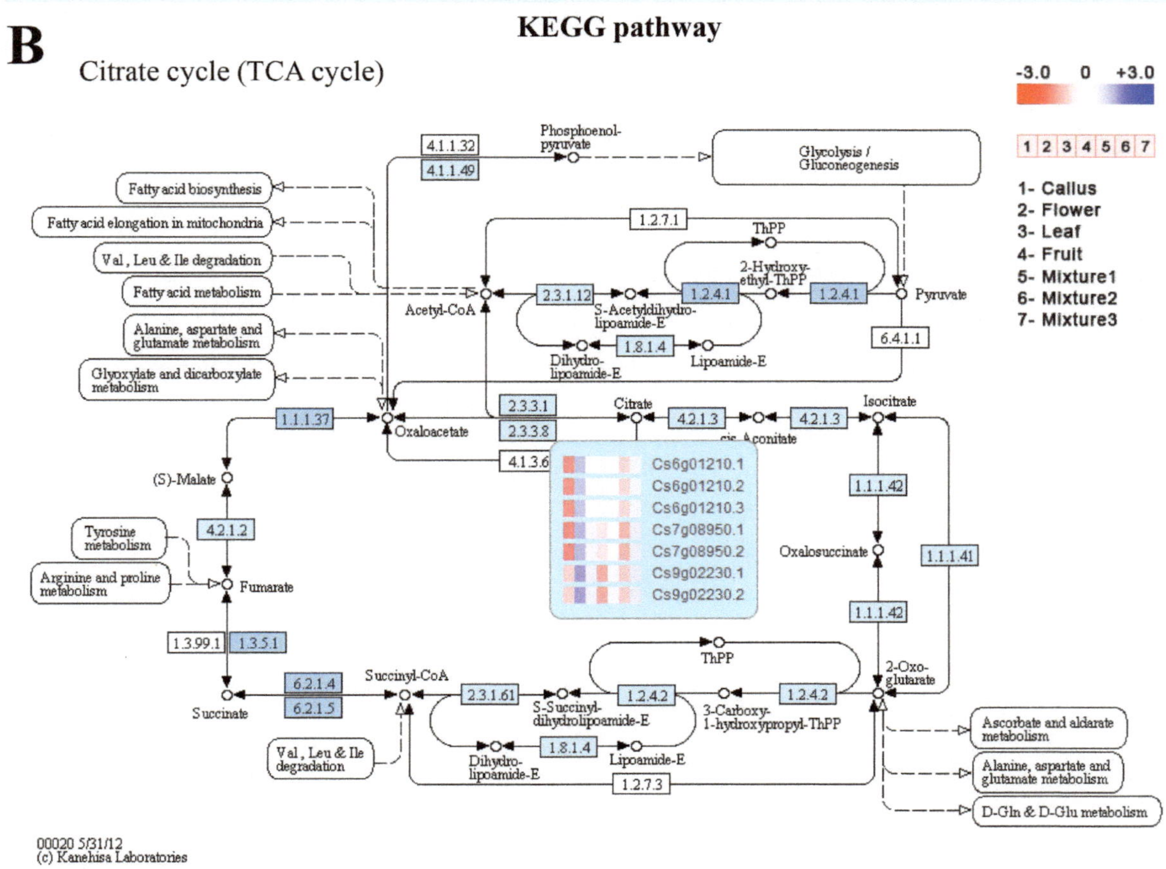
B
KEGG pathway
Citrate cycle (TCA cycle)
-3.0 0 +3.0
1 2 3 4 5 6 7
1- Callus
2- Flower
3- Leaf
4- Fruit
5- Mixture1
6- Mixture2
7- Mixture3
Phosphoenol-pyruvate
Glycolysis / Gluconeogenesis
4.1.1.32
4.1.1.49
Fatty acid biosynthesis
Fatty acid elongation in mitochondria
Val, Leu & Ile degradation
Fatty acid metabolism
Alanine, aspartate and glutamate metabolism
Glyoxylate and dicarboxylate metabolism
1.2.7.1
ThPP
2-Hydroxy-ethyl-ThPP
Acetyl-CoA
2.3.1.12
S-Acetyldihydro-lipoamide-E
1.2.4.1
1.2.4.1
Pyruvate
6.4.1.1
Dihydro-lipoamide-E
1.8.1.4
Lipoamide-E
2.3.3.1
2.3.3.8
Citrate
4.2.1.3
4.2.1.3
Isocitrate
cis-Aconitate
1.1.1.37
Oxaloacetate
4.1.3.6
Cs6g01210.1
Cs6g01210.2
Cs6g01210.3
Cs7g08950.1
Cs7g08950.2
Cs9g02230.1
Cs9g02230.2
(S)-Malate
4.2.1.2
Tyrosine metabolism
Arginine and proline metabolism
Fumarate
1.1.1.42
Oxalosuccinate
1.1.1.41
1.1.1.42
1.3.99.1
1.3.5.1
ThPP
2-Oxo-glutarate
Succinyl-CoA
6.2.1.4
6.2.1.5
Succinate
2.3.1.61
S-Succinyl-dihydrolipoamide-E
1.2.4.2
1.2.4.2
3-Carboxy-1-hydroxypropyl-ThPP
Val, Leu & Ile degradation
Dihydro-lipoamide-E
1.8.1.4
Lipoamide-E
1.2.7.3
Ascorbate and aldarate metabolism
Alanine, aspartate and glutamate metabolism
D-Gln & D-Glu metabolism
00020 5/31/12
(c) Kanehisa Laboratories

Figure 3. The predicted protein interactions and KEGG pathway in CAP. (**A**) The predicted PPI network of Cs8g02750.1 and Cs4g05680.2. (**B**) Citrate cycle (TCA cycle) metabolic pathway in sweet orange.

Author Contributions

Conceived and designed the experiments: LLC DC. Performed the experiments: JW YL DC. Analyzed the data: JWC BHH FX SL. Contributed reagents/materials/analysis tools: QX XXD. Wrote the paper: LLC DC.

References

1. FAO (2011) Food and Agricultural Organization of the United Nations: FAOSTAT [online]. Available: http://faostat.fao.org/default.aspx.
2. Arumuganathan K, Earle ED (1991) Nuclear DNA content of some important plant species. Plant Mol Biol Rep 9: 208–218.
3. Xu Q, Chen LL, Ruan X, Chen D, Zhu A, et al. (2013) The draft genome of sweet orange (*Citrus sinensis*). *Nat Genet* 45: 59–66.
4. Kanehisa M, Goto S, Sato Y, Furumichi M, Tanabe M (2012) KEGG for integration and interpretation of large-scale molecular data sets. Nucleic Acids Res 40: D109–114.
5. Harris MA, Clark J, Ireland A, Lomax J, Ashburner M, et al. (2004) The Gene Ontology (GO) database and informatics resource. Nucleic Acids Res 32: D258–261.
6. Zdobnov EM, Apweiler R (2001) InterProScan–an integration platform for the signature-recognition methods in InterPro. Bioinformatics 17: 847–848.
7. Boeckmann B, Bairoch A, Apweiler R, Blatter MC, Estreicher A, et al. (2003) The SWISS-PROT protein knowledgebase and its supplement TrEMBL in 2003. Nucleic Acids Res 31: 365–370.
8. Thimm O, Blasing O, Gibon Y, Nagel A, Meyer S, et al. (2004) MAPMAN: a user-driven tool to display genomics data sets onto diagrams of metabolic pathways and other biological processes. Plant J 37: 914–939.
9. Attwood TK, Coletta A, Muirhead G, Pavlopoulou A, Philippou PB, et al. (2012) The PRINTS database: a fine-grained protein sequence annotation and analysis resource - its status in 2012. Database (Oxford) 2012: bas019.
10. Punta M, Coggill PC, Eberhardt RY, Mistry J, Tate J, et al. (2012) The Pfam protein families database. Nucleic Acids Res 40: D290–301.
11. Wilson D, Pethica R, Zhou Y, Talbot C, Vogel C, et al. (2009) SUPERFAMILY – sophisticated comparative genomics, data mining, visualization and phylogeny. Nucleic Acids Res 37: D380–386.
12. Stein LD (2013) Using GBrowse 2.0 to visualize and share next-generation sequence data. Brief Bioinform 14: 162–171.
13. Tweedie S, Ashburner M, Falls K, Leyland P, McQuilton P, et al. (2009) FlyBase: enhancing *Drosophila* gene ontology annotations. Nucleic Acids Res 37: D555–559.
14. Rogers A, Antoshechkin I, Bieri T, Blasiar D, Bastiani C, et al. (2008) WormBase 2007. Nucleic Acids Res 36: D612–617.
15. Hong EL, Balakrishnan R, Dong Q, Christie KR, Park J, et al. (2008) Gene Ontology annotations at SGD: new data sources and annotation methods. Nucleic Acids Res 36: D577–581.
16. Duan J, Li R, Cheng D, Fan W, Zha X, et al. (2010) SilkDB v2.0: a platform for silkworm (*Bombyx mori*) genome biology. Nucleic Acids Res 38: D453–456.
17. Burge C, Karlin S (1997) Prediction of complete gene structures in human genomic DNA. J Mol Biol 268: 78–94.
18. Blanco E, Abril JF (2009) Computational gene annotation in new genome assemblies using GeneID. Methods Mol Biol 537: 243–261.
19. Majoros WH, Pertea M, Salzberg SL (2004) TigrScan and GlimmerHMM: two open source *ab initio* eukaryotic gene-finders. Bioinformatics 20: 2878–2879.
20. Ding YD, Chang JW, Guo J, Chen DJ, Li S, et al. (2014) Prediction and functional analysis of sweet orange protein-protein interaction network. Submitted.
21. Lopes CT, Franz M, Kazi F, Donaldson SL, Morris Q, et al. (2010) Cytoscape web: an interactive web-based network browser. *Bioinformatics* 26: 2347–2348.
22. Finn RD, Clements J, Eddy SR (2011) HMMER web server: interactive sequence similarity searching. Nucleic Acids Res 39: W29–37.
23. Lorenz R, Bernhart SH, Höner Zu Siederdissen C, Tafer H, Flamm C, et al. (2011) ViennaRNA Package 2.0. Algorithms Mol Biol 6: 26.
24. Buchan DW, Minneci F, Nugent TC, Bryson K, Jones DT (2013) Scalable web services for the PSIPRED protein analysis workbench. Nucleic Acids Res 41: W349–357.
25. Lamesch P, Berardini TZ, Li D, Swarbreck D, Wilks C, et al. (2012) The Arabidopsis Information Resource (TAIR): improved gene annotation and new tools. Nucleic Acids Res 40: D1202–1210.

16

B Chromosomes of *Aegilops speltoides* Are Enriched in Organelle Genome-Derived Sequences

Alevtina Ruban[1,2], Jörg Fuchs[2], André Marques[3], Veit Schubert[2], Alexander Soloviev[1], Olga Raskina[4], Ekaterina Badaeva[5], Andreas Houben[2]*

1 Russian State Agrarian University – Moscow Timiryazev Agricultural Academy, Department of Genetics and Biotechnology, Moscow, Russia, **2** Leibniz-Institute of Plant Genetics and Crop Plant Research (IPK), Chromosome Structure and Function Laboratory, Gatersleben, Germany, **3** Laboratory of Plant Cytogenetics and Evolution, Department of Botany, Universidade Federal de Pernambuco, Recife, Brazil, **4** Institute of Evolution, University of Haifa, Laboratory of Plant Molecular Cytogenetics, Haifa, Israel, **5** Engelhardt Institute of Molecular Biology Russian Academy of Sciences, Laboratory of Molecular Karyology, Moscow, Russia

Abstract

B chromosomes (Bs) are dispensable components of the genome exhibiting non-Mendelian inheritance. Chromosome counts and flow cytometric analysis of the grass species *Aegilops speltoides* revealed a tissue-type specific distribution of the roughly 570 Mbp large B chromosomes. To address the question whether organelle-to-nucleus DNA transfer is a mechanism that drives the evolution of Bs, *in situ* hybridization was performed with labelled organellar DNA. The observed B-specific accumulation of chloroplast- and mitochondria-derived sequences suggests a reduced selection against the insertion of organellar DNA in supernumerary chromosomes. The distribution of B-localised organellar-derived sequences and other sequences differs between genotypes of different geographical origins.

Editor: Khalil Kashkush, Ben-Gurion University, Israel

Funding: This work was supported by the DFG Germany (HO 1779/14-1, http://www.dfg.de/). The funders had no role in study design, data collection and analysis, decision to publish, or preparation of the manuscript.

Competing Interests: The authors have read the journal's policy and have the following conflicts: Andreas Houben also acts as an Academic Editor for PLOS ONE. There are no patents, products in development or marketed products to declare. This does not alter their adherence to all the PLoS ONE policies on sharing data and materials.

* E-mail: houben@ipk-gatersleben.de

Introduction

B chromosomes (Bs) are optional additions to the basic set of standard A chromosomes (As), and occur in all eukaryotic groups. They differ from the As in inheritance, and Bs are not required for normal growth and development of the host organism. Due to the dispensable nature of Bs, they can be present or absent among individuals of the same population in a species. It is widely accepted that B chromosomes derived from the A chromosomes and/or from sex chromosomes. However, there is also evidence suggesting that Bs can be spontaneously generated in response to the new genome conditions following interspecific hybridisation (for reviews, see [1–3]).

Sequence characterisation of the B chromosome of rye provided a unique opportunity for the analysis of the origin and evolution of this enigmatic genome component [4,5]. In contrast to the prevalent view that Bs do not harbor genic sequences, analyses showed that rye Bs are rich in partly transcriptionally active gene-derived sequences [6]. This allowed us to trace their origin from duplicated fragments of multiple A chromosomes [4]. Beside the amplification of B-located satellite repeats [5], this selfish chromosome has accumulated significantly greater amounts of chloroplast- and of mitochondrion-derived sequences than the A chromosomes [4]. Almost all parts of the chloroplast and mitochondrial genomes are found on the Bs, indicating that all sequences are transferable. The higher amount of organelle-derived DNA inserts in B than in A chromosomes and an increased mutation frequency of B-located organellar DNA suggests a reduced selection against the insertion of organellar DNA in supernumerary chromosomes. To study whether the in rye observed B-specific accumulation of organelle-derived DNA can also be found in other species we analysed the B chromosomes of the grass *Aegilops speltoides*.

Aegilops speltoides Tausch (syn. *Triticum speltoides* (Tausch) Gren.) is an annual diploid species (2n = 2x = 14, genome type: S) which belongs to section *Sitopsis* (Triticeae, Poaceae). This species is mainly distributed in the Fertile Crescent and also occurs in south-eastern part of the Balkan peninsula [7]. The submetacentric Bs of *Ae. speltoides* are about 2/3 of the average length of the A chromosomes [8]. The Bs of this species are absent in the roots but stably present in the aerial tissue of the same individual [9], and a maximum number of eight Bs per cell was reported [10]. The maintenance of Bs in natural populations is possible by their transmission at higher than Mendelian frequencies. Accumulation of Bs occurs due to the non-disjunction of B chromatids and the preferential transport of both B chromatids into the generative nucleus during the first mitosis in the male gametophyte [9]. *Aegilops* Bs probably originated from the standard set of As as a consequence of interspecific hybridisation and/or chromosomal rearrangement events. Proposed potential donors of the Bs are the A chromosomes 1, 4 and 5 of the *Ae. speltoides* genome [11,12]. Consequently, the Bs are also characterized by a number of A chromosome-localised repeats like *Spelt1*, pSc119.2 tandem repeats, 5S rDNA and *Ty3-gypsy* retroelements [11–14].

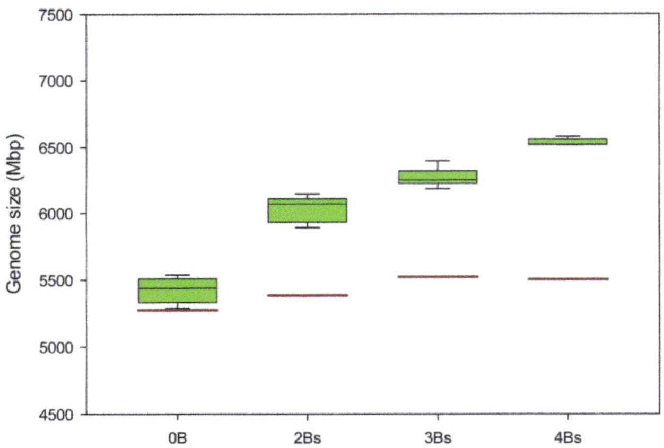

Figure 1. Size determination of the B chromosome. Boxplot representing the genome size distribution in plants without and with B chromosomes of *Ae. speltoides* from Katzir (Israel) measured by flow cytometry. In plants without Bs no difference was found between nuclei isolated from leaf (green boxes) and root (brown line) tissues, while it differed remarkably in plants with Bs. The box boundaries indicate the 75th and 25th percentiles and the error bars the 90th and 10th percentiles of the five replicates.

Materials and Methods

Plant material and plant cultivation

Plants from natural populations of *Aegilops speltoides* ssp. *aucheri* (Boiss.) Chennav. from Katzir, Israel (TS 89 Mediterranean, 233–250 m, 32829′N, 35805′E), Technion, Haifa, Israel (2.36 Mediterranean, 224 m, 32846′N, 35800′E), Ramat Hanadiv, Israel, (2.46 Mediterranean, 100–125 m, 32833′N, 34856′E) and Tartus, Syria (PI 487238 Mediterranean, 600 m, 35807′N, 36807′E) were cultivated first under greenhouse conditions (16 h light, 22°C day/16°C night) and were further cultivated in a garden (Gatersleben, Germany) under natural condition. The plant material was provided by USDA, Aegean Agricultural Research Institute (Turkey) and the Institute of Evolution, Haifa (Israel).

Flow cytometric genome size determination

The genome size was determined using a CyFlow Space (Partec) or a FACSAria Flow Cytometer (BD Biosciences) following the protocol of Dolezel et al. [15]. The analysis is based on the mean of five independent measurements of individuals with known number of Bs. For all measurements, *Secale cereale* L. (Genebank Gatersleben, accession number R 737) with an estimated genome size of 16.19 pg/2C was taken as internal reference standard.

Chromosome preparations derived from shoot meristems

Young secondary shoots before stem elongation were used for chromosome preparation. After an ice-water treatment of 24 hours the material was fixed in 3:1 (ethanol:acetic acid) and stored at 4°C until use from 4 days to several weeks. The fixative was changed after first hour of fixation and then each few weeks during storage. The basal parts of shoots were excised and washed first in ice-cold water and next twice in 1× citric buffer. Meristematic tissue was treated with an enzyme mixture (0.7% cellulase R10, 0.7% cellulase, 1.0% pectolyase, and 1.0% cytohelicase in 1× citric buffer) for 50 min at 37°C. Afterwards material was washed in 1× citric buffer and twice in ice-cold water. The shoot meristem was fragmented in 7 µl of 60% freshly prepared acetic acid into smaller pieces with the help of a needle on a slide. After, another 7 µl of 60% acetic acid was added the specimen was kept for 2 min at room temperature. Next, a homogenization step was performed with an additional 7 µl 60% acetic acid and the slide was placed on a 50°C hot plate for 2 min. The material was spread by hovering a needle over the drop without touching the hot slide. After spreading of cells, the drop was surrounded by 200 µl of ice-cold, freshly prepared 3:1 fixative. More fixative was added and the slide was shortly washed in fixative, then dipped in 60% acetic acid for 10 min and rinsed 5 times in 96% ethanol. A quality check of the air dried slides was performed by phase contrast microscopy. The slides were stored until use in 96% ethanol at 4°C.

Fluorescence *in situ* hybridisation (FISH) and microscopy

The following probes were used: barley BACs encoding chloroplast DNA (BAC clone ChHB 100G01) and mitochondrial DNA (BAC clone MnHB 0205G01) [4,5] and the *Arabidopsis*-type telomere repeat [16]. Amplicons for the 5S ribosomal DNA, which include the coding as well as the flanking spacer region, were generated by PCR as described by Fukui et al. [17]. The *Spelt-1* tandem repeat probe was generated by PCR according to Salina et al. [18]. FISH probes were labelled with ChromaTide Texas Red-12-dUTP or Alexa Fluor 488-5-dUTP (http://www.invitrogen.com) by nick translation. FISH was performed as described by Ma et al. [16]. Fluorescence signals were observed by standard epifluorescence microscopy or to achieve a resolution of ~100 nm by structure illumination microscopy (SIM) using an Elyra super resolution microscope (Zeiss). All images were collected in grey scale and pseudocoloured.

Results and Discussion

Tissue type-specific distribution of B chromosomes

Since B chromosomes of *Ae. speltoides* are known to possess 5S rDNA sites, plants carrying B chromosomes were screened based on additional hybridisation signals after FISH on interphase nuclei of leaf tissue using a corresponding hybridisation probe. This allowed an unambiguous identification of B-positive plants at an early stage of development as the Bs of this species are absent in roots (Figure S1). The tissue-type specific distribution pattern of Bs was confirmed by flow cytometric analysis of 0B, 2B, 3B and 4B plants from the Katzir (Israel) population (Table S1). The average size of a single unreplicated B chromosome was estimated to be ~570 Mbp, ranging from 550 to 600 Mbp, for (1C). Hence, one B equals ~10% of the genome size of a 0B plants (5,400 Mbp). The genome size of nuclei isolated from roots was similar in all analysed plants, while it increased in leaf nuclei depending on the number of Bs (Figure 1). An analysis of Bs in differentiated tissues has not been reported before. A comparable situation of tissue-type specific B chromosome distribution is also known for species as, *Agropyron cristatum* [19], *Poa alpina* [20], or *Aegilops mutica* [21]. The absence of Bs in some organs or tissues could be caused by a specific elimination process of Bs during an early stage of embryo differentiation.

The Bs are polymorphic and enriched in organellar-derived DNA sequences

To determine whether the Bs of *Ae. speltoides* from Katzir accumulated mitochondria- and plastid-derived sequences as demonstrated for the Bs of rye [4], we hybridised labelled BACs encoding barley plastid and mitochondrial DNA with *Ae. speltoides*

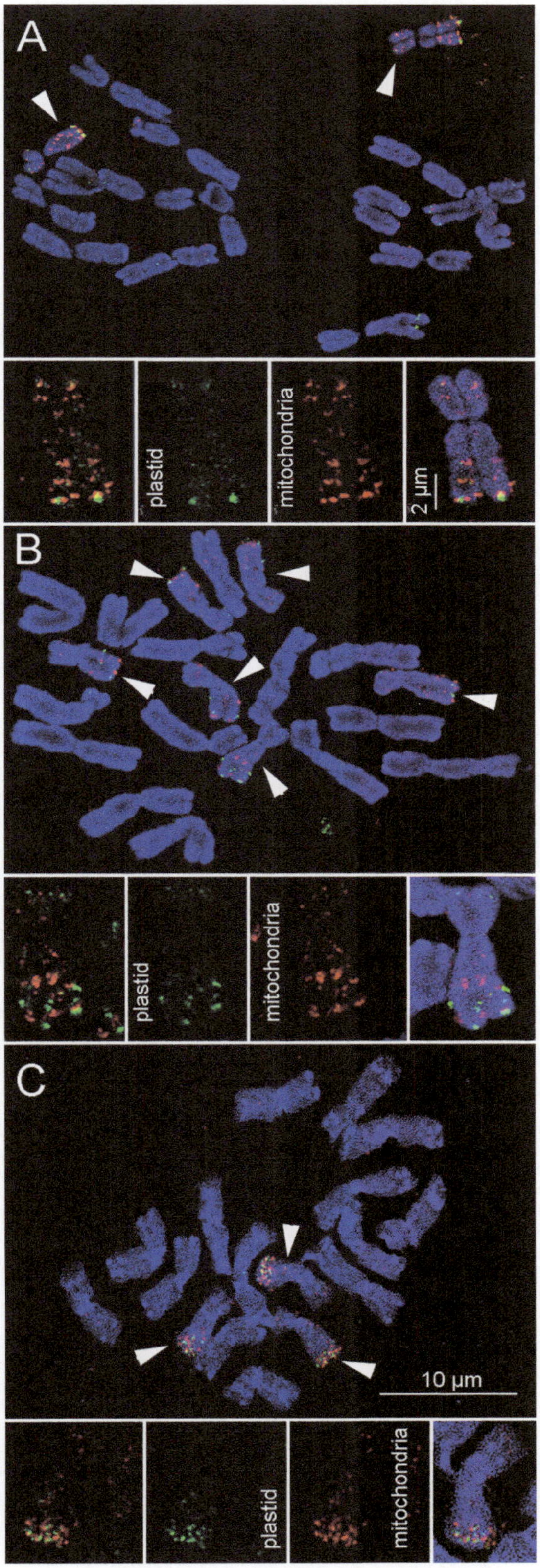

Figure 2. B chromosomes of *Aegilops speltoides* are enriched in organelle genome-derived sequences. Distinct chromosomal localisation of plastid (in green) and mitochondria (in red) derived sequences in three *Ae. speltoides* populations after applying SIM. (A) Somatic metaphase cells with 2 Bs from Katzir (Israel), (B) with 6 Bs from Tartus (Syria), and (C) with 3 Bs from the Technion (Israel) population. Note the geographical origin-dependent hybridisation patters along the Bs (arrowed).

mitotic chromosomes. Both types of probes revealed B-enriched hybridisation signals (Figure 2A). Multiple mitochondria-derived DNA insertions were found along both arms of the B chromosomes except at the pericentromere. The global distribution of plastid DNA-specific signals was similar but of less amount, likely due to a lower abundance. The A chromosomes revealed only minor insertions of mitochondrial origin mainly (Figure S2).

To address whether Bs present in geographically distinct population of *Ae. speltodies* are polymorphic we analysed the chromosomal distribution patterns of organellar DNA in Bs derived from different accessions. As B chromosomes are dispensable, it is expected to observe polymorphisms among populations. The distribution of B-localised organellar-derived signals differed between the tested genotypes. Bs from Tartus showed less organellar-derived signals, although exhibiting a similar distribution as for the Bs from Katzir (Figure 2B). In contrast, Bs from the Technion population displayed a strong accumulation in the subtelomeric region of the long arm for both types of probes (Figure 2C). A partial colocalisation of both mitochondria and plastid DNA-specific signals was found in all Bs independently of their geographical origin. In contrast, a comparative analysis of rye Bs coming from geographically distinct populations did not identify polymorphisms regarding the distribution of accumulated organellar DNA and different B-specific repeats [22]. We conclude that the Bs of *Ae. speltoides* accumulated a significant amount of organellar DNA and that differences of the molecular composition exist between the Bs of different geographical origin. These differences between populations are also reflected by the distribution of FISH signals in interphase nuclei (Figure S3).

The observed polymorphism was further tested by applying the *Triticeae*-specific tandem repeat *Spelt-1*, which is a dynamic component of the *Ae. speltoides* genome [11,18], as well as of the 5S rDNA. Bs of the Technion and Katzir population revealed 5S rDNA signals localised in the terminal part of short arm and in the long arm near the *Spelt-1* cluster. In some plants of the Katzir population the 5S rDNA locus in the short arm is absent (Figure 3). *Ae. speltoides* Bs are submetacentric, but in one plant from the Katzir population we found a metacentric type of Bs. Likely due to terminal deletion in the long arm, former interstitial 5S rDNA and *Spelt-1* clusters became terminal (Figure 3). Apparently, the repeat clusters of the B are hot spots of chromosomal rearrangements. Comparable intraspecific differences in the patterns of repeat clusters were reported previously for the A chromosomes of *Ae.*

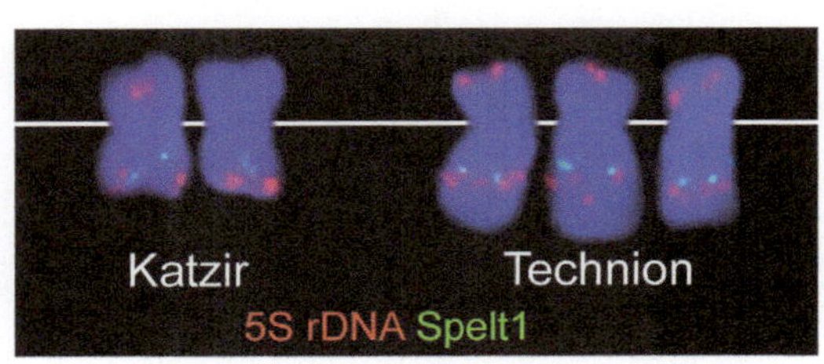

Figure 3. Localisation of 5S rDNA and *Spelt-1* tandem repeat along selected Bs. Deficient-B type from the Katzir population and standard Bs from the Technion population.

speltoides [13]. The localisation of *Arabidopsis*-type telomeric repeats is identical on the A- and B-chromosomes of all populations analysed (Figure S4).

What mechanism could account for the accumulation of organellar DNA on the Bs of *Ae. speltoides*? The first possibility is that insertion into B chromosome DNA has fewer deleterious genetic consequences than their counterparts on As. Insertions into A chromosomes may disrupt gene function with lethal consequences. In contrast, Bs which are not required for growth and development can tolerate more mutations. The second possibility is that the nuclear integration of organellar sequences may be dependent on the formation of double strand breaks and if Bs are particularly prone to double strand breaks, this could facilitate the preferred integration of promiscuous DNA in Bs.

Plastid DNA fragments are very numerous in some tissues such as the developing pollen gametophyte [23] or after stress [24]. Hence, uncontrolled insertion would be expected to result in the accumulation of plastid- and, by analogy, mitochondria-derived sequences. Alternatively, the mechanisms that prevent nuclear genome expansion may be impaired on the Bs. Transfer of organellar DNA to the nucleus is very frequent [25], but much of the promiscuous DNA also is rapidly lost again within one generation [26]. If this elimination mechanism (e.g. illegitimate recombination) is impaired in Bs, the high turnover rates that prevent such sequences from accumulating on the A chromosomes would be absent. Thus, the dynamic equilibrium between frequent integration and rapid elimination of organellar DNA could be impaired for B chromosomes. Alternatively, a B-specific amplification process increased the copy number of integrated organellar DNA.

The observed partial overlapping of plastid and mitochondria DNA-specific hybridisation patterns suggests that both types of promiscuous DNA are chromosomally inserted under similar constrains in *Ae. speltoides* Bs. Future analyses of other B-bearing species such as maize will be needed to address the question whether organelle-to-nucleus DNA transfer is an important mechanism that drives the evolution of B chromosomes [27].

Supporting Information

Figure S1 FISH of isolated *Ae. speltoides* nuclei labelled with 5S rDNA. (A) Nucleus of a plant without Bs and (B) with Bs (arrowed). (B) The large 5S rDNA signals are of A chromosome origin, while the arrowed minor signals are B chromosome derived. Scale bar equals 10 μm.

Figure S2 Localisation of mitochondrial- and plastid-derived sequences on *Ae. speltoides* metaphase chromosomes of a plant without Bs from the Ramat Hanadiv population. Scale bar equals 10 μm.

Figure S3 Localisation of mitochondria (in red)- and plastid (in green) derived sequences on *Ae. speltoides* interphase nuclei. (A) Nucleus of a plant with 2Bs from Katzir. (B) Nucleus of a plant with 3Bs from the Technion population. Scale bar equals 10 μm.

Figure S4 Localisation of *Spelt-1* tandem repeat (in green) and *Arabidopsis*-type telomeric repeat (in red) sequences on *Ae. speltoides* metaphase chromosomes. The Bs are marked with arrows. Scale bar equals 10 μm.

Acknowledgments

We are grateful to Katrin Kumke and Joachim Bruder for technical support.

Author Contributions

Conceived and designed the experiments: EB AH. Performed the experiments: AR JF AM VS. Analyzed the data: AR JF AH. Contributed reagents/materials/analysis tools: AS OR. Wrote the paper: AR EB AH.

References

1. Jones N, Houben A (2003) B chromosomes in plants: escapees from the A chromosome genome? Trends Plant Sci 8: 417–423.
2. Houben A, Banaei-Moghaddam AM, Klemme S (2013) Biology and Evolution of B Chromosomes. In: Greilhuber J, Dolezel J, Wendel JF, editors. Plant Genome Diversity Volume 2: Springer Vienna. pp.149–165.
3. Jones RN, Rees H (1982) B-Chromosomes. New York: Academic Press. viii + 266 p.
4. Martis MM, Klemme S, Banaei-Moghaddam AM, Blattner FR, Macas J, et al. (2012) Selfish supernumerary chromosome reveals its origin as a mosaic of host genome and organellar sequences. Proceedings of the National Academy of Sciences of the United States of America 109: 13343–13346.
5. Klemme S, Banaei-Moghaddam AM, Macas J, Wicker T, Novák P, et al. (2013) High-copy sequences reveal distinct evolution of the rye B-chromosome. New Phytologist 199: 550–558.
6. Banaei-Moghaddam AM, Meier K, Karimi-Ashtiyani R, Houben A (2013) Formation and expression of pseudogenes on the B chromosome of rye. Plant Cell (In press).
7. Van Slageren MW (1994) Wild Wheats: a monograph of *Aegilops* L. and *Amblyopyrum* (Jaub. et Spach) Eig (*Poaceae*). Wageningen: Wageningen Agricultural University and ICARDA, Aleppo, Syria. 514 p.
8. Simchen G, Zarchi Y, Hillel J (1971) Supernumerary chromosomes in second outbreeding species of wheat group. Chromosoma 33: 63–69.
9. Mendelson D, Zohary D (1972) Behavior and transmission of supernumerary chromosomes in *Aegilops speltoides*. Heredity 29: 329–339.
10. Raskina O, Belyayev A, Nevo E (2004) Quantum speciation in *Aegilops*: molecular cytogenetic evidence from rDNA cluster variability in natural populations. Proceedings of the National Academy of Sciences of the United States of America 101: 14818–14823.
11. Belyayev A, Raskina O (2013) Chromosome evolution in marginal populations of *Aegilops speltoides*: causes and consequences. Annals of Botany 111: 531–538.
12. Friebe B, Jiang J, Gill B (1995) Detection of 5S rDNA and other repeated DNA on supernumerary B-chromosomes of *Triticum* species (*Poaceae*). Plant Systematics and Evolution 196: 131–139.
13. Raskina O, Brodsky L, Belyayev A (2011) Tandem repeats on an eco-geographical scale: outcomes from the genome of *Aegilops speltoides*. Chromosome Research 19: 607–623.
14. Hosid E, Brodsky L, Kalendar R, Raskina O, Belyayev A (2012) Diversity of long terminal repeat retrotransposon genome distribution in natural populations of the wild diploid wheat *Aegilops speltoides*. Genetics 190: 263–274.
15. Doležel J, Greilhuber J, Lucretti S, Meister A, Lysák MA, et al. (1998) Plant genome size estimation by flow cytometry: Inter-laboratory comparison. Annals of Botany 82: 17–26.
16. Ma L, Vu GTH, Schubert V, Watanabe K, Stein N, et al. (2010) Synteny between *Brachypodium distachyon* and *Hordeum vulgare* as revealed by FISH. Chromosome Research 18: 841–850.
17. Fukui K, Kamisugi Y, Sakai F (1994) Physical mapping of 5S rDNA loci by direct-cloned biotinylated probes in barley chromosomes. Genome 37: 105–111.
18. Salina EA, Pestsova EG, Adonina IG, Vershinin AV (1998) Identification of a new family of tandem repeats in *Triticeae* genomes. Euphytica 100: 231–237.
19. Baenziger H (1962) Supernumerary chromosomes in diploid and tetraploid forms of crested wheatgrass. Canadian Journal of Botany 40: 549–561.
20. Müntzing A (1948) Accessory chromosomes in *Poa alpina*. Heredity 2: 49–61.
21. Ohta S (1995) Distinct numerical variation of B-chromosomes among different tissues in *Aegilops mutica* Boiss. Japanese Journal of Genetics 70: 93–101.

22. Marques A, Banaei-Moghaddam AM, Klemme S, Blattner FR, Niwa K, et al. (2013) B chromosomes of rye are highly conserved and accompanied the development of early agriculture. Annals of Botany 112: 527–534.
23. Sheppard AE, Ayliffe MA, Blatch L, Day A, Delaney SK, et al. (2008) Transfer of plastid DNA to the nucleus is elevated during male gametogenesis in tobacco. Plant Physiology 148: 328–336.
24. Wang D, Lloyd AH, Timmis JN (2012) Environmental stress increases the entry of cytoplasmic organellar DNA into the nucleus in plants. Proc Natl Acad Sci U S A 109: 2444–2448.
25. Timmis JN, Ayliffe MA, Huang CY, Martin W (2004) Endosymbiotic gene transfer: Organelle genomes forge eukaryotic chromosomes. Nature Reviews Genetics 5: 123–135.
26. Sheppard AE, Timmis JN (2009) Instability of plastid DNA in the nuclear genome. PLoS Genet 5: e1000323.
27. Houben A, Banaei-Moghaddam AM, Klemme S, Timmis JN (2013) Evolution and biology of supernumerary B chromosomes. Cellular and Molecular Life Sciences: 1–12.

17

Variation, Evolution, and Correlation Analysis of C+G Content and Genome or Chromosome Size in Different Kingdoms and Phyla

Xiu-Qing Li[1]*, Donglei Du[2]

1 Molecular Genetics Laboratory, Potato Research Centre, Agriculture and Agri-Food Canada, Fredericton, New Brunswick, Canada, **2** Quantitative Methods Research Group, Faculty of Business Administration, University of New Brunswick, Fredericton, New Brunswick, Canada

Abstract

C+G content (GC content or G+C content) is known to be correlated with genome/chromosome size in bacteria but the relationship for other kingdoms remains unclear. This study analyzed genome size, chromosome size, and base composition in most of the available sequenced genomes in various kingdoms. Genome size tends to increase during evolution in plants and animals, and the same is likely true for bacteria. The genomic C+G contents were found to vary greatly in microorganisms but were quite similar within each animal or plant subkingdom. In animals and plants, the C+G contents are ranked as follows: monocot plants>mammals>non-mammalian animals>dicot plants. The variation in C+G content between chromosomes within species is greater in animals than in plants. The correlation between average chromosome C+G content and chromosome length was found to be positive in Proteobacteria, Actinobacteria (but not in other analyzed bacterial phyla), Ascomycota fungi, and likely also in some plants; negative in some animals, insignificant in two protist phyla, and likely very weak in Archaea. Clearly, correlations between C+G content and chromosome size can be positive, negative, or not significant depending on the kingdoms/groups or species. Different phyla or species exhibit different patterns of correlation between chromosome-size and C+G content. Most chromosomes within a species have a similar pattern of variation in C+G content but outliers are common. The data presented in this study suggest that the C+G content is under genetic control by both *trans-* and *cis-* factors and that the correlation between C+G content and chromosome length can be positive, negative, or not significant in different phyla.

Editor: Zhang Zhang, Beijing Institute of Genomics, Chinese Academy of Sciences, China

Funding: The research was supported by LXQ's research funding from Agriculture and Agri-Food Canada potato genomics project, Canadian Food Inspection Agency CHA-P-1101 project, and New Brunswick EARI Growing Forward Program. The funders had no role in study design, data collection and analysis, decision to publish, or preparation of the manuscript.

Competing Interests: The author has declared that no competing interests exist.

* E-mail: Xiu-Qing.Li@agr.gc.ca

Introduction

Base composition is a fundamental property of genomes and has a strong influence on gene function and regulation. It is known from Chargaff's rule [1] that in cellular DNA, the amount of adenine (A) is approximately equal to that of thymine (T) and the amount of cytosine (C) is approximately equal to that of guanine (G). It is also known from Watson and Crick's pair theory [2] of double-stranded DNA molecules that A pairs with T and C with G. Since the A-T base pair has two hydrogen bonds and the G-C base pair has three hydrogen bonds, the G-C interaction is stronger than A-T. For this reason, C+G content (also called GC content, CG content, or G+C content) is one of the main parameters used to describe DNA base composition.

C+G content can be measured by several methods, including melting–annealing profiles, CsCl gradient buoyant densities, flow cytometry, and genome sequencing, each of which has its advantages and disadvantages. Although C+G content is one of the major factors affecting melting–annealing profiles, the melting–annealing speed of DNA is not 100% correlated with GC content, because the speed is also influenced by the degree of repetition of sequences such as satellite sequences [3–5]. For the DNA density bands on the CsCl gradient liquid, it is difficult to achieve a very high resolution in small centrifuge tubes. Flow cytometry measurement is correlated with sequencing data in rice, but the C+G content measured by flow cytometry after staining with DAPI (4′,6-diamidino-2-phenylindole) is consistently higher than the content in the sequencing data [6]. Although the genome-sequencing approach has the highest resolution and is the most precise of all these methods, it is limited by the number of species for which complete genome sequences are available, and its degree of accuracy is affected by the degree of completeness of genome assembly. Shotgun sequencing, together with the assignment of scaffolds to specific chromosomes, can partly address the gap issue. Currently, most of the complete genomes available are from archaeans and bacteria, and a few are from animals and plants. That is likely why the evaluation of base composition in animals has focused mainly on DNA segments and transcribed regions [7–9]. In plants, most base compositions are determined on the basis of annealing temperature, and only recently were two dicots and one monocot compared [10]. A general comparison of genomic base composition between major kingdoms (domains of life) is still lacking.

Previous research found that C+G content is positively correlated with chromosome (genome) size in bacteria [11] and some other prokaryotes [12], in keeping with the finding that long coding sequences are usually GC-rich [13,14]. Since each bacterium usually has a single chromosome, it is unclear whether C+G content correlates positively with chromosome size or genome size and whether C+G content is regulated differently between chromosomes in the same species. One potential scenario is that a higher C+G content makes the chromosome stronger and therefore allows it to be longer. It is unknown whether positive correlation between C+G content and genome or chromosomes size exist in other kingdoms or phyla. However, in a previous study using pooled data for 1 plant, 4 animals, 3 protists, and 7 fungi, no significant correlation were found between the genomic GC contents and genome sizes [15]. It is possible that no correlations were found between the 15 genome sizes and the 15 C+G contents concerned because different mechanisms operate in different kingdoms. To determine whether C+G content is correlated with genome size or chromosome size, it is necessary to analyze many more species and compare the species within each kingdom and subkingdom. Since complete or nearly complete chromosome sequences are available for fungi, protists and some higher eukaryotes, we can examine the correlation between C+G contents and chromosome lengths for those particular species.

We speculate that C+G content and genome size may or may not be positively correlated depending on the kingdom or phylum, but that the species in each subkingdom share similar pattern for the relationship between chromosome length and C+G contents, likely because of shared evolutionary history, similar functional needs, and similar molecular mechanisms. In the present study, we analyzed the base compositions, particularly the C+G contents, of nearly all major kingdoms and phyla by using complete and nearly complete genomes from the GenBank, RefSeq, EMBL, DDBJ, and PDB databases obtained through searches of National Center for Biotechnology Information (NCBI) (http://www.ncbi.nlm.nih.gov/nucleotide/).

Results

General Description of Genome Size and C+G Content Variation at the Kingdom and Subkingdom Levels

The average genome sizes for the six kingdoms/subkingdoms that were analyzed are listed here in order from smallest to largest: Archaea (1.86 Mb)<Bacteria (2.61 Mb) <protists (3.16 Mb)< fungi (18.44 Mb) <plants (538.84 Mb)<animals (1,837.44 Mb) (Table 1; Figure 1). Microorganisms and higher organisms showed very different patterns of C+G content when the genomes were ranked by size within each large phylogenetic group (Table 1; Figure 1). The genome-level average C+G contents were ranked according to the percentage of the total C+G nucleotides in the total genomic nucleotides of the kingdom or subkingdom, with the results being as follows: protists (26.42%) <plants (41.10%)< animals (41.20%) <Archaea (44.88%)<fungi (47.96%)<bacteria (50.76%) (Table 1).

The protists for which genome size and C+G content were analyzed in this study are from only two phyla, Alveolata and Cryptophyta. It is unclear at this stage whether the genome size and average C+G content (25.7%) obtained is representative of all protists. However, the order shown above for other kingdoms should be very reliable because of the relatively large numbers of species in each kingdom. Although eukaryote species have multiple chromosomes, prokaryotes usually (except three archaeal species) have just one.

In terms of average genome size, we obtained the following ranking: Gram-positive bacteria (2.53 Mb)<Gram-negative bacteria (2.67 Mb); dicot plants (368.78 Mb)<monocot plants (793.93 Mb); and non-mammalian animals (674.06 Mb)<non-primate mammalian animals (2,190.40 Mb) <primate animals (2,739.35 Mb) (Table 1). These results appear to point to an evolutionary increase in genome size.

The ranking of the average genomic C+G content is very similar to that of average genome size in the different subkingdoms of bacteria, plants, and animals. The order found for genome C+G content is as follows: Gram-positive bacteria<Gram-negative bacteria; dicot plants<monocot plants; non-mammalian animals<mammalian animals (Table 1). In higher organisms, the C+G content was lowest in dicot plants and highest in monocot plants, with animals having an intermediate content (Figure 1).

The C+G content, showed considerable variation within each kingdom among microorganisms (Figure 1), was relatively similar among species of dicot plants and monocot plants, and highly similar among species of mammalian animals (Figure 1). With regard to subkingdoms of plants and animals, the C+G content was dramatically different between dicot and monocot plants and also clearly different between non-mammalian animals and mammalian animals, although the difference between non-primates mammals and primates was relatively minor (Figure 1).

Genome Size and C+G Content Statistical Distribution and Normal Distribution Probability

When C+G content was plotted against genome size, the distribution of the actual chromosome lengths and C+G contents within each kingdom was found to be very scattered (Figure 2). A clear positive correlation between C+G content and chromosome length was observed, as well as some outliers, in bacteria (Figure 2). In the other kingdoms (Archaea, Fungi, Protists, Plants, and Animals), no linear correlation was found between the C+G content and chromosome length (Figure 2).

Like the kingdoms, the subkingdoms did not show a simple linear correlation between C+G content and chromosome size (Figure 3). Gram-positive bacteria and Gram-negative bacteria appear to show clear positive correlations between C+G content and chromosome size (Figure 3). Again, the C+G content and chromosome length data did not show a simple linear relationship, which suggests that the Pearson correlation coefficient model does not apply in this situation.

The probability distribution plots of C+G content and chromosome length for different kingdoms showed that none of the six kingdoms were well modelled by the normal distribution regardless of chromosome length (Figure 4) or C+G content (Figure 5). Interestingly, in each kingdom, the probability plot showed that most chromosomes fit normal distribution fairly well but the longest chromosomes clearly do not fit (Figure 4). Also, in each kingdom, with regard to the chromosome C+G content, the majority of chromosomes fit an approximately normal distribution (despite partitioning into more than one group in some cases); the lowest and highest C+G contents were clearly outliers (Figure 5).

When all the chromosomes within each subkingdom were pooled and plotted together, average chromosome C+G content and chromosome sequence length showed the following patterns: positive linear correlation for Gram-positive bacteria, Gram-negative bacteria, and fungi, as well as dicot and monocot plants; negative correlation for non-mammalian animals and primate animals; and no significant correlations for Archaea and non-primate mammalian animals (Table 2). When all animal chromosomes were pooled and analyzed together, a slight negative correlation was found between C+G content and chromosome size

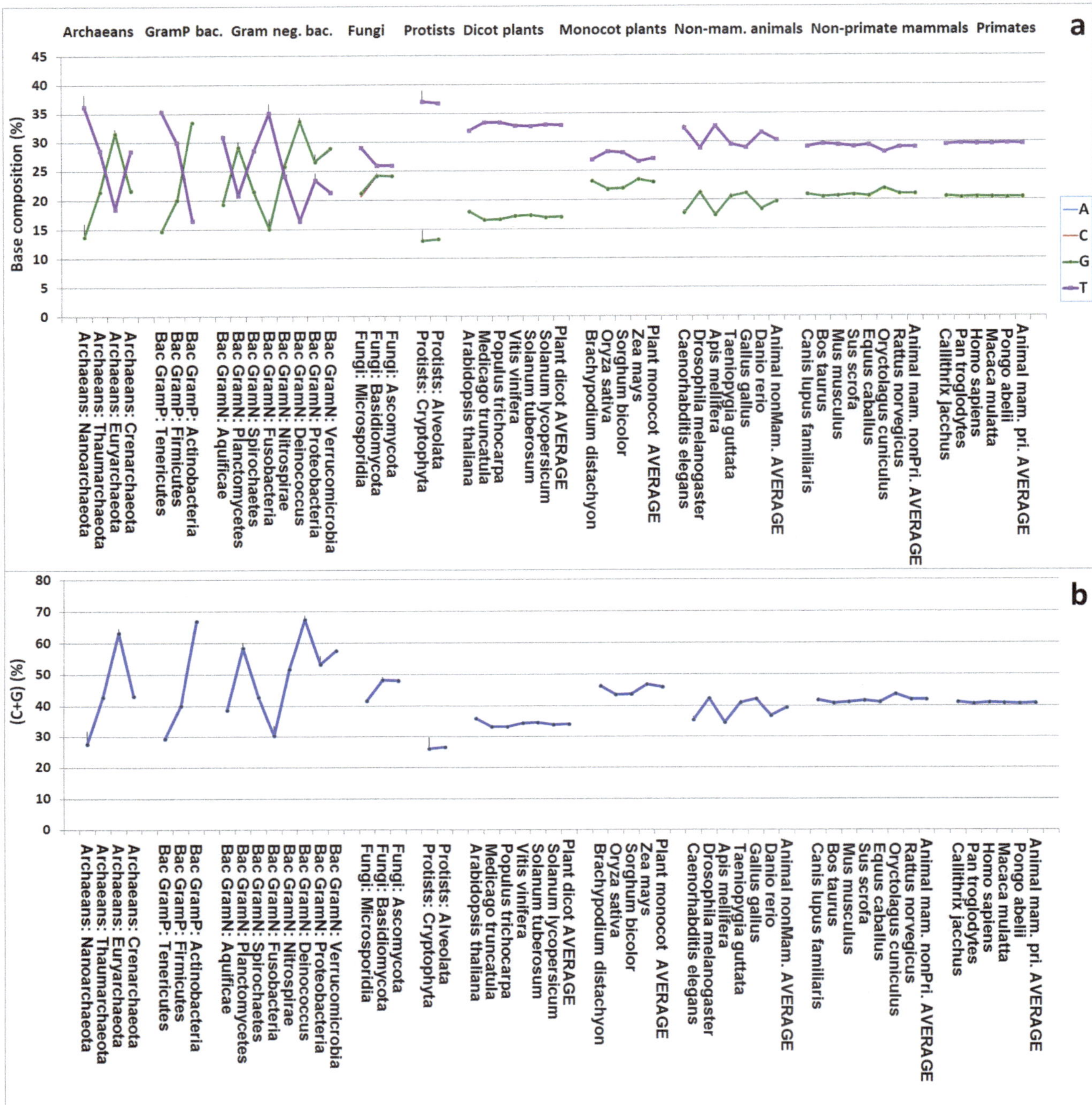

Figure 1. The base compositions and C+G contents of different kingdoms and large groups. The information is provided by phylum for archaeans, gram-positive bacteria, gram-negative bacteria, fungi, and protists and by species for dicot plants, monocot plants, non-mammalian animals, non-primate mammalian animals, and primate animals. Note that in panel (**a**), the A and T contents essentially overlap and are indistinguishable, and a similar situation exist for the C and G contents.

(Table 2). A positive correlation was found between C+G content and chromosome size (Table 2) for pooled dicot and monocot plant chromosomes.

At the small phylum or species level, in some cases such as in the Basidiomycota (fungi), the C+G content and chromosome length were found to be normally distributed, but no clear-cut correlations were found (Figure 6). Although most chromosomes fit normal distribution, outliers were frequently found, such as in *Zea mays*, *Drosophila melanogaster*, and *Taeniopygia guttata* (Figure 6).

In summary, Pearson models of linear regression were not suitable for calculating correlation coefficients. We decided to use the Spearman model to calculate the monotonic relationship between C+G content and chromosome size for each kingdom (Table 2), subkingdom (Table 2), and large species groups (Table 3).

Table 1. Average genome size and average genomic C+G content in different kingdoms and subkingdoms.

Kingdom or subkingdom	Number of genomes	Average genome size (Mb)	Average genome C+G (%)
Kingdoms:			
Archaeans	58	1.86	44.88
Bacteria	430	2.61	50.76
Fungi	11	18.44	47.96
Protists	21	3.16	26.42
Plants	10	538.84	41.10
Animals	18	1,837.44	41.20
Subkingdoms:			
Gram-positive bacteria	184	2.53	49.40
Gram-negative bacteria	246	2.67	51.71
Dicot plants	6	368.78	34.15
Monocot plants	4	793.93	45.93
Non-mammalian animals	6	674.06	39.44
Non-primate mammalian animals	7	2,190.40	41.97
Primate animals	5	2,739.35	40.85

General Comparative Analysis of Chromosome Size and C+G Content at the Kingdom and Phylum Levels

C+G contents were ranked (in terms of the average C+G content per chromosome or large scaffold) among kingdoms, yielding the following results: protists <plants<animals<Archaea<bacteria<fungi (Table 2). In microorganisms, chromosome size or length in terms of nucleotide number was ranked from smallest to largest: protists>fungi>archaeans>Gram-positive bacteria>Gram-negative bacteria (Table 2). Animal and plant chromosome sizes were less precise because some chromosomes were not complete; however, the assembled chromosomes were already very large, with average sizes ranging from 29.90 Mb in dicot plants to 92.93 Mb in non-primate mammalian animals. These were likely large enough to be used in estimating the approximate C+G contents of the chromosomes (Table 1).

Chromosome Analysis Within Archaeans

Archaeal genome size varied considerably. The average genome sizes of the four phyla Nanoarchaeota, Thaumarchaeota, Crenarchaeota and Euryarchaeota were 0.14, 1.97, 2.07, and 3.39 Mb, respectively (Table 3). The average chromosome size in Euryarchaeota was 1.69 Mb, because each of its genomes had two chromosomes (Table 3). In archaeans, chromosome size was equal to genome size in all species except in the phylum Euryarchaeota in which there were two chromosomes. The genomes in the phylum Nanoarchaeota phylum were the smallest and they had very low C+G contents (27.6%), with an (A+T)/(C+G) ratio of 2.62 (Table 3). The phylum Euryarchaeota had the highest C+G content (63.3% on average) and the lowest (A+T)/(C+G) ratio (0.58) (Table 3).

For the pooled chromosomes of all the archaeal species, the Spearman correlation between C+G content and chromosome length was weak ($R = 0.21$) and not significant (Table 2). When each subkingdom was analyzed separately, a highly significant Spearman R value (−0.52) was obtained for the Crenarchaeota (Table 3). However, when we plotted the data, two found two distinct groups (Figure 3). The genomes located in the right lower corner were from three species of Sulfolobus. When we analyzed the two groups separately, the Spearman R value was for the Sulfolobus group was 0.01, and the value for the major Crenarchaeota group was 0.04 (Figure 3). Therefore, there was no correlation between C+G content and genome size in the two subgroups of Crenarchaeota.

Three species from the phylum Euryarchaeota (*Halorubrum lacusprofundi*, *Haloarcula hispanica*, and *Haloarcula marismortui*) each had one larger chromosome (2.7–3.1 Mb, called Chromosome 1) and one smaller chromosome (288–525 kb, called Chromosome 2). Both of these chromosomes had the highest C+G contents among archean species (Table 3; Figure 1); their contents were much higher than those of the species of Thaumarchaeota and Crenarchaeota. The smaller chromosomes had relatively lower C+G contents than did the larger chromosomes in all three species. Chromosome length and C+G content showed a positive correlation when all chromosomes in these three species were analyzed together (Figure 7a,b). However, this correlation does not necessarily mean that C+G content actually increased when chromosome size increased. We verified this by conducting separate analyses of Chromosome 1 and Chromosome 2 from all the Euryarchaeota species and found that C+G content decreased when the size of Chromosome 1 increased; a similar pattern was also observed for Chromosome 2 (Figure 7c,d).

From the above analysis, it is more plausible to conclude that there was likely no moderate or strong correlation between C+G content and genome size in the 61 archaeal genomes analyzed in this study. This issue may need to be revisited once a much larger number of archaeal species have been sequenced, with a focus on whether there is a negative correlation between C+G content and chromosome size in Euryarchaeota.

Chromosome Analysis Within Gram-positive Bacteria

In total, 184 genomes of Gram-positive bacteria were analyzed and the general correlation between C+G content and genome size was found to be very strong ($R = 0.75$; Table 2 for correlation; Figure 3 for distribution). The genomes were from three phyla–Tenericutes, Firmicutes, and Actinobacteria–with average genome sizes of 0.86, 3.36, and 5.16 Mb, respectively, and average C+G

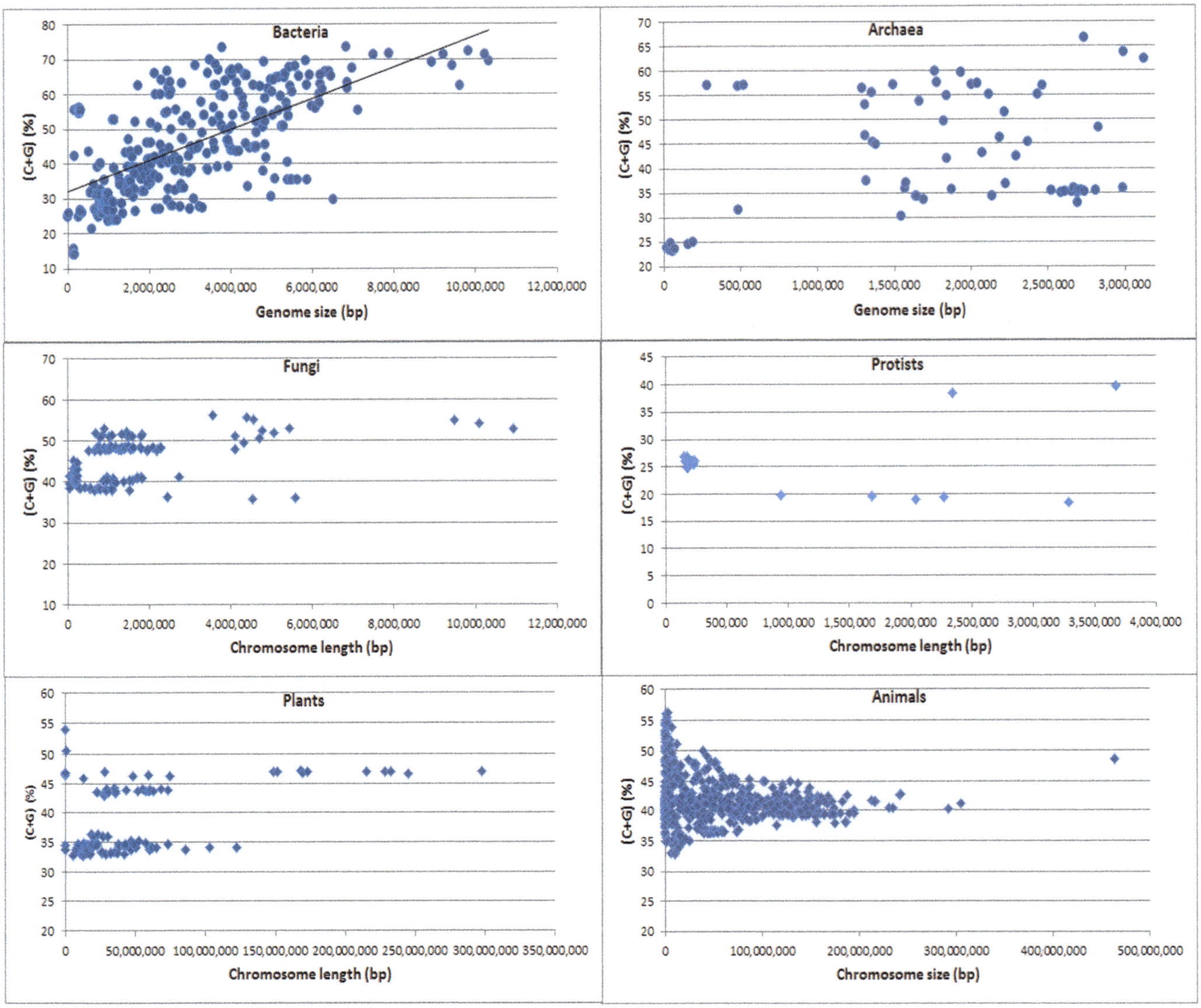

Figure 2. Distribution of values of chromosome size and C+G content in each kingdom of living organisms. The chromosome number for bacteria, archaea, fungi, protists, plants, and animals is 430, 61, 139, 21, 115, and 565, respectively. Note that the C+G content and chromosome or genome size in bacteria is positively correlated but there is no such a correlation or linear relationship in other kingdoms.

contents of 29.3%, 40.1%, and 67.0%, respectively. In terms of the average genome size of Gram-positive bacteria, there was a clear increase in C+G content with increasing genome size (Figure 1). When all the chromosomes were ranked from smallest to largest (figure not shown), the first 88 genomes (small) were from the phylum Tenericutes and the remaining genomes were from the other two phyla. Although the Tenericutes genome showed an approximately fivefold increase in size among its species, there was no tendency toward increasing C+G content ($R = 0.11$; Table 3). No correlation was found for Firmicutes, either (Table 3). The only positive correlation between C+G content and genome size was found among the species of the Actinobacteria group (Table 3). Among the Gram-positive bacteria, there were at least two evolutionary patterns: genome size was strongly and positively associated with genomic C+G content in the Actinobacteria group but no correlation was found in the other two phyla (Table 3).

Chromosome Analysis Within Gram-negative Bacteria

Eight phyla or large groups of Gram-negative bacteria, namely (in order of average genome size) Aquificae, Planctomycetes, Spirochaetes, Fusobacteria, Nitrospirae, Deinococcus, Proteobacteria, and Verrucomicrobia, were analyzed (Table 3). When all the genomes of the Gram-negative bacteria regardless of phylum were ranked from smallest to largest, there was clearly a tendency toward increasing C+G content with increasing genome size ($R = 0.50$) (Table 2; Figure 3). When genome-size ranking was done for each phylum or large phylogenetic group, marked increases in C+G content and genome size were observed in the phylum Proteobacteria. While this relationship may also characterize the phyla Nitrospirae and Verrucomicrobia, it may not be significant owing to the small number of genomes. However, no such relationship was found for Aquificae, Spirochaetes, Planctomycetes and Deinococcus, and the results for Fusobacteria were not clear. The C+G contents ranked from only 14.55% in

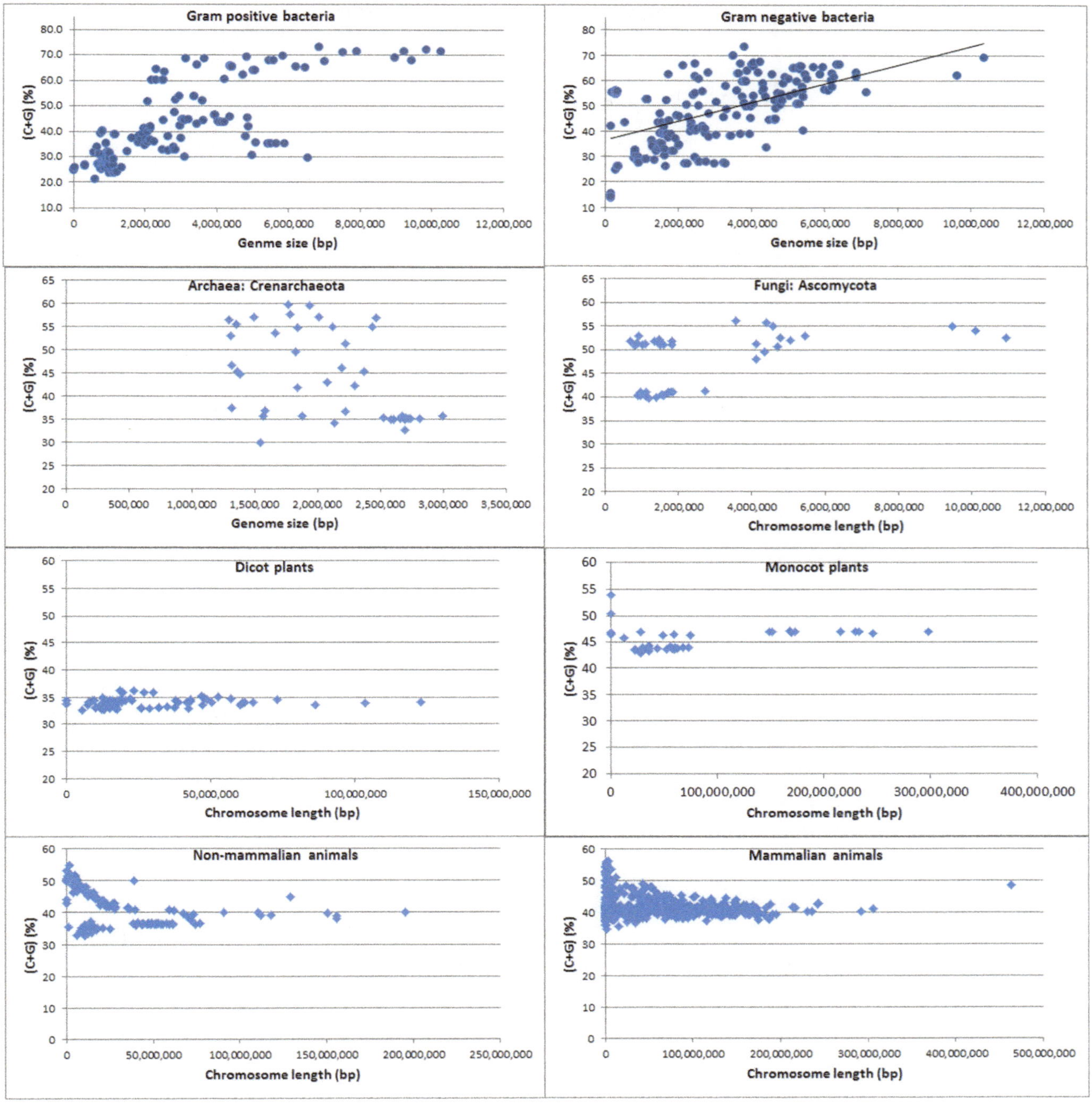

Figure 3. Distribution of values of chromosome size and the C+G content in each subkingdom. The chromosome number for Gram-positive bacteria, Gram-negative bacteria, Crenarchaeota, and Ascomycota is 184, 246, 42, and 61, respectively. The chromosome or large scaffold number for dicot plants, monocot plants, non-mammalian animals, and mammalian animals is 74, 41, 120, and 445, respectively. Note that C+G content and chromosome or genome size in both Gram-positive and Gram-negative bacteria are positively correlated but there is linear relationship in other subkingdoms. Note that in Crenarchaeota (archaea) there are two groups in the distribution. The lower right corner group is Sulfolobus species. There is no correlation between C+G content and genome size within each of the two distribution group of Crenarchaeota.

Carsonella ruddii (isolate Thao2000, CP003544, 157.5 kb) to 69.19% in *Myxococcus stipitatus* (CP004025, 10,350 kb). This corresponds to a 67-fold difference in genome size and a 4.8-fold difference in C+G content between the two species. The C+G content increase appears to have continued at the same rate during the evolution of this phylum Proteobacteria (data not shown). These results suggest that genome size and C+G content have coevolved in some bacterial phyla but not in others.

Chromosome Analysis Within Fungi

The three fungal phyla (Microsporidia, Basidiomycota, and Ascomycota) showed phylum-level C+G content increases in

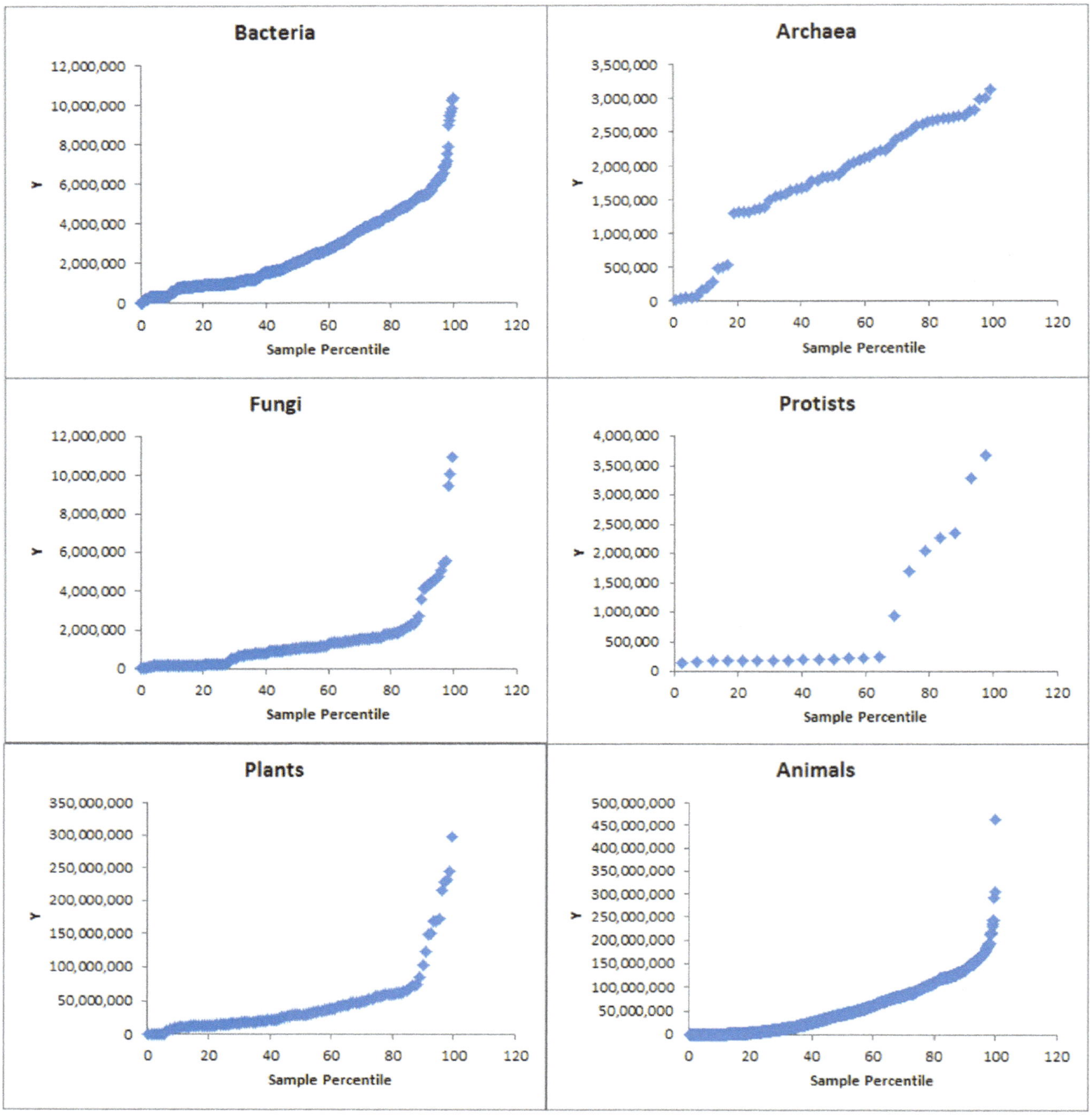

Figure 4. Normal probability plots of genome/chromosome size (bp, the Y-axis). Note that the large genomes/chromosomes (on the right of each plot) do not fit a normal distribution.

association with genome size increases (Figure 1). There was a moderate positive correlation between C+G content increase and genome size increase in Ascomycota ($R = 0.47$) but not in Microsporidia or Basidiomycota (Table 3).

Chromosome Analysis Within Protists

Among the two large groups or phyla (Alveolata and Cryptophyta) of protists analyzed, there was a very large difference in genome size, but the C+G contents were quite similar. There was no correlation between C+G content and chromosome length in these two phyla of protists (Table 3). When we pooled and plotted all the chromosomes, we found a weak negative correlation but with at least two outliers corresponding to exceptionally high C+G contents (Figure 2). The correlation between C+G content and chromosome length in protists should be re-examined after more protist genomes are sequenced.

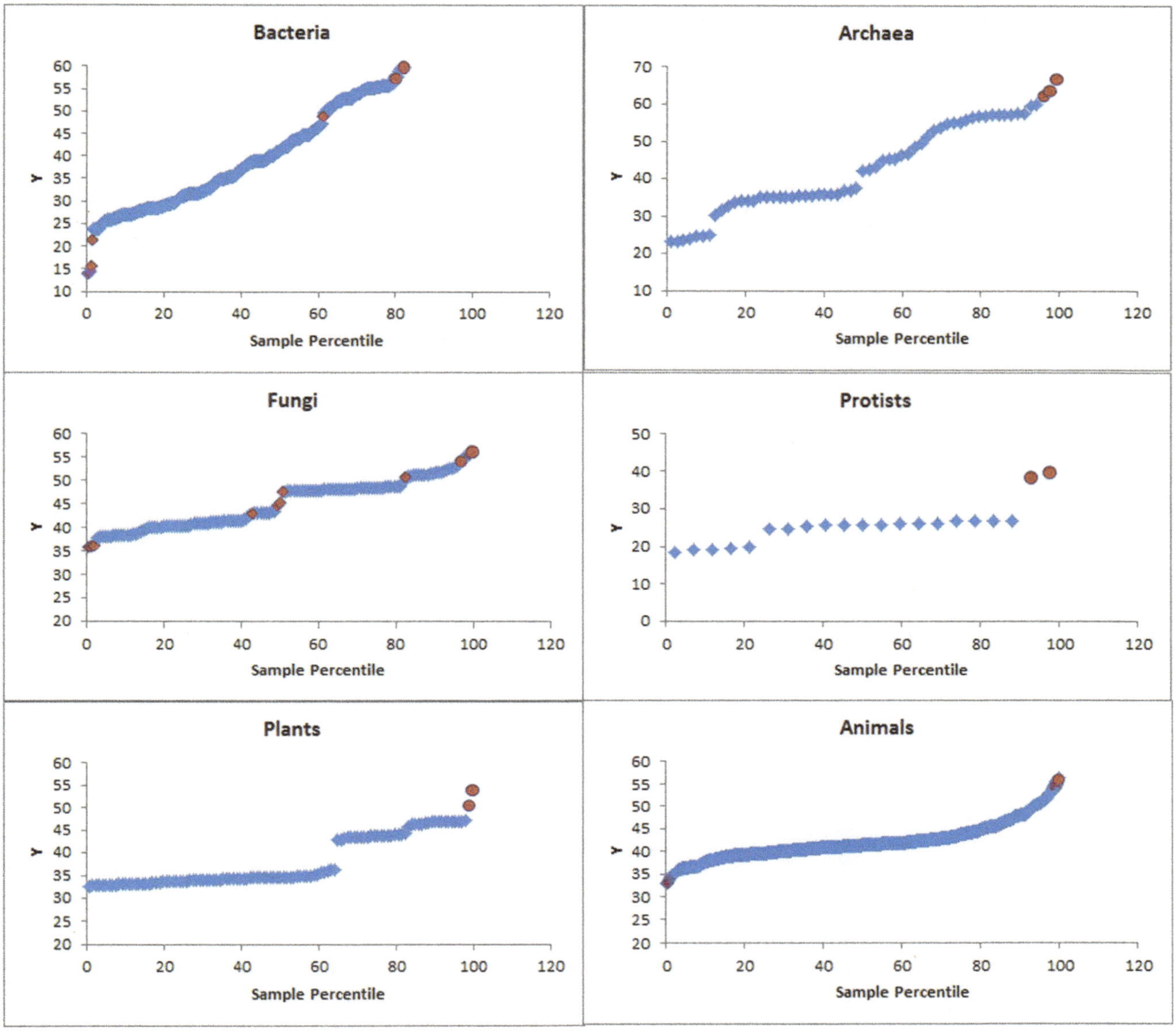

Figure 5. Normal probability plots of C+G contents (%, the Y-axis). Note that the lowest and the highest C+G contents (on the left and right, respectively, of each plot) do not fit a normal distribution. Red dots: outliers.

Chromosome Analysis Within Dicot Plants

Six dicot plants, namely *Arabidopsis thaliana*, *Medicago truncatula*, *Populus trichocarpa*, *Vitis vinifera*, *Solanum tuberosum*, and *Solanum lycopersicum*, were analyzed (Table 2; Figure 1). The average genome size was 0.37 Mb when the unanchored scaffolds were not counted. In the dicot group, *A. thaliana* had the smallest genome (119.0 Mb), *S. lycopersicum* the largest (719.0 Mb), and *S. tuberosum* (Group Phureja) the second largest (578.4 Mb) when the unanchored scaffolds were not counted. Given that the current draft genome (PGSC_DM_v3_2.1.10) of *S. tuberosum* (Group Phureja) has 11% unanchored sequences, actual chromosome size may be a few percent larger. All the dicot plant species had much smaller C+G contents than did the mammalian animals (Figure 1).

Genomic C+G content was remarkably similar in these plants, with exception of *A. thaliana* (Figure 8a). When all the plant chromosomes were pooled together, there was a positive but relatively weak correlation between C+G content and genome size (R = 0.39; Table 2). When each species was analyzed separately, potato (*S. tuberosum*) was found to have a strong positive correlation between C+G content and chromosome length ($R = 64$; Table 3). Tomato (*S. lycopersicum*) showed a moderate correlation ($R = 0.49$) but it was not significant (Table 3). The other plant species did not show a significant correlation. Since the complete genome sequence is not available for either the potato or the tomato, the correlation between C+G content and chromosome size should be re-examined in future to determining whether the correlation is indeed positive, once the genome assembly in these two closely related species has been enhanced further.

Table 2. Correlation between chromosome size and chromosome C+G content in different kingdoms and subkingdoms.

Kingdom or subkingdom	Number of chromosomes	Average chromosome size (Mb)[a]	Average chromosome C+G (%)	Spearman correlation between chr size and C+G (*R*)
Kingdoms:				
Archaeans	61	1.77	43.09	0.21 NS
Bacteria	430	2.61	43.71	0.61 **
Fungi	139	1.46	45.09	0.46 **
Protists	21	0.90	25.63	−0.29 NS
Plants	115	46,86	38.11	0.39 **
Animals	565	58,54	42.06	−0.19 **
Subkingdoms:				
Gram-positive bacteria	184	2.53	40.05	0.75 **
Gram-negative bacteria	246	2.67	46.45	0.50 **
Dicot plants	74	29.90	34.07	0.26 *
Monocot plants	41	77.46	45.41	0.32 *
Non-mammalian animals	120	33.70	41.55	−0.43 **
Non-primate mammalian animals	165	92.93	42.03	−0.09 NS
Primate animals	280	48.91	42.29	−0.20 **
Mammals (pooled)				−0.14 **

[a]The smaller chromosome/scaffold size in primate animals relative to non-primate mammalian animals is likely due to the incompleteness of genome assembly.
*, **, and NS: Significant (P<0.05), highly significant (P<0.01) and not significant (P>0.05) according to the "algorithm AS 89" test (see Methods for more details).

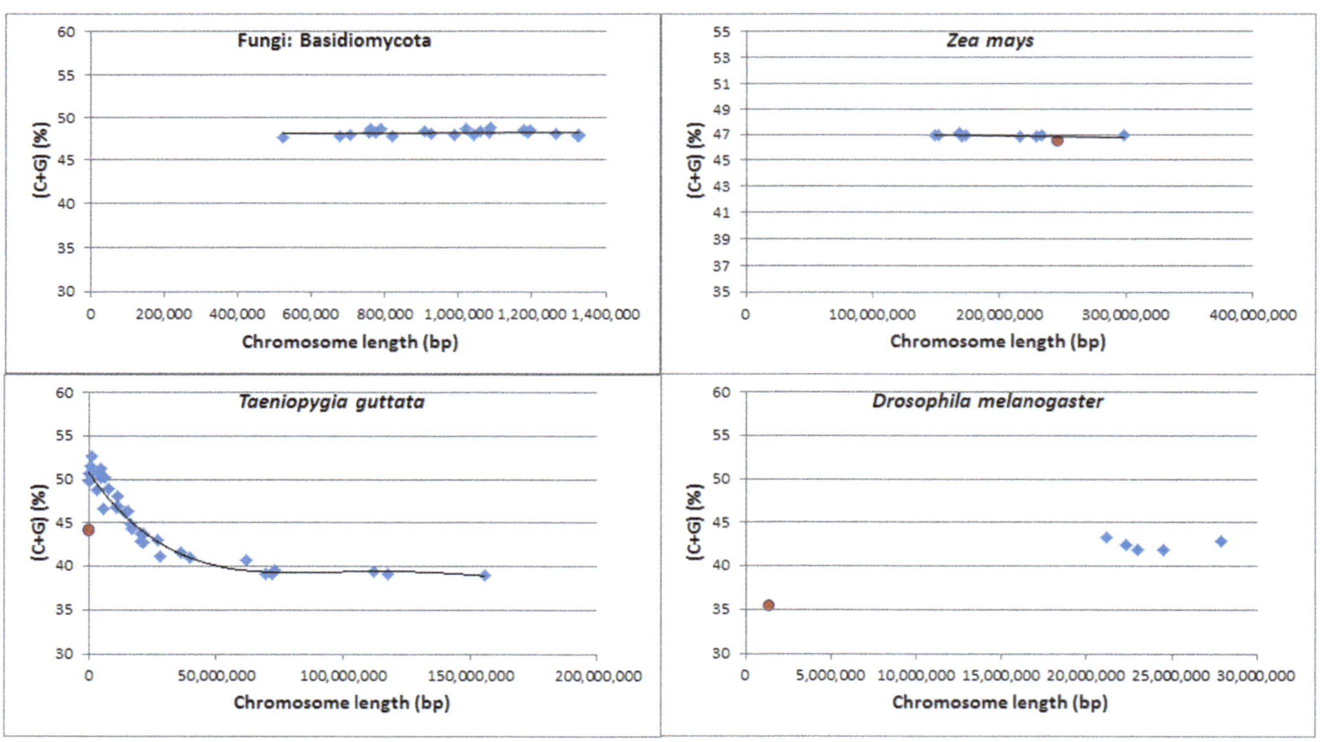

Figure 6. Distribution of C+G contents and genome/chromosome sizes showing some outliers. Blue diamonds: genome or chromosomes which appear to be normally distributed. Red round dots: outliers.

Table 3. Genome sizes, C+G contents, and correlations between genome or chromosome size and C+G content in different phyla and species analyzed.

Kingdom, phylum, or species[a]	No. of chrs[b]	Average genome size[c]	Genome C+G content[d] (%)	Genome (A+T)/(C+G) ratio	Spearman correlation (chr size and C+G)[e] (R)
Archaeans					
Crenarchaeota	42	2,072,521	43.2	1.32	−0.52[f] **
Euryarchaeota	6	3,388,400	63.3	0.58	0.60 NS
Nanoarchaeota	8	135,584	27.6	2.62	0.62 NS
Thaumarchaeota	5	1,970,970	43.0	1.33	0.50 NS
Bacteria: Gram-positive					
Actinobacteria	37	5,162,387	67.0	0.49	0.73 **
Firmicutes	59	3,360,830	40.1	1.49	0.03 NS
Tenericutes	88	857,257	29.3	2.41	0.11 NS
Bacteria: Gram-negative					
Aquificae	14	1,638,805	38.7	1.59	−0.06 NS
Deinococcus	7	3,064,306	67.3	0.49	0.18 NS
Fusobacteria	7	2,499,267	30.2	2.31	0.43 NS
Nitrospirae	4	2,821,645	51.6	0.94	0.80 NS
Planctomycetes	33	1,892,222	58.4	0.71	0.33 NS
Proteobacteria	130	3,195,588	53.3	0.88	0.84 **
Spirochaetes	47	1,937,957	43.0	1.32	0.27 NS
Verrucomicrobia	4	3,664,906	57.5	0.73	0.80 NS
Fungi					
Ascomycota	61	17,487,539	48.2	1.08	0.47 **
Basidiomycota	42	18,762,089	48.2	1.08	0.05 NS
Microsporidia	36	2,218,723	41.8	1.39	0.14 NS
Protists					
Alveolata	7	8,125,950	26.5	2.77	0.14 NS
Cryptophyta	14	540,299	26.0	2.84	−0.27 NS
Dicot plants					
Arabidopsis thaliana	5	118,960,141	36	1.78	−0.20 NS
Medicago truncatula	8	245,176,270	33.2	2.02	−0.31 NS
Populus trichocarpa	20	260,960,130	33.3	2	−0.06 NS
Solanum lycopersicum [g]	10	718,969,627	34	1.95	0.49 NS
Solanum tuberosum	12	578,393,875	34.7	1.88	0.64 *
Vitis vinifera	19	290,237,009	34.4	1.9	0.37 NS
Monocot plants					
Brachypodium distachyon	10	224,012,348	46.3	1.16	−0.47 NS
Oryza sativa	12	370,733,456	43.6	1.3	0.47 NS
Sorghum bicolor	9	549,132,399	43.8	1.28	0.32 NS
Zea mays	10	2,031,824,535	46.9	1.13	−0.44 NS
Animals: non-mammalian					
Apis mellifera	14	164,585,183	34.7	1.88	0.27 NS
Caenorhabditis elegans	6	100,269,912	35.4	1.82	−0.54 NS
Danio rerio	26	1,474,835,229	36.8	1.72	0.27 NS
Drosophila melanogaster	6	120,290,946	42.4	1.36	0.26 NS
Gallus gallus	33	1,169,412,079	42.1	1.37	−0.81 **
Taeniopygia guttata	35	1,014,970,640	41.0	1.44	−0.91 **
Animals: non-primate mammalian					
Bos taurus	9	1,068,912,767	40.9	1.44	0.02 NS
Canis lupus familiaris	11	898,613,247	41.8	1.39	−0.08 NS

Table 3. Cont.

Kingdom, phylum, or species[a]	No. of chrs[b]	Average genome size[c]	Genome C+G content[d] (%)	Genome (A+T)/(C+G) ratio	Spearman correlation (chr size and C+G)[e] (R)
Equus caballus	32	2,335,454,483	41.2	1.42	−0.10 NS
Mus musculus	26	1,135,801,574	41.2	1.42	0.21 NS
Oryctolagus cuniculus	23	2,603,978,348	43.7	1.29	−0.50 **
Rattus norvegicus	45	5,058,733,204	42	1.38	−0.23 NS
Sus scrofa	19	2,231,281,778	41.7	1.4	0.01 NS
Animals: primate					
Callithrix jacchus	23	2,177,235,585	41.1	1.43	−0.31 NS
Homo sapiens	190	2,804,441,965	40.9	1.44	−0.12 NS
Macaca mulatta	23	2,871,002,222	40.9	1.45	−0.48 *
Pan troglodytes	25	2,752,354,403	40.7	1.46	−0.56 **
Pongo abelii	26	3,093,520,335	40.7	1.46	−0.43 **

[a]From archaeans to protists: by phylum or large group; from plants to animals: by species.
[b]The number of chromosomes. Because archaeans and bacteria (except for the three archaean species in the Euryarchaeota group, as shown in Figure 7) have one chromosome per cell, the number equals the genome number (except for the Euryarchaeota group of archaeans. Euryarchaeota. The average genome size in Euryarchaeota was 3,388,400 bp).
[c]From archaeans to protists: average genome size of all the species in the phylum or large group; from animals to plants: genome size of the species.
[d]The overall C+G content when all the genomes or chromosomes were pooled.
[e]The size and C+G content of each chromosome were calculated first, and then sizes and C+G contents were used to calculate the Spearman R value.
[f]R = 0.04 after removing *Sulfolobus* spp.
[g]*Solanum lycopersicum* (tomato) chromosomes 8 and 12 were omitted from this calculation because NCBI did not have the complete sequences of those chromosomes at the time of downloading.
*, **, and NS: Significant ($P<0.05$), highly significant ($P<0.01$) and not significant ($P>0.05$) according to the "algorithm AS 89" test (see Methods for more details).

Chromosome Analysis Within Monocot Plants (Cereals)

Four monocot plants were analyzed (Figure 1). These cereals, namely *Brachypodium distachyon*, *Oryza sativa*, *Sorghum bicolor*, and *Zea mays*, were ranked in order of genome size from smallest to largest (Table 3). The average genome size was 0.79 Gb when the unanchored scaffolds were not counted. In the monocot group, *B. distachyon* had the smallest genome (0.22 Gb) and *Z. mays* the largest (2.03 Gb) when the unanchored scaffolds were not counted. The C+G contents in these monocot plants, despite variation among species (Figure 8b), were much higher than those in the dicot plants and were also higher than those in the non-mammalian, non-primate mammalian, and primate animals (Figure 1).

When we pooled all the monocot plant chromosomes together, we found a weak positive correlation between C+G content and genome or chromosome size ($R = 0.32$; Table 2). When each species was analyzed separately, none of the species showed a significant correlation between the C+G content and chromosome length (Table 3). Although the C+G content in *B. distachyon* differed among the chromosomes (Figure 8b), very similar C+G contents were found among the chromosomes within species in other three cereals, regardless of chromosome size (Figure 8b). The correlation between C+G content and chromosomes size was moderately positive in *O. sativa* but moderately negative in *B. distachyon* and *Zea mays* (Table 3).

Chromosome Analysis Within Non-mammalian Animals

The six non-mammalian animal species with complete or nearly complete genome sequences had an average genome size of 674 Mb and an average genome C+G content of 49.4% (Table 1). There were clear differences in genome size among the six genomes. *Caenorhabditis elegans* had the smallest genome (100.27 Mb) and *Danio rerio* the largest (1,474.84 Mb, or 1.47 Gb).

For these species, at the genome level, there was no clear correlation between C+G content and average genome size (Figure 1). However, when all the 120 chromosomes of these six animals were pooled, a negative correlation was found between C+G content and chromosome size ($R = -0.43$; Table 2). Interestingly, two species, *Taeniopygia guttata* and *Gallus gallus*, showed a very strong negative correlation between C+G content and chromosome size ($R = 0.91$ and $R = 0.81$, respectively; Table 3). Despite the general negative correlation between C+G content and chromosome size in *G. gallus*, two large chromosomes actually had quite high C+G contents. This suggests that chromosome-level *cis*-regulation may exist, in addition to the overall cell-level regulation (Figure S1).

Chromosome Analysis Within Non-primate Mammalian Animals

This analysis encompassed seven non-primate mammalian species whose sequence scaffolds are clearly assigned to specific chromosomes, even though the contiging is often incomplete (Figure 1). The average genome size was 2.19 Gb. There were clear-cut differences in genome size among the seven genomes. *Canis lupus familiaris* had the smallest genome (0.90 Gb) and *Rattus norvegicus* the largest (5.06 Gb). However, there was no clear correlation between C+G content and genome size (Figure 1). Within species, C+G content and chromosome size showed a moderate negative correlation ($R = -0.50$) in *Oryctolagus cuniculus* but there was no significant correlation in the other non-primate mammalian animals (Table 3).

In Figure 9a, the genome sizes of four nearly complete genomes are ranked from smallest to largest among species, and the

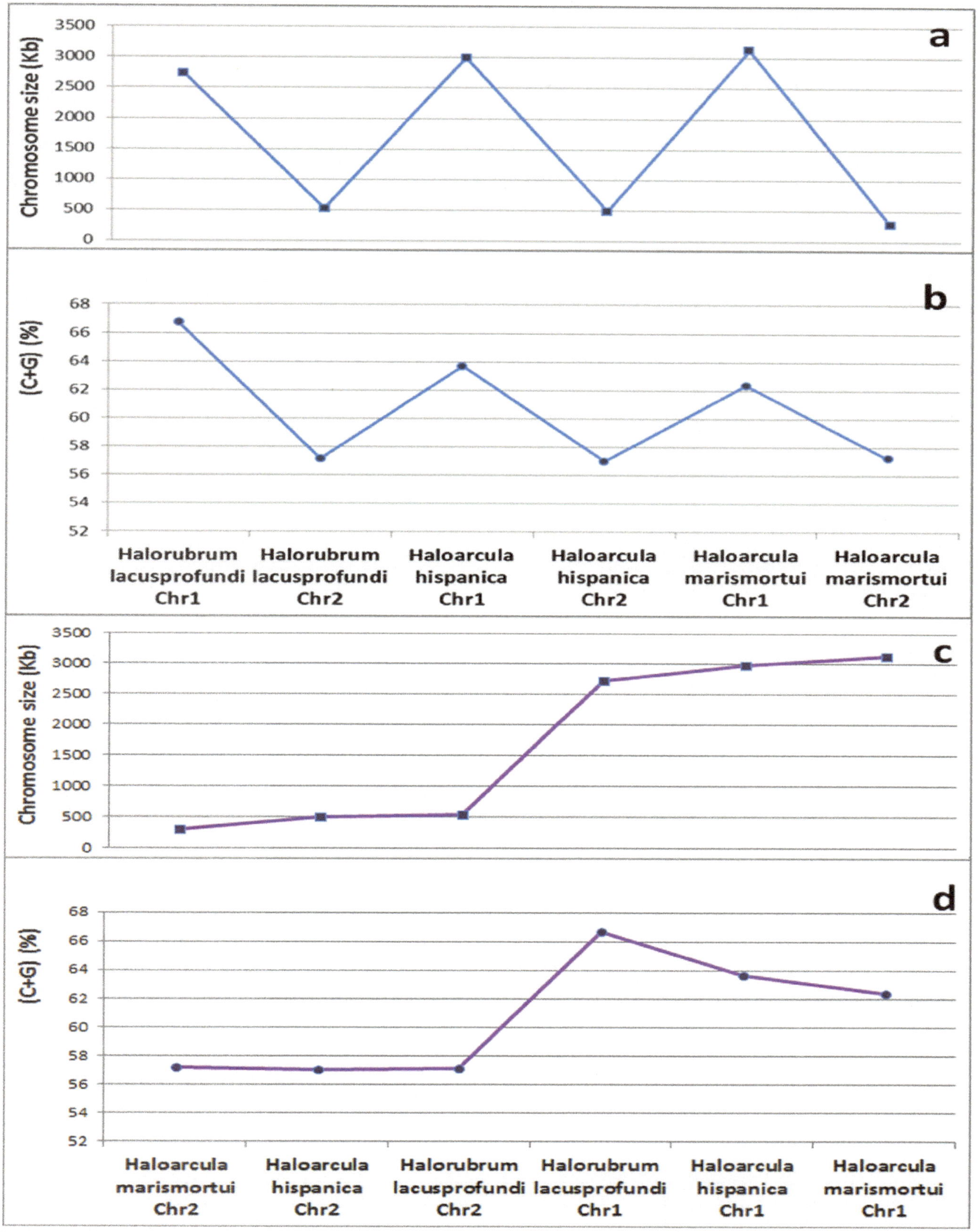

Figure 7. The chromosome sizes and C+G contents of three Archaea species, which each have one large chromosome (Chr1) and one small chromosome (Chr2). The three species (*Haloarcula hispanica, Haloarcula marismortui*, and *Halorubrum lacusprofundi*) belong to the

Euryarchaeota. Note that although the C+G contents appear to be correlated with chromosome sizes in (**a**) and (**b**), the C+G contents are actually negatively correlated with chromosome size within each type of chromosome (Chr1 and Chr2), as shown in (**c**) and (**d**).

chromosome sizes are arranged from smallest to largest within each species. *Mus musculus* and *Equus caballus*, two species with relatively small genomes, showed similar pattern of C+G content variation when chromosome sizes were arranged from smallest to largest (Figure 9a, first two species). The larger chromosomes within each species tended to have smaller C+G contents. Interestingly, *Oryctolagus cuniculus* and *R. norvegicus*, two species with relatively large genome sizes, had very similar patterns of C+G content among their chromosomes: both the largest and smallest chromosomes within each species tended to have larger C+G contents than the chromosomes of intermediate size (Figure 9a, last two species).

Chromosome Analysis Within Primate Animals

Five primate species (Table 2) were selected for analysis because they had nearly complete genome sequences with clear grouping of scaffolds for each chromosome (Figure 1). Their genomes were all very large, with an overall average genome size of 2.74 Gb. The average C+G content in primates (40.9%) was higher than that in non-mammalian animals (39.44%) but lower than that in non-primate mammals (41.97%) (Table 1). The genome size differences among the five primates were less clear-cut than those found among the non-primate animals. *Callithrix jacchus* had the smallest genome (2.18 Gb) and *Pongo abelii* the largest (3.09 Gb) when the unanchored (unlocated) scaffolds were not counted. The C+G contents of these primate animals were remarkably similar and quite consistent relative to those of other groups of living organisms (Figure 1).

Within each species, C+G content and chromosome size showed a consistently negative correlation, which reached a high level of significance in *Pan troglodytes*, *Macaca mulatta*, and *Pongo abelii* (Table 3). Four species (*C. jacchus*, *P. troglodytes*, *M. mulatta*, and *P. abelii*) had the most clearly defined chromosomes or pseudomolecules within the primate group. When these four genomes were ranked from smallest to largest among species and the chromosome sizes were arranged from smallest to largest within each species (Figure 9b), three species (*Pan troglodytes*, *M. mulatta*, and *Pongo abelii*) had very similar C+G content variation patterns (Figure 9b). The smallest chromosome group within each species tended to have the largest C+G content, and the largest two or three chromosomes also had relatively high C+G contents in comparison with the intermediate chromosomes (Figure 9b). The C+G content variation patterns for these species produced an imperfect U shape in Figure9b. These corresponding patterns in

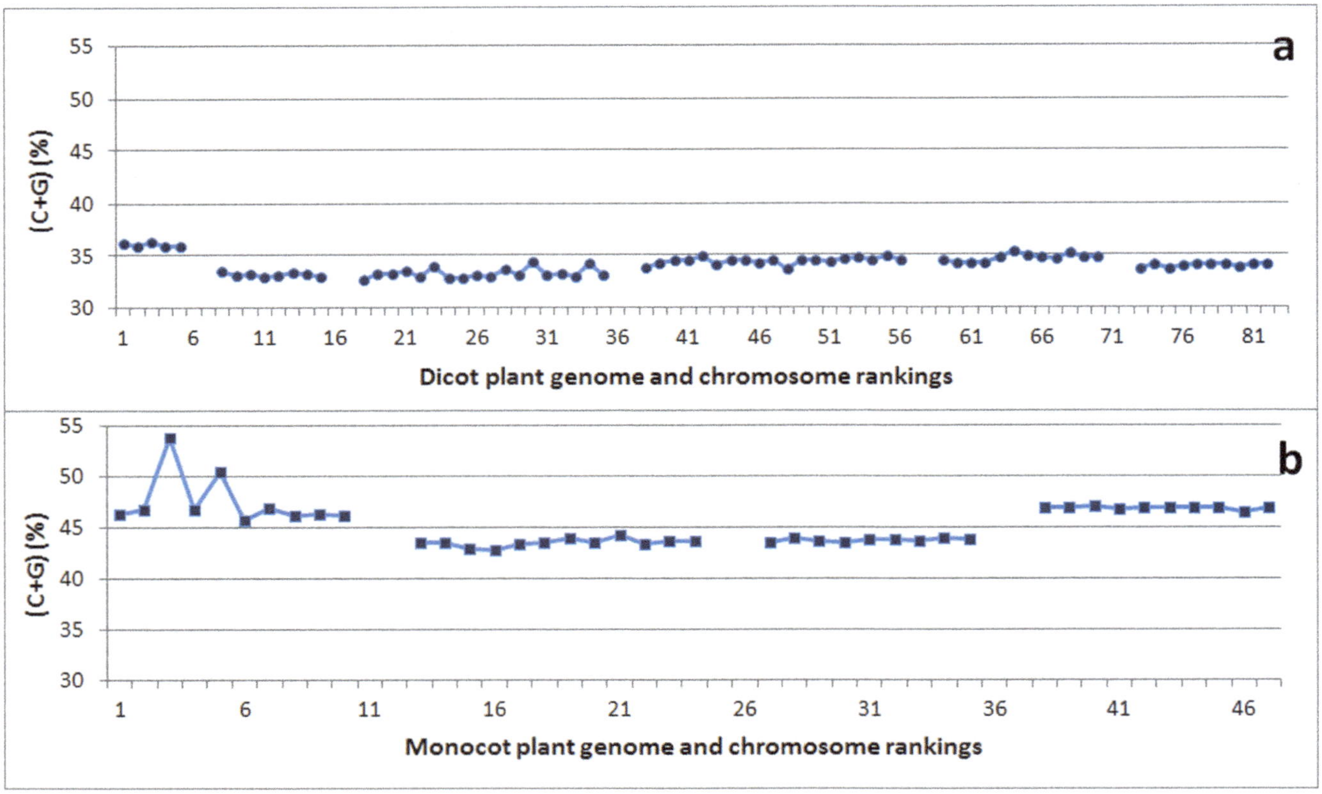

Figure 8. The C+G contents (%) per chromosome in dicot and monocot plants, ranked by genome size among species and by chromosome size within each species. Panel (**a**) shows dicot plants, namely (in order from left to right) *Arabidopsis thaliana*, *Medicago truncatula*, *Populus trichocarpa*, *Vitis vinifera*, *Solanum tuberosum*, and *Solanum lycopersicum*. For *S. lycopersicum*, chromosomes 6 and 8 were analyzed together as one file, and chromosome 10 and 12 also as one file because they were together in the downloaded files. Panel (**b**) shows monocot plants, namely (in order from left to right): *Brachypodium distachyon*, *Oryza sativa*, *Sorghum bicolor*, and *Zea mays*. Note that there is very little variation in C+G content between chromosomes within each dicot or monocot species, with the exception of *B. distachyon*.

primates showed some similarity to, but a more pronounced U shape than the patterns in the non-primates *Oryctolagus cuniculus* and *Rattus norvegicus* (Figure 9a, last two species). This finding may provide insight into the evolutionary relationship between primates and some non-primates. Further research is required to understand the similarity in the relationship between C+G variation patterns and chromosome sizes.

Discussion

Positive, Negative or No Correlation Between C+G Content and Genome Size or Chromosome Length Depending on Kingdoms or Phyla

In this general analysis, in which all chromosomes within each large group were pooled, we found the following relationship between C+G content and chromosome size: strong positive correlation in Gram-positive bacteria and Gram-negative bacteria; strong negative correlation in some animal species (among chromosomes within species in *Taeniopygia guttata* and *Gallus gallus*); and some positive correlations in fungi and plants (Tables 2 and 3). In plants, given the general difference in C+G contents between dicot and monocot plants, a separate analysis is required within each large group.

Most of the plant and animals species in this genome study were also used in our previous analysis of RNA poly(A) site selection [16]. Even through fungi, protists, plants and animals have very different genomic C+G contents, as showed in the present study, the poly(A) tail starting position of the poly(A) site is usually an adenosine in every species [16]. However, the preference for C over G at the mRNA poly(A) tail attachment position is very different in animals and plants: there is almost no difference between C and G in animals, a moderate bias toward C versus G in dicot plants, and a strong preference for C over G in monocot plants [16]. A large-scale analysis is required in order to obtain an overview of genome characteristics (or genome biology traits).

In our analysis, we found that C+G content and genome or chromosome size were positively correlated, negatively correlated, or not correlated in a kingdom-, phylum-, or species-dependent manner. In bacteria, the correlation was positive in 2 of the 11 phyla (within currently available genomes), a finding that may explain the positive correlation previously reported for bacteria [11,12]. However, we found a lack of correlation for 9 of 11 bacterial phyla and an existence of negative correlation between chromosome size and C+G contents in animal species (Table 3).

Unlike the situation for prokaryotes and fungi, in which C+G content and chromosome size were either positively correlated or not correlated, negative correlations were found in animal species (Table 3). Similarly, gene direction evolved from mainly same-direction neighbours in archaeans, bacteria and protozoa (protists) to opposite-direction neighbours in fungi, chlorophyta (protists) and some animals [17]. These findings point to the need for large-scale analyses of different kingdoms in genome biology research. Both gene direction and the correlation between C+G content and genome/chromosome size provide evidence of evolutionary change associated with increasing biological complexity.

Independent Emergence of Strong Correlation Between C+G Content and Genome Size or Chromosome Length During Evolution

Bacteria and Archaea diverged likely since a very early cellular life form appeared on Earth. The first phylogenetically distinct branches for their evolutionary descent from a common ancestor

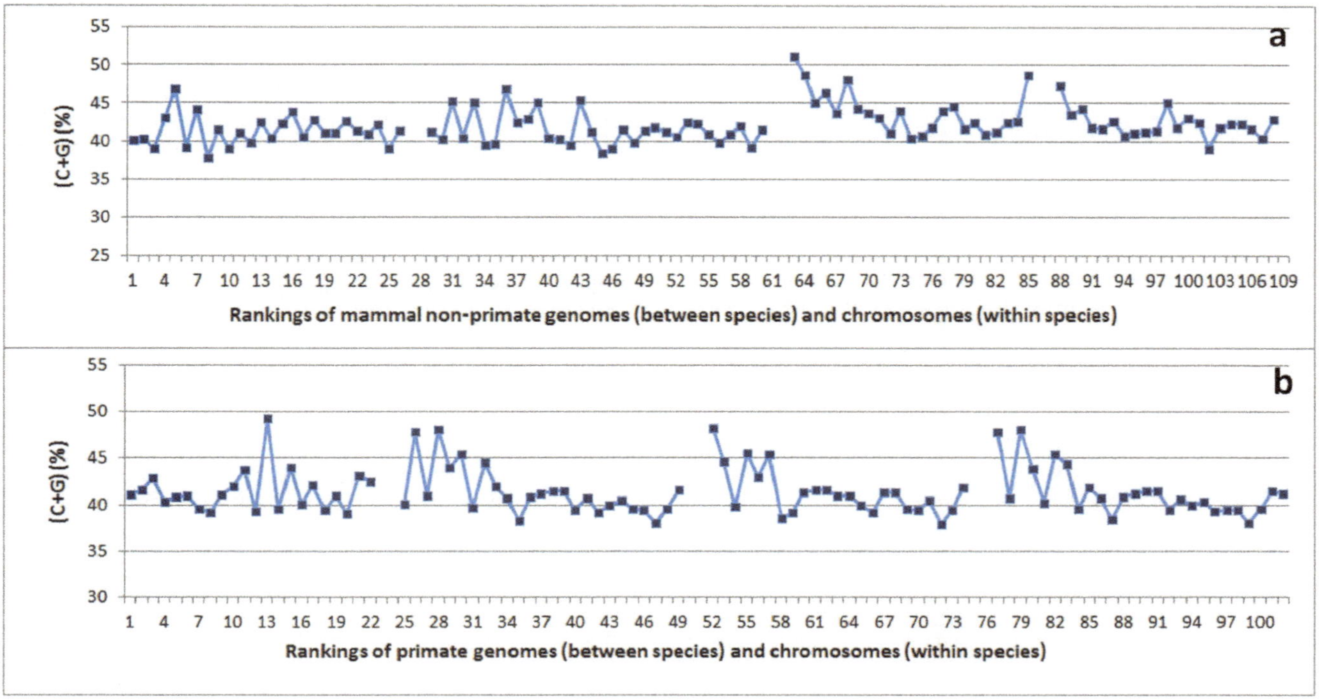

Figure 9. The C+G content (%) per chromosome in mammals, ranked by genome size among species and by chromosome size within each species, from smallest to largest. Panel (**a**) shows non-primate mammalian animals, namely (in order from left to right) *Mus musculus, Equus caballus, Oryctolagus cuniculus*, and *Rattus norvegicus*. Panel (**b**) shows primate animals, namely (in order from left to right) *Callithrix jacchus, Pan troglodytes, Macaca mulatta*, and *Pongo abelii*. Note that the patterns are quite similar among the last three species of primate animals.

have not been identified. However, in some phylogenetic trees [18], bacterial phyla Aquificae and Deinococcusare are among early bacterial branches in terms of the relative branching order, and archean phylum Crenarchaeota also likely branched out earlier than most other archaeal phyla [19]. A quite different tree was derived recently based on an alignment of 24 marker genes; however FastTree was used and the bootstrap support value was very low [20]. Therefore, controversy still exists regarding which clades represent the earliest branches from the common ancestors of the Bacteria or Archaea. If Aquificae and Deinococcusare represent two of the basal branches on the phylogenetic tree for bacteria, the data in the present study may suggest that the early evolutionary branches in bacteria lack any strong correlation between C+G content and genome size. Taken together, Aquificae bacteria, archaeans, and fungi all lacked a strong correlation between C+G content and genome or chromosome size. Therefore, it is likely that the genomes of the basal branches on the tree of life were characterized by a lack of strong correlation between C+G content and genome size. The strong correlations found, such as the strong positive correlation in Actinobacteria and Proteobacteria and the strong negative correlation in some animals, likely emerged independently later after the divergence from their last common ancestor.

Different Genetic Controls for the Gram Staining Trait and the Correlation Between C+G Content and Genome Size in Bacteria

In Gram-positive bacteria, the C+G content and genome size showed a very strong positive correlation in the phylum Actinobacteria but there was no detectable correlation in the phyla Firmicutes and Tenericutes (Table 3). In Gram-negative bacteria, the positive correlation between C+G content and genome size was very strong in Proteobacteria but almost no correlation was found in Aquificae (Table 3). It is known that most Gram-positive bacteria fall in the same large clade on the bacterial phylogenetic tree [21]. The results clearly indicate that the correlation between C+G content and genome size and the Gram staining reaction trait are two different genetically controlled traits.

The phyla Aquificae and Deinococcus are phylogenetically distinct from most other Gram-negative bacteria, and Deinococcus is located on a branch that is dominated by Gram-positive dominated bacteria [21]. In a recent phylogenetic study [20], Aquificae and Deinococcus were grouped together on a large branch that is located between Gram-negative and Gram-positive bacteria. These two bacterial phyla can be in the same situation with regard to the absence of correlation between C+G content and genome size, regardless of whether this pattern has existed since they emerged or is the result of mutation. It provides further evidence that the correlation between C+G content and genome size and the Gram staining colour are two separate traits and controlled by different genes.

Evidence of Both *Cis*- and *Trans*-regulation of Genome Size and C+G Content

The probability plot of genome size in Figure 4 suggests that most genomes or chromosomes fit a normal or nearly normal distribution but the largest genomes and chromosomes stand apart. The results indicate that most chromosome sizes fall within the normal distribution range, likely under a common unknown *trans*-acting regulation that operate on these chromosomes at the cellular level, but some large chromosomes cannot be explained by the common regulation. It is likely that the largest chromosomes were created through some specific events, which may include processes such as chromosome fusion, translocation, *cis*-acting transposition and *cis*-acting gene amplification.

The probability plot of C+G content in Figure 5 suggests that most genomes have the same normal or nearly normal distribution for C+G content. The existence of outliers is very clear in the case of *Drosophila melanogaster* (Figure 6), *Gallus gallus* (Figure S1), *Zea mays* (Figure 6) and *Taeniopygia guttata* (Figure 6). In *Drosophila melanogaster*, the high similarity (41.8% to 43.3%) of C+G content in Chromosomes 2L, 2R, 3L, and X points to a cellular-level influence on C+G content. Chromosome 4 (C+G content: 35.5%) in *Drosophila melanogaster* is an exception, which provides evidence for the existence of local GC-poor mechanism (i.e., *cis*-factor) specifically applying to this chromosome.

Other evidence in support of *cis*- regulation for the variation of C+G content between chromosomes within species comes from the U-shape co-variation pattern between chromosome size and C+G content in animals (Figure 9). These findings have enhanced our knowledge of the correlation between C+G content and genome or chromosome size. Further research is required to investigate the molecular mechanisms behind these *trans*- and *cis*-regulations of C+G contents. Potential cellular-level *in-trans* control can include enzymatic regulation, cellular-level gene regulation on transposition, and the influence of substrate concentration. Candidate local *in-cis* factors may include, for example, specific transposition, gene amplification, chromosome number and structural variation.

The dynamics of genome-size variation in many eukaryotes can be largely explained by transposable elements [22]. For example, SINE (short interspersed elements) represents one of the drivers of genome evolution [23]. In maize, 49% to 78% of the genome is made up of retrotransposons (transposons via RNA intermediates) [24]. In wheat, about 90% of the genome consists of repeated sequences and 68% consists of transposable elements [25]. Around 42% of the human genome is made up of retrotransposons, whereas DNA transposons only account for about 2% to 3% [26]. It is possible that *Bytmar1*, a transposon that is rich in GC and is found in the deep-sea hydrothermal crab, encodes three transposase isoforms from a single open reading frame [27]. It follows that the C+G content should be species-dependent. The transposon type may also partly explain why the correlation between C+G content and chromosome size was positive, negative, or very weak depending on the non-mammal animal species considered in this study. Various mammal animals, particularly primates were found to have a negative correlation between C+G content and chromosome size. Does this mean that a *cis*-acting, GC-poor transposon contributed greatly to primate chromosome size? Clearly, more research is required to clarify this matter. It is also very interesting that even though cereals (monocots) were found to have much higher C+G contents than dicot plants, cereals did not show any strong correlation between C+G content and chromosome size in this study. This finding of an animal–plant difference in the correlation between C+G content and chromosome size may stimulate further comparative research on the regulation of base composition in mammalian animals and plants with a view to determining whether C+G content or genome/chromosome size serves as the main driving force during the interactive evolution between the two genome characteristics.

Somatic Genome Variation of C+G Content

Somatic genome variation (i.e., somagenetic variation) may have partly contributed to the C+G content difference we found between species and between chromosomes in the same species, in our analysis of animals and plants. In flax [28,29] and certain vertebrates [30–32], studies have shown that the genome varies during development and in response to the environment [28,29,33,34]. Viral infection can also promote genome rearrangement in plants [35]. Even in the same plant species (*Nicotiana sylvestris*), somagenetic variation can be stable [36] or unstable over sexual generations [37] depending on the type of genome variation that occurs. Therefore, the genome sequences we analyzed only provide a snapshot of some genomes; at least minor variation in C+G contents is expected if other DNA samples from the same species are sequenced.

In addition, it is known that genome abnormalities and mutation directions can be influenced greatly by dNTP concentrations and non-proportional disturbance in the dNTP pool as well as by temperature range conditions [38–41]. DNA polymerase stability can be influenced by environmental factors such as UV radiation [42]. Deamination of DNA can also change base composition [43]. These influences may at least partly explain the C+G content variation and the patterns of correlation between C+G content and genome size in archaeans and bacteria. Further research is required to clarify whether this non-transposition influence also exists in eukaryotes.

Speciation, Progressive Changes and Sudden Changes

It is interested to note that the large groups of living organisms have distinct C+G contents or correlations between C+G content and chromosome size (Figure 1). According to Darwin's theory [44], evolution is "a slow and gradual process–one species first giving rise to two or three varieties, these being slowly converted into species," and a consequence of natural selection is "Divergence of Character and the Extinction of less-improved forms." The C+G contents analyzed in this study of living organisms were therefore likely favoured and preserved during evolution, and the species with intermediate C+G contents likely became extinct. Natural selection undoubtedly played an important role in determining the C+G contents found in living organisms, and the extinction of some unfit species contributed to the gaps between the large groups existing today (Figure 1). However, some large groups such as the mammal subkingdom have a very long history (250 million years) [45], but all the sequenced genomes still have very similar C+G contents (Figure 1). This may indicate that the evolution path from variety to species may not be as smooth as Darwin believed. A sudden major gene mutation resulting from a process involving transposition, deamination, dNTP pool balance, or DNA polymerase might accelerate the process. The rate of change slows when the new species or group is nearly equilibrated in terms of cellular function, biochemical resource usage, body part cooperation, and adaptation to the environment, and therefore the C+G contents are largely similar among the species of the new group. Both progressive changes and sudden changes contribute to speciation. Further research at the molecular level is required to verify why the C+G contents differ so much between large groups of living organisms.

C+G Contents in Whole Genomes and Specific Function Units

Although the genomic C+G content varies greatly among kingdoms, subkingdoms, and species (Table 3), the C+G contents in certain functional units were often found to be less variable. The longest coding sequences-exons in various vertebrates and prokaryotes tended to be GC-rich [14]. The RNA poly(A) tail starting position of poly(A) site is usually an adenosine in every species of microorganisms, plants and animals studied to date [16]. Uracil (U) content was found to be higher in the mRNA 3′ untranslated region (3′UTR) than in the pre-mRNA 3′ cleaved-off region (3′COR), and always higher than that of the whole genome in every subkingdom of plants and animals [46]. The genomic T content appears to be too low in monocotyledonous genomes and must be enriched to a certain degree to permit the 3′COR and 3′UTR to function properly in monocotyledonous cells [46]. The functional units, such as 3′COR and 3′UTR, on genome are found to be relatively stable while the whole genomes diverge much faster in terms of the base composition [46]. Further research is required to understand the roles of genomic base composition in evolution of living organisms.

This study analyzed C+G contents and their correlations with chromosome size in different kingdoms of living organisms. Positive, negative, or insignificant correlations were found depending on the kingdom, large phylogenetic group or species considered. Certain co-variation patterns between chromosome-size and C+G content were conserved between species within groups of animals and plants. This suggests that different molecular and evolutionary mechanisms, e.g., genetic control by both *trans*- and *cis*-factors, are responsible for the different patterns of C+G content and genome or chromosome size in different kingdoms, groups, and species.

Methods

Genome and Chromosome Sequences

The completely sequenced genomes and various incomplete but assembled genomes in the GenBank, RefSeq, EMBL, DDBJ, and PDB databases were downloaded from the National Center for Biotechnology Information (NCBI) website (http://www.ncbi.nlm.nih.gov). For bacterial genomes, because of limitations in terms of computational power, only the 508 most recently sequenced genomes (non-redundant ones) were analyzed. For archaeans and bacteria, only the complete sequences of the genomes were used. For species in other kingdoms, the chromosomes were essentially described either as complete sequences or as whole-genome shotgun sequences grouped into specific chromosomes. Although the human genome sequence had not yet been completed and did not meet these criteria, it was included in this analysis because of its obvious importance and its interest for readers. If the same strain was sequenced independently by different research groups and there was more than one genome sequence in the NCBI, we randomly chose one of the sequences if sequence length was identical. The genome and chromosome ID list is provided in Table S1.

C+G Content Calculation

The genome or chromosome sizes and the base compositions were calculated using PERL scripts. The base composition characteristics included the percentages of A, C, G, T, C+G, and A+T, and the (A+T)/(C+G) ratio. Genome base composition was calculated from the overall content of each specific base in the overall amount of bases in that genome. The average chromosome base composition was calculated by taking the average value of all the chromosomes after calculating the base composition of each chromosomes or large scaffold. The genome C+G content and the average chromosome C+G content were very similar but often not exactly the same.

Correlation Analysis

The genome size and C+G content were plotted first to see whether there was a linear correlation. If these plots did not show a linear relationship between the C+G content and genome or chromosome size, Pearson correlation analysis and linear regression analysis were not used as they would likely not be suitable. In such a case, we used Spearman correlation analysis to test for monotonic relationship. The Spearman correlation analysis was conducted using the CORREL function in the Excel 2010 data analysis package after ranking the data from the smallest to the largest. The Spearman R values along with the p-values were calculated and verified using the COR.TEST function in the statistical software of R program (http://www.r-project.org/). The p-values were computed using the algorithm AS 89 for $n < 1290$ and exact = TRUE; otherwise, the asymptotic t approximation was used [47,48]. Probability plots of genome size and C+G content were created to examine the type of distribution (not for correlation analysis), and outliers were identified using the regression function in the Excel 2010 data analysis package.

Calculations and statistical analysis were carried out mainly at the first three of the following levels: 1) genome-level comparison of kingdoms and large groups based on the genome sizes and C+G contents of the genomes; 2) chromosome-level comparison of kingdoms and large groups; 3) chromosome-level analysis within each group and species; and 4) chromosome-type-level analysis for some archaean species, because three of them had one larger chromosome and one smaller chromosome.

We ranked the strength of the Spearman correlation R using the grouping by Swinscow and Campbell [41]: R = 0.0–0.19 very weak (which we called it "no detectable correlation or "no correlation"), 0.2–0.39 weak, 0.40–0.59 moderate, 0.6–0.79 strong and 0.8–1 very strong. However, we used this ranking of moderate, strong, and very strong R values only if the R values were significantly different from zero ($P<0.05$).

Supporting Information

Figure S1 The chromosome C+G contents in *Gallus gallus*. Note that, on average, C+G content and chromosome size are negatively correlated in this species, but four chromosomes behaved differently.

Acknowledgments

The author would like to thank Mr. Martin Lague, of Agriculture and Agri-Food Canada, for the technical assistance in writing the PERL scripts used for file downloading and reformatting.

Author Contributions

Conceived and designed the experiments: XQL. Performed the experiments: XQL. Analyzed the data: XQL DD. Contributed reagents/materials/analysis tools: XQL. Wrote the paper: XQL. Wrote and tested most of the Perl and R programs: DD. Revised and finalized the manuscript: XQL DD.

References

1. Chargaff E, Lipshitz R, Green C (1952) Composition of the desoxypentose nucleic acids of four genera of sea-urchin. J Biol Chem 195: 155–160.
2. Watson JD, Crick FHC (1953) Molecular structure of nucleic acids: A structure for deoxyribose nucleic acid. Nature 171: 737–738.
3. Cionini PG, Bassi P, Cremonini R, Cavallini A (1985) Cytological localization of fast renaturing and satellite DNA sequences in *Vicia faba*. Protoplasma 124: 106–111.
4. Appels R, Dennis ES, Smyth DR, Peacock WJ (1981) Two repeated DNA sequences from the heterochromatic regions of rye (*Secale cereale*) chromosomes. Chromosoma 84: 265–277.
5. Nellåker C, Li F, Uhrzander F, Tyrcha J, Karlsson H (2009) Expression profiling of repetitive elements by melting temperature analysis: Variation in HERV-W gag expression across human individuals and tissues. BMC Genomics 10: 532.
6. Šmarda P, Bureš P, Šmerda J, Horová L (2012) Measurements of genomic GC content in plant genomes with flow cytometry: A test for reliability. New Phytol 193: 513–521.
7. Mooers AØ, Holmes EC (2000) The evolution of base composition and phylogenetic inference. Trends Ecol Evol 15: 365–369.
8. Gardiner K (1996) Base composition and gene distribution: Critical patterns in mammalian genome organization. Trends Genet 12: 519–524.
9. Presber W, Schroeder C, Krueger DH (1976) The evolution of mutation rates via DNA base composition and mutator genes: A proposition. J Theor Biol 59: 353–360.
10. Šmarda P, Bureš P (2012) The variation of base composition in plant genomes. In: Wendel JF, Greilhuber J, Doležel J, Leitch IJ, editors. Plant Genome Diversity: Springer-Verlag Wien. pp. 209–235.
11. Nishida H (2012) Evolution of genome base composition and genome size in bacteria. Front Microbiol 3: 420.
12. Musto H, Naya H, Zavala A, Romero H, Alvarez-Valín F, et al. (2006) Genomic GC level, optimal growth temperature, and genome size in prokaryotes. Biochem Biophys Res Commun 347: 1–3.
13. Pozzoli U, Menozzi G, Fumagalli M, Cereda M, Comi GP, et al. (2008) Both selective and neutral processes drive GC content evolution in the human genome. BMC Evol Biol 8: 99.
14. Oliver JL, Marín A (1996) A relationship between GC content and coding-sequence length. J Mol Evol 43: 216–223.
15. Mitchell D (2007) GC content and genome length in Chargaff compliant genomes. Biochem Biophys Res Commun 353: 207–210.
16. Li XQ, Du D (2013) RNA polyadenylation sites on the genomes of microorganisms, animals, and plants. PLoS ONE 8: e79511.
17. Li XQ, Du D (2012) Gene direction in living organisms. Sci Rep 2: 982.
18. Allers T, Mevarech M (2005) Archaeal genetics - The third way. Nat Rev Genet 6: 58–73.
19. Woese CR, Kandler O, Wheelis ML (1990) Towards a natural system of organisms: Proposal for the domains Archaea, Bacteria, and Eucarya. Proc Natl Acad Sci U S A 87: 4576–4579.
20. Lang JM, Darling AE, Eisen JA (2013) Phylogeny of bacterial and archaeal genomes using conserved genes: supertrees and supermatrices. PLoS ONE 8: e62510.
21. Wu D, Hugenholtz P, Mavromatis K, Pukall R, Dalin E, et al. (2009) A phylogeny-driven genomic encyclopaedia of Bacteria and Archaea. Nature 462: 1056–1060.
22. Pritham EJ (2009) Transposable elements and factors influencing their success in eukaryotes. J Hered 100: 648–655.
23. Schmitz J (2012) SINEs as driving forces in genome evolution. Genome Dyn 7: 92–107.
24. Sanmiguel P, Bennetzen JL (1998) Evidence that a recent increase in maize genome size was caused by the massive amplification of intergene retro-transposons. Ann Bot 82: 37–44.
25. Li W, Zhang P, Fellers JP, Friebe B, Gill BS (2004) Sequence composition, organization, and evolution of the core Triticeae genome. Plant J 40: 500–511.
26. Lander ES, Linton LM, Birren B, Nusbaum C, Zody MC, et al. (2001) Initial sequencing and analysis of the human genome. Nature 409: 860–921.
27. Halaimia-Toumi N, Casse N, Demattei MV, Renault S, Pradier E, et al. (2004) The GC-rich transposon *Bytmar1* from the deep-sea hydrothermal crab, *Bythograea thermydron*, may encode three transposase isoforms from a single ORF. J Mol Evol 59: 747–760.
28. Cullis CA (1986) Phenotypic consequences of environmentally induced changes in plant DNA. Trends Genet 2: 307–309.
29. Cullis CA (2005) Mechanisms and control of rapid genomic changes in flax. Ann Bot 95: 201–206.
30. Smith JJ, Antonacci F, Eichler EE, Amemiy CT (2009) Programmed loss of millions of base pairs from a vertebrate genome. Proc Natl Acad Sci U S A 106: 11212–11217.
31. Smith JJ, Baker C, Eichler EE, Amemiya CT (2012) Genetic consequences of programmed genome rearrangement. Curr Biol 22: 1524–1529.
32. Smith JJ, Kuraku S, Holt C, Sauka-Spengler T, Jiang N, et al. (2013) Sequencing of the sea lamprey (*Petromyzon marinus*) genome provides insights into vertebrate evolution. Nat Genet 45: 415–421.
33. Li XQ (2008) Nature, extent and developmental regulation of somagenetic variations. CAB Reviews: Perspect Agri, Vet Sci Nutri Natur Resourc 3: No.031.
34. Li XQ (2009) Developmental and environmental variation in genomes. Heredity 102: 323–329.
35. Yao Y, Bilichak A, Golubov A, Kovalchuk I (2011) Local infection with oilseed rape mosaic virus promotes genetic rearrangements in systemic Arabidopsis tissue. Mut Res 709–710: 7–14.
36. Li XQ, Chetrit P, Mathieu C, Vedel F, De Paepe R, et al. (1988) Regeneration of cytoplasmic male sterile protoclones of *Nicotiana sylvestris* with mitochondrial variations. Curr Genet 13: 261–266.
37. Li XQ (1983) Analysis of the evolution and nature of DNA content changes induced by androgenesis in *Nicotiana sylvestris* Spegaz and Comes, during plant develpment. DEA Thesis, Univ Paris-Sud: 24 pages (in French).

38. Davidson MB, Katou Y, Keszthelyi A, Sing TL, Xia T, et al. (2012) Endogenous DNA replication stress results in expansion of dNTP pools and a mutator phenotype. EMBO J 31: 895–907.
39. Ouspenski II, Elledge SJ, Brinkley BR (1999) New yeast genes important for chromosome integrity and segregation identified by dosage effects on genome stability. Nucleic Acids Res 27: 3001–3008.
40. Chabes A, Stillman B (2007) Constitutively high dNTP concentration inhibits cell cycle progression and the DNA damage checkpoint in yeast *Saccharomyces cerevisiae*. Proc Natl Acad Sci U S A 104: 1183–1188.
41. Zheng H, Wu H (2010) Gene-centric association analysis for the correlation between the guanine-cytosine content levels and temperature range conditions of prokaryotic species. BMC Bioinformatics 11: S7.
42. Skoneczna A, McIntyre J, Skoneczny M, Policinska Z, Sledziewska-Gojska E (2007) Polymerase eta is a short-lived, proteasomally degraded protein that is temporarily stabilized following UV irradiation in *Saccharomyces cerevisiae*. J Mol Biol 366: 1074–1086.
43. Jiang C, Zhao Z (2006) Directionality of point mutation and 5-methylcytosine deamination rates in the chimpanzee genome. BMC Genomics 7: 316.
44. Darwin CR (1876) The Origin of Species by Means of Natural Selection, or the Preservation of Favoured Races in the Struggle for Life. London: John Murray (6th edition with additions and corrections). 460 p.
45. Okada N, Sasaki T, Shimogori T, Nishihara H (2010) Emergence of mammals by emergency: Exaptation. Genes Cells 15: 801–812.
46. Li XQ (2014) Comparative analysis of the base compositions of the pre-mRNA 3′ cleaved-off region and the mRNA 3′ untranslated region relative to the genomic base composition in animals and plants. PLoS ONE (Accepted with revision).
47. Best DJ, Roberts DE (1975) Algorithm AS 89: The upper tail probabilities of Spearman's rho. Appl Statist 24: 377–379.
48. Hollander M, Wolfe DA (1973) Nonparametric statistical methods. New York: John Wiley & Sons.

18

Selection Strategies for the Development of Maize Introgression Populations

Eva Herzog[1], Karen Christin Falke[2], Thomas Presterl[3], Daniela Scheuermann[3], Milena Ouzunova[3], Matthias Frisch[1]*

1 Institute of Agronomy and Plant Breeding II, Justus Liebig University, Giessen, Germany, **2** Institute for Evolution and Biodiversity, University of Münster, Münster, Germany, **3** KWS Saat AG, Einbeck, Germany

Abstract

Introgression libraries are valuable resources for QTL detection and breeding, but their development is costly and time-consuming. Selection strategies for the development of introgression populations with a limited number of individuals and high-throughput (HT) marker assays are required. The objectives of our simulation study were to design and compare selection strategies for the development of maize introgression populations of 100 lines with population sizes of 360–720 individuals per generation for different DH and S_2 crossing schemes. Pre-selection for complete donor chromosomes or donor chromosome halves reduced the number of simultaneous backcross programs. The investigated crossing and selection schemes differed considerably with respect to their suitability to create introgression populations with clearly separated, evenly distributed target donor chromosome segments. DH crossing schemes were superior to S_2 crossing schemes, mainly due to complete homozygosity, which greatly reduced the total number of disjunct genome segments in the introgression populations. The S_2 crossing schemes were more flexible with respect to selection and provided economic alternatives to DH crossing schemes. For the DH crossing schemes, increasing population sizes gradually over backcross generations was advantageous as it reduced the total number of required HT assays compared to constant population sizes. For the S_2 crossing schemes, large population sizes in the final backcross generation facilitated selection for the target segments in the final backcross generation and reduced fixation of large donor chromosome segments. The suggested crossing and selection schemes can help to make the genetic diversity of exotic germplasm available for enhancing the genetic variation of narrow-based breeding populations of crops.

Editor: Lewis Lukens, University of Guelph, Canada

Funding: Funding from the German Federal Ministry of Education and Research (BMBF Grant 0315951) is gratefully acknowledged. [http://www.bmbf.de/en/index.php] The funders had no role in study design, data collection and analysis, decision to publish, or preparation of the manuscript.

Competing Interests: The authors have the following interests: Thomas Presterl, Daniela Scheuermann and Milena Ouzunova are employed by KWS Saat AG. There are no patents, products in development or marketed products to declare related to KWS SAAT AG and the subject matter of the publication. This does not alter the authors' adherence to all the PLOS ONE policies on sharing data and materials, as detailed online in the guide for authors.

* E-mail: matthias.frisch@uni-giessen.de

Introduction

Introgression libraries are valuable resources for the identification of alleles of agricultural interest in exotic germplasm. They facilitate the introduction of new genetic variation into elite breeding germplasm by providing favorable chromosome segments from wild or exotic species in an adapted genetic background [1,2]. Ideally, an introgression library consists of a set of homozygous introgression lines (ILs) which carry short marker-defined chromosome segments from an exotic donor in a common genetic background. The concept was first described in tomato [3]. In the mean time, introgression libraries have been developed for the model species *Arabidopsis thaliana* [4,5], and in many agriculturally important crops, such as rice [6,7], barley [8,9], wheat [10,11], maize [12,13] and rye [14].

Introgression libraries are usually developed by marker-assisted backcrossing followed by selfing or production of double haploid (DH) lines. The backcross process for their development is costly and labor-intensive if complete coverage of the donor genome by short evenly distributed target chromosome segments is to be achieved. Often additional backcross programs have to be run for the developed ILs in order to close gaps in donor genome coverage, or to shorten donor chromosome segments by additional recombination events [3,9]. In spite of the high resource requirements, only incomplete donor genome coverage has been achieved for most of the reported introgression libraries [9,14].

In previous simulation studies on introgression libraries, two generations of selfing were investigated for line development [15,16]. Recent genetic studies in maize were based on ILs that underwent two to five generations of selfing [17–19]. The use of DH technology has to our knowledge not yet been investigated in simulation studies on the development of introgression libraries. However, *in vivo* induction of maternal haploids is currently a routine method of DH production in commercial maize breeding programs. The main advantage of the DH technology is that complete homozygosity can be obtained after only two generations. Inspite of this time-saving, the production of DH lines is still considerably more costly than conventional selfing [20]. Moreover, a current drawback of *in vivo* induction of maternal haploids in maize is that on average only one viable DH line can be derived from one backcross individual. It is therefore of economic interest to compare this method with S_2 crossing schemes which require the same number of generations to evaluate the benefits of DH lines.

A possible approach to tackle the high costs required for the development of ideal introgression libraries would be to resort to introgression populations which are not perfect in appearance, but carry some additional donor segments outside the actual target segments. Such introgression populations could be developed with fewer individuals and marker assays. Complete coverage of the donor genome is desirable in order to capture the whole wealth of alleles of agricultural interest in the exotic donor. It is therefore one component of a minimum standard which introgression populations should meet. A second component are short, evenly distributed target donor chromosome segments in a clean adapted background, as they facilitate the use of the ILs in the following breeding process.

The design of the crossing scheme and the selection strategy are the most important factors that influence the distribution of donor chromosome segments in the introgression population. Falke et al. [16] suggested for the development of ideal introgression libraries that a chromosome-based selection strategy which pre-selects individuals carrying the donor alleles on complete chromosomes in generation BC_1 saves resources. Adapting and advancing this concept to crossing schemes with small population sizes might be an efficient approach to develop introgression populations with a limited number of marker assays.

The objectives of our simulation study were (1) to design selection strategies and crossing schemes for the development of maize introgression populations with limited resources, (2) to compare these selection strategies with respect to the distribution and length of donor chromosome segments and the required investments in terms of time, individuals and marker assays, (3) to give guidelines for the optimal experimental design for constructing introgression populations.

Materials and Methods

Software

All simulations were conducted in R version 3.0.0 [21] with the software package SelectionTools, which is available from http://www.uni-giessen.de/population-genetics/downloads.

Genetic Model

A genetic model of maize with 10 equally sized chromosomes of 200 cM length was used for the simulations. Genetic markers for selection were equally spaced. The distance between two adjacent marker loci was 1 cM. All markers were polymorphic between donor and recipient. It was assumed that markers were analyzed with high-throughput (HT) assays. One HT assay comprised genotyping one individual at all marker loci in the linkage map. Recombination was modelled assuming no interference in crossover formation [22]. Each simulation of an introgression population of 100 ILs was replicated 1,000 times in order to reduce sampling effects and to obtain results with high numerical accuracy and a small standard error.

Crossing Schemes

Four crossing schemes were investigated: BC_2DH, BC_3DH, BC_2S_2, BC_3S_2. Each crossing scheme started with the cross of a homozygous donor and a homozygous recipient to create one F_1 individual. The F_1 individual was backcrossed to the recipient to create a BC_1 population of size n_{BC1}. From the BC_1 population, the best individuals with the highest values of selection indices for the respective selection strategy were selected. Each of the selected BC_1 individuals was backcrossed to the recipient to create BC_2 sub-populations of size n_{BC2}. From these BC_2 sub-populations, the best individuals with the highest values of the respective selection indices were selected. For the DH crossing schemes, *in vivo* induction of maternal haploids was assumed with a success rate of one viable DH line per backcross individual. For the BC_2DH schemes, one DH line was thus created from each of the selected BC_2 individuals. For the BC_2S_2 crossing schemes, the selected BC_2 individuals were selfed to create a fixed number of S_1 individuals. Each of the S_1 individuals was selfed again and one S_2 individual was created. For the BC_3 crossing schemes, each of the selected BC_2 individuals was backcrossed to the recipient to create BC_3 sub-populations of size n_{BC3}. From these BC_3 sub-populations, the best individuals with the highest values of the respective selection indices were selected. The generations S_1, S_2 or DH of the BC_3 crossing schemes were carried out as described for the BC_2 crossing schemes.

Evaluation of Selection Candidates

The final introgression populations should consist of 100 ILs which guarantee an acceptable resolution of QTL detection in maize, and which can be immediately used in further breeding steps. Each IL should ideally carry a 20 cM chromosome segment from the donor to provide a complete and even coverage of the donor genome without overlap. The 20 cM chromosome segments are hereafter simply referred to as "target segments". To determine the selection index for an individual with respect to a given target segment, we denote with t_c the donor genome proportion of the chromosome on which the target segment is located, with t_h the donor genome proportion of the chromosome half on which the target segment is located and with t_s the donor genome proportion of the target segment itself. The values for the genetic background b_c, b_h, b_s correspond to t_c, t_h, t_s and denote the recipient genome proportion outside the respective chromosome region. Depending on the selection strategy, t and b are used to define selection indices.

Selection Strategies

We considered generations $g=\{BC_1,BC_2,BC_3,DH,S_1,S_2\}$ for selection. Generation DH was the generation in which homozygous diploid DH lines were available for selection. In each generation g, the genome was divided into selection regions that could either be 10 complete chromosomes, 20 chromosome halves or 100 target segments. For selection for complete donor chromosomes, a fixed number n_{sel} of best individuals for each of the chromosomes $c=1,2,...,10$ with the highest values for selection index $i=t_c+b_c$ were selected. For selection for donor chromosome halves, a fixed number n_{sel} of best individuals for each of the chromosome halves $h=1,2,...,20$ with the highest values for selection index $i=t_h+b_h$ were selected. For selection for donor target segments, a fixed number n_{sel} of best individuals for each of the target segments $s=1,2,...,100$ with the highest values for selection index $i=t_s+b_s$ were selected.

Selection for complete donor chromosomes, donor chromosome halves and donor target segments were combined to form different selection strategies. Selection for complete donor chromosomes in a backcross generation is denoted by a C in the strategy name, selection for donor chromosome halves is denoted by an H, and selection for donor target segments is denoted by an S. For example, for strategy CH, selection for complete donor chromosomes was conducted in generation BC_1 while selection for donor chromosome halves was conducted in generation BC_2. An overview of the investigated selection strategies is presented in Table 1. The investigated combinations of crossing scheme and selection strategy are listed in the first column of Table 2. For all

Table 1. Definition of the selection index i in generations BC_1, BC_2, BC_3, DH, S_1, S_2 for different selection strategies for developing introgression populations.

	Generation				
Strategy	BC_1	BC_2	BC_3	S_1	**DH/**S_2
C	t_c+b_c	–	–	–	t_s-b_s
H	t_h+b_h	–	–	–	t_s-b_s
CC	t_c+b_c	t_c+b_c	–	–	t_s-b_s
HH	t_h+b_h	t_h+b_h	–	–	t_s+b_s
CH	t_c+b_c	t_h+b_h	–	–	t_s+b_s
CCC	t_c+b_c	t_c+b_c	t_c+b_c	–	t_s+b_s
HHH	t_h+b_h	t_h+b_h	t_h+b_h	–	t_s+b_s
CHH	t_c+b_c	t_h+b_h	t_h+b_h	–	t_s+b_s
HHS	t_h+b_h	t_h+b_h	t_s+b_s	–	t_s+b_s

Selection for complete donor chromosomes (C), selection for donor chromosome halves (H) and selection for donor target segments (S) were combined to form different selection strategies (left column). t_c, t_h and t_s denote the donor genome proportions of the chromosome on which the target segment is located, of the chromosome half on which the target segment is located and of the target segment itself. b_c, b_h and b_s correspond to t_c, t_h, t_s and denote the recipient genome proportion outside the respective chromosome region.

selection strategies, the best 100 ILs for selection index $i=t_s+b_s$ were selected in generation DH or S_2, depending on the crossing scheme.

Population Sizes and Simulation Series

We investigated population sizes of $n_{tot}=360-720$ individuals per backcross generation. This should be within a range which can be realized in practical maize breeding programs. Variations in population size were investigated to determine both the effect on preserving the target segments up to line development as well as on recovering the genotype of the recipient outside the target segments.

In the first series of simulations, basic crossing schemes were investigated. Selection was carried out in generation BC_1 for basic crossing schemes with two backcross generations, and in generations BC_1 and BC_2 for basic crossing schemes with three backcross generations. The total population size per generation was kept constant at $n_{tot}=360$ individuals in every generation g.

In the second series of simulation, crossing schemes with high selection intensity were investigated. Population size was doubled compared to the basic crossing schemes ($n_{tot}=720$) in every generation g, while the number of selected individuals was the same as for the basic crossing schemes. The crossing schemes with high selection intensity are denoted by BC_3-CC', BC_3-HH' and BC_3-CH' (Table 2). In the first and second series of simulations, all backcross individuals generated in the final backcross generation were used for line development for both DH and S_2 crossing schemes. One IL was derived from one backcross individual.

In the third series of simulations, crossing schemes with selection in the final backcross generation were investigated. n_{tot} was doubled to 720 individuals in the final backcross generation for the DH crossing schemes BC_2DH-CC, BC_2DH-HH, BC_2DH-CH, $BC_3DH-CCC$, $BC_3DH-HHH$, $BC_3DH-CHH$. This increase in population size was necessary to enable selection and to keep n_{tot} at 360 individuals in generation DH. For the corresponding S_2 schemes, n_{tot} was kept at 360 individuals also in the final backcross generation.

In the fourth series of simulations, crossing schemes with increasing population sizes were investigated. Selection was conducted in the final backcross generation. The crossing schemes with increasing population sizes are denoted by BC_3-HHH^* and BC_3-HHS^*. The details concerning the total population size n_{tot} and population sizes in the sub-populations n_g for all investigated combinations of crossing scheme and selection strategy are summarized in Table 2. Schematic representations of the crossing schemes $BC_3DH-HHH^*$ and $BC_3S_2-HHS^*$ are given in Figure 1 and Figure 2 for illustration.

Measures

To evaluate and compare introgression populations originating from different crossing and selection schemes, the following measures were determined: (a) the genome coverage of the donor O in percent, which is defined as the proportion of the donor genome which is covered by the introgression population, irrespective of whether by the target segments or other donor segments in the genetic background, (b) the depth of donor genome coverage T, which is defined as the average number of ILs in which each donor allele appears in the introgression population, (c) the number of disjunct genome segments in the introgression population S, (d) the resolution of the introgression population R in cM, which is defined as the total genome length of the genetic model in cM divided by S, (e) the average number of donor segments per IL N, (f) the average length of donor segments per IL L in cM, (g) the average total donor genome proportion of the introgression population D_t in percent, (h) the average donor genome proportion of the chromosomes carrying the respective target segments D_c in percent, (i) the average donor genome proportion of the target segments D_s in percent.

Results

High values for the donor genome coverage O around 99% were observed for all crossing schemes (Table 3). However, the resulting introgression populations differed substantially in the values for the number of disjunct genome segments S, the total donor genome proportion D_t, the donor genome proportion of the carrier chromosomes D_c and the donor genome proportion of the target segments D_s. BC_3 crossing schemes resulted in 2–3% lower values for D_t than BC_2 crossing schemes, even if the number of

Table 2. Subdivision of the total population sizes n_{tot} into sub-population sizes n_g in generations $g=BC_1,BC_2,BC_3,S_1,DH,S_2$ for different crossing and selection schemes for developing introgression populations.

	Generation				
Scheme	BC_1	BC_2	BC_3	S_1	**DH/S_2**
Basic crossing schemes					
BC_2DH-C	$1\times1\times360$	$1\times10\times36$	–	–	$10\times36\times1$
BC_2DH-H	$1\times1\times360$	$1\times20\times18$	–	–	$20\times18\times1$
BC_3DH-CC	$1\times1\times360$	$1\times10\times36$	$10\times1\times36$	–	$10\times36\times1$
BC_3DH-HH	$1\times1\times360$	$1\times20\times18$	$20\times1\times18$	–	$20\times18\times1$
BC_3DH-CH	$1\times1\times360$	$1\times10\times36$	$10\times2\times18$	–	$20\times18\times1$
Crossing schemes with high selection intensity					
BC_3DH-CC'	$1\times1\times720$	$1\times10\times72$	$10\times1\times72$	–	$10\times72\times1$
BC_3DH-HH'	$1\times1\times720$	$1\times20\times36$	$20\times1\times36$	–	$20\times36\times1$
BC_3DH-CH'	$1\times1\times720$	$1\times10\times72$	$10\times2\times36$	–	$20\times36\times1$
Crossing schemes with selection in the final BC generation					
BC_2DH-CC	$1\times1\times360$	$1\times10\times72$	–	–	$10\times36\times1$
BC_2DH-HH	$1\times1\times360$	$1\times20\times36$	–	–	$20\times18\times1$
BC_2DH-CH	$1\times1\times360$	$1\times10\times72$	–	–	$10\times(2\times18)\times1$
$BC_3DH-CCC$	$1\times1\times360$	$1\times10\times36$	$10\times1\times72$	–	$10\times36\times1$
$BC_3DH-HHH$	$1\times1\times360$	$1\times20\times18$	$20\times1\times36$	–	$20\times18\times1$
$BC_3DH-CHH$	$1\times1\times360$	$1\times10\times36$	$10\times2\times36$	–	$20\times18\times1$
Crossing schemes with increasing population sizes					
$BC_3DH-HHH^*$	$1\times1\times180$	$1\times20\times18$	$20\times1\times27$	–	$20\times18\times1$
$BC_3DH-HHS^*$	$1\times1\times180$	$1\times20\times18$	$20\times1\times30$	–	$20\times(5\times3)\times1$
Basic crossing schemes					
BC_2S_2-C	$1\times1\times360$	$1\times10\times36$	–	$10\times36\times1$	$10\times36\times1$
BC_2S_2-H	$1\times1\times360$	$1\times20\times18$	–	$20\times18\times1$	$20\times18\times1$
BC_3S_2-CC	$1\times1\times360$	$1\times10\times36$	$10\times1\times36$	$10\times36\times1$	$10\times36\times1$
BC_3S_2-HH	$1\times1\times360$	$1\times20\times18$	$20\times1\times18$	$20\times18\times1$	$20\times18\times1$
BC_3S_2-CH	$1\times1\times360$	$1\times10\times36$	$10\times2\times18$	$20\times18\times1$	$20\times18\times1$
Crossing schemes with high selection intensity					
BC_3S_2-CC'	$1\times1\times720$	$1\times10\times72$	$10\times1\times72$	$10\times72\times1$	$10\times72\times1$
BC_3S_2-HH'	$1\times1\times720$	$1\times20\times36$	$20\times1\times36$	$20\times36\times1$	$20\times36\times1$
BC_3S_2-CH'	$1\times1\times720$	$1\times10\times72$	$10\times2\times36$	$20\times36\times1$	$20\times36\times1$
Crossing schemes with selection in the final BC generation					
BC_2S_2-CC	$1\times1\times360$	$1\times10\times36$	–	$10\times1\times36$	$10\times36\times1$
BC_2S_2-HH	$1\times1\times360$	$1\times20\times18$	–	$20\times1\times18$	$20\times18\times1$
BC_2S_2-CH	$1\times1\times360$	$1\times10\times36$	–	$10\times2\times18$	$20\times18\times1$
BC_3S_2-CCC	$1\times1\times360$	$1\times10\times36$	$10\times1\times36$	$10\times1\times36$	$10\times36\times1$
BC_3S_2-HHH	$1\times1\times360$	$1\times20\times18$	$20\times1\times18$	$20\times1\times18$	$20\times18\times1$
BC_3S_2-CHH	$1\times1\times360$	$1\times10\times36$	$10\times2\times18$	$20\times1\times18$	$20\times18\times1$
Crossing schemes with increasing population sizes					
$BC_3S_2-HHH^*$	$1\times1\times180$	$1\times20\times18$	$20\times1\times27$	$20\times1\times18$	$20\times18\times1$
$BC_3S_2-HHS^*$	$1\times1\times180$	$1\times20\times18$	$20\times1\times23$	$20\times5\times4$	$100\times1\times4$

The total population size in generation g is defined as $n_{tot}=n_{pop}\times n_{sel}\times n_g$. n_{pop}: number of sub-populations in generation $g-1$; n_{sel}: number of individuals selected from the sub-populations in generation $g-1$; n_g: population size per sub-population in generation g.

generations of selection was the same. For example, the basic crossing scheme BC_3DH-CC resulted in a D_t of only 5.0%, while crossing scheme BC_2DH-CC with selection in the final backcross generation resulted in a D_t of 7.8%. An additional generation of selection in BC_2 schemes only resulted in minor improvements of D_t of 0.4–1.4% compared to the basic crossing schemes without selection. For example, scheme BC_2DH-CC improved D_t only by 0.5% compared to scheme BC_2DH-C.

The DH crossing schemes had in most cases better values for T, D_t, D_c, D_s and especially S than the S_2 crossing schemes (Table 3).

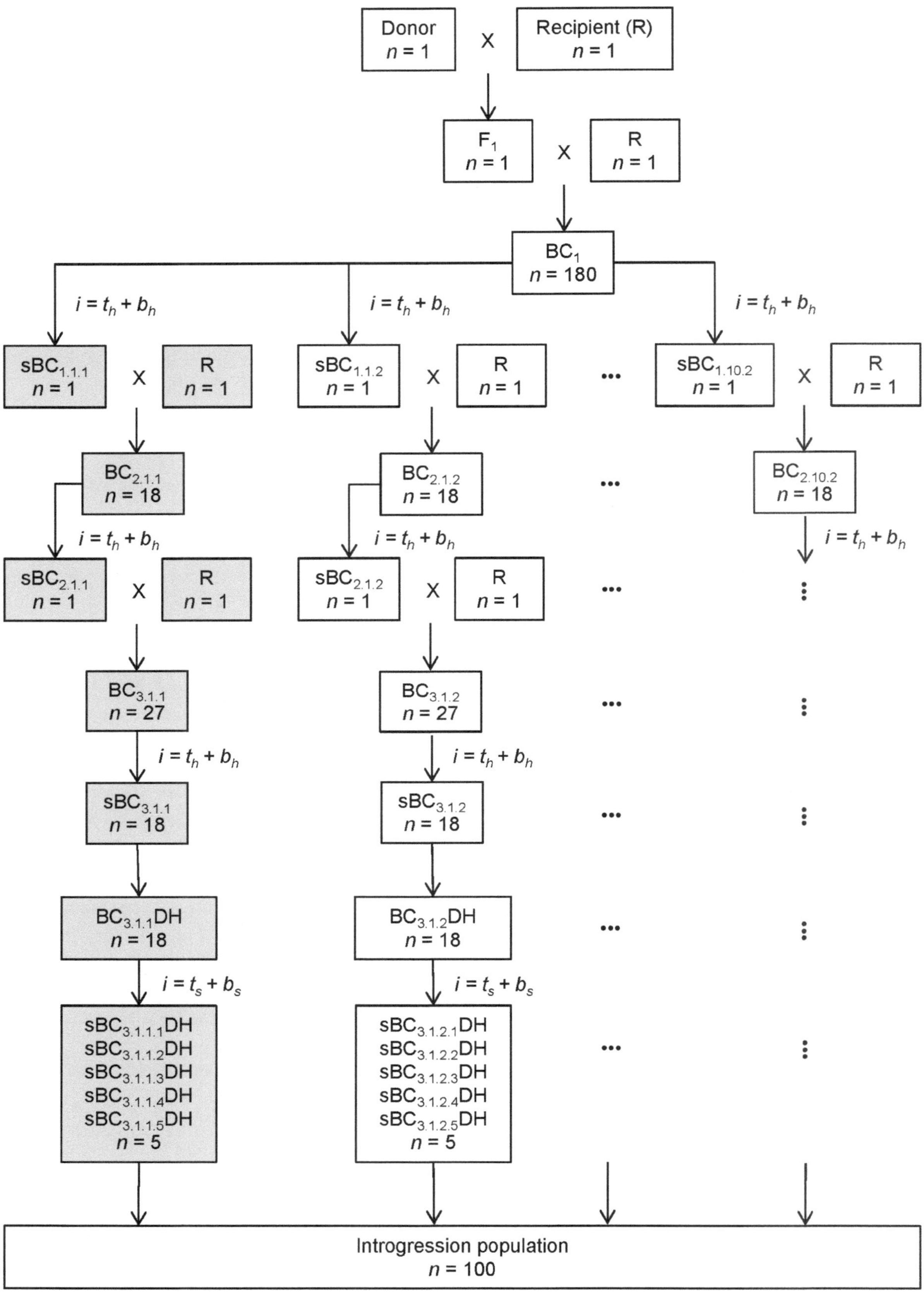

Donor n = 1
X
Recipient (R) n = 1
F1 n = 1
X
R n = 1
BC1 n = 180
i = th + bh
sBC1.1.1 n = 1
R n = 1
sBC1.1.2 n = 1
sBC1.10.2 n = 1
BC2.1.1 n = 18
BC2.1.2 n = 18
BC2.10.2 n = 18
sBC2.1.1 n = 1
sBC2.1.2 n = 1
BC3.1.1 n = 27
BC3.1.2 n = 27
sBC3.1.1 n = 18
sBC3.1.2 n = 18
BC3.1.1DH n = 18
BC3.1.2DH n = 18
i = ts + bs
sBC3.1.1.1DH sBC3.1.1.2DH sBC3.1.1.3DH sBC3.1.1.4DH sBC3.1.1.5DH n = 5
sBC3.1.2.1DH sBC3.1.2.2DH sBC3.1.2.3DH sBC3.1.2.4DH sBC3.1.2.5DH n = 5
Introgression population n = 100

Figure 1. Schematic representation of crossing scheme $BC_3DH-HHH^*$**.** Crossing scheme $BC_3DH-HHH^*$ is characterized by increasing population sizes in the backcross generations and selection for donor chromosome halves in the final backcross generation. The parts highlighted in gray represent one branch of the crossing scheme. Sub-populations are indexed by BC_g, $BC_{g.c.h}$ and $BC_{g.c.h.s}$, where g is the respective backcross generation, c is the respective chromosome, h is the respective chromosome half, s is the respective target segment; $sBC_{g.c.h}$ and $sBC_{g.c.h.s}$ denote individuals selected for the respective selection regions.

Very high values of $S > 1000$ segments were observed for the basic crossing schemes BC_2S_2-C and BC_2S_2-H. These crossing schemes had on average $N = 1$ additional donor segment per IL compared to the corresponding DH crossing schemes. However, they were also characterized by incomplete homozygosity (Figure 3B). The S_2 crossing schemes with selection in the final backcross generation required 360 individuals and HT assays less then the corresponding DH crossing schemes (Tables 2 and 3). Nevertheless, the differences between DH and S_2 crossing schemes then diminished. For example, scheme BC_3S_2-HHH resulted in similar values for most measures as the corresponding scheme $BC_3DH-HHH$ (Table 3).

The differences in the total donor genome proportion D_t between selection for complete donor chromosomes and selection for donor chromosome halves ranged only between 0.1–0.7% for the same number of backcross generations and generations of selection. However, substantial differences were observed for the donor genome proportion of the carrier chromosomes D_c and the donor genome proportion of the target segments D_s. For selection for complete donor chromosomes, high values for D_c of up to 48% were observed. They were clearly visible in the graphical genotypes for schemes BC_2S_2-CC and BC_3DH-CC (Figure 3B and C). For selection for donor chromosome halves, the values for D_c were much lower and did not exceed 42% (Table 3). Without selection in the final backcross generation, selection for donor chromosome halves resulted in substantially reduced values for D_s. For example, the basic crossing schemes BC_3DH-HH and BC_3S_2-HH resulted in values for D_s of only 94% and 90%. Moreover, the ranges for D_s for these crossing schemes were substantially greater (Figure 4 for S_2 crossing schemes, for DH data not shown).

The basic crossing schemes BC_3DH-CH and BC_3S_2-CH which combined selection for complete donor chromosomes and selection for donor chromosome halves resulted in similarly low values for D_s of 93.7% and 89.9% as selection for donor chromosome halves only (Table 3). In addition, the combined strategies CH and CHH resulted in high values for D_c of up to 45.9%. The low values for D_s and the high values for D_c were reflected in the graphical genotype of scheme BC_3DH-CH, e.g. in ILs 47 and 54 (Figure 3A).

Doubling population sizes n_{tot} from 360 to 720 individuals in the crossing schemes with high selection intensity reduced the total donor genome proportion D_t from 5.0–5.1% to 3.6–3.8% compared to the basic DH crossing schemes, and from 5.3–5.7% to 4.3–4.4% compared to the basic S_2 crossing schemes (Table 3). The donor genome proportion of the carrier chromosomes D_c was reduced by about 4.2–7.5% for the DH crossing schemes, and by about 0.9–6.9% for the S_2 crossing schemes. The reduction of the donor genome proportion of the target segments D_s in combination with increased ranges that was observed with selection for donor chromosome halves in the basic crossing schemes was not observed in the crossing schemes with high selection intensity (Table 3 and Figure 4). D_s was increased by 5.2% for crossing scheme BC_3DH-HH' and by 8.6% for crossing scheme BC_3S_2-HH' compared to the basic crossing schemes BC_3DH-HH and BC_3S_2-HH. However, these improvements were only achieved with 2160 HT assays compared to 1080 HT asssays in the basic crossing schemes (Table 3).

The crossing schemes with selection in the final backcross generation resulted in values for D_t that were 1.1–1.2% higher for the DH crossing schemes und 0.6–1.4% higher for the S_2 crossing schemes compared to the crossing schemes with high selection intensity. The ranges of D_s for selection for donor chromosome halves were about the same size as for the crossing schemes with high selection intensity (Figure 4). The average values for D_s were 0.5% lower for scheme $BC_3DH-HHH$ and 0.3% lower for scheme BC_3S_2-HHH (Table 3). The number of required HT assays was reduced by 360 for the DH crossing schemes and by 720 for S_2 crossing schemes compared to the crossing schemes with high selection intensity. For the crossing schemes with selection in the final backcross generation, selection for donor chromosome halves was the most advantageous selection strategy with respect to the genetic background and to the target segments. Most notably, the crossing schemes $BC_3DH-HHH$ and BC_3S_2-HHH resulted in the lowest values for the donor genome proportion of the carrier chromosomes D_c. Compared to the most efficient basic crossing schemes BC_2DH-C and BC_3DH-CC, the crossing schemes BC_2DH-HH and $BC_3DH-HHH$ resulted in small improvements of both the genetic background and D_s. However, in both cases 720 additional HT assays had to be invested. For the S_2 crossing schemes with selection in the final backcross generation, high values of D_c of 38.1–48.3% were observed. Large donor chromosome segments on the carrier chromosomes were also visible in the graphical genotypes for schemes BC_2S_2-CC and BC_3S_2-HHH (Figure 3B and D). The high values for D_c were associated with a considerable reduction of the number of disjunct genome segments S of >200 segments for the BC_2 crossing schemes and of 100–200 segments for the BC_3 crossing schemes compared to the basic S_2 crossing schemes (Table 3).

The crossing schemes with increasing population sizes reduced the number of required HT assays for DH crossing schemes by 360 in comparison to the crossing schemes with selection in the final backcross generation and constant population sizes. The crossing schemes $BC_3DH-HHH^*$ and $BC_3DH-HHS^*$ resulted in similar values for most measures as the crossing scheme $BC_3DH-HHH$. However, D_c and D_s were slightly reduced for crossing scheme $BC_3DH-HHS^*$. Compared to the most efficient basic crossing scheme BC_3DH-CC, crossing scheme $BC_3DH-HHH^*$ required 360 additional HT assays, but reduced D_c by 1.9% and increased D_s by 0.6%. The crossing scheme $BC_3S_2-HHH^*$ resulted with 38.0% in a much higher D_c than the crossing scheme $BC_3S_2-HHS^*$ with 30.4%. For crossing scheme $BC_3S_2-HHS^*$, the average D_s was only 96.2% and the range for D_s was higher than for the crossing schemes $BC_3S_2-HHH^*$ and BC_3S_2-HHH (Figure 4). However, D_t and D_c were the lowest for all investigated crossing schemes, with the exception of the crossing schemes with high selection intensity and $n_{tot} = 720$ (Table 3). The clear-cut separation of the target segments is also visible in the graphical genotype (Figure 3F).

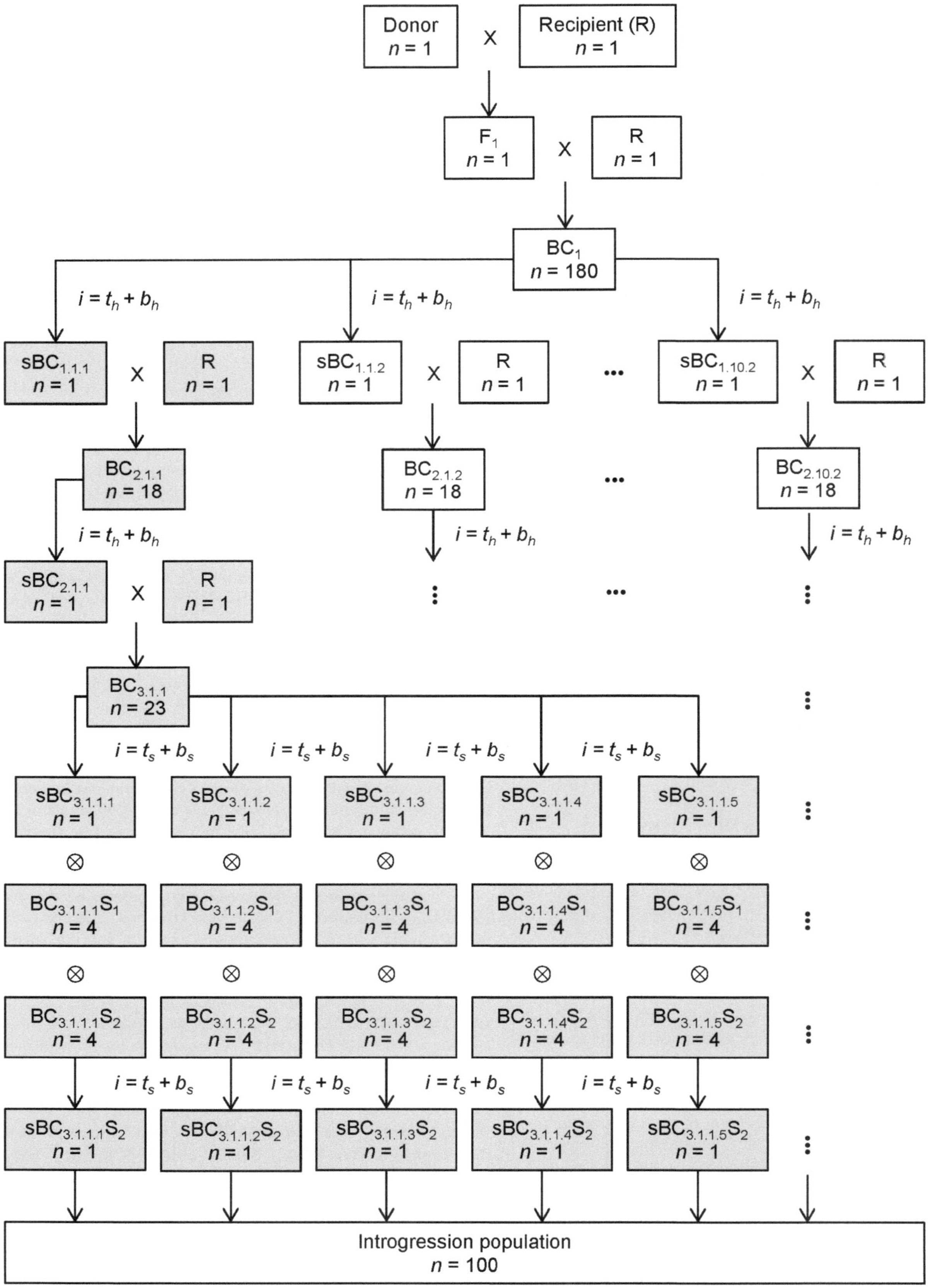

Donor
n = 1
X
Recipient (R)
n = 1
F1
n = 1
R
n = 1
BC1
n = 180
i = th + bh
sBC1.1.1
n = 1
sBC1.1.2
n = 1
sBC1.10.2
n = 1
BC2.1.1
n = 18
BC2.1.2
n = 18
BC2.10.2
n = 18
sBC2.1.1
n = 1
BC3.1.1
n = 23
i = ts + bs
sBC3.1.1.1
sBC3.1.1.2
sBC3.1.1.3
sBC3.1.1.4
sBC3.1.1.5
BC3.1.1.1S1
BC3.1.1.2S1
BC3.1.1.3S1
BC3.1.1.4S1
BC3.1.1.5S1
n = 4
BC3.1.1.1S2
BC3.1.1.2S2
BC3.1.1.3S2
BC3.1.1.4S2
BC3.1.1.5S2
sBC3.1.1.1S2
sBC3.1.1.2S2
sBC3.1.1.3S2
sBC3.1.1.4S2
sBC3.1.1.5S2
Introgression population
n = 100

Figure 2. Schematic representation of crossing scheme $BC_3S_2-HHS^*$. Crossing scheme $BC_3S_2-HHS^*$ is characterized by increasing population sizes in the backcross generations and selection for target segments in the final backcross generation. The parts highlighted in gray represent one branch of the crossing scheme. Sub-populations are indexed by BC_g, $BC_{g.c.h}$ and $BC_{g.c.h.s}$, where g is the respective backcross generation, c is the respective chromosome, h is the respective chromosome half, s is the respective target segment; $sBC_{g.c.h}$ and $sBC_{g.c.h.s}$ denote individuals selected for the respective selection regions.

Discussion

Measures for Characterizing Introgression Populations

Measures for the description of introgression populations should allow to distinguish between introgression populations of different structure. Complete donor genome coverage O is desirable in order to make the complete genetic variation of the donor available for the breeding process. However, high values for O can also be caused by donor segments outside the target segments which could not be removed from the genetic background. O is therefore only informative if interpreted in relation to measures which reflect the distribution of the donor genome in the introgression population. A distinctive description of introgression populations is possible with the total donor genome proportion D_t, the donor genome proportion on the carrier chromosomes D_c and the donor genome proportion of the actual target segments D_s.

A high total donor genome proportion D_t is often associated with a high number of disjunct genome segments S. S determines the resolution R, which is an important parameter for the accuracy of QTL detection. However, if S is greater than the number of ILs, the problem of overparameterization arises with classical linear model approaches. This issue has only in part been resolved by using statistical methods which pre-select a reduced number of ILs for the linear model [23].

High values for the donor genome proportion on the carrier chromosomes D_c and the depth of donor genome coverage T reflect undesired donor segments attached to the actual target segments. Such large donor segments which overlap between ILs have been reported to increase the risk of false-positive effects in QTL detection and reduce the power of QTL detection [24]. This is mainly a problem if linkage maps with large distances between adjacent markers of 10 cM or more are employed, because QTLs located between the last marker of the target segment and the next marker outside the target segment are incorrectly assigned to the target segments. With dense marker maps which are now available this problem should be overcome. However, large donor segments also increase the risk of linkage drag in the breeding process and often require further steps of separation [24].

Low values for the donor genome proportion of the target segments D_s indicate a loss of target segments and potentially useful alleles. This is a problem that arises with small population sizes as were investigated in the present study [16]. Even if the missing target segments are present in the genetic background of other ILs, this might impair QTL detection and the further use of the ILs for the breeding progress.

We therefore argue that short non-overlapping target segments in a clean recipient background are advantageous also with dense marker maps. For 20 cM target segments and a genomic model of 10 equally sized chromosomes of 200 cM length, this corresponds to $D_t = 1\%$, $D_c = 10\%$ and $D_s = 100\%$ in the ideal case. The effort and time required for developing introgression populations with such characteristics is beyond the scope of most breeding programs. With the limited population sizes and number of HT assays investigated in this study, these ideal values could not be achieved with two or three backcross generations (Table 3). We therefore considered those crossing and selection schemes as efficient which with a given limited resource input resulted in the highest coverage of target segments D_s in combination with low overlap of target segments reflected in D_c and T and a low total donor genome proportion D_t.

With respect to QTL detection, it can be expected that the optimal values for the suggested measures will depend on the statistical method and the genetic architecture of the trait. They could be determined for a given statistical method by including QTLs of different number and effect in future simulation studies. We plan further investigations in this area of research.

Crossing Schemes

BC_3 crossing schemes had 2–3% lower values for the total donor genome proportion D_t than BC_2 crossing schemes (Table 3), even if no selection for the genetic background was conducted in generation BC_3. Selection in generation BC_2, as was investigated with the crossing schemes BC_2-CC, BC_2-HH and BC_2-CH, only resulted in a reduction of D_t of 0.4–1.4% compared to the basic crossing schemes BC_2-C and BC_2-H (Table 3). An explanation for this comparatively small reduction is that the limiting factor for the reduction of D_t is the number of recombinations during meiosis. Hence, even though BC_2 crossing schemes have a time advantage, the effect of a third backcross generation cannot be compensated by investing in additional marker analyses. We therefore conclude that BC_3 crossing schemes result in introgression populations with an improved structure, and that the time investment in the additional backcross generation is worthwhile.

DH crossing schemes were for most measures superior to the corresponding S_2 crossing schemes. The differences were most pronounced in the number of disjunct genome segments S. Even though the S_2 schemes on average had a slightly higher number of donor segments per IL N, it seems that the very high values for S that were observed especially in the BC_2-S_2 crossing schemes mainly had to be attributed to incomplete homozygosity (Figure 3B). It can be expected that introgression populations with $S > 1000$ segments in 100 ILs (Table 3) are not suitable for effective QTL detection. We therefore conclude that the DH method is essential for short crossing schemes with only two backcross generations.

A drawback of the DH method is that with current protocols of *in vivo* DH induction of maternal haploids, only a very limited number of viable DH lines can be derived from one backcross individual. We expect that our assumption of one DH line per backcross individual is a conservative, but realistic estimate. In contrast, with selfing, many progenies can be derived from one selected backcross individual. In the S_2 crossing schemes, it is consequently comparatively cheap and easy to conduct selection in the final backcross generation. For the DH schemes, selection in the final backcross generation could only be conducted if population size in this generation was higher than the desired number of final DH lines. As a result, the S_2 crossing schemes with selection in the final backcross generation required 360 HT assays less than the corresponding DH schemes (Table 3). Moreover, the selected fractions of best backcross individuals were much greater for the DH than for the S_2 crossing schemes (Table 2). This resulted in a lower selection intensity for both the selection region of the final backcross generation and the genetic background in

Table 3. Measures evaluated for introgression populations resulting from different crossing and selection schemes.

Scheme	O	T	S	R	N	L	D_t	D_c	D_s	HT
Basic crossing schemes										
BC_2DH-C	99.9	8.3	691	2.9	6.2	28.7	8.3	38.9	98.7	720
BC_2DH-H	100.0	8.9	751	2.7	6.4	29.2	8.9	37.3	94.1	720
BC_3DH-CC	99.2	5.1	457	4.4	3.8	29.8	5.0	35.2	97.8	1080
BC_3DH-HH	99.8	5.2	487	4.1	3.9	29.8	5.1	33.3	94.2	1080
BC_3DH-CH	99.6	5.2	469	4.3	3.8	30.5	5.1	35.1	93.7	1080
Crossing schemes with high selection intensity										
BC_3DH-CC'	99.3	3.6	389	5.1	3.0	27.8	3.6	27.7	98.8	2160
BC_3DH-HH'	99.9	3.8	406	4.9	3.0	29.7	3.8	29.1	99.4	2160
BC_3DH-CH'	99.6	3.9	399	5.0	3.0	30.0	3.8	29.9	99.0	2160
Crossing schemes with selection in the final BC generation										
BC_2DH-CC	99.9	7.8	676	3.0	5.9	28.7	7.8	41.0	98.9	1440
BC_2DH-HH	100.0	8.1	716	2.8	6.0	29.0	8.1	38.9	99.3	1440
BC_2DH-CH	99.9	8.5	679	2.9	6.0	30.4	8.5	43.7	98.8	1440
$BC_3DH-CCC$	99.1	4.9	457	4.4	3.7	30.2	4.8	35.8	98.0	1800
$BC_3DH-HHH$	99.9	4.8	464	4.3	3.6	31.0	4.7	33.9	98.9	1800
$BC_3DH-CHH$	99.6	4.8	450	4.4	3.5	31.4	4.7	35.0	98.5	1800
Crossing schemes with increasing population sizes										
$BC_3DH-HHH^*$	99.9	5.0	492	4.1	3.8	29.7	5.0	33.3	98.4	1440
$BC_3DH-HHS^*$	99.8	4.9	484	4.1	3.8	29.4	4.8	32.5	97.5	1440
Basic crossing schemes										
BC_2S_2-C	100.0	11.4	1021	2.0	7.3	26.4	9.3	41.4	97.6	720
BC_2S_2-H	100.0	11.4	1073	1.9	7.4	25.9	9.3	36.3	90.4	720
BC_3S_2-CC	99.3	6.9	684	2.9	4.5	27.7	5.7	39.2	97.0	1080
BC_3S_2-HH	99.9	6.3	702	2.8	4.4	26.3	5.3	33.0	90.3	1080
BC_3S_2-CH	99.7	6.4	681	2.9	4.3	26.9	5.3	35.0	89.9	1080
Crossing schemes with high selection intensity										
BC_3S_2-CC'	99.5	5.1	585	3.4	3.6	26.6	4.3	32.3	98.8	2160
BC_3S_2-HH'	99.9	5.1	591	3.4	3.5	27.8	4.4	32.1	98.9	2160
BC_3S_2-CH'	99.7	5.2	581	3.4	3.5	28.2	4.4	33.4	98.4	2160
Crossing schemes with selection in the final BC generation										
BC_2S_2-CC	99.3	10.5	795	2.5	6.5	27.2	8.3	48.3	98.0	1080
BC_2S_2-HH	99.9	9.7	785	2.5	6.2	27.0	7.9	42.8	98.8	1080
BC_2S_2-CH	99.8	9.7	761	2.6	6.1	27.6	7.9	45.9	98.4	1080
BC_3S_2-CCC	98.3	7.1	588	3.4	4.2	30.5	5.7	44.6	97.5	1440
BC_3S_2-HHH	99.7	5.9	510	3.9	3.7	31.0	5.0	38.1	98.6	1440
BC_3S_2-CHH	99.3	6.0	509	3.9	3.7	31.3	5.0	39.5	98.2	1440
Crossing schemes with increasing population sizes										
$BC_3S_2-HHH^*$	99.7	5.8	508	3.9	3.8	30.7	4.9	38.0	98.7	1440
$BC_3S_2-HHS^*$	99.8	5.1	596	3.4	3.7	26.1	4.3	30.4	96.2	1400

O: donor genome coverage in percent; T: depth of donor genome coverage; S: number of disjunct genome segments; R: resolution; N: number of donor segments per IL; L: length of donor segments per IL in cM; D_t: total donor genome proportion in percent; D_c: donor genome proportion of carrier chromosomes in percent; D_s: donor genome proportion of target segments in percent; HT: the required number of HT assays. Measures are arithmetic means over 1,000 replications.

the DH crossing schemes. We therefore suggest that a comparison of DH and S_2 crossing schemes should take the distinctive features of both methods into account. The evaluation of efficiency should also be based on the number of required HT assays. Considering this, S_2 crossing schemes which exploit their selection advantages represent economic and easy-to-handle alternatives to DH crossing schemes.

Selection Strategies for Small and Constant Population Sizes

For a given genetic model and crossing scheme, the selection strategy is the most important factor that influences the structure of the resulting introgression population. In the following paragraphs, different aspects such as the length of the selection

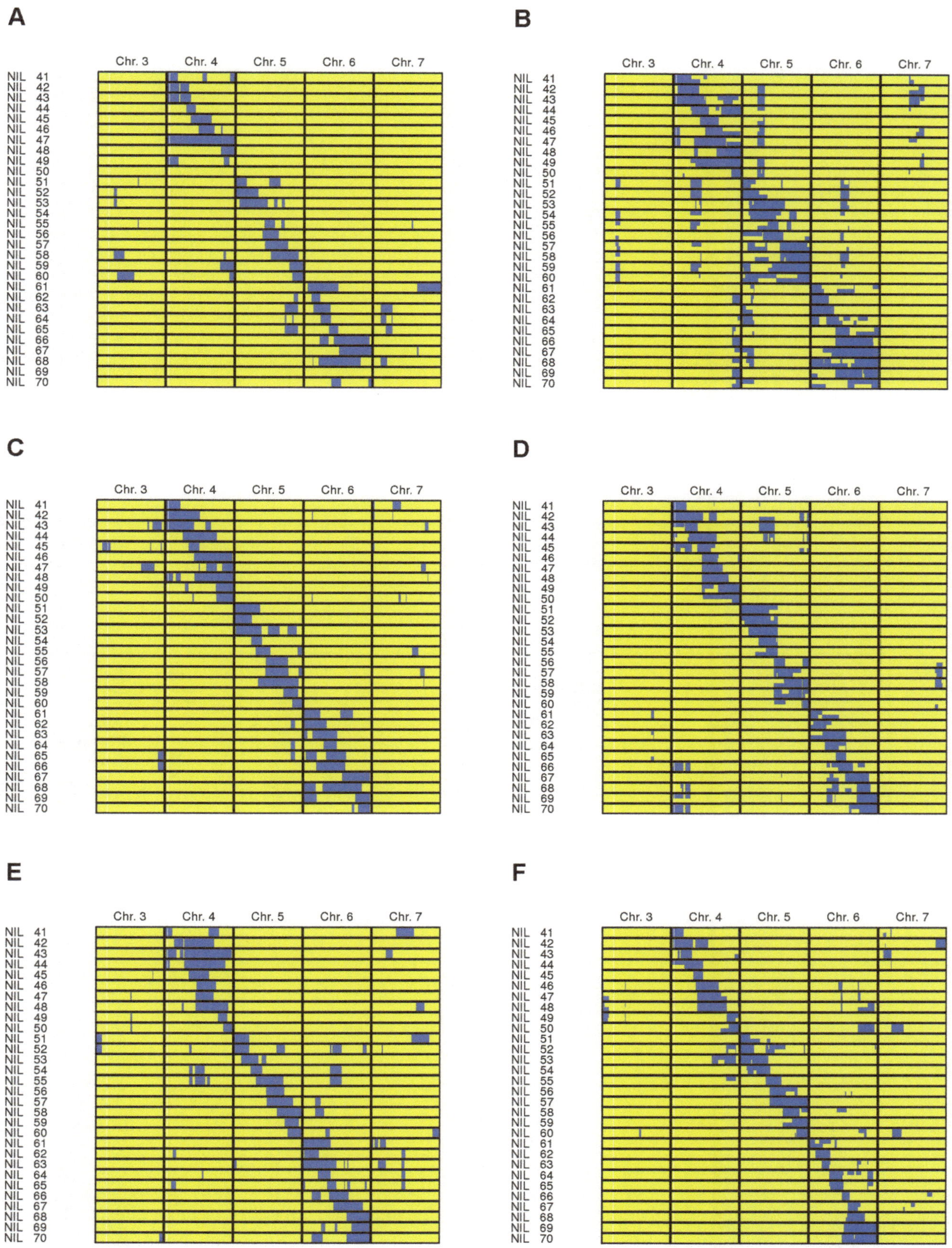
A
Chr. 3
Chr. 4
Chr. 5
Chr. 6
Chr. 7
NIL 41
NIL 42
NIL 43
NIL 44
NIL 45
NIL 46
NIL 47
NIL 48
NIL 49
NIL 50
NIL 51
NIL 52
NIL 53
NIL 54
NIL 55
NIL 56
NIL 57
NIL 58
NIL 59
NIL 60
NIL 61
NIL 62
NIL 63
NIL 64
NIL 65
NIL 66
NIL 67
NIL 68
NIL 69
NIL 70
B
Chr. 3
Chr. 4
Chr. 5
Chr. 6
Chr. 7
NIL 41
NIL 42
NIL 43
NIL 44
NIL 45
NIL 46
NIL 47
NIL 48
NIL 49
NIL 50
NIL 51
NIL 52
NIL 53
NIL 54
NIL 55
NIL 56
NIL 57
NIL 58
NIL 59
NIL 60
NIL 61
NIL 62
NIL 63
NIL 64
NIL 65
NIL 66
NIL 67
NIL 68
NIL 69
NIL 70
C
Chr. 3
Chr. 4
Chr. 5
Chr. 6
Chr. 7
NIL 41
NIL 42
NIL 43
NIL 44
NIL 45
NIL 46
NIL 47
NIL 48
NIL 49
NIL 50
NIL 51
NIL 52
NIL 53
NIL 54
NIL 55
NIL 56
NIL 57
NIL 58
NIL 59
NIL 60
NIL 61
NIL 62
NIL 63
NIL 64
NIL 65
NIL 66
NIL 67
NIL 68
NIL 69
NIL 70
D
Chr. 3
Chr. 4
Chr. 5
Chr. 6
Chr. 7
NIL 41
NIL 42
NIL 43
NIL 44
NIL 45
NIL 46
NIL 47
NIL 48
NIL 49
NIL 50
NIL 51
NIL 52
NIL 53
NIL 54
NIL 55
NIL 56
NIL 57
NIL 58
NIL 59
NIL 60
NIL 61
NIL 62
NIL 63
NIL 64
NIL 65
NIL 66
NIL 67
NIL 68
NIL 69
NIL 70
E
Chr. 3
Chr. 4
Chr. 5
Chr. 6
Chr. 7
NIL 41
NIL 42
NIL 43
NIL 44
NIL 45
NIL 46
NIL 47
NIL 48
NIL 49
NIL 50
NIL 51
NIL 52
NIL 53
NIL 54
NIL 55
NIL 56
NIL 57
NIL 58
NIL 59
NIL 60
NIL 61
NIL 62
NIL 63
NIL 64
NIL 65
NIL 66
NIL 67
NIL 68
NIL 69
NIL 70
F
Chr. 3
Chr. 4
Chr. 5
Chr. 6
Chr. 7
NIL 41
NIL 42
NIL 43
NIL 44
NIL 45
NIL 46
NIL 47
NIL 48
NIL 49
NIL 50
NIL 51
NIL 52
NIL 53
NIL 54
NIL 55
NIL 56
NIL 57
NIL 58
NIL 59
NIL 60
NIL 61
NIL 62
NIL 63
NIL 64
NIL 65
NIL 66
NIL 67
NIL 68
NIL 69
NIL 70

Figure 3. Graphical genotypes of introgression populations resulting from six different crossing schemes. A: BC_3DH-CH; B: BC_2S_2-CC; C: BC_3DH-CC; D: BC_3S_2-HHH; E: $BC_3DH-HHH^*$; F: $BC_3S_2-HHS^*$. The graphical genotypes display the chromosomes 3 to 7 of ILs 41–70 and are examples from one simulation run. Chromosome segments which stem from the donor are displayed in blue, whereas chromosome segments which stem from the recipient are displayed in yellow. The graphical genotypes illustrate the differences between the alternative crossing schemes with respect to their suitability to create introgression populations with complete donor genome coverage and clearly separated, evenly distributed target donor chromosome segments.

regions, the number of generations of selection and the required population sizes for effective selection are discussed.

Selection strategies which pre-select individuals carrying complete donor chromosomes reduce the number of simultaneous backcross programs to the number of chromosomes [16]. They are therefore suitable for breeding programs with limited resources. However, for long chromosomes of 200 cM length, selection for complete donor chromosomes preserved large donor chromosome segments on the carrier chromosomes up to line development (Figure 3C and B). This was reflected in high values for the proportion of donor genome on the carrier chromosomes D_c of up to 48% (Table 3). The selection regions for selection in the backcross generations were therefore reduced to donor chromosome halves for selection strategies H, HH and HHH. In all four series of simulations, selection for donor chromosome halves resulted in the desired reduction of D_c compared to selection for complete donor chromosomes (Table 3). Other measures for the genetic background were approximately equivalent. We therefore conclude that for crop species with long chromosomes such as maize, wheat or rapeseed, selection for donor chromosome halves reduces the length of the donor segments attached to the actual target segments and the risk of linkage drag.

However, for crossing schemes without selection in the final backcross generation and constant population sizes of $n_{tot}=360$ individuals, selection for donor chromosome halves resulted in a considerable reduction of the donor genome proportion of the target segments D_s of up to 7%. Moreover, the estimated values for D_s were less reliable for these crossing schemes, *e.g.*, in schemes BC_2S_2-H and BC_3S_2-HH (Figure 4). These findings have to be attributed to the small population sizes n_g in the sub-populations and the structure of the selection index *i*. In generation DH or S_2, population sizes were reduced to $n_g=18$ individuals with selection for donor chromosome halves (Table 2). Without selection in the final backcross generation, around 50% of the ILs developed from the backcross individuals are expected to carry no donor allele at a given locus within the respective target segment. The probability to find five ILs with complete donor target segments for the introgression population was therefore even further reduced. As the selection index $i=t_s+b_s$ weighed the target segments and the genetic background equally, a clean genetic background sometimes outweighed a reduced D_s and led to the observed loss of target segments in these small sub-populations. We therefore conclude that a sufficiently large population size is the crucial

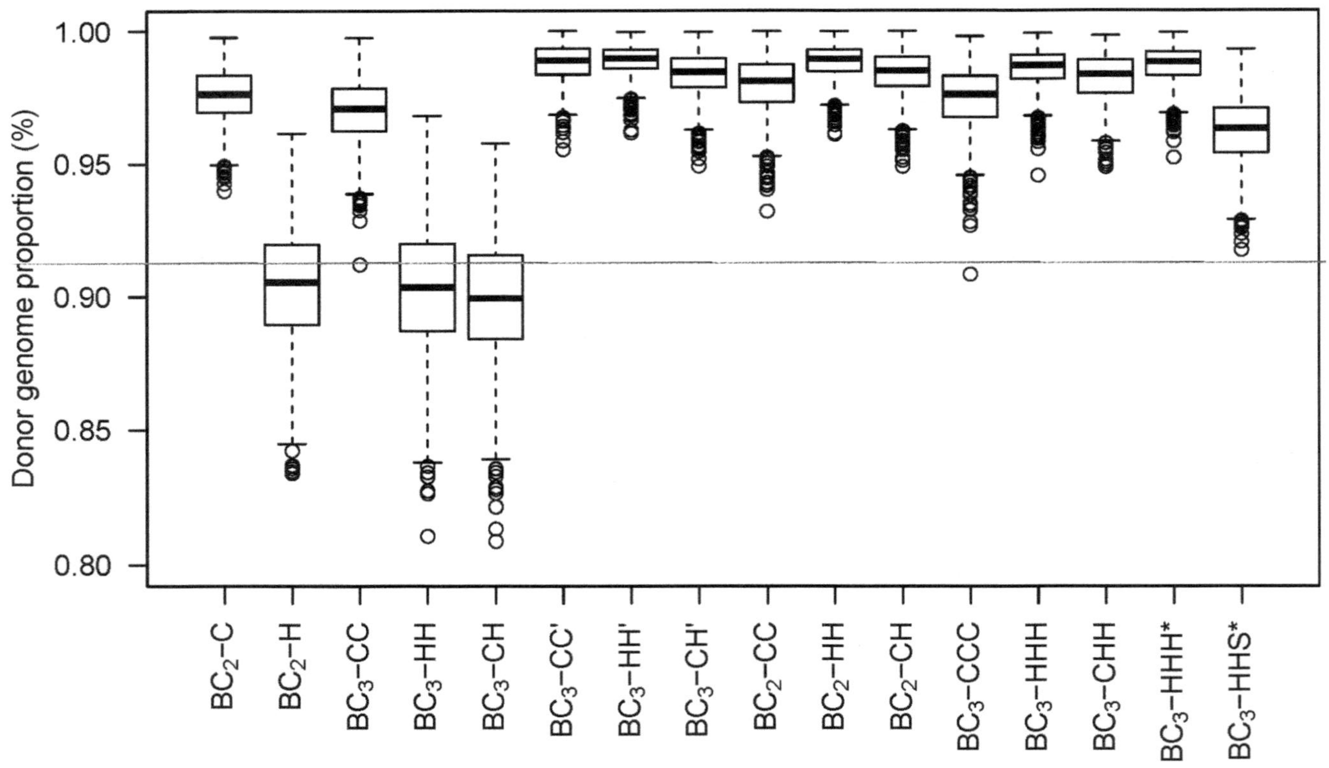

Figure 4. Donor genome proportion of target segments D_s for all investigated S_2 crossing schemes. The boxplots represent the distribution over 1,000 replications of the simulations. The basic crossing schemes BC_2S_2-H, BC_3S_2-HH and BC_3S_2-CH which select for donor chromosome halves are characterized by higher ranges for D_s.

factor for the successful application of selection for donor chromosome halves.

A loss of target segments caused by small population sizes was also observed for the basic combined selection strategy CH which selected for complete donor chromosomes in generation BC_1 and for donor chromosome halves in generation BC_2. In addition, the combined strategies CH and CHH resulted in high values for the donor genome proportion on the carrier chromosomes D_c of up to 45% (Table 3). This can be explained by the efficient selection for complete donor chromosomes from the comparatively large BC_1 population of $n_g = n_{tot} = 360$ individuals (Table 2). The pre-selected complete donor chromosomes are in large part preserved up to line development. The combination of missing target segments with large donor chromosome segments on carrier chromosomes was also reflected in the graphical genotype for scheme BC_3DH-CH, *e.g.*, in IL 47 and 54 (Figure 3A). The selection strategies CH and CHH therefore combine the drawbacks of both selection for complete donor chromosomes and selection for chromosome halves. They are not suitable for crossing schemes with small and constant population sizes, in which the population sizes n_g in the sub-populations are subsequently reduced over the backcross generations. We conclude that for for small breeding programs with a constant population size of $n_{tot} = 360$ and a limited number of HT assays for selection, selection strategies which only select for complete donor chromosomes in the backcross generations should be employed in both DH and S_2 crossing schemes to avoid the loss of target segments.

Finding more Carriers of Donor Target Segments for Line Development

To employ selection for donor chromosome halves effectively for reducing the donor genome proportion of the carrier chromosomes D_c without losing the target segments, it is necessary to increase the frequency of carriers of donor target segments for line development. Using larger population sizes is a straightforward solution for this problem, which in addition can improve the overall structure of introgression populations. The crossing schemes with high selection intensity and double population sizes of $n_{tot} = 720$ individuals resulted in small improvements of the total donor genome proportion D_t of about 1–1.5% compared to the basic crossing schemes (Table 3). The desired increase in the donor proportion of the target segments D_s was achieved. For selection for donor chromosome halves, D_s was increased by 5.2–8.6%. Selection for donor chromosome halves was then even superior to selection for complete donor chromosomes. Moreover, the donor genome proportion on the carrier chromosomes D_c was reduced by up to 7.5%, indicating an improved separation of target segments. The observed improvements were greater for the DH than for the S_2 crossing schemes. Nevertheless, the comparatively small improvements of the introgression populations required 1080 additional HT assays. We therefore conclude that such large population sizes are only suitable for breeding programs with access to DH technology, less stringent resource restrictions and high requirements with respect to the genetic background. If the requirements concerning the structure of the introgression population are not that high, it might be more economic to increase population size only in the final backcross generation and/or to invest in additional HT assays only in this generation.

For crossing schemes with selection in the final backcross generation, the total donor genome proportion D_t was similar to the values of the basic crossing schemes, and about 1% higher than for the crossing schemes with higher selection intensity (Table 3). However, the average values for the donor genome proportion of the target segments D_s were similar to the crossing schemes with higher selection intensity (Table 3) and the ranges were effectively reduced (Figure 4). Moreover, the number of required HT assays was reduced by 360 for the DH crossing schemes and by 720 for S_2 crossing schemes compared to the crossing schemes with higher selection intensity (Table 3). The decision for doubling population sizes requires the same resources as would be required for generating an additional introgression population. This large effort seems not to be justified by the relatively small improvements compared to the basic crossing schemes. We therefore conclude that selection in the final backcross generation is the more efficient solution for both DH and S_2 crossing schemes.

Selection for donor chromosome halves was the best strategy with selection in the final backcross generation for both DH and S_2 crossing schemes (Table 3). However, for the DH schemes, only small improvements for schemes BC_2DH-HH and $BC_3DH-HHH$ were observed compared to the most efficient basic crossing schemes BC_2DH-C and BC_3DH-CC (Table 3). For these small improvements, 720 additional individuals and HT assays had to be invested. For the S_2 schemes, considerable reductions in S of 174 and 236 segments were observed for schemes BC_2S_2-HH and BC_3S_2-HHH with selection in the final backcross generation compared to the basic crossing schemes BC_2S_2-C and BC_2S_2-CC. D_t was only slightly reduced. However, the donor genome proportion on carrier chromosomes D_c was in general very high for the crossing schemes with selection in the final backcross generation with 38–48%. This indicates a fixation of the selection regions of the final backcross generation (Figure 3B and D). In schemes BC_2S_2-CC and BC_3S_2-HH, complete donor chromosomes and donor chromosome halves still appear as blocks around the target segments. These blocks lead to an overlap of donor segments between ILs that reduces the effective resolution of the introgression population for QTL detection. The overlap also hampers the further use of the ILs in the breeding process, as further steps of separation of the target segments by backcrossing are required. We therefore conclude that the crossing schemes with selection in the final backcross generation have the potential to improve the resulting introgression populations at moderate cost. However, for the DH crossing schemes, the number of required HT assays and individuals has to be reduced. For the S_2 crossing schemes, the fixation of large donor chromosome segments has to be avoided. Optimizations of the respective crossing schemes are presented in the following.

Increasing Population Sizes Over Backcross Generations

With constant population sizes of $n_{tot} = 360$ individuals, the population size in generation BC_1 was large in relation to the genetic gains that could be achieved by selecting a comparatively small fraction of 10 or 20 individuals (Table 2). Starting with smaller population sizes in generation BC_1 and gradually increasing population sizes in the following backcross generations was therefore an efficient option to reduce the overall number of required individuals and HT assays for selection in the final backcross generation. Larger population sizes in generation BC_3 also enabled selection for target segments, which was investigated as an option to avoid the fixation of large donor chromosome segments especially for the S_2 crossing schemes.

The schemes $BC_3DH-HHH^*$ and $BC_3DH-HHS^*$ resulted in similar values for all measures (Table 3). However, D_c and D_s were slightly lower for scheme $BC_3DH-HHS^*$. We therefore conclude that selection for target segments already in the final backcross generation is not efficient for DH crossing schemes. In comparison to the best but also very expensive scheme $BC_3DH-HHH$ with selection in the final backcross generation, scheme $BC_3DH-HHH^*$ can be considered equivalent, but required 360 individuals and HT assays less. In comparison to the more economic basic crossing scheme BC_3DH-CC, scheme $BC_3DH-HHH^*$ improved D_c and D_s and thus the separation of target segments. This is also visible in the graphical genotype (Figure 3). The investment in the additional 360 HT assays seems therefore worthwhile (Table 3).

The scheme $BC_3S_2-HHS^*$ resulted in better values than the schemes BC_3S_2-HHH and $BC_3S_2-HHH^*$. Most notably, it resulted in a much lower D_c of 30% compared to 38%. Scheme $BC_3S_2-HHS^*$ resulted in a D_s that was 2.4% lower compared to schemes BC_3S_2-HHH and $BC_3S_2-HHH^*$ and the ranges for D_s were higher (Figure 4). Nevertheless, it resulted in the lowest values of D_t and D_c and the best separation of target genes of all investigated DH and S_2 crossing schemes with comparable population sizes. The comparatively high value of S of 596 segments in combination with reduced values for D_t can in this case be explained by a greatly improved separation of target segments compared to the other S_2 schemes with selection in the final backcross generation. The improved separation of target segments is also visible in the graphical genotype (Figure 3F). This was achieved with 40 HT assays less (Table 3). We therefore expect that scheme $BC_3S_2-HHS^*$ will result in an improved power of QTL detection, and recommend selection for target segments in the final backcross generation for S_2 crossing schemes.

Compared to the best but expensive comparable DH crossing scheme $BC_3DH-HHH$, the S_2 crossing scheme $BC_3S_2-HHS^*$ resulted in similar values and required 400 HT assays less. Overall, we conclude that increasing population sizes over backcross are advantageous and economic for both DH and S_2 crossing schemes. Moreover, crossing scheme $BC_3S_2-HHS^*$ can provide a cheap alternative to comparable DH crossing schemes.

Conclusions

Our study has shown that introgression populations with complete coverage of the donor genome and reasonably clean recipient background can be developed with a limited number of backcross individuals and HT assays. It has provided further insight on how different crossing and selection schemes influence the structure of the resulting introgression populations. The guidelines which have been derived for maize are transferable to other crop species with similar number and length of chromosomes. For crops with different genome size, some considerations are discussed in the following.

Rapeseed is a crop with a large genome of 19 chromosomes, for which efficient protocols of microspore culture are available for DH production. For the large genome of rapeseed, it can be expected that the values for the total donor genome proportion D_t will be lower than those observed for the smaller genome of maize. With the investigated selection index i, the selection pressure on the carrier chromosomes will be reduced with increasing genome size and number of chromosomes. It might therefore be an interesting option for rapeseed to put more weight on the background markers on the carrier chromosomes to achieve an efficient reduction of D_c. As with microspore culture many DH lines can usually be derived from one backcross individual, the advantages of DH production should be more pronounced than for maize. However, the optimal selection strategies for DH crossing schemes in rapeseed should then be similar to those for selfing in maize.

Sugar beet is a crop with a small genome of 9 chromosomes, for which the guidelines for selfing should be most relevant. In smaller genomes, equivalent values of D_t can usually be reached with smaller population sizes and with fewer backcrosses. However, the average length of the chromosomes in cM is also much shorter than in maize. This implies that fewer crossovers occur per meiosis, and that it might require more individuals and backcross generations to effectively separate the target segments. The combined effects of genome size and chromosome length will also depend on the desired number and length of the target segments.

Simulations can considerably facilitate the planning process for the development of introgression populations in different crop species. The derived guidelines can help breeders and geneticists to enhance the genetic variation of narrow based breeding populations of crops.

Author Contributions

Conceived and designed the experiments: EH KCF TP DS MO MF. Performed the experiments: EH KCF. Analyzed the data: EH. Wrote the paper: EH MF.

References

1. Zamir D (2001) Improving plant breeding with exotic genetic libraries. Nat Rev Genet 2: 983–989.
2. McCouch S (2004) Diversifying selection in plant breeding. PLoS Biol 2: e347.
3. Eshed Y, Zamir D (1994) A genomic library of *Lycopersicon pennellii* in *L. esculentum*: a tool for fine mapping of genes. Euphytica 79: 175–179.
4. Keurentjes JJ, Bentsink L, Alonso-Blanco C, Hanhart CJ, Blankestijn-De Vries H, et al. (2007) Development of a near-isogenic line population of *Arabidopsis thaliana* and comparison of mapping power with a recombinant inbred line population. Genetics 175: 891–905.
5. Törjék O, Meyer RC, Zehnsdorf M, Teltow M, Strompen G, et al. (2008) Construction and analysis of 2 reciprocal Arabidopsis introgression line populations. J Hered 99: 396–406.
6. Lin S, Sasaki T, Yano M (1998) Mapping quantitative trait loci controlling seed dormancy and heading date in rice, *Oryza sativa* L., using backcross inbred lines. Theor Appl Genet 96: 997–1003.
7. Cheema KK, Bains NS, Mangat GS, Das A, Vikal Y, et al. (2008) Development of high yielding IR64× *Oryza rufipogon* (Griff.) introgression lines and identification of introgressed alien chromosome segments using SSR markers. Euphytica 160: 401–409.
8. Matus I, Corey A, Filichkin T, Hayes P, Vales M, et al. (2003) Development and characterization of recombinant chromosome substitution lines (RCSLs) using *Hordeum vulgare* subsp. *spontaneum* as a source of donor alleles in a *Hordeum vulgare* subsp. *vulgare* background. Genome 46: 1010–1023.
9. Schmalenbach I, Körber N, Pillen K (2008) Selecting a set of wild barley introgression lines and verification of QTL effects for resistance to powdery mildew and leaf rust. Theor Appl Genet 117: 1093–1106.
10. Liu S, Zhou R, Dong Y, Li P, Jia J (2006) Development, utilization of introgression lines using a synthetic wheat as donor. Theor Appl Genet 112: 1360–1373.
11. Pumphrey MO, Bernardo R, Anderson JA (2007) Validating the *Fhb1* QTL for Fusarium head blight resistance in near-isogenic wheat lines developed from breeding populations. Crop Sci 47: 200–206.
12. Ribaut JM, Ragot M (2007) Marker-assisted selection to improve drought adaptation in maize: the backcross approach, perspectives, limitations, and alternatives. J Exper Bot 58: 351–360.
13. Szalma S, Hostert B, LeDeaux J, Stuber C, Holland J (2007) QTL mapping with near-isogenic lines in maize. Theor Appl Genet 114: 1211–1228.
14. Falke K, Sušić Z, Hackauf B, Korzun V, Schondelmaier J, et al. (2008) Establishment of introgression libraries in hybrid rye (*Secale cereale* L.) from an Iranian primitive accession as a new tool for rye breeding and genomics. Theor Appl Genet 117: 641–652.
15. Syed N, Pooni H, Mei M, Chen Z, Kearsey M (2004) Optimising the construction of a substitution library in *Arabidopsis thaliana* using computer simulations. Mol Breeding 13: 59–68.
16. Falke KC, Miedaner T, Frisch M (2009) Selection strategies for the development of rye introgression libraries. Theor Appl Genet 119: 595–603.
17. Welcker C, Sadok W, Dignat G, Renault M, Salvi S, et al. (2011) A common genetic determinism for sensitivities to soil water deficit and evaporative

demand: meta-analysis of quantitative trait loci and introgression lines of maize. Plant Physiol 157: 718–729.
18. Belcher AR, Zwonitzer JC, Santa Cruz J, Krakowsky MD, Chung CL, et al. (2012) Analysis of quantitative disease resistance to southern leaf blight and of multiple disease resistance in maize, using near-isogenic lines. Theor Appl Genet 124: 433–445.
19. Mano Y, Omori F (2013) Flooding tolerance in interspecific introgression lines containing chromosome segments from teosinte (*Zea nicaraguensis*) in maize (*Zea mays subsp. mays*). Ann Bot-London 112: 1125–1139.
20. Lübberstedt T, Frei UK (2012) Application of doubled haploids for target gene fixation in backcross programmes of maize. Plant Breeding 131: 449–452.
21. R Core Team (2013) R: A Language and Environment for Statistical Computing. R Foundation for Statistical Computing, Vienna, Austria. URL http://www.R-project.org/.
22. Stam P (1979) Interference in genetic crossing over and chromosome mapping. Genetics 92: 573–594.
23. Mahone GS, Borchardt D, Presterl T, Frisch M (2012) A comparison of tests for QTL mapping with introgression libraries containing overlapping and nonoverlapping donor segments. Crop Sci 52: 2198–2205.
24. Falke K, Frisch M (2011) Power and false-positive rate in QTL detection with near-isogenic line libraries. Heredity 106: 576–584.

De Novo Reconstruction of Consensus Master Genomes of Plant RNA and DNA Viruses from siRNAs

Jonathan Seguin[1,4], Rajendran Rajeswaran[1], Nachelli Malpica-López[1], Robert R. Martin[2,3], Kristin Kasschau[3], Valerian V. Dolja[3], Patricia Otten[4], Laurent Farinelli[4], Mikhail M. Pooggin[1]*

1 University of Basel, Department of Environmental Sciences, Institute of Botany, Basel, Switzerland, **2** United States Department of Agriculture–Agricultural Research Service, Horticultural Crops Research Laboratory, Corvallis, Oregon, United States of America, **3** Oregon State University, Department of Botany and Plant Pathology, Center for Genome Research and Biocomputing, Corvallis, Oregon, United States of America, **4** Fasteris SA, Plan-les-Ouates, Geneva, Switzerland

Abstract

Virus-infected plants accumulate abundant, 21–24 nucleotide viral siRNAs which are generated by the evolutionary conserved RNA interference (RNAi) machinery that regulates gene expression and defends against invasive nucleic acids. Here we show that, similar to RNA viruses, the entire genome sequences of DNA viruses are densely covered with siRNAs in both sense and antisense orientations. This implies pervasive transcription of both coding and non-coding viral DNA in the nucleus, which generates double-stranded RNA precursors of viral siRNAs. Consistent with our finding and hypothesis, we demonstrate that the complete genomes of DNA viruses from *Caulimoviridae* and *Geminiviridae* families can be reconstructed by deep sequencing and *de novo* assembly of viral siRNAs using bioinformatics tools. Furthermore, we prove that this 'siRNA omics' approach can be used for reliable identification of the consensus master genome and its microvariants in viral quasispecies. Finally, we utilized this approach to reconstruct an emerging DNA virus and two viroids associated with economically-important red blotch disease of grapevine, and to rapidly generate a biologically-active clone representing the wild type master genome of *Oilseed rape mosaic virus*. Our findings show that deep siRNA sequencing allows for *de novo* reconstruction of any DNA or RNA virus genome and its microvariants, making it suitable for universal characterization of evolving viral quasispecies as well as for studying the mechanisms of siRNA biogenesis and RNAi-based antiviral defense.

Editor: Hanu Pappu, Washington State University, United States of America

Funding: The work was supported by European Cooperation in Science and Technology [grant SERI No. C09.0176 to L.F. and M.M.P.]; Swiss National Science Foundation [grant 31003A_143882/1 to M.M.P.]; Vinoculate, Inc. [contract 2010-744 to V.V.D.], USDA-NIFA [subcontract 2009-04401 to V.V.D.]; USDA-NIFA-SCRI [2009-51181-06027 subaward to R.R.M.]; and Bard [award IS-4314-10C to V.V.D.]. The funders had no role in study design, data collection and analysis, decision to publish, or preparation of the manuscript.

Competing Interests: The work in VVD lab is partially supported by a contract with Vinoculate, Inc. that also holds exclusive rights to an OSU patent "Closterovirus Vectors and Methods". No. 8,415,147 Issued April 9, 2013; OSU Ref. No. 06-57; Klarquist Ref. No. 245-79793-10. Jonathan Seguin, Patricia Otten and Laurent Farinelli are employed by Fasteris SA. There are no further patents, products in development or marketed products to declare. This does not alter the authors' adherence to all the PLOS ONE policies on sharing data and materials, as detailed online in the guide for authors.

* E-mail: Mikhail.Pooggin@unibas.ch

Introduction

Owing to error-prone replication, viruses accumulate microvariants which deviate from a consensus master genome by one or more SNPs (single-nucleotide polymorphisms) and/or indels (insertions/deletions) and comprise a viral quasispecies that can rapidly evolve in changing environment [1]. Resistance-breaking strains often emerge from such microvariants and from recombination events involving distinct viral strains or viruses. Existing methods of viral diagnostics using antibodies and PCR often fail to identify new pathogenic strains and are not applicable for emerging viruses with unknown genomes. Therefore, next generation deep sequencing approaches and *de novo* assembly of virus genomes from sequencing reads hold a great promise for universal diagnostics of viral pathogens and reliable characterization of causative agent(s) of any given disease [2,3]. In a pioneering work, Kreuze *et al.* [2] have demonstrated that a complete genome of a known plant RNA virus can be reconstructed *de novo* from multiple contigs of short interfering RNAs (siRNAs) which are generated in infected plants by the evolutionarily conserved RNA silencing/RNA interference (RNAi) machinery [4–6]. This and the follow-up studies have proven that deep siRNA sequencing and bioinformatics are applicable for identification and at least partial genome reconstruction of plant viruses and viroids [7–14] as well as insect viruses [15,16]. Here we extend these findings by demonstrating that the complete genomes of plant DNA viruses of two major families – *Caulimoviridae* and *Geminiviridae* – can be reconstructed without a reference genome as a single contig or a few overlapping contigs of viral siRNAs. Furthermore, we show that bioinformatics analysis of viral siRNA population allows for the identification of the master genome and its microvariants in viral quasispecies. We also used this technology to reconstruct a newly emerged single-stranded DNA virus and two viroids associated with the red blotch disease of grapevines in the United States. Thus, deep siRNA sequencing can be used for identification and reconstruction of the consensus master genome of any plant virus or viroid, and for studying virus diversity and evolution. Moreover, our analysis of siRNAs derived from DNA viruses and viroids contributes to further understanding the mechanisms of siRNA biogenesis and RNAi-based antiviral defense and raises new questions for future research.

Results and Discussion

Growing evidence indicates that 21–24 nt virus-derived siRNAs are produced from double-stranded (ds) RNA precursors covering the entire viral genome sequences [4,17,18]. Accordingly, complete or near complete genomes of new RNA virus strains have been reconstructed using a reference strain sequence as a scaffold for assembly of overlapping siRNA contigs which were generated by the short sequence read assembler Velvet [2,7,8,10,11–13]. Similar approaches have succeeded in detection and partial genome reconstruction of plant DNA viruses [2,9,10,19]. Because DNA viruses do not replicate through dsRNA intermediates, viral siRNA precursors are likely produced by the host Polymerase II-mediated, bidirectional transcription of a circular viral DNA beyond the poly(A) sites, thereby including both coding (mRNA) and non-coding (promoter) sequences [4,17,18]. To validate this hypothesis and test if complete genomes of DNA viruses could also be reconstructed as single contigs of viral siRNAs we used the small RNA (sRNA) deep-sequencing libraries obtained from *Arabidopsis* plants infected with *Cauliflower mosaic virus* (CaMV) [18] and *Cabbage leaf curl virus* (CaLCuV) [17] (Datasets S1*A* and S1*B*), which represent the *Caulimoviridae* and the *Geminiviridae* families, respectively. Bioinformatic analysis of the redundant sRNA reads revealed that the hotspots of viral siRNA production cover only portions of the viral genome in both cases. The non-redundant reads, however, cover the entire circular genomes of CaMV and CaLCuV in the sense and antisense orientations without gaps (Figure 1; Dataset S2), albeit the density of non-redundant reads is somewhat higher in the hotspot regions (Dataset S2). This suggests that both coding and non-coding regions of circular viral dsDNA are transcribed in both orientations to generate dsRNA precursors of viral siRNAs. Since genetic evidence revealed that the silencing-related, DNA-dependent RNA polymerases Pol IV and Pol V, or RNA-dependent RNA polymerases RDR1, RDR2 and RDR6 are not required for the biogenesis of CaMV- and CaLCuV-derived siRNAs [17,18], these dsRNA precursors are likely generated by Pol II-mediated sense and antisense transcription. However, potential involvement of RDR3, RDR4 or RDR5 in viral siRNA biogenesis was not ruled out yet. In conclusion, similar to RNA viruses, the entire genomes of DNA viruses from the families *Caulimoviridae* and *Geminiviridae* are densely covered with non-redundant viral siRNA species and therefore can potentially be reconstructed as single contigs of the viral siRNAs.

To *de novo* assemble viral siRNAs, we tested different algorithms using Velvet followed by Oases or Metavelvet for assembling redundant or non-redundant reads and Seqman for merging the resulting contigs. In some cases, we also used mapping to the plant genome as a filtering step before Seqman to separate the viral siRNA contigs from the plant sRNA contigs (Figure 2). As a result, with both Oases and Metavelvet, the complete 8,031 nt genome of CaMV was reconstructed as a single terminally-redundant contig (Figure 1). The bipartite genome of CaLCuV was assembled as one terminally-redundant contig covering 2,512 nt DNA-B and two contigs covering 2,583 nt DNA-A (Figure 1). In the latter case, the filtering step was required. Because DNA-A and DNA-B of CaLCuV share a near identical common region of 195 nts (with 7 SNPs), during *de novo* assembly the DNA-A siRNA contig gets split in two contigs within this region. Generally, Oases generated longer contigs, while Metavelvet more precise contigs. Non-redundant reads assembled in longer contigs. SNPs and short indels that occurred in some of the contigs could be identified and corrected by SNP calling with redundant reads (see Materials and Methods for further details of the bioinformatics analysis). Thus, using Oases or Metavelvet followed by Seqman, the complete viral genomes were assembled *de novo* as single contigs of non-redundant siRNAs. This is unlike most of the above mentioned reports of virus or viroid reconstruction, in which multiple contigs of redundant siRNAs generated by Velvet were assembled using a reference genome as a scaffold. Furthermore, we found that the filtering step before Velvet, which was applied in some previous studies in efforts to remove host small RNAs interfering with assembly of viral siRNAs, often generates gaps in viral siRNA contigs. In contrast, the filtering step before Seqman used in our study enables reconstruction of complete genomes, especially in the case of DNA viruses. Moreover, we found that SNP calling with redundant siRNA reads can be applied to correct potential errors in *de novo* assembly algorithms.

To determine if such 'siRNA omics' (siRomics) approach is applicable for identification of a master genome in viral 'quasispecies cloud', we sequenced sRNAs from *Arabidopsis* infected with *Oilseed rape mosaic virus* (ORMV), the RNA virus for which an available cDNA clone was not infectious [20] because of potential cloning errors or because it represented a defective microvariant from the ORMV quasispecies. Using Oases followed by Seqman, the 6,303 nt ORMV genome was reconstructed *de novo* as a single contig from two independent sRNA libraries (Dataset S1*C* and Dataset S2). This reconstructed genome differed from the available cDNA sequence at three positions (G-to-A at position 12, A-to-T at position 231, and G-to-T at position 5612; Figure 1). SNP calling using the two sRNA libraries confirmed these mismatches in 96.5–97.7% (A12), 99.5–99.9% (T231) and 88.6–93.8% (T5612) viral redundant reads and highlighted the overall variation in the ORMV quasispecies (Dataset S3*A*). We corrected these mismatches (presumably cloning errors) in the cDNA clone and tested it for infectivity. Strikingly, the resulting clone was fully biologically active, causing the disease symptoms indistinguishable from those of the wild type ORMV sap (Figure 3). Thus, a common problem of virology, often taking years to overcome [21], was solved in one step. Our unpublished results suggest that G at position 12 in the original cDNA clone had a drastic impact on ORMV infectivity, possibly because it affects initiation of positive stand synthesis during the viral replication process; the nucleotide substitutions at the positions 231 and 5612 likely represent viable variants in the virus quasispecies (N.M.L., R.R., and M.M.P., in preparation).

Interestingly, SNP calling revealed that ORMV, CaMV and CaLCuV do not differ drastically in the frequency of SNPs or the average degree of deviation (in %) from the master genome nucleotides (Dataset S3*A-D*). This implies that distinct replication mechanisms of these viruses, involving the viral RDR (ORMV), the viral reverse transcriptase (CaMV), or the host DNA polymerase (CaLCuV), may have a comparable error rate. The error rate of the host DNA polymerase possessing proof-reading activity might be as high as the error rates of the viral RDR and reverse transcriptase lacking any proof-reading activity, because it replicates viral DNA via a rolling circle mechanism involving a viral Rep protein (recently reviewed by Pooggin [22]). Alternatively, comparable accumulation of the microvariants in all the three viruses may reflect their adaptation to the experimental host plant *Arabidopsis thaliana.* Note, that for identification of the SNPs listed in Dataset S3, we used a quite conservative cut-off, 10% non-redundant reads, to account for both an error of Illumina sequencing of sRNAs (0.1–0.5%) and sequence-specific biases that may lead to overrepresentation of certain sRNAs in redundant

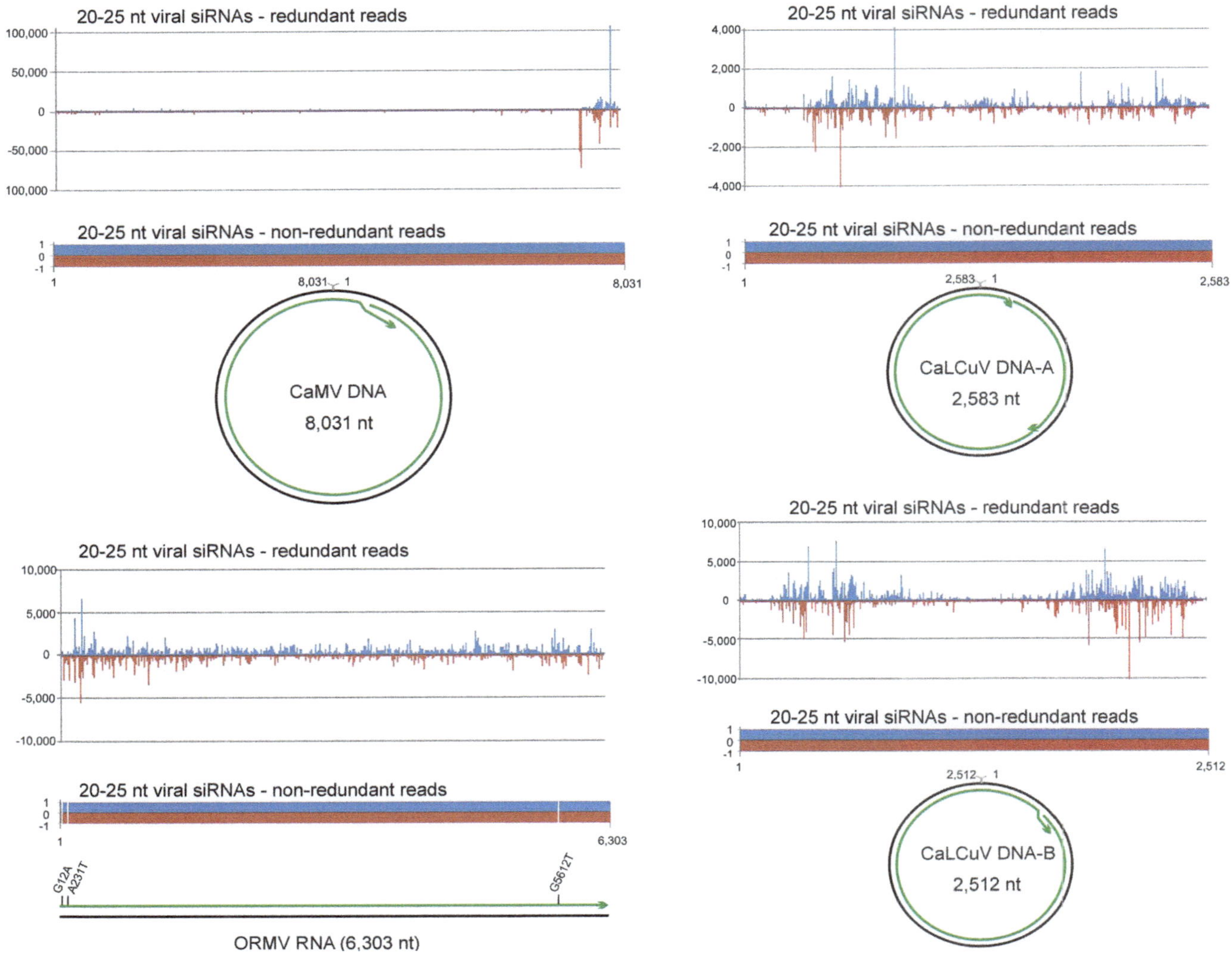

Figure 1. Maps of viral siRNAs and their contigs. The graphs plot the number of 20–25 nt viral siRNA reads (redundant and non-redundant) at each nucleotide position of the genomes of CaMV, ORMV and CaLCuV (DNA-A and DNA-B); Bars above the axis represent sense reads starting at respective positions; those below the antisense reads ending at respective positions. Circular DNA genomes of CaMV and CaLCuV and linear RNA genome of ORMV are shown below the graphs, with the siRNA contigs covering the genomes depicted as green lines with arrowheads. Mismatches between the ORMV contig and the reference genome are indicated.

reads. Since host RDR activity may amplify CaLCuV siRNAs and thereby contribute to the observed deviations from the CaLCuV master genome, we compared the viral microvariant accumulation in CaLCuV-infected wild-type plants and *rdr1/2/6* triple mutant plants with diminished RDR activities [17]: no drastic difference was observed in the frequency of SNPs or the average degree of deviation from the master genome nucleotides (Dataset S3*C-D*). Our findings for CaLCuV are consistent with the observations that geminiviruses have high mutation frequency and evolve as fast as RNA viruses (see [23] and references therein).

To evaluate the potential of siRomics for diagnostics of an unknown disease, we deep sequenced sRNAs from grapevines (*Vitis vinifera* cv. Pinot noir) grown at vineyards in Oregon, some of which exhibited severe leaf red blotch disease symptoms of unknown etiology, and from control plants with green, healthy-looking leaves. *De novo* reconstruction revealed that both infected and healthy-looking vines harbored *Grapevine yellow speckle viroid 1* (GYSVd-1) and *Hop stunt viroid* (HSVd), whose small circular RNA genomes were assembled as single terminally-redundant contigs and verified by redundant sRNA coverage (Figure 4; Dataset S1*D* and Dataset S2). Previously, these two viroids have been identified in grapevines in Oregon and elsewhere (see Materials and Methods). In addition, the disease-affected vines harbored a virus with a 3,206 nt circular genome reconstructed *de novo* from four contigs and validated by redundant sRNA coverage (Figure 4A; Dataset S1*D* and Dataset S2). Phylogenomic analysis showed that the virus is distantly related to circular DNA genomes of plant geminiviruses. The only closely related genome at the time of our analysis was that of a recently identified DNA virus from a declining Cabernet franc vineyard in New York State [24] (named 'grapevine geminivirus' (GVGV); the NCBI Genbank accession NC_017918). Despite their occurrence across the continent in distinct grapevine cultivars, the New York and Oregonian isolates differed by only 11 SNPs (Dataset S3*E*). Notably, all these 11 nucleotides in the Oregonian isolate are supported by at least 90% redundant reads in three independent red blotch leaf samples and therefore represent quite stable nucleotide positions in the master genome, unlike some of the other positions identified by SNP

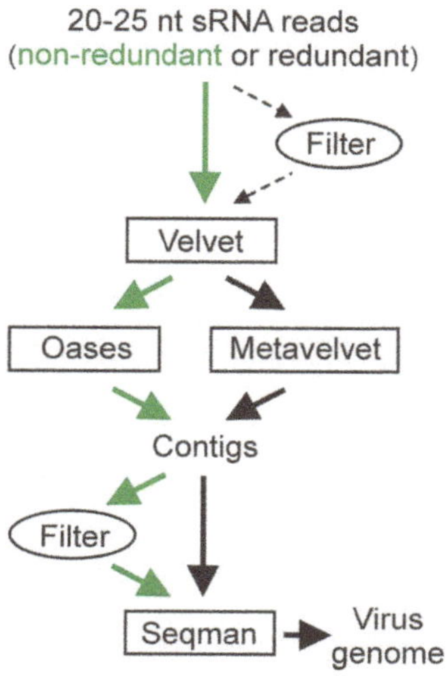

Figure 2. Bioinformatics algorithms for *de novo* reconstruction of viral/viroid genomes from siRNAs. The *de novo* assembler programs are boxed. Green arrows indicate the algorithm which in many cases generated the longest contigs.

calling (Dataset S3*B*). We designed PCR primers specific to the virus (5′-TGCAAGTGGACATACGTTTA and 5′-GGGATCC-CATCAATTGTTCT) and confirmed its presence in DNA samples from 12 of 16 symptomatic vines from the same vineyard, but not from any of 18 symptomless vines. Intriguingly, the most recent reports describe *Grapevine red blotch-associated virus* (GRBaV) and *Grapevine red leaf-associated virus* (GRLaV) that severely affect vineyards in States of California [25] and Washington [26], respectively. Both GRBaV and GRLaV sequences were found to be very similar to the geminivirus from New York [25,26]. Thus, the virus that we identified in Oregon appears to be geographically wide-spread and associated with the emerging disease that threatens the high cash-value crop, grapevine. Whether this virus causes red blotch disease alone or in complex with the viroids identified in our study remains to be investigated. Interestingly, GRLaV was found to be associated with two RNA viruses and four viroids including HSVd and GYSVd-1 [26]. Thus, both or one of the latter two viroids may contribute to the disease.

Unlike most RNA viruses that generate predominantly 21-nt siRNAs (e.g. ORMV [4]; Dataset S1*C*), DNA viruses that are transcribed in the nucleus spawn 21-, 22- and 24-nt siRNAs (e.g. CaMV and CaLCuV; Dataset S1*A* and Dataset S1*B*), which are processed by distinct Dicer-like (DCL) enzymes from long dsRNA precursors [4,17,18]. We found that the grapevine geminivirus also induces formation of predominantly 21-, 22- and 24-nt siRNAs (Dataset S1*D*), thus resembling other geminiviruses. Furthermore, both coding and non-coding regions of this virus are covered with viral siRNA species in both orientations without gaps (Figure 4; Dataset S2). This suggests that bi-directional readthrough transcription of circular viral DNA may generate dsRNA precursors covering the entire virus genome, like it was proposed in the case of CaLCuV [17]. Analysis of 5′-nucleotides of the grapevine geminivirus-derived siRNAs and the host sRNAs (Dataset S1*D*) revealed similar biases to 5′U in 21-nt sRNAs and 5′A in 24-nt sRNAs. This indicates that, similar to plant miRNAs and siRNAs [27,28], viral siRNAs are also sorted by Argonaute (AGO) proteins based on 5′-nucleotide identity to form silencing complexes. Bioinformatics analysis of siRNAs derived from the grapevine viroids HSVd and GYSVd-1 (Dataset S1*C*) revealed

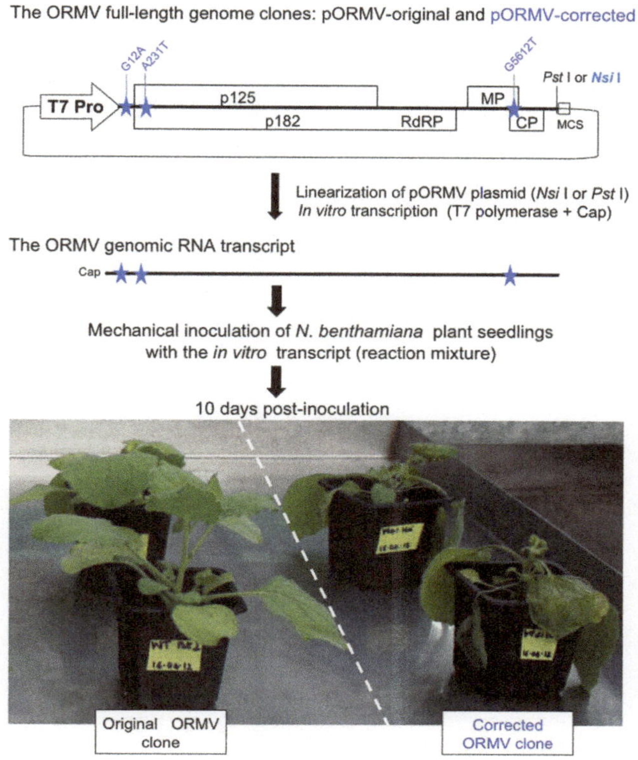

Figure 3. Test of the original and the corrected ORMV clones for infectivity. The plasmid containing the full-length ORMV genome sequence (original or corrected) behind the T7 promoter is depicted schematically: the restriction site *Pst* I or *Nsi* I, respectively, just downstream of the genome (located in multiple cloning site; MCS) was used for linearization of the plasmid, followed by run-off transcription by T7 polymerase in the presence of a cap analog. The resulting *in vitro* transcript (ORMV genomic RNA) was taken for mechanical inoculation of *N. benthamiana* plants. The picture shows the inoculated plants at 10 days post-inoculation.

that these nucleus-localized viroids produce predominantly 21-, 22- and 24-nt siRNAs reminiscent to those produced by DNA viruses transcribed in the nucleus. In further similarity to DNA viruses, siRNA species of each size-class and polarity densely cover the entire circular RNA genomes of these viroids (Dataset S2; Figure 4). This implies that viroid siRNAs are processed from dsRNA intermediates of Pol II-mediated replication of circular single-stranded viroid RNA [29]. The hotspots of viroid/virus siRNA production that map to similar locations for each siRNA size-class and polarity (Dataset S2) may result from preferential internal processing of the dsRNA precursors by distinct DCLs and/or stabilization of siRNAs with certain nucleotide compositions by AGO proteins.

In summary, using deep siRNA sequencing and bioinformatics, dubbed siRomics, we developed a pipeline for nucleotide sequencing of the entire genomes of DNA and RNA viruses or viroids and for identification of the consensus master genome and its microvariants in viral or viroid quasispecies. Moreover, we demonstrated utility of siRomics for the universal diagnostics of known and emerging viral diseases, as well as for rapid generation of the biologically-active clones of problematic viruses. In addition to highlighting the potential of siRomics for applied research, our findings contribute to further understanding the siRNA silencing mechanisms targeting DNA viruses as well as RNA viruses and viroids.

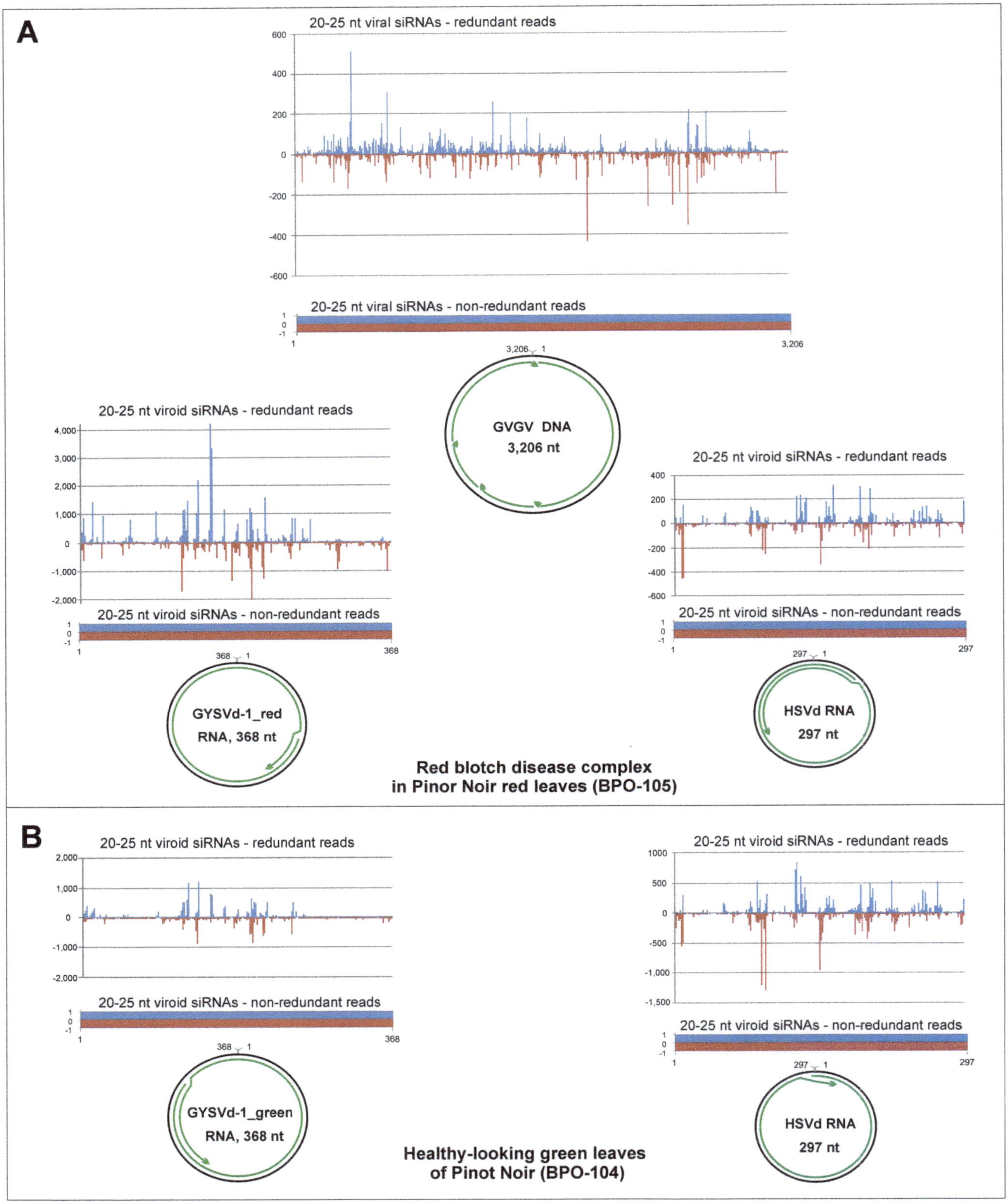

Figure 4. Maps of viral and viroid siRNAs and their contigs from red blotch disease-infected and healthy-looking leaves of grapevine. (**A**) Red blotch disease-infected leaves. (**B**) Healthy-looking green leaves. The graphs plot the number of 20–25 nt viral or viroid siRNA reads (redundant and non-redundant) at each nucleotide position of the genomes of the grapevine geminivirus (GVGV; also named GRBaV and GRLaV) and the viroids HSVd, GYSVd1_red and GYSVd1_green; Bars above the axis represent sense reads starting at respective positions; those below represent antisense reads ending at respective positions. The circular DNA genome of GVGV and the circular RNA genomes of HSVd, GYSVd1_red and

GYSVd1_green are shown below the graphs, with the siRNA contigs covering the genomes depicted as green lines with arrowheads.

Materials and Methods

Plants and viruses

Growth conditions and virus infections of *Arabidopsis thaliana* Col-0 plants were described in detail previously [4]. Briefly, seedlings were infected either by biolistic inoculation with DNA clones of *Cauliflower mosaic virus* (CaMV; the NCBI Genbank accession V00140) and *Cabbage leaf curl virus* (CaLCuV; U65529.2 for DNA-A and U65530.2 for DNA-B) or by mechanical inoculation with sap from *Oilseed rape mosaic virus* (ORMV)-infected *Nicotiana benthamiana*. A previously constructed plasmid containing ORMV cDNA downstream of the T7 promoter (kindly provided by Dr. Fernando Ponz) was modified using synthetic DNA fragments and suitable restriction sites to correct the cloning errors and obtain the reconstructed wild type ORMV genome clone (deposited to the Genbank as KF137561). The resulting and the original clones were linearized downstream of the ORMV sequence and used as templates for *in vitro* transcription reactions (MEGAscript T7 kit, Ambion) to produce a capped viral genomic RNA. The reaction mixtures were used for mechanical inoculation. Symptom development at day 10 post-inoculation is shown in Figure 3.

Samples of the red leaves displaying leafroll-like disease symptoms (named 'red blotch' disease) and healthy-looking green leaves of grapevine cv. Pinot noir plants were collected in a privately owned vineyard near Newberg, Oregon, USA in summer, 2011. The samples were scion clone 777 grafted onto rootstock 44–53 and collected with permission of the owner.

Deep sequencing and bioinformatics analysis of viral/viroid siRNAs

Total RNA from infected and control tissue samples was extracted with Trizol and used for Illumina sequencing of 19–30 nt RNAs as described for CaLCuV by Aregger *et al.* [17]. The resulting small RNA (sRNA) libraries (detailed in Dataset S1) were taken for bioinformatics analysis and for *de novo* reconstruction of the viral genomes using the algorithms summarized in Figure 2. The results of bioinformatics analysis of the viral and host sRNA populations are summarized in Datasets S1, S2 and S3. To reconstruct viral and viroid genomes, the non-redundant or redundant sRNA reads ranging from 20 to 25 nts were assembled into contigs using Velvet 1.2.07 (www.ebi.ac.uk/~zerbino/velvet) [30] followed by Oases 0.2.08 (www.ebi.ac.uk/~zerbino/oases) [31] or Metavelvet 1.2.01 (metavelvet.dna.bio.keio.ac.jp) [32]. Number and size of the resulting contigs varied depending on the choice of Velvet *k*-mer values (13 through 23). 100% coverage of a virus genome could be achieved either with single *k*-mers or certain combinations thereof, as exemplified for CaMV, CaLCuV and ORMV in Dataset S4. SNP calling and correction of errors in viral contigs/genomes was done using Integrative Genomics Viewer (IGV; www.broadinstitute.org/igv) [33]. Oases and Metavelvet contigs obtained with all *k*-mer values or their selected combinations were merged using the Seqman module of Lasergene DNASTAR 8.1.2 Core Suite (DNAStar, Madison, WI). If required, the filtering step before Seqman (or Velvet) was done by mapping contigs (or sRNA sets) to the *Arabidopsis thaliana* genome (TAIR9) or *Vitis vinifera* genome (PRJNA33471) using Burrows-Wheeler Aligner (BWA) 0.5.9 [34]. The *de novo* reconstructed viral genomes were scanned for SNPs and indels using IGV with redundant reads. Finally, single-base resolution maps of viral sRNAs on the virus genomes were created using BWA and a sRNA map tool MISIS (www.fasteris.com/apps; [35]). Reads mapping to several positions on the reference sequence were attributed at random to one of them. To account for a circular virus/viroid genome the first 25 bases of the genome sequence were added to its 3′-end. For each reference genome and each sRNA size (20 to 25 nt), MISIS counted total number of reads, reads in forward and reverse orientation (Dataset S1) and thus generated single-base resolution maps (Dataset S2), where for each position starting from the 5′ end of the reference genome, the number of matches starting at this position in forward (first base of the read) and reverse (last base of the read) orientation for each read length is given. The reads mapped to the last 25 bases of the extended genome sequence were added to the reads mapped to the first 25 bases. MISIS generated two types of counts tables, one with zero mismatches and another with up to two mismatches. Comparison of the two tables was informative for identification and correction of potential mismatches between a reference sequence and the master genome sequence as well as for initial identification of SNPs and short indels in viral quasispecies (see Dataset S2, for all the viruses and viroids analyzed in this study). The positions of SNPs and the degree of deviation (in %) from the master genome nucleotide at each position in viral and viroid quasispecies were identified by IGV analysis of redundant and non-redundant sRNAs mapped to the reference genome with up to 2 mismatches. For identification of the SNPs listed in Dataset S3, we set an arbitrary cutoff value of 10% non-redundant reads: in other words, ten or more percent of the reads support the deviation from the master genome nucleotide for each SNP.

The analysis of siRNAs and complete genome contigs revealed that the infectious DNA-B clone of CaLCuV differs from its reference sequence U65530.2 by a single nucleotide deletion at the last position of the reference (making the genome 1 nt shorter), the infectious clone of CaMV differs from its reference sequence V00140 by two substitutions (C6175A and T6281C), while the original non-infectious ORMV clone differs from its reference sequence (NC_004422; named *Youcai mosaic virus*) at several positions. These apparent sequencing errors were confirmed by re-sequencing of the three clones. The original ORMV clone (confirmed by re-sequencing) differs at the three positions (Dataset S3*A*) from the reconstructed wild-type ORMV genome (KF137561) described in this study. *Hop stunt viroid* (HSVd; deposited to the Genbank as KF137565), which we reconstructed from each of the two green and three red leaf samples of grapevine cv. Pinot noir (Dataset S1*D*), is 100% identical to the sequences of other HSVd isolates, e.g. from grapevine cultivars Lumunage and Thompson Seedless in China (DQ371455, DQ371459) and a citrus tree in Tunisia (GU825977). In addition, the green leaves contained a variant of this viroid with the two SNPs supported by ca. 60% reads (Dataset S3*H*). *Grapevine yellow speckle viroid 1* (GYSVd-1) reconstructed from the two green leaf samples (GYSVd1_green; deposited to the Genbank as KF137564) is most closely related to the GYSVd-1 isolate from Germany (X87911; two SNPs), while GYSVd-1 reconstructed from the three red leaf samples (GYSVd1_red; deposited to the Genbank as KF137563) to the GYSVd-1 isolate from Japan (AB028466; two SNPs). The genome sequence of grapevine geminivirus (GVGV) reconstructed from the red leaves was deposited to the Genbank as KF137562.

Acknowledgments

We thank Thomas Boller for supporting the research of M.M.P. group at the Botanical Institute, and Fernando Ponz and Manfred Heinlein for providing ORMV materials.

Supporting Information

Dataset S1 Counts of viral and endogenous sRNAs in the sRNA deep-sequencing libraries from mock-inoculated and CaMV-infected *Arabidopsis* (Table S1A), CaLCuV-infected *Arabidopsis* (Table S1B), ORMV-infected *Arabidopsis* (Table S1C), and healthy-looking green and red blotch disease-infected leaves of grapevine cv. Pinot noir plants (Table S1D).

Dataset S2 MISIS-generated, single-base resolution maps of 20–25 nt viral siRNAs from CaMV (BPO-20, BPO-22)-, CaLCuV (BPO-57)- and ORMV (BPO-38, BPO-44)- infected *Arabidopsis* plants and of 20–25 nt viral (GVGV) and viroid (HSVd, GYSVd1_red, GYSVd1_green) siRNAs from red blotch disease-infected red leaves (BPO-105) and healthy-looking green leaves (BPO-104) of grapevine cv. Pinot noir plants. The numbering of nucleotide positions are according to the NCBI Genbank reverence sequences of CaMV (V00140; note that two corrections C6175A and T6281C were introduced in this sequence based on the sRNA and DNA sequencing), CaLCV DNA-A (U65529.2), CaLCuV DNA-B (U65530.2; the last nucleotide of this reference sequence, position 2513, was deleted based on the sRNA and DNA sequencing), ORMV (KF137561), GVGV (KF137562), HSVd (KF137565), GYSVd1_green (KF137564) and GYSV-d1_red (KF137563). Note that the positions of 5′-terminal nucleotide of sense sRNAs and 3′-terminal nucleotide of antisense sRNAs along the reference sequence are given, and the read counts are given for each sRNA of 20-, 21-, 22-, 23-, 24- and 25-nt classes mapped to the forward strand (X20, X21, X22, X23, X24, X25) and the reverse strand (X20_rev, X21_rev, X22_rev, X23_rev, X24_rev, X25_rev) with zero mismatches, along with the total counts of 20–25 nt sRNAs mapped on the forward (total_forward) and reverse (total_reverse) strands and on both strands (total). The last column contains the total number of 20–25 nt sRNA mapped to the reference sequence with up to two mismatches.

Dataset S3 SNPs in viral and viroid quasispecies. **Table S3A**: SNPs at positions of the mismatches between the reconstructed wild-type ORMV genome and the original ORMV genome clone as well as SNPs in the wild type ORMV viral quasispecies; **Table S3B**: SNPs in CaMV; **Table S3C**: SNPs in CaLCuV DNA-A; **Table S3D**: SNPs in CaLCuV DNA-B; **Table S3E**: SNPs at the positions of the mismatches between the GVGV-Oregon and the GVGV-New York genomes as well as other SNPs in the GVGV-Oregon quasispecies; **Table S3F**: SNPs in the GYSVd-1 (red) and green) quasispecies; **Table S3G**: SNPs in the GYSVd-1 (green) quasispecies; **Table S3H**: SNPs in the HSVd quasispecies.

Dataset S4 Analysis of the contigs generated by Velvet and Oases using non-redundant 20–25 nt sRNA libraries from CaMV (BPO-20 and BPO-21)-, CaLCuV (BPO-57)- and ORMV (BPO38 and BPO44)-infected *Arabidopsis*. Coverage (in %) of the viral genome sequences with the siRNA contigs is calculated for single k-mer values and combination thereof.

Author Contributions

Conceived and designed the experiments: MMP LF VVD. Performed the experiments: JS RR NML RRM KK PO. Analyzed the data: MMP JS PO VVD. Wrote the paper: MMP VVD.

References

1. Domingo E, Sheldon J, Perales C (2012) Viral quasispecies evolution. Microbiol Mol Biol Rev 76: 159–216.
2. Kreuze JF, Perez A, Untiveros M, Quispe D, Fuentes S, et al. (2009) Complete viral genome sequence and discovery of novel viruses by deep sequencing of small RNAs: a generic method for diagnosis, discovery and sequencing of viruses. Virology 388: 1–7.
3. Hagen C, Frizzi A, Gabriels S, Huang M, Salati R, et al. (2012) Accurate and sensitive diagnosis of geminiviruses through enrichment, high-throughput sequencing and automated sequence identification. Arch Virol. 157: 907–15.
4. Blevins T, Rajeswaran R, Shivaprasad PV, Beknazariants D, Si-Ammour A, et al. (2006) Four plant Dicers mediate viral small RNA biogenesis and DNA virus induced silencing. Nucleic Acids Res 34: 6233–6246.
5. Donaire L, Wang Y, Gonzalez-Ibeas D, Mayer KF, Aranda MA, et al. (2009) Deep-sequencing of plant viral small RNAs reveals effective and widespread targeting of viral genomes. Virology 392: 203–214.
6. Llave C (2010) Virus-derived small interfering RNAs at the core of plant-virus interactions. Trends Plant Sci 15: 701–707.
7. Al Rwahnih M, Daubert S, Golino D, Rowhani A (2009) Deep sequencing analysis of RNAs from a grapevine showing Syrah decline symptoms reveals a multiple virus infection that includes a novel virus. Virology 387: 395–401.
8. Cuellar WJ, Cruzado RK, Fuentes S, Untiveros M, Soto M, et al. (2011) Sequence characterization of a Peruvian isolate of Sweet potato chlorotic stunt virus: further variability and a model for p22 acquisition. Virus Res 157: 111–115.
9. Zhang Y, Singh K, Kaur R, Qiu W (2011) Association of a novel DNA virus with the grapevine vein-clearing and vine decline syndrome. Phytopathology 101: 1081–90.
10. Hagen C, Frizzi A, Kao J, Jia L, Huang M, et al. (2011) Using small RNA sequences to diagnose, sequence, and investigate the infectivity characteristics of vegetable-infecting viruses. Arch Virol 156: 1209–1216.
11. Li R, Gao S, Hernandez AG, Wechter WP, Fei Z, Ling KS (2012) Deep sequencing of small RNAs in tomato for virus and viroid identification and strain differentiation. PLoS One 7: e37127.
12. Fuentes S, Heider B, Tasso RC, Romero E, Zum Felde T, et al. (2012) Complete genome sequence of a potyvirus infecting yam beans (Pachyrhizus spp.) in Peru. Arch Virol 157: 773–776.
13. Giampetruzzi A, Roumi V, Roberto R, Malossini U, Yoshikawa N, et al. (2012) A new grapevine virus discovered by deep sequencing of virus- and viroid-derived small RNAs in Cv Pinot gris. Virus Res 163: 262–268.
14. Wu Q, Wang Y, Cao M, Pantaleo V, Burgyan J, et al. (2012) Homology-independent discovery of replicating pathogenic circular RNAs by deep sequencing and a new computational algorithm. Proc Natl Acad Sci U S A 109: 3938–3943.
15. Wu Q, Luo Y, Lu R, Lau N, Lai EC, et al. (2010) Virus discovery by deep sequencing and assembly of virus-derived small silencing RNAs. Proc Natl Acad Sci U S A 107: 1606–1611.
16. Vodovar N, Goic B, Blanc H, Saleh MC (2011) In silico reconstruction of viral genomes from small RNAs improves virus-derived small interfering RNA profiling. J Virol 85: 11016–11021.
17. Aregger M, Borah BK, Seguin J, Rajeswaran R, Gubaeva EG, et al. (2012) Primary and secondary siRNAs in geminivirus-induced gene silencing. PLoS Pathog 8: e1002941.
18. Blevins T, Rajeswaran R, Aregger M, Borah BK, Schepetilnikov M, et al. (2011) Massive production of small RNAs from a non-coding region of Cauliflower mosaic virus in plant defense and viral counter-defense. Nucleic Acids Res 39: 5003–5014.
19. Loconsole G, Saldarelli P, Doddapaneni H, Savino V, Martelli GP, et al. (2012) Identification of a single-stranded DNA virus associated with citrus chlorotic dwarf disease, a new member in the family Geminiviridae. Virology 432: 162–172.
20. Mansilla C, Sánchez F, Padgett HS, Pogue GP, Ponz F (2009) Chimeras between oilseed rape mosaic virus and tobacco mosaic virus highlight the relevant role of the tobamoviral RdRp as pathogenicity determinant in several hosts. Mol Plant Pathol 10: 59–68.
21. Kurth EG, Peremyslov VV, Prokhnevsky AI, Kasschau KD, Miller M, et al. (2012) Virus-derived gene expression and RNA interference vector for grapevine. J Virol 86: 6002–6009.

22. Pooggin MM (2013) How can plant DNA viruses evade siRNA-directed DNA methylation and silencing? Int J Mol Sci 14: 15233–15259.
23. Duffy S, Holmes EC (2009) Validation of high rates of nucleotide substitution in geminiviruses: phylogenetic evidence from East African cassava mosaic viruses. J Gen Virol 90: 1539–1547.
24. Krenz B, Thompson JR, Fuchs M, Perry KL (2012) Complete genome sequence of a new circular DNA virus from grapevine. J Virol 86: 7715.
25. Al Rwahnih M, Dave A, Anderson MM, Rowhani A, Uyemoto JK, et al. (2013) Association of a DNA virus with Grapevines affected by Red Blotch disease in California. Phytopathology 103: 1069–1076.
26. Poojari S, Alabi OJ, Fofanov VY, Naidu RA (2013) A leafhopper-transmissible DNA virus with novel evolutionary lineage in the family geminiviridae implicated in grapevine redleaf disease by next-generation sequencing. PLoS One 8: e64194.
27. Mi S, Cai T, Hu Y, Chen Y, Hodges E, et al. (2008) Sorting of small RNAs into Arabidopsis argonaute complexes is directed by the 5′ terminal nucleotide. Cell 133: 116–127.
28. Havecker ER, Wallbridge LM, Hardcastle TJ, Bush MS, Kelly KA, et al. (2010) The Arabidopsis RNA-directed DNA methylation argonautes functionally diverge based on their expression and interaction with target loci. Plant Cell 22: 321–334.
29. Gómez G, Pallás V (2013) Viroids: a light in the darkness of the lncRNA-directed regulatory networks in plants. New Phytol 198: 10–15.
30. Zerbino DR, Birney E (2008) Velvet: algorithms for de novo short read assembly using de Bruijn graphs. Genome Res 18: 821–829.
31. Schulz MH, Zerbino DR, Vingron M, Birney E (2012) Oases: robust de novo RNA-seq assembly across the dynamic range of expression levels. Bioinformatics 28: 1086–1092.
32. Namiki T, Hachiya T, Tanaka H, Sakakibara Y (2012) MetaVelvet: an extension of Velvet assembler to de novo metagenome assembly from short sequence reads. Nucleic Acids Res 40: e155.
33. Thorvaldsdóttir H, Robinson JT, Mesirov JP (2013) Integrative Genomics Viewer (IGV): high-performance genomics data visualization and exploration. Brief Bioinform 14: 178–192.
34. Li H, Durbin R (2009) Fast and accurate long-read alignment with Burrows-Wheeler transform. Bioinformatics 26: 1754–1760.
35. Seguin J, Otten P, Baerlocher L, Farinelli L, Pooggin MM (2014) MISIS: A bioinformatics tool to view and analyze maps of small RNAs derived from viruses and genomic loci generating multiple small RNAs. J Virol Methods 195: 120–122.

20

Combined Analysis of the Chloroplast Genome and Transcriptome of the Antarctic Vascular Plant *Deschampsia antarctica* Desv

Jungeun Lee, Yoonjee Kang, Seung Chul Shin, Hyun Park*, Hyoungseok Lee*

Division of Life Sciences, Korea Polar Research Institute, Incheon, South Korea

Abstract

Background: Antarctic hairgrass (*Deschampsia antarctica* Desv.) is the only natural grass species in the maritime Antarctic. It has been researched as an important ecological marker and as an extremophile plant for studies on stress tolerance. Despite its importance, little genomic information is available for *D. antarctica*. Here, we report the complete chloroplast genome, transcriptome profiles of the coding/noncoding genes, and the posttranscriptional processing by RNA editing in the chloroplast system.

Results: The complete chloroplast genome of *D. antarctica* is 135,362 bp in length with a typical quadripartite structure, including the large (LSC: 79,881 bp) and small (SSC: 12,519 bp) single-copy regions, separated by a pair of identical inverted repeats (IR: 21,481 bp). It contains 114 unique genes, including 81 unique protein-coding genes, 29 tRNA genes, and 4 rRNA genes. Sequence divergence analysis with other plastomes from the BEP clade of the grass family suggests a sister relationship between *D. antarctica*, *Festuca arundinacea* and *Lolium perenne* of the Poeae tribe, based on the whole plastome. In addition, we conducted high-resolution mapping of the chloroplast-derived transcripts. Thus, we created an expression profile for 81 protein-coding genes and identified *ndhC*, *psbJ*, *rps19*, *psaJ*, and *psbA* as the most highly expressed chloroplast genes. Small RNA-seq analysis identified 27 small noncoding RNAs of chloroplast origin that were preferentially located near the 5′- or 3′-ends of genes. We also found >30 RNA-editing sites in the *D. antarctica* chloroplast genome, with a dominance of C-to-U conversions.

Conclusions: We assembled and characterized the complete chloroplast genome sequence of *D. antarctica* and investigated the features of the plastid transcriptome. These data may contribute to a better understanding of the evolution of *D. antarctica* within the Poaceae family for use in molecular phylogenetic studies and may also help researchers understand the characteristics of the chloroplast transcriptome.

Editor: Szabolcs Semsey, Niels Bohr Institute, Denmark

Funding: This work was supported by Functional Genomics on Polar Organisms grant (PE13020) and the Basic Research Program (PE13120) funded by Korea Polar Research Institute (KOPRI). The funders had no role in study design, data collection and analysis, decision to publish, or preparation of the manuscript.

Competing Interests: The authors have declared that no competing interests exist.

* E-mail: soulaid@kopri.re.kr (HL); hpark@kopri.re.kr (HP)

Introduction

Chloroplasts are plant-specific organelles that conduct photosynthesis, providing essential energy for the synthesis of starch, fatty acids, pigments, and amino acids [1,2]. Chloroplasts contain DNA and their own genetic information. In higher plants, chloroplast genomes exist as circular DNA, with the size ranging from 120 kb to 150 kb, and generally have a highly conserved quadripartite organization composed of two copies of inverted repeats (IRs), which separate the large single copy (LSC) and small single copy (SSC) regions [3,4]. In vascular plants, chloroplast genomes usually contain 110–130 unique genes encoding 4 rRNAs, 30–31 tRNAs, and 80–90 proteins; these encode ribosomal proteins and RNA polymerase subunits involved in protein synthesis, thylakoid proteins, and the Rubisco large subunit for photosynthesis, as well as protein subunits for an NADH dehydrogenase complex, which mediates redox reactions [2,5]. Advances in high-throughput sequencing technologies have resulted in the full sequences of organelle genomes from a growing number of organisms [6]. Currently, plastid genome resources with >420 records have been established. These provide a vast amount of high-resolution information that can be exploited in phylogenetic and ecological studies, making it possible to track the evolutionary history of a species after obtaining the full sequence of its chloroplast genome.

The grass family (Poaceae), which occurs in nearly every terrestrial habitat, is one of the most diverse angiosperm families, including approximately 10,000 species over 700 genera. To date, 38 chloroplast genomes of grass species [32 from the BEP (Bambusoideae, Ehrhartoideae, Pooideae) clade and 6 from the PACMAD (Panicoideae, Arundinoideae, Chloridoideae, Micrairoideae, Aristidoideae, and Danthonioideae) clade] have been deposited into the GenBank database, and recent studies have tried to reconstruct the phylogeny of the subfamilies and genera in

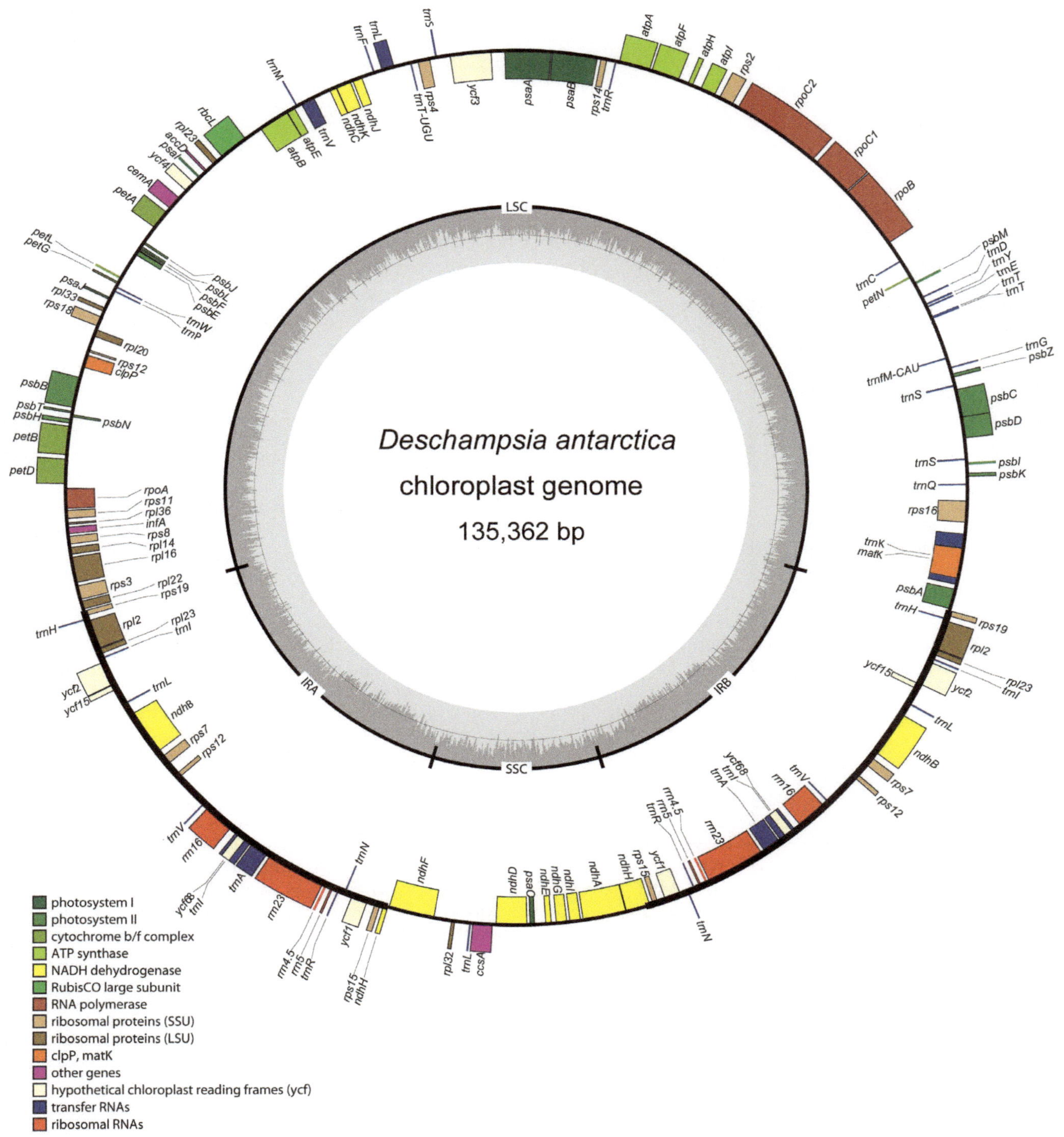

Figure 1. Map of the chloroplast genome of *Deschampsia antarctica*. Genes lying outside of the outer circle are transcribed clockwise, while those inside the circle are transcribed counterclockwise. Genes belonging to different functional groups are color coded. The innermost darker gray corresponds to GC, while the lighter gray corresponds to AT content. IR, inverted repeat; LSC, large single copy region; SSC, small single copy region.

the Poaceae family using whole sequences of chloroplast genomes [7,8].

Extremophile plants have evolved tolerance overcoming unfavorable environmental conditions, such as freezing temperatures, drought, high salinity, and high UV radiance. The genetic information on such species provides clues for the evolutionary or geological history of the species, as well as resources for genetic engineering. Antarctic hairgrass (*Deschampsia antarctica* Desv.) is the only native grass species that thrives in the harsh environment of Antarctica [9]. As an extremophile, it may be useful as a source of genes associated with stress tolerance [10]. It has also been suggested as an ecological marker of global warming because of its

Table 1. Genes present in the *Deschampsia antarctica* chloroplast genome.

Products		Genes
1	Photosystem I	*psaA, B, C, I, J, ycf3*[a], *ycf4*
2	Photosystem II	*psbA, B, C, D, E, F, H, I, J, K, L, M, N, T, Z*
3	Cytochrome b6/f	*petA, B*[b], *D*[b], *G, L, N*
4	ATP synthase	*atpA, B, E, F*[b], *H, I*
5	Rubisco	*rbcL*
6	NADH oxidoreductase	*ndhA*[b], *B*[b, c], *C, D, E, F, G, H*[c], *I, J, K*
7	Large subunit ribosomal proteins	*rpl2*[b, c], *14, 16*[b], *20, 22, 23*[c], *32, 33, 36*
8	Small subunit ribosomal proteins	*rps2, 3, 4, 7*[c], *8, 11, 12*[b, c, d], *14, 15*[c], *16*[b], *18, 19*[c]
9	RNAP	*rpoA, rpoB, C1, C2*
10	Other proteins	*accD, ccsA, cemA, clpP, matK, infA*
11	Proteins of unknown function	*ycf1*[c, e], *ycf2*[c], *ycf15*[c], *ycf68*[c]
12	Ribosomal	*rrn23*[c], *16*[c], *5*[c], *4.5*[c]
13	Transfer RNAs	*trnA(UGC)*[b, c], *C(GCA), D(GUC), E(UUC), F(GAA), G(UCC), H(GUG)*[c], *I(CAU)*[c], *I(GAU)*[b, c], *K(UUU)*[b], *L(UAA)*[b], *L(UAG), L(CAA)*[c], *fM(CAU), M(CAU), N(GUU)*[c], *P(UGG), Q(UUG), R(ACG)*[c], *R(UCU), S(GCU), S(GGA), S(UGA), T(GGU), T(UGU), V(UAC)*[b], *V(GAC)*[c], *W(CCA), Y(GUA)*

[a]Gene containing two introns.
[b]Gene containing a single intron.
[c]Two gene copies in the IRs.
[d]Gene divided into two independent transcription units.
[e]Pseudogene.

successful adaptation to climate change and its rapid spread [10,11]. Despite the importance of this terrestrially isolated plant, its phylogenetic position is still controversial [12–14], and available genetic resources are limited.

Table 2. Genes containing introns in the *Deschampsia antarctica* chloroplast genome and the length of the exons and introns.

Gene	Location	Length (bp)				
		Exon I	Intron I	Exon II	Intron II	Exon III
rps16	LSC	40	830	209		
atpF	LSC	159	802	408		
ycf3	LSC	126	749	228	728	159
petB	LSC	6	760	642		
petD	LSC	9	686	525		
rpl16	LSC	9	893	402		
*rps12**	LSC	117	-	231		
rpl2	IR	393	660	432		
ndhB	IR	777	712	756		
ndhA	SSC	549	1012	540		
trnK-UUU	LSC	38	2486	33		
trnL-UAA	LSC	37	537	50		
trnV-UAC	LSC	39	605	37		
trnI-GAU	IR	42	801	35		
trnA-UGC	IR	38	811	35		

**rps12* is trans-spliced gene with 59 end exon located in the LSC region and the duplicated 39 end exon located in IR regions.

Here, we obtained the complete chloroplast genome sequence of *D. antarctica* by high-throughput sequencing and *de novo* assembly. By comparison with the chloroplast genomes from other representative members of the BEP clade, we explored the deep-phylogenetic relationship of *D. antarctica* to other grass species at the genomic level. In addition, using combinatorial analysis of the RNA-seq data, we conducted high-resolution mapping of the chloroplast-derived transcripts to a reference chloroplast genome to demonstrate transcriptome profiles of the coding and noncoding genes and the posttranscriptional processing by RNA editing in the chloroplasts of *D. antarctica*. These data may contribute to a better understanding of the evolution of *D. antarctica* within the Poaceae family and the characteristics of the chloroplast transcriptome.

Methods

Ethics Statement

This study including sample collection and experimental research conducted on these materials was according to the law on activities and environmental protection to Antarctic approved by the Minister of Foreign Affairs and Trade of the Republic of Korea.

Plant Materials

Deschampsia antarctica Desv. (Poaceae) plants growing under natural conditions were collected in the vicinity of the Korean King Sejong Antarctic Station (62°14′29″S, 58°44′18″W) on the Barton Peninsula of King George Island and then transferred to the lab and grown hydroponically, supplemented with 0.5× Murashige and Skoog (MS) medium containing 2% sucrose under a 16:8 h light:dark cycle with a light intensity of 150 μmol m^{-2} s^{-1} at 15°C, a temperature that results in high Rubisco activity in *D. antarctica* [15].

Table 3. The codon–anticodon recognition pattern and codon usage in the *Deschampsia antarctica* chloroplast genome.

Amino acid	Codon	No.*	tRNA	Amino acid	Codon	No.	tRNA
Phe	UUU	790		Tyr	UAU	599	
Phe	UUC	448	trnF-GAA	Tyr	UAC	211	trnY-GUA
Leu	UUA	790	trnL-UAA	Stop	UAA	48	
Leu	UUG	445	trnL-CAA	Stop	UAG	20	
Leu	CUU	492		His	CAU	371	
Leu	CUC	226		His	CAC	164	trnH-GUG
Leu	CUA	363	trnL-UAG	Gln	CAA	572	trnQ-GUU
Leu	CUG	150		Gln	CAG	235	
Ile	AUU	874		Asn	AAU	647	
Ile	AUC	379	trnI-GAU	Asn	AAC	274	trnN-GUU
Ile	AUA	562	trnI-CAU	Lys	AAA	865	trnK-UUU
Met	AUG	522	trn(f)M-CAU	Lys	AAG	367	
Val	GUU	473		Asp	GAU	619	
Val	GUC	182	trnV-GAC	Asp	GAC	209	trnD-GUC
Val	GUA	505	trnV-UAC	Glu	GAA	807	trnE-UUC
Val	GUG	196		Glu	GAG	372	
Ser	UCU	458		Cys	UGU	215	
Ser	UCC	328	trnS-GGA	Cys	UGC	106	trnC-GCA
Ser	UCA	296	trnS-UGA	Stop	UGA	17	
Ser	UCG	161		Trp	UGG	430	trnW-CCA
Pro	CCU	375		Arg	CGU	312	trnR-ACG
Pro	CCC	243		Arg	CGC	152	
Pro	CCA	291	trnP-UGG	Arg	CGA	311	
Pro	CCG	151		Arg	CGG	152	
Thr	ACU	507		Arg	AGA	436	trnR-UCU
Thr	ACC	236	trnT-GGU	Arg	AGG	221	
Thr	ACA	331	trnT-UGU	Ser	AGU	349	
Thr	ACG	173		Ser	AGC	176	trnS-GCU
Ala	GCU	593		Gly	GGU	539	
Ala	GCC	242		Gly	GGC	225	trnG-GCC
Ala	GCA	413	trnA-UGC	Gly	GGA	653	trnG-UCC
Ala	GCG	202		Gly	GGG	359	

*Numerals indicate the frequency of usage of each codon in 23430 in codons in 81 potential protein-coding genes.

DNA and RNA Sequencing

Total genomic DNA was extracted from leaf tissues using the DNeasy Plant Mini Kit (Qiagen, Valencia, CA, USA) according to the manufacturer's instructions. Total RNA was extracted from whole plants using the RNeasy Plant Mini Kit (Qiagen). For the small noncoding RNA library, total RNA was extracted from leaves using the mirVana Kit (Ambion, Austin, TX, USA). The quality of the RNA and DNA was checked on a Bioanalyzer 2100 (Agilent, Santa Clara, CA, USA). The libraries were prepared and sequenced according to the manufacturer's instructions (Illumina, San Diego, CA, USA). The DNA library was constructed using TruSeq DNA sample preparation kits and a single lane of an Illumina HiSeq2000 sequencer (PE, 2×101 bp). For the mRNA library, multiplex libraries were obtained using TruSeq RNA sample preparation kits, and the samples were sequenced in one lane of an Illumina HiSeq2000 sequencer (PE, 2×101 bp). The small RNA library was constructed using the TruSeq Small RNA Sample Prep Kit; the resulting single end library was sequenced in one lane of an Illumina GAIIX sequencer (SE, 1×35 bp). The files containing the sequences and quality scores of reads were deposited in the NCBI Short Read Archive, and the accession numbers are SRX465632 (genomic DNA-Seq), SRX465633 (mRNA-Seq), and SRX465634 (Small RNA-Seq).

Genome Assembly, Annotation, and Sequence Analysis

After trim of low quality reads and adapters, the raw reads were aligned to 330 publicly available chloroplast genomes downloaded from NCBI organelle genome resources. *De novo* assembly was done with the collected chloroplast-related reads by Celera Assembler 6.1 (Celera Genomics, Alameda, USA). The assembled contigs were ordered with reference chloroplast genomes of two ryegrass species, *Lolium multiforum* (NC_019651) and *Festuca altissima* (JX871939), which were identified as the top-hit species when the input reads were blasted against the nr database. The gaps were filled by realignment of input reads using Geneious R6 v6.1.5 (Biomatters Ltd., Auckland, New Zealand) and PCR-based Sanger sequencing using primers designed for gap-flanking regions (Table S1). The sequences from the junction and highly variable region were validated by Sanger sequencing. The complete plastome was annotated using the online software DOGMA with default parameters [16]. Repeat sequences were analyzed using REPuter [17].

Phylogenetic Analysis

Complete plastome sequences of nine Poaceae species (accession numbers are listed in Table S2) were aligned using the LAGAN program within the mVISTA online suite of computational tools [18]. Default parameters were applied, and the annotation framework of the perennial ryegrass chloroplast genome was used. The percentage identity between each plastome, all relative to that of *D. antarctica*, was subsequently visualized using an mVISTA plot [19]. The plastome-based phylogeny was reconstructed for the nine Poaceae species using the whole plastome alignment generated by LAGAN. The phylogenetic tree was constructed through the method of maximum parsimony, as implemented by MEGA 5.2 [20]. Sites with gaps or missing data were excluded from the analysis, and statistical support was achieved through bootstrapping using 1000 replicates.

Transcriptome and Small Noncoding RNA Analysis

We analyzed in-house RNA-seq data libraries generated from two sets of RNAs (mRNA and small RNA), obtained as described above. For transcriptome analysis, we analyzed combined data sets of mRNAs and small RNAs. The reads of the combined data sets were mapped to the complete chloroplast genome, and the filtered reads were collected using the Bowtie 2.0 program with mismatch ≤2 bp [21]. The filtered reads were remapped according to the genome annotation using Cufflinks to calculate the fragments per kilobase of exon per million fragments mapped (FPKM) values of the transcripts and TopHat for alignment of transcript variants [22]. For small noncoding RNA analysis, we collected the reads in the size range of 20–24 nt from the small RNA data set. The size-filtered reads were mapped using Bowtie 2.0 with the criterion of zero mismatch. To search for RNA-editing sites in the chloroplast genome, putative target sites were predicted using two independent methods: 1) the PREP-chloroplast [23] search program using the chloroplast-genome sequence and 2) SAMtools/BCFtools, which calls single-nucleotide polymorphisms (SNPs) and indels by

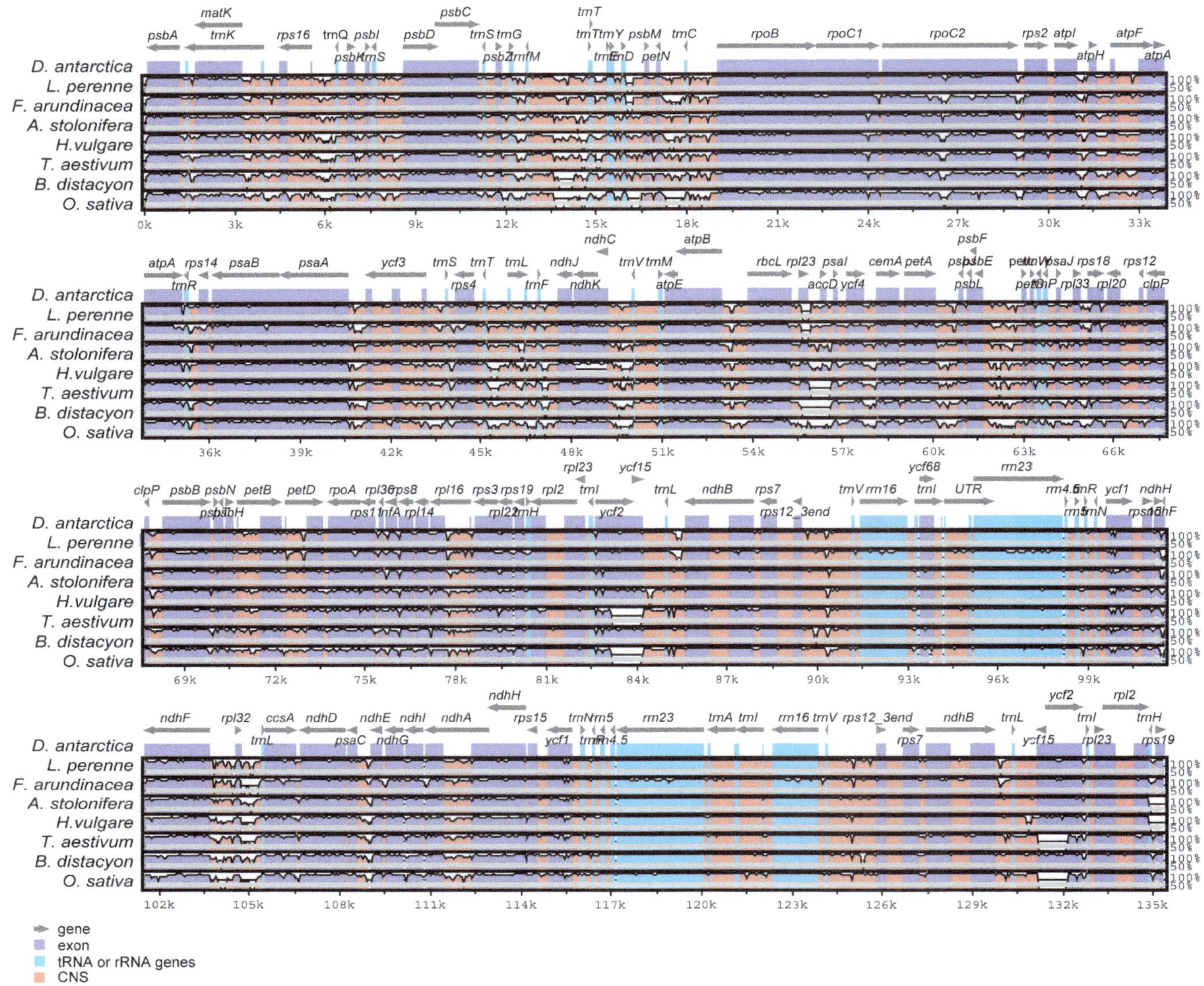

Figure 2. Sequence alignment of eight Poaceae chloroplast genomes. The top line shows genes in order (transcriptional direction indicated by arrows). The sequence similarity of the aligned regions between *Deschampsia antarctica* and the other seven species is shown as horizontal bars indicating the average percent identity between 50% and 100% (shown on the *y*-axis of the graph). The *x*-axis represents the coordinate in the chloroplast genome. Genome regions are color coded as protein-coding (exon), tRNA or rRNA, and conserved noncoding sequences (CNS).

comparing transcripts against references [24]. After prediction, the candidate sites were manually examined in the transcriptome data using the Integrative Genomics Viewer (IGV) genome browser.

Results

Chloroplast Genome Assembly and Validation

Illumina paired-end sequencing produced 153,346,825 raw reads with a sequence length of 101 bp and a total base number of 15,488,029,325. After quality trim and alignment of the raw reads against the publicly available chloroplast genomes reported in NCBI, we collected 1,985,544 chloroplast-related paired reads with 191,735,269 bases. The subsequent *de novo* assembly resulted in 18 large contigs >3 kb (max: 50,269 bp, min: 3,046 bp). To order the contigs, the chloroplast genomes of *L. multiforum*, and *F. altissima* were used as references because these species were identified as the top-hit species when the input reads were blasted against the nr database. The resulting gaps were filled by alignment of the input reads using the Geneious program and PCR-based Sanger sequencing. The sequences from the junction regions (LSC–IRA, LSC–IRB, SSC–IRA, SSC–IRB) and the regions with high interspecific variability were validated by Sanger sequencing. The final *D. antarctica* chloroplast genome sequence has been submitted to GenBank (Accession No. KF887484).

Genome Organization and Gene Content

The size of the *D. antarctica* chloroplast genome was 135,362 bp, similar in range as other Poaceae species, with a typical quadripartite structure (Figure 1). The LSC and SSC regions were 79,881 bp and 12,519 bp in size, respectively, separated by a pair of inverted repeats (IRa and IRb), which were both 21,481 bp in length. The GC content of the *D. antarctica* chloroplast genome was 38.3%, consistent with other reported Poaceae chloroplast genomes. The GC contents of the LSC and SSC regions were

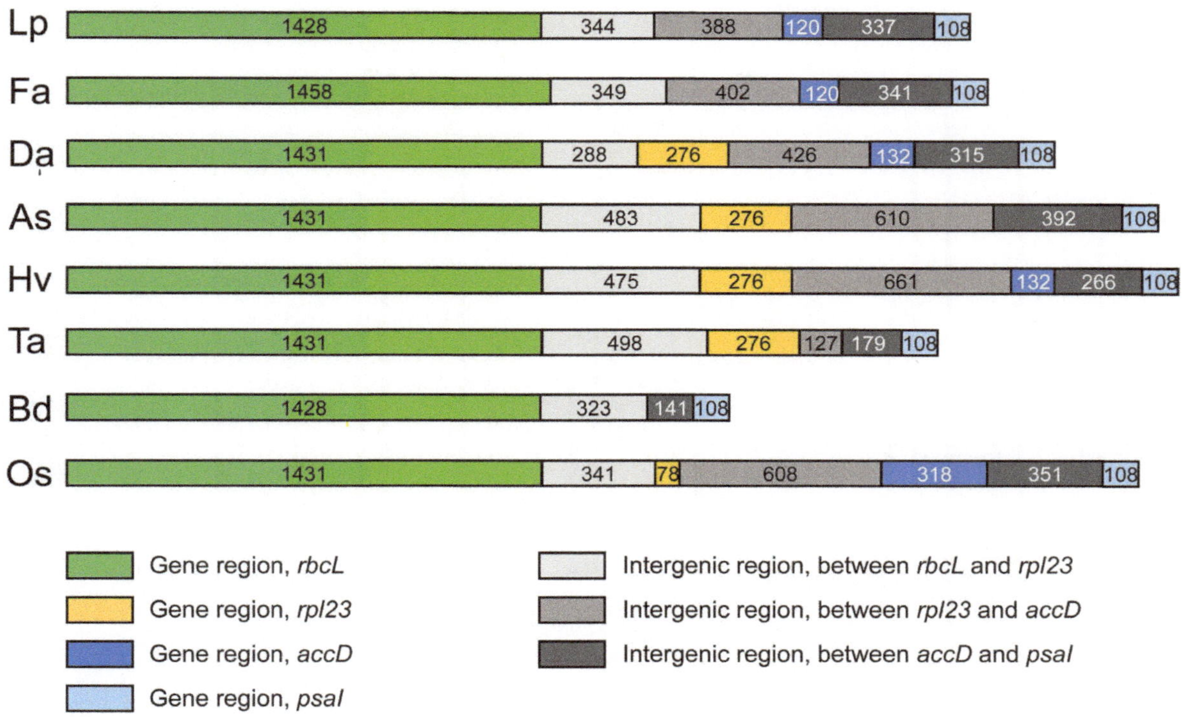

Figure 3. Comparison of the *rbcL-psaI* region among eight Poaceae species. The genes and intergenic regions between *rbcL* and *psaI* are indicated by boxes, with the length presented in bp. (Lp: *Lolium perenne*, Fa: *Festuca arundinacea*, As: *Agrostis stolonifera*, Hv: *Hordeum vulgare*, Ta: *Triticum aestivum*, Bd: *Brachypodium distachyon*, Os: *Oryza sativa* subsp. *japonica*).

36.3% and 32.4%, respectively, whereas that of the IR region was 43.85%.

The *D. antarctica* chloroplast genome contained 81 unique protein-coding genes, 12 of which were duplicated in the IR, including *rps7*, *rps12*, *rps15*, *rps19*, *rpl2*, *rpl23*, *ycf1*, *ycf2*, *ycf15*, *ycf68*, *ndhB*, and partial *ndhH*. Additionally, 29 unique tRNA genes, representing all 20 amino acids, were distributed throughout the genome (1 in the SSC region, 20 in the LSC region, and 8 in the IR region). Four rRNA genes were also identified, with complete duplication in the IR regions. Altogether, the *D. antarctica* chloroplast genome contained 114 unique genes (Table 1). Among them, 14 genes contained a single intron (9 protein-coding genes and 5 tRNA genes), while *ycf3* contained two introns. Of the 15 genes with introns, 10 were located in the LSC (7 protein-coding genes and 3 tRNAs; 9 contained one intron and 1 contained two introns), 1 in the SSC (a protein-coding gene with a single intron), and 4 in the IR region (2 protein coding genes and 2 tRNAs, all 4 containing a single intron) (Table 2). The *rps12* gene is a trans-spliced gene with a 5′-end exon located in the LSC region and duplicated 3′-end exons located in the IR region. The *trnK-UUU* gene contained the largest intron (2,486 bp), which included the *matK* gene.

On the basis of the sequences of protein-coding genes and tRNA genes within the chloroplast genome, the frequency of codon usage was deduced (Table 3). Among these codons, 2,466 (11.22%) encode for leucine, while 321 (1.46%) encode for

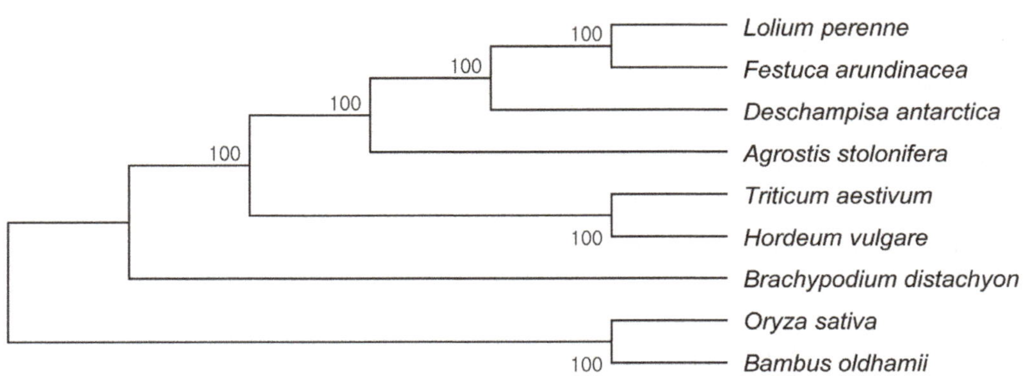

Figure 4. Maximum parsimony analysis of nine Poaceae species based on the whole plastome sequence. The plastome sequences of *Oryza sativa* and *Bambusa oldhamii* were included as outgroup species. The phylogenetic tree was drawn using MEGA5, and bootstrap support was achieved using 1,000 replicates.

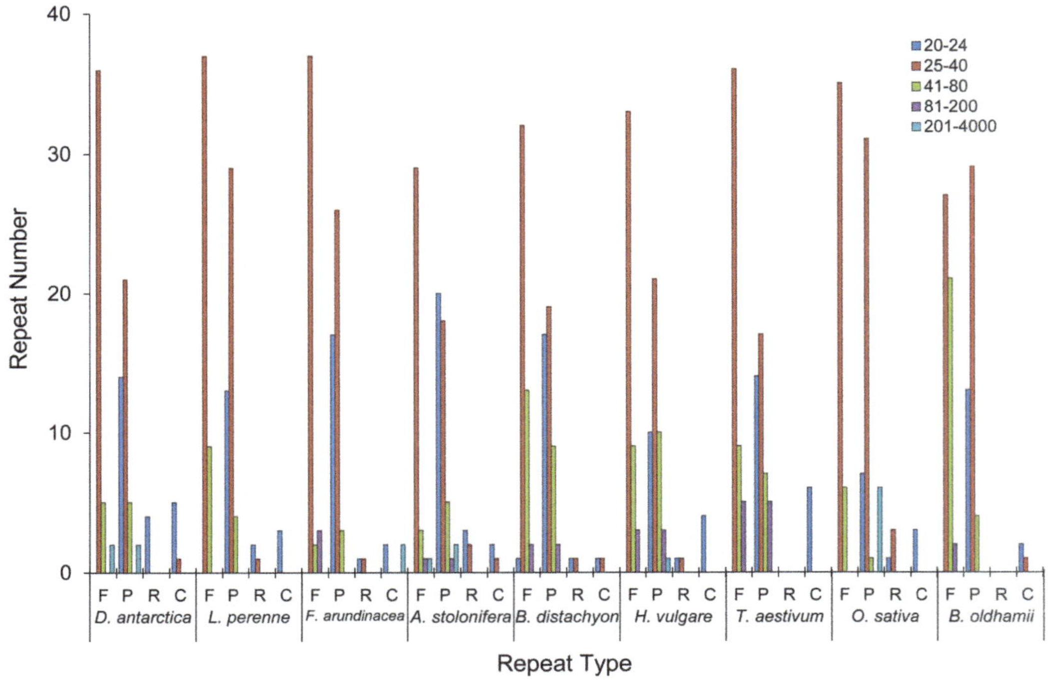

Figure 5. Repeat analysis in the *Deschampsia antarctica* chloroplast genome. Repeat sequences are compared among eight chloroplast genomes in the Poaceae family. To identify repeat sequences, the REPuter program was used. Repeats with length >20 bp and sequence identity *e*-value $<10^{-3}$ were selected and categorized to four types based on their orientations (F: forward, P: palindromic, R: reverse).

cysteine, which are the most and least used amino acids, respectively. The codon usage is biased toward a high representation of A and T at the third codon position, which is similar to a previous report [25].

Comparison with Other Poaceae Chloroplast Genomes

The availability of multiple complete Poaceae chloroplast genomes provides an opportunity to compare sequence variation within the family at the genome-level. The sequence identity of seven Poaceae chloroplast genomes was plotted using the mVISTA program, with the annotation of *D. antarctica* as a reference (Figure 2, percent identity plot, as summarized in Table S3). The whole aligned sequences indicate that the Poaceae chloroplast genomes are rather conservative, although some divergent regions were found between these genomes. Similar to other plant species, the coding region is more conservative than the noncoding counterpart. Of all genes, *ycf1* appears to be the most divergent pseudogene. In addition, *rpl32*, *ycf2*, and *rpoC2* also displayed high sequence divergence. The noncoding regions showed a higher sequence divergence than the coding regions among the eight Poaceae chloroplast genomes. In the alignment sequences, several intergenic regions were found to display high divergence, including *trnG(UCC)-trnfM(CAU)*, *trnY(GUA)-trnD(GUC)*, *ndhF-rpl32*, and *rpl32-trnL(UAG)*. In addition, the intron sequences from *trnK(UUU)*, *trnL(UAA)*, and *ndhA* showed high sequence divergence.

The length variation was also examined among *D. antarctica* and the eight Poaceae chloroplast genomes. The most interesting region with length variation was the *rbcL-psaI* region, which contains four gene regions and three intergenic regions (Figure 3). The variation of gene region was detected in the presence of an *rpl23* translocation product and an *accD* pseudogene in the region between *rbcL* and *psaI*. The *rpl23* gene was absent from *L. perenne*, *F. arundinacea*, and *Brachypodium distachyon*, and was present in the five other analyzed Poaceae species, including *D. antarctica*. Remnants of the *accD* gene were detected in *D. antarctica*, *L. perenne*, *F. arundinacea*, and *Hordeum vulgare*. This pseudogene was identified in rice but was not predicted in the other species according to DOGMA. The variation in size of the intergenics regions was also detected among species of the Pooideae subfamily. Three intergenic regions occurred between the *rbcL* and *psaI* genes. The intergenic region between *rbcL* and *rpl23* ranged from 288 bp (*D. antarctica*) to 498 bp (*Triticum aestivum*). Between *rpl23* and *accD*, it ranged from 0 bp (*B. distachyon*) to 661 bp (*H. vulgare*), and between *accD* and *psaI*, it ranged from 141 bp (*B. distachyon*) to 392 bp (*Agrostis stolonifera*). In cases when a particular gene was absent, the boundaries of the intergenic regions were determined based on homologies between the species.

Phylogenomic Analysis

Phylogenomic analysis of representatives from the Pooideae subfamily, including *D. antarctica*, produced a single, well-supported tree using maximum parsimony (Figure 4). The tree is well congruent with respect to species, and the two outgroup species belonging to the BEP clade (*Bambusa oldhamii* from Bambusoideae and *Oryza sativa* subsp. *japonica* from Ehrhartoideae) are basal to the remaining species in a separate resolved clade.

Repeat Sequence Analysis

Repeat regions of DNA are an important factor in genome recombination and rearrangement. We identified 69 repeats in *D. antarctica*, including 43 forward, 24 palindromic, and 2 reverse repeats with a length >20 bp and a sequence identity *e*-value $<10^{-3}$, using the REPuter program (Table S4). Among the 69 repeats, 58 (84%) were 25–80 bp in length, 51 (63%) were 25–

Table 4. RNA Expression of protein coding genes in the *Deschampsia antarctica* chloroplast genome.

locus ID	gene name	locus	FPKM	locus ID	gene name	locus	FPKM
DeanCp027	*ndhC*	48846–49209	87311	DeanCp032	*accD*	56279–56411	326
DeanCp037	*psbJ*	60890–61013	19529	DeanCp073	*rps15*	100781–101054	314
DeanCp064	*rps19*	79951–80233	13440	DeanCp076	*rpl32*	104531–104714	309
DeanCp043	*psaJ*	64095–64224	10915	DeanCp025	*ndhJ*	47535–48015	288
DeanCp002	*psbA*	83–1145	10274	DeanCp016	*atpI*	30197–30941	283
DeanCp042	*petG*	63234–63348	9557	DeanCp074	*ndhH*	101191–101431	254
DeanCp051	*psbN*	70154–70286	7916	DeanCp004	*rps16*	4488–5567	250
DeanCp011	*petN*	17020–17110	7370	DeanCp080	*ndhE*	109031–109337	202
DeanCp018	*atpF*	32063–33432	6499	DeanCp036	*petA*	59093–60056	188
DeanCp039	*psbF*	61282–61402	6371	DeanCp065	*rpl2*	80495–81980	185
DeanCp038	*psbL*	61143–61260	4812	DeanCp069	*ndhB*	85606–87851	165
DeanCp010	*psbM*	16638–16743	3750	DeanCp070	*rps7*	88150–88621	159
DeanCp006	*psbI*	7354–7465	3335	DeanCp060	*rpl14*	76697–77069	152
DeanCp050	*psbT*	69989–70106	3089	DeanCp023	*ycf3*	41199–43189	152
DeanCp040	*psbE*	61412–61664	3020	DeanCp068	*ycf15*	83879–84167	149
DeanCp052	*psbH*	70389–70611	2998	DeanCp045	*rps18*	65145–65658	144
DeanCp020	*rps14*	35628–35940	2224	DeanCp058	*infA*	75691–76042	144
DeanCp017	*atpH*	31349–31595	2138	DeanCp046	*rpl20*	65815–66175	144
DeanCp005	*psbK*	6761–6947	1970	DeanCp024	*rps4*	44159–44765	142
DeanCp009	*psbZ*	11675–11864	1948	DeanCp048	*clpP*	67131–67782	141
DeanCp030	*rbcL*	53858–55292	1782	DeanCp081	*ndhG*	109549–110080	136
DeanCp007	*psbD*	8635–9697	1333	DeanCp034	*ycf4*	57149–57707	134
DeanCp033	*psaI*	56726–56837	1308	DeanCp003	*matK*	1685–3221	133
DeanCp041	*petL*	62963–63059	1282	DeanCp061	*rpl16*	77186–78490	132
DeanCp049	*psbB*	68293–69820	1197	DeanCp059	*rps8*	76143–76554	128
DeanCp071	*ycf68*	93397–93832	1096	DeanCp063	*rpl22*	79429–79873	113
DeanCp079	*psaC*	108276–108522	1052	DeanCp035	*cemA*	58162–58861	110
DeanCp019	*atpA*	33523–35047	971	DeanCp075	*ndhF*	101464–103684	102
DeanCp054	*petD*	72341–73561	956	DeanCp077	*ccsA*	105547–106507	87
DeanCp057	*rpl36*	75472–75586	956	DeanCp082	*ndhI*	110198–110741	85
DeanCp044	*rpl33*	64666–64867	922	DeanCp055	*rpoA*	73770–74796	74
DeanCp001	*rps12*	66870–89475	840	DeanCp014	*rpoC2*	24536–28943	66
DeanCp021	*psaB*	36086–38291	676	DeanCp015	*rps2*	29236–29947	62
DeanCp022	*psaA*	38316–40569	558	DeanCp083	*ndhA*	110838–112939	58
DeanCp008	*psbC*	9644–11066	513	DeanCp078	*ndhD*	106654–108157	57
DeanCp084	*ndhH*	112940–114122	510	DeanCp072	*ycf1*	99622–100414	52
DeanCp053	*petB*	70745–72153	495	DeanCp056	*rps11*	74860–75292	48
DeanCp031	*rpl23*	55577–55853	419	DeanCp062	*rps3*	78636–79356	44
DeanCp047	*rps12*	125837–126080	402	DeanCp067	*ycf2*	82674–83874	39
DeanCp066	*rpl23*	81998–82280	400	DeanCp090	*ycf2*	131435–132638	39
DeanCp029	*atpB*	51509–53006	385	DeanCp013	*rpoC1*	22302–24333	31
DeanCp026	*ndhK*	48118–48856	342	DeanCp012	*rpoB*	19034–22265	29
DeanCp028	*atpE*	51099–51513	329				

40 bp in length, and 10 (21%) were 41–80 bp in length. The repeats were mostly located in the intergenic sequences (54%), followed by coding sequences (37%) and intronic sequences (9%). The structure of the repeats in the other seven Poaceae species was also analyzed using REPuter. The majority of repeats in Poaceae species within the size range of 25–80 bp commonly are forward or palindromic (Figure 5). The total number of repeats varied among species (*D. antarctica*: 69, *L. perrene*: 72, *F. arundinacea*: 59, *A. stolonifera*: 50, *B. distachyon*: 60, *H. vulgare*: 67, *T. aestivum*: 79, *O. sativa*: 78, *B. oldhamii*: 74). The repeat pattern in *D. antarctica* was

Table 5. RNA editing sites in the *Deschampsia antarctica* chloroplast genome.

Gene	length	location from start	codon change	amino acid change	Nucleotide change	Number of reads*
matK	1536	1258	CAU>UAU	His>Tyr	C>U	U;11 (28.9%), C; 27 (71.1%)
rpoB	3231	398	CGC>CAC	Arg>His	G>A	A:4 (25%), G:10 (62.5%), U:1(6.3%),
rpoC1	2031	603	GAA-GAU	Glu>Asp	A>U	A:19 (74.1%), U:7 (25.9%)
rpoC1	2031	612	GCG-GCA	Ala->Ala	G>A	A:7 (24.1%), G:22 (75.9%)
rpoC2	4407	650	AUA>AGA	Ile>Arg	U>G	G:2 (20.0%. U:8 (80%)
atpA	1524	334	UUG>CUG	Leu>Leu	U>C	C:13 (43.3%), U:17(56.7%)
atpA	1524	367	AUA>GUA	Ile>Val	A>G	A:23 (76.7%), G:7(23.3%)
atpA	1524	933	GAA>GAC	Glu>Asp	A>C	A:41(54.7%), C:34(45.3%)
atpA	1524	1148	UCA>UUA	Ser>Leu	C>U	C:2(2.9%), U:66 (97.1%)
ycf3	513	44	UCC>UUC	Ser>Phe	C>U	U:19 (100%)
rps4	606	588	UAU>UAA	Tyr>stop	U>A	A:55 (66.3%),U:28 (33.7%)
rps4	606	580	GUG>CUG	Val>Leu	G>C	G:31(36%), C:55 (64%)
rps4	606	370	AAU>GAU	Asn>Asp	A>G	G:3 (42.9%), A:4 (57.1%)
ndhJ 3′ UTR			AAU>AAC	Asn>Asn	A>C	A:4 (30.8%), C: 9 (69.2%)
ndhJ	480	480	UGA>UGG	stop>Trp	A>G	A:4 (30.8%), G: 9 (64.3%)
ndhK	738	125	CCA>CUA	Pro>Leu	C>U	C:2(9.5%), U:19 (90.5%)
ndhC	363	13	CAC>UAC	His>Tyr	C>U	C:3(50%). U:3 (50%)
psbL	117	111	UUC>UUU	Phe>Phe	C>U	C:2 (15.4%), U: 10 (76.9%), G: 1(7.7%)
petL	96	56	CCA>CUA	Pro>Leu	C>U	U:2 (100%)
rpl20	360	308	UCA>UUA	Ser>Leu	C>U	C:5 (45.5%), U:6 (54.5%)
psbB	5127	867	AGC>AGU	Ser>Ser	C>U	C:25 (83.3%), U:5 (16.7%)
petB	648	611	CCA>CUA	Pro>Leu	C>U	U:19 (100%)
rpoA	1026	527	UCC>UUC	Ser>Phe	C>U	C:2 (18.2%), U:9 (81.8%)
rps8	411	182	UCA>UUA	Ser>Leu	C>U	C:1 (7.7%), U:12 (92.3%)
rpl16	411	250	GGC>ACG	Gly>Ser	G>A	G:2 (33.3%), A:4 (66.7%)
rps3	720	30	UUC>UUU	Phe>Phe	C>U	C:6 (30%), U:14 (70%)
ndhD	1503	878	UCA>UUA	Ser>Leu	C>U	C:1 (9.1%), U:10 (90.9%)
ndhG	531	347	CCA>CUA	Pro>Leu	C>U	U: 29 (100%)
ndhA	1089	722	GCA>GUA	Ala>Val	C>U	C: 9 (81.8%), U: 2 (18.2%)
ndhA	1089	474	UCA>UUA	Ser>Leu	C>U	C: 2 (10.5%), U: 17 (89.5%)

*indicates the number of reads with an alternate base and the number of reads with the same base as the reference.

more similar with *L. perenne* and *F. arundinacea* in the Poeae tribe than with *B. oldhamii* from the Bambusoideae. For example, repeats in the size range of 41–80 bp represent ≤20% of the total number of repeats in species of the Pooideae subfamily, whereas they represent >28% of the total in *O. sativa* and *B. oldhamii.*

Expression Analysis

We performed an expression analysis of the 81 chloroplast protein-coding genes using in-house RNA-seq data from leaf tissues of *D. antarctica* (Lee et al., unpublished data). The short reads were mapped to the *D. antarctica* chloroplast genome, and the numbers of reads corresponding to coding genes were calculated and normalized according to gene length (Table 4). The most abundant genes were *ndhC, psbJ, rps19, psaJ,* and *psbA*, with FPKM value >10,000. Thirteen genes (*ccsA, ndhI, rpoA, rpoC2, rps2, ndhA, ndhD, ycf1, rps11, rps3, ycf2, rpoC1,* and *rpoB*) had low expression, with FPKM value <100.

A total of 247,904 reads mapped to the protein coding region. Among these, 89,675 (36.2%) and 73,054 reads (29.5%) were generated from genes encoding components of the cyclic electron transfer system and photosystem II (PSII) complex, respectively. In addition, among the 18 highly expressed genes (FPKM value >2,000), 10 genes were found to encode subunits of the PSII complex (*psbA, psbB, psbE, psbF, psbH, psbI, psbL, psbM, psbN,* and *psbT*). In contrast, *rpoA, rpoB, rpoC1,* and *rpoC2*, which encode plastid RNA polymerase, showed very low expression.

RNA Editing

RNA editing is a sequence-specific posttranscriptional modification resulting in conversion, insertion, and deletion of nucleotides in a precursor RNA. Such modifications are observed across organisms. In plants, RNA editing has been reported to occur with C-to-U or U-to-C (rare) conversions in mitochondria and plastids [26].

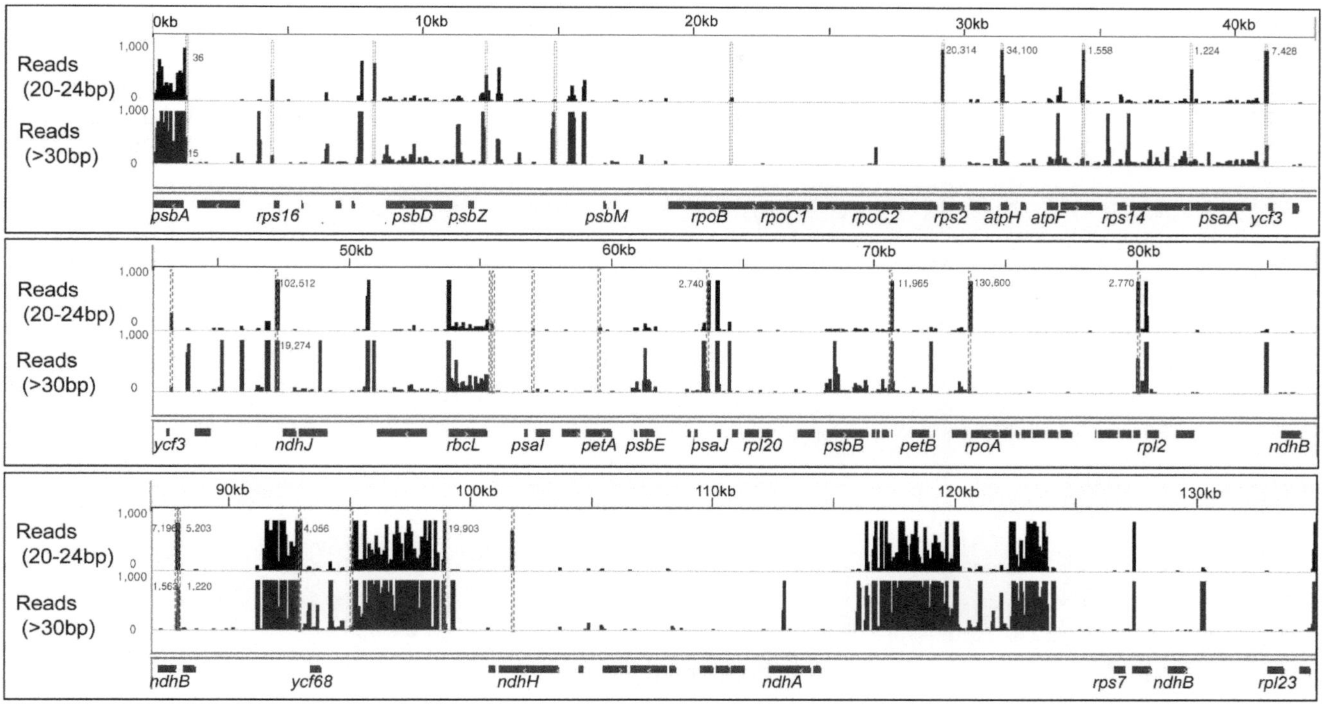

Figure 6. Distribution of plastid small RNAs in the *Deschampsia antarctica* chloroplast genome. The reads from small RNA-seq were divided into two groups according to the length (20–24 nt and >30 nt) and aligned to the *D. antarctica* chloroplast genome with 100% identity. The distributions of reads were compared between the two groups. In total, 12,753,636 reads were distributed unevenly in the chloroplast genome with high density in the coding regions of *psbA* and *rbcL*, intergenic regions, and inverted repeat regions in which most of the rRNA genes exist. The 27 loci enriched with 20–24 nt RNAs are indicated in red, along with the number of reads. The *y*-axis shows the number of reads (from 0 to 1000).

In the *Deschampsia* chloroplast genome, we first predicted 37 RNA-editing sites out of 16 genes using the PREP-chloroplast program (Table S5). Using another method, we aligned read sequences from the RNA-seq data using variant searching tools comparing transcripts against a reference genome and confirmed 30 editing sites. The 30 nucleotide substitutions occur in 23 genes in the *D. antarctica* chloroplast genome, which results in 25 non-synonymous amino acid changes (Table 5). Of the substitutions, 17 (54.8%) were C-to-U conversions, resulting in 14 non-synonymous amino acid changes. In contrast, only 1 edit was a U-to-C conversion with synonymous base change. Although RNA editing of plant plastids has been shown to be conversions of C to U and U to C, we observed different versions of edits, including 3 A-to-Cs, 3 A-to-Gs, 3 G-to-As, 1 G- to- C, 1 U-to-A, 1 A-to-U, and 1 U-to-G in 13 sites.

We calculated the ratio between the number of reads with an alternate base and the number of reads with the same base as the reference. The percentages of the conversion rates of each edit varied with the locus (16–100%) (Table 5). However, some edits with C-to-U conversion in several genes showed very high editing rates (>90%), especially for *atpA*, *ycf3*, *ndhK*, *petB*, *rpoA*, *rps8*, *ndhD*, *ndhG*, and *ndhA*, suggesting that the edited RNAs for these gene are common forms in the processed RNA pools in *D. antarctica*.

Discovery of Plastid Small Noncoding RNA in *D. antarctica*

Numerous small noncoding RNAs have been identified in the nuclear genomes of bacteria and eukaryotes. Small noncoding RNAs are also transcribed from mitochondria and plastid genomes [27–29]. In this study, we screened for small noncoding RNAs from our deep sequencing data in the small RNA library generated from *D. antarctica* leaf tissues. The reads between 20 and 24 nt in length were mapped to the chloroplast genome with 100% identity. In total, 12,753,636 reads were distributed unevenly throughout the chloroplast genome (Figure 6), including coding regions of *psbA* and *rbcL*, intergenic regions, regions encoding several tRNA genes, and inverted repeat regions in which most of the rRNA genes exist. To exclude RNA fragments that may have been generated from abundant RNA species, we compared the distribution of reads that were 20–24 nt in length with those longer than 30 nt. As a result, we identified 27 loci where short noncoding RNAs (sRNAs) of 20–24 nt length with unique sequences were abundantly expressed (Table 6).

The *D. antarctica* plastid sRNAs were not evenly distributed throughout the genome. The relative positions of the sRNAs showed that 19 of 27 (71%) were located in the noncoding regions (18 in intergenic regions and 1 in an intronic region). In particular, 30% and 11%, respectively, of the intergenic sRNAs were located at the 5′- and 3′-ends of genes (>100 bp from the start or termination codons) (Figure 7). Fifteen (55.6%) sRNAs were located within −150 to +50 bp from the start codon of genes, suggesting that proximity to the 5′-ends of genes is important.

To determine if the identified sRNAs are evolutionarily conserved, we compared the sequences of 27 sRNAs in *D. antarctica* with the sRNAs reported for other plant species by multiple sequence alignment [28,29]. In total, we found that 13 sRNAs have orthology with the plastid sRNAs found in *Arabidopsis*, rice, or barley (Figures 8, Figure S1, and Table 6). Among the pairs identified, four sRNAs (*psbH-petB*, *atpH* 5′end, *ndhB* 5′end, and *petD_rpoA*) showed >90% sequence homology, and their

Table 6. Distribution of small RNAs in the chloroplast genome of *Deschampsia antarctica*.

Loca -tion	start	end	F/R*	core sequence	length	reads number	At	Os	Hv
psbA 5′ end	1229	1209	R	AACAAGCCTTCTATTATCTA	20	36	+		
trnK-rps16	4378	4356	R	TGTCGTGCCAATCCAACATAAGCC	23	819		+	+
psbI-psbD	8129	8150	F	TTCCTTAGACTTAGACCGCGC	21	1200			
trnG-trnfM	12314	12333	F	ACCGTATCCCTTACTATTCT	20	1056			
trnT 3′ end	14786	14806	F	GGTTCAAATCCGATAAAGGGC	21	148			
rpoB CDS	21313	21332	F	CGTCGTATATCGCGGAAGCT	20	166			
rpoC2-rps2	29130	29147	F	ATTTCAAGCTATTTCGGA	18	20314			+
atpH 5′ end	31305	31325	F	ATTGTATCCTTAACCATTTCT	21	34100	+	+	+
atpA CDS	34274	34297	F	TTATGTACCGCGAACGGCATA	21	1558			
psaB 5′ end	38291	38315	R	AGGAGGATTTGAAAGGCATTA	21	1224			
ycf3 3′ end	41108	41088	R	TTCATTATATCGCTTTCTTCT	21	7428			+
ycf3 5′ end	43251	43232	R	TTTGTTTTTATGTTATTTTG	20	450	+		+
trnF-ndhJ	47285	47265	R	CTTTGTATCGCGCGCATGACT	21	102512			
rbcL 3′ end_1	55406	55426	F	CTCGGCTCAATCTTTTTTAGA	21	111	+	+	
rbcL 3′ end_2	55424	55431	F	AAAAAAAAGATTGAGCCGAAT	21	160	+	+	
psaI-ycf4	57004	57024	F	TGAATAGAAAGTCAATGTATC	21	120			
petA CDS	59574	59553	R	TTTCACTATATTTCTTACCGGG	22	230			
trnP 5′ end	63758	63739	R	AGGGATGTAGCGCAGCTTGG	20	2740			
psbH-petB	70702	70721	F	GGTAGTTCGACCGCGGAATT	20	11965	+	+	+
petD-rpoA	73658	73677	F	TTATTATGATCCATTTCGCG	20	130600		+	+
rps19 CDS	80081	80098	F	ATGAATCGCGATTGTATG	18	2770			
ndhB 5′ end 1	87859 (127454)	87839 (127474)	R	ACTAATTCATGATCTGGCATG	21	7196	+	+	+
ndhB 5′ end 2	87863 (127450)	87843 (127470)	R	AGTTACTAATTCATGATCTGG	21	5203	+	+	+
rrn16 3′ end	92921 (122393)	92941 (122373)	F	GGTGCGGCTGGATCACCTCCT	21	4056			
trnA intron	95093 (120221)	95113 (120201)	F	CTTAGCGGATACTATGATAGC	21	982			
trnR 3′ end	98927 (116387)	98946 (116368)	F	GTGTCGGGGGTTCGAATCCC	20	19903		+	
ndhF CDS	101690	101709	F	ATAACCGCGATTATATGACC	20	1149			

*F/R: Direction of transcripts (F: forward, R: reverse).

locations within the genome were the same in all of the species examined, suggesting these plastid sRNAs may be evolutionarily conserved across angiosperms (Figure 8).

Discussion

We obtained the completed sequence of the chloroplast genome of *D. antarctica* using whole genome sequencing data from total genomic DNA from leaves. As previous studies have reported, aligning all the reads against the plastid genome database allow the rapid and efficient assembly of the chloroplast genome [8,30,31]. By this method, we identified 1.2% of the total genomic reads as chloroplast-related sequences.

The chloroplast genome of *D. antarctica* has the typical features found in the genomes of other Poaceae species. The size of its genome and the ratio of GC content is 135,362 bp and 38.3%, respectively, similar to other Poaceae species. The subfamily Pooideae, which includes one-third of all grass species, has been divided into 13 tribes [14], but recent analyses have demonstrated wide variations between them. For example, neither Poeae nor Aveneae are monophyletic, and the components of these two groups are intermixed within a clade [13,32]. Traditional morphological phylogenetic studies placed *Deschampsia* within the

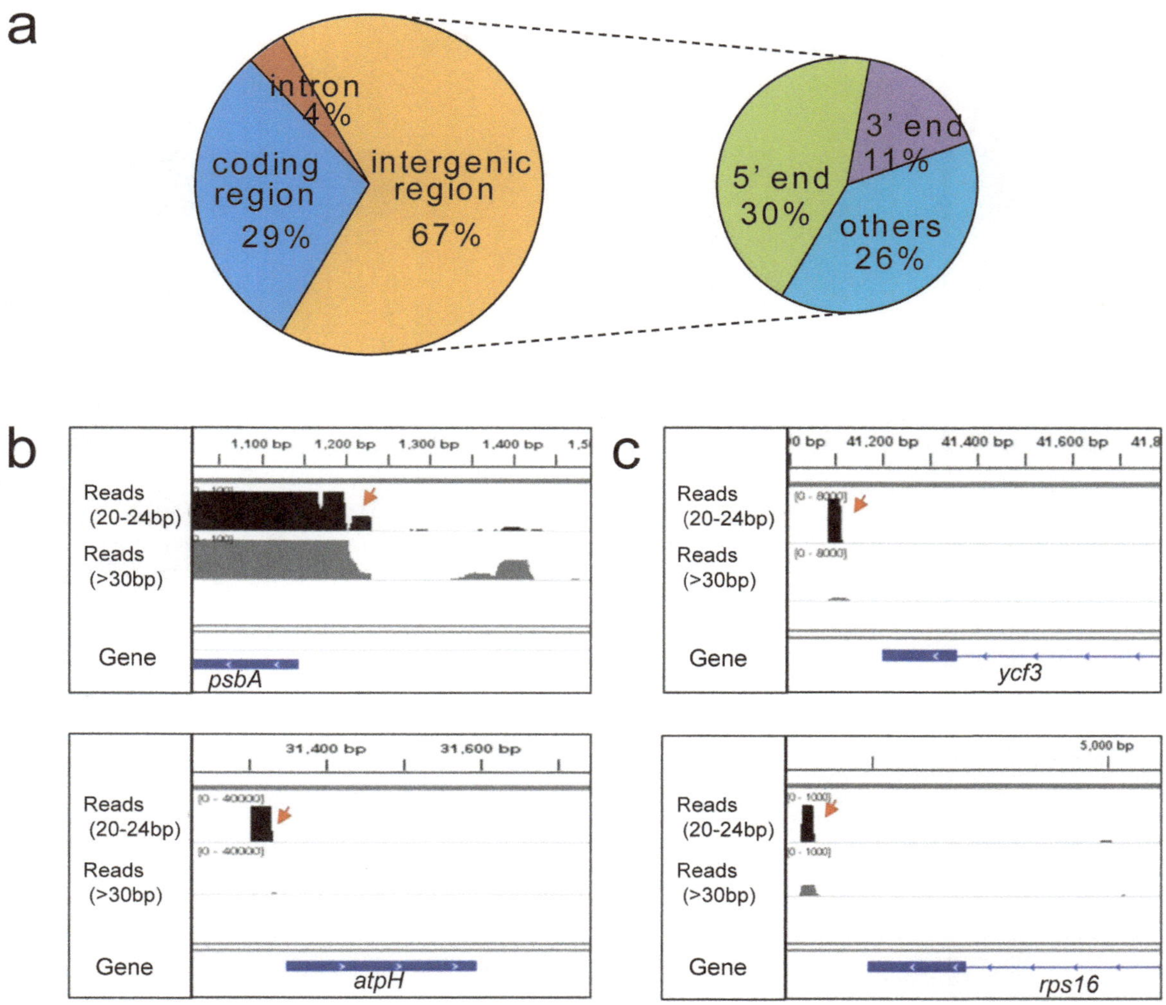

Figure 7. Relative locations of small RNAs in the *Deschampsia antarctica* chloroplast genome. a Relative locations of plastid small RNAs according to the gene structure; **b** examples of small RNAs located proximal to the 5′ ends of the coding genes; **c** examples of small RNAs located proximal to the 3′ end of the coding genes.

tribe Aveneae. However, molecular studies inferred alternative phylogenetic positions of *Deschampsia* (i.e., Aveneae or Poeae), depending on the target sequences used for examination or the parameters used for grouping [12,13,32–35]. In this study, we revised the phylogenetic position of *D. antarctica* using complete sequences of chloroplast DNA. A comparative analysis based on both whole plastome and open reading frame sequences of coding genes suggest that *D. antarctica* is more closely related with species in the Poeae tribe than the Aveneae tribe. This is in agreement with the results of Davis and Soreng [13], Catalan et al. [33], and Nadot et al. [34], in which *Deschampsia* forms a closer relationship with species of the Poeae than with those of Aveneae, as suggested by Souto et al. [12] and Hsiao et al. [35]. However, in our genome structure analysis, we found an interesting region (*rbcL–psaI*) where both the *rpl23* translocation product and *accD* pseudogene were found. This appears to be specific to *Deschampsia*, since other Poeae or Aveneae species have kept only one remnant of *accD* or *rpl23* in the region, suggesting that this region could be molecular evidence for an intermixed lineage of *Deschampsia*.

For the transcriptome analysis of the chloroplast genome, we utilized RNA-seq data from libraries generated by two preparation methods (mRNA-seq and small RNA-seq). We found that a significant proportion of the reads from RNA-seq data represent the organelle derived sequences, suggesting that the eukaryotic RNA-seq results are very good resources for a functional study of genes in organelles.

The transcriptome analysis of *D. antarctica* plastid RNAs revealed several interesting aspects of RNA metabolism. First, a search of the variant transcripts revealed numerous RNA-editing sites in the *D. antarctica* chloroplast genome. RNA editing has been observed in the chloroplasts of extant descendants of early land plants other than liverworts and mosses. In angiosperm plastids, RNA editing is mostly restricted to a C-to-U conversion, and the conversion occurs at about 30 different positions, whereas hornworts and fern plastids extensively edit U-to-C as well as C-to-U at >300 different positions [36]. A comparative analysis of eight land plants, including hornworts, ferns, and seed plants, suggested that chloroplast RNA editing is of monophyletic origin and evolved as a system to generate new variations [37]. Our transcriptome analysis revealed *in situ* editing sites beyond those predicted by computational tools (Table 4 vs. Table S5). According to the variant transcript search, the major form of

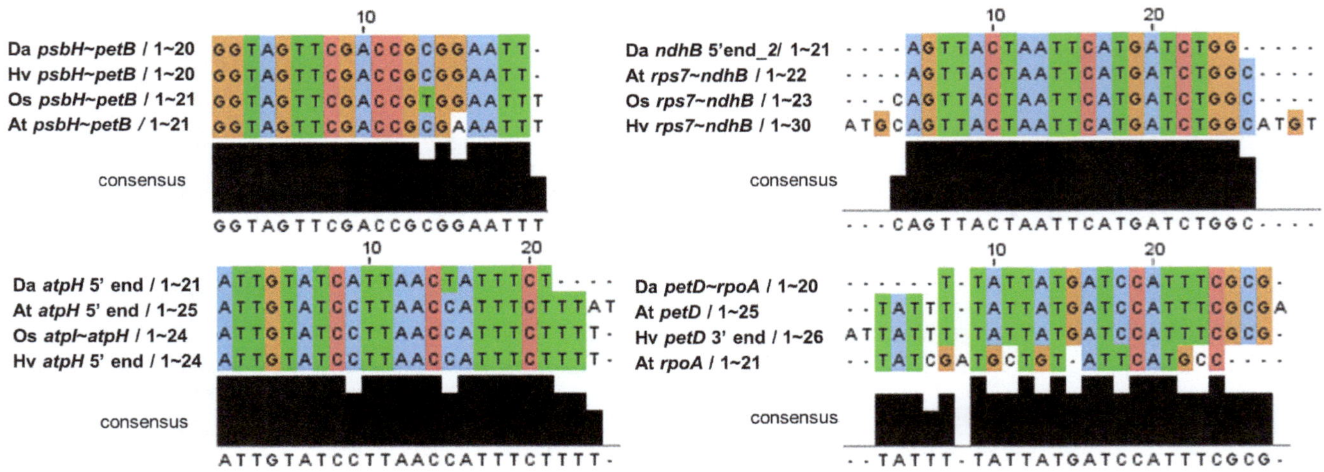

Figure 8. Sequence conservation among orthologs of plastid small RNAs. To determine if the identified sRNAs are evolutionarily conserved, *Deschampsia antarctica* sRNAs were compared with the plastid sRNAs identified in *Arabidopsis*, rice, or barley [28,29]. The sequence aligments of sRNAs which have >90% sequence homology are shown. The multiple sequence alignments were performed with ClustalW2 algorithm (http://www.ebi.ac.uk/Tools/msa/clustalw2/) and visualized with Jalview program [41]. The consensus sequences between ortholog sRNAs were shown at the bottom of each alignment.

RNA-editing is C-to-U conversion (54.8%), and the conversion rate of C-to-U edits (>90%) is much higher than those of other edits. Some edits with C-to-U conversion in several genes, such as *atpA*, *ycf3*, *ndhK*, *petB*, *rpoA*, *rps8*, *ndhD*, *ndhG*, and *ndhA*, have been reported in other species [37], indicating that these edits are functionally conserved in plants. Comparison between the whole genome DNA and transcriptome data also showed that various versions of edits exist and that their respective conversion rates differ. The difference in conversion rates among edits might be the result of tissue-specific, gene-specific, or developmental stage-specific RNA-editing patterns. Considering that mitochondrial RNA editing occurs with developmental and tissue specificity in plants [38–40], exploring whether tissue-disparity exists in plastid RNA-editing and the regulatory mechanisms that underlie it would be worthwhile.

We identified 27 plastid small noncoding RNAs in the *D. antarctica* chloroplast genome by high-resolution mapping of the transcriptome data. In *Arabidopsis*, rice, maize, and barley, small RNAs are expressed in plastids and their sequences correlate with the termini of processed mRNA [28,29]. These studies also suggested that the small RNAs are footprints of the RNA-binding pentatricopeptide repeat (PPR) proteins, which protect RNAs from exonucleolytic degradation. Our results support this hypothesis. We observed a large amount of small RNAs expressed in the *D. antarctica* plastid, and these RNAs were not randomly distributed but were located in intergenic regions preferentially near the 5′- or 3′-ends of coding regions. This suggests that many small RNAs are evolutionarily conserved in their sequences and locations, which might have resulted from the functionally conserved gene regulatory system of higher plants.

Conclusions

Using Illumina high-throughput sequencing technology, we obtained the complete sequence of the *D. antarctica* chloroplast genome. This is the first chloroplast genome sequenced from a plant species endemic to Antarctica. Sequence divergence analysis with other plastomes of the BEP clade in the grass family suggests a sister relationship between *D. antarctica* and two species of the Poeae tribe, *F. anrundinacea* and *L. perenne*. In addition, we conducted high-resolution mapping of the chloroplast-derived transcripts resulting from RNA-seq data. As a result, we could make an expression profile for 81 protein-coding genes and proposed *ndhC*, *psbJ*, *rps19*, *psaJ*, and *psbA* as the most highly expressed chloroplast genes in *D. antarctica*. Analysis of small RNA-seq revealed that 27 small noncoding RNAs are preferentially located close to the 5′- or 3′-ends of genes. Also, >30 RNA-editing sites were found in the *D. antarctica* chloroplast genome, with a predominance of C-to-U conversions. These will be very useful for molecular phylogeny studies of the evolution of Antarctic plants and for transcriptome studies specific to plant organelles.

Acknowledgments

The authors wish to thank Dr. Sanghee Kim (Korea Polar Research Institute) for providing support of analysis software.

Author Contributions

Conceived and designed the experiments: JL HL HP. Performed the experiments: JL YK. Analyzed the data: JL YK HL SCS. Wrote the paper: JL HL HP.

References

1. Neuhaus HE, Emes MJ (2000) Nonphothosynthetic metabolism in plastids. Annual Review of Plant Physiology and Plant Molecular Biology 51: 111–140.
2. Wicke S, Schneeweiss G, dePamphilis C, Müller K, Quandt D (2011) The evolution of the plastid chromosome in land plants: gene content, gene order, gene function. Plant Molecular Biology 76: 273–297.
3. Chumley TW, Palmer JD, Mower JP, Fourcade HM, Calie PJ, et al. (2006) the complete chloroplast genome sequence of *Pelargonium* × *hortorum*: organization and evolution of the largest and most highly rearranged chloroplast genome of land plants. Molecular Biology and Evolution 23: 2175–2190.
4. Yang M, Zhang X, Liu G, Yin Y, Chen K, et al. (2010) The complete chloroplast genome sequence of *Date Palm Phoenix dactylifera L.* PLoS ONE 5: e12762.
5. Bock R (2007) Structure, function, and inheritance of plastid genomes. In: Bock R, editor. Cell and Molecular Biology of Plastids: Springer Berlin Heidelberg. pp. 29–63.
6. Moore M, Dhingra A, Soltis P, Shaw R, Farmerie W, et al. (2006) Rapid and accurate pyrosequencing of angiosperm plastid genomes. BMC Plant Biology 6: 17.
7. Wu Z-Q, Ge S (2012) The phylogeny of the BEP clade in grasses revisited: Evidence from the whole-genome sequences of chloroplasts. Molecular Phylogenetics and Evolution 62: 573–578.
8. Hand ML, Spangenberg GC, Forster JW, Cogan NOI (2013) Plastome sequence determination and comparative analysis for members of the *Lolium-Festuca* grass species complex. G3: Genes|Genomes|Genetics 3: 607–616.
9. Alberdi M, Bravo LA, Gutiérrez A, Gidekel M, Corcuera LJ (2002) Ecophysiology of Antarctic vascular plants. Physiologia Plantarum 115: 479–486.
10. Lee J, Noh E, Choi H-S, Shin S, Park H, et al. (2013) Transcriptome sequencing of the Antarctic vascular plant *Deschampsia antarctica* Desv. under abiotic stress. Planta 237: 823–836.
11. Xiong FS, Mueller EC, Day TA (2000) Photosynthetic and respiratory acclimation and growth response of Antarctic vascular plants to contrasting temperature regimes. American Journal of Botany 87: 700–710.
12. Souto DPF, Catalano SA, Tosto D, Bernasconi P, Sala A, et al. (2006) Phylogenetic relationships of *Deschampsia antarctica* (Poaceae): Insights from nuclear ribosomal ITS. Plant Systematics and Evolution 261: 1–9.
13. Davis JI, Soreng RJ (2007) A preliminary phylogenetic analysis of the grass subfamily pooideae (Poaceae), with attention to structural features of the plastid and nuclear genomes, including an intron loss in GBSSI. Aliso: A Journal of Systematic and Evolutionary Botany 23: Article 27.
14. GPWG (2001) Phylogeny and subfamilial classification of the grasses (Poaceae). Annals of the Missouri Botanical Garden 88: 373–457.
15. Pérez-Torres E, García A, Dinamarca J, Alberdi M, Gutiérrez A, et al. (2004) The role of photochemical quenching and antioxidants in photoprotection of *Deschampsia antarctica*. Functional Plant Biology 31: 731–741.
16. Wyman SK, Jansen RK, Boore JL (2004) Automatic annotation of organellar genomes with DOGMA. Bioinformatics 20: 3252–3255.
17. Kurtz S, Schleiermacher C (1999) REPuter: fast computation of maximal repeats in complete genomes. Bioinformatics 15: 426–427.
18. Brudno M, Malde S, Poliakov A, Do CB, Couronne O, et al. (2003) Glocal alignment: finding rearrangements during alignment. Bioinformatics 19: i54–i62.
19. Frazer KA, Pachter L, Poliakov A, Rubin EM, Dubchak I (2004) VISTA: computational tools for comparative genomics. Nucleic Acids Research 32: W273–W279.
20. Tamura K, Peterson D, Peterson N, Stecher G, Nei M, et al. (2011) MEGA5: molecular evolutionary genetics analysis using maximum likelihood, evolutionary distance, and maximum parsimony methods. Molecular Biology and Evolution 28: 2731–2739.
21. Langmead B, Trapnell C, Pop M, Salzberg S (2009) Ultrafast and memory-efficient alignment of short DNA sequences to the human genome. Genome Biology 10: R25.
22. Trapnell C, Roberts A, Goff L, Pertea G, Kim D, et al. (2012) Differential gene and transcript expression analysis of RNA-seq experiments with TopHat and Cufflinks. Nat Protocols 7: 562–578.
23. Mower JP (2009) The PREP suite: predictive RNA editors for plant mitochondrial genes, chloroplast genes and user-defined alignments. Nucleic Acids Research 37: W253–W259.
24. Li H, Handsaker B, Wysoker A, Fennell T, Ruan J, et al. (2009) The Sequence Alignment/Map format and SAMtools. Bioinformatics 25: 2078–2079.
25. Zhang Y, Nie X, Jia X, Zhao C, Biradar SS, et al. (2012) Analysis of codon usage patterns of the chloroplast genomes in the Poaceae family. Australian Journal of Botany 60: 461–470.
26. Germain A, Hotto AM, Barkan A, Stern DB (2013) RNA processing and decay in plastids. Wiley Interdisciplinary Reviews: RNA 4: 295–316.
27. Lung B, Zemann A, Madej MJ, Schuelke M, Techritz S, et al. (2006) Identification of small non-coding RNAs from mitochondria and chloroplasts. Nucleic Acids Research 34: 3842–3852.
28. Ruwe H, Schmitz-Linneweber C (2012) Short non-coding RNA fragments accumulating in chloroplasts: footprints of RNA binding proteins? Nucleic Acids Research 40: 3106–3116.
29. Zhelyazkova P, Hammani K, Rojas M, Voelker R, Vargas-Suárez M, et al. (2012) Protein-mediated protection as the predominant mechanism for defining processed mRNA termini in land plant chloroplasts. Nucleic Acids Research 40: 3092–3105.
30. Zhang T, Zhang X, Hu S, Yu J (2011) An efficient procedure for plant organellar genome assembly, based on whole genome data from the 454 GS-FLX sequencing platform. Plant Methods 7: 38.
31. Wang W, Messing J (2011) High-throughput sequencing of three *Lemnoideae* (Duckweeds) chloroplast genomes from total DNA. PLoS ONE 6: e24670.
32. Davis JI, Soreng RJ (2000) Phylogenetic structure in Poaceae subfamily Pooideae as inferred from molecular and morphological characters: misclassification versus reticulation. In: Jacobs SWL, Everett J, editors. Grasses: systematics and evolution. Collingwood, Victoria, Austrailia: CSIRO Publishing. pp. 61–74.
33. Catalán P, Kellogg EA, Olmstead RG (1997) Phylogeny of Poaceae subfamily pooideae based on chloroplast ndhF gene sequences. Molecular Phylogenetics and Evolution 8: 150–166.
34. Nadot S, Bajon R, Lejeune B (1994) The chloroplast gene rps4 as a tool for the study of Poaceae phylogeny. Plant Systematics and Evolution 191: 27–38.
35. Hsiao C, Chatterton NJ, Asay KH, Jensen KB (1995) Molecular phylogeny of the Pooideae (Poaceae) based on nuclear rDNA (ITS) sequences. Theoretical and Applied Genetics 90: 389–398.
36. Stern DB, Goldschmidt-Clermont M, Hanson MR (2010) Chloroplast RNA metabolism. Annual Review of Plant Biology 61: 125–155.
37. Tillich M, Lehwark P, Morton BR, Maier UG (2006) The evolution of chloroplast RNA editing. Molecular Biology and Evolution 23: 1912–1921.
38. Grosskopf D, Mulligan RM (1996) Developmental- and tissue-specificity of RNA editing in mitochondria of suspension-cultured maize cells and seedlings. Current Genetics 29: 556–563.
39. Howad W, Kempken F (1997) Cell type-specific loss of *atp6* RNA editing in cytoplasmic male sterile Sorghum bicolor. Proceedings of the National Academy of Sciences 94: 11090–11095.
40. Picardi E, Horner DS, Chiara M, Schiavon R, Valle G, et al. (2010) Large-scale detection and analysis of RNA editing in grape mtDNA by RNA deep-sequencing. Nucleic Acids Research.
41. Waterhouse AM, Procter JB, Martin DMA, Clamp M, Barton GJ (2009) Jalview Version 2-a multiple sequence alignment editor and analysis workbench. Bioinformatics 25: 1189–1191.

Permissions

All chapters in this book were first published in PLOS ONE, by The Public Library of Science; hereby published with permission under the Creative Commons Attribution License or equivalent. Every chapter published in this book has been scrutinized by our experts. Their significance has been extensively debated. The topics covered herein carry significant findings which will fuel the growth of the discipline. They may even be implemented as practical applications or may be referred to as a beginning point for another development.

The contributors of this book come from diverse backgrounds, making this book a truly international effort. This book will bring forth new frontiers with its revolutionizing research information and detailed analysis of the nascent developments around the world.

We would like to thank all the contributing authors for lending their expertise to make the book truly unique. They have played a crucial role in the development of this book. Without their invaluable contributions this book wouldn't have been possible. They have made vital efforts to compile up to date information on the varied aspects of this subject to make this book a valuable addition to the collection of many professionals and students.

This book was conceptualized with the vision of imparting up-to-date information and advanced data in this field. To ensure the same, a matchless editorial board was set up. Every individual on the board went through rigorous rounds of assessment to prove their worth. After which they invested a large part of their time researching and compiling the most relevant data for our readers.

The editorial board has been involved in producing this book since its inception. They have spent rigorous hours researching and exploring the diverse topics which have resulted in the successful publishing of this book. They have passed on their knowledge of decades through this book. To expedite this challenging task, the publisher supported the team at every step. A small team of assistant editors was also appointed to further simplify the editing procedure and attain best results for the readers.

Apart from the editorial board, the designing team has also invested a significant amount of their time in understanding the subject and creating the most relevant covers. They scrutinized every image to scout for the most suitable representation of the subject and create an appropriate cover for the book.

The publishing team has been an ardent support to the editorial, designing and production team. Their endless efforts to recruit the best for this project, has resulted in the accomplishment of this book. They are a veteran in the field of academics and their pool of knowledge is as vast as their experience in printing. Their expertise and guidance has proved useful at every step. Their uncompromising quality standards have made this book an exceptional effort. Their encouragement from time to time has been an inspiration for everyone.

The publisher and the editorial board hope that this book will prove to be a valuable piece of knowledge for researchers, students, practitioners and scholars across the globe.

List of Contributors

Silvia Dal Santo, Giovanni Battista Tornielli, Marianna Fasoli, Luca Venturini, Mario Pezzotti and Sara Zenoni
Department of Biotechnology, University of Verona, Verona, Italy

Alessandro Vannozzi
Department of Agronomy, Food, Natural Resources, Animals and Environment, University of Padova, Legnaro, Padova, Italy

Barry G. Hall
Bellingham Research Institute, Bellingham, Washington, United States of America

Jeffrey C. Glaubitz, Terry M. Casstevens, Fei Lu, James Harriman and Robert J. Elshire
Institute for Genomic Diversity, Cornell University, Ithaca, New York, United States of America

Qi Sun
Biotechnology Resource Center Bioinformatics Facility, Cornell University, Ithaca, New York, United States of America

Edward S. Buckler
Institute for Genomic Diversity, Cornell University, Ithaca, New York, United States of America

USDA Agricultural Research Service, Ithaca, New York, United States of America

Yanxin Zhao, Manjun Cai, Xiaobo Zhang, Yurong Li, Jianhua Zhang, Hailiang Zhao, Fei Kong,
Yonglian Zheng and Fazhan Qiu

National Key Laboratory of Crop Genetic Improvement, Huazhong Agricultural University, Wuhan, China

Clio Der Sarkissian, Paul Brotherton, Jennifer E. L. Templeton, Bastien Llamas, Julien Soubrier1, Alan Cooper and Wolfgang Haak
Australian Centre for Ancient DNA, School of Earth and Environmental Sciences, University of Adelaide, Adelaide, South Australia, Australia

Oleg Balanovsky
Vavilov Institute for General Genetics, Russian Academy of Sciences, Moscow, Russia

Research Centre for Medical Genetics, Russian Academy of Medical Sciences, Moscow, Russia

Vyacheslav Moiseyev and Valery Khartanovich
Peter the Great Museum of Anthropology and Ethnography (Kunstkamera) RAS, St Petersburg, Russia

Thomas P. Howard III, Andrew P. Hayward, Anthony Tordillos, Christopher Fragoso,
Maria A. Moreno and Stephen L. Dellaporta

Department of Molecular, Cellular and Developmental Biology, Yale University, New Haven, Connecticut, United States of America

Joe Tohme
International Center for Tropical Agriculture, Cali, Colombia

Albert P. Kausch and John P. Mottinger
Department of Cell and Molecular Biology, University of Rhode Island, Kingston, Rhode Island, United States of America

Awdhesh Kumar Mishra, Mehanathan Muthamilarasan, Yusuf Khan, Swarup Kumar Parida and Manoj Prasad
National Institute of Plant Genome Research, Aruna Asaf Ali Marg, New Delhi

Yong-Yi Shen
State Key Laboratory of Genetic Resources and Evolution, Kunming Institute of Zoology, the Chinese Academy of Sciences, Kunming, China

Joint Influenza Research Centre (SUMC/HKU), Shantou University Medical College, Shantou, China

Kun Dai
Xinjiang Institute of Ecology and Geography, the Chinese Academy of Sciences, Urumqi, China

Xue Cao
State Key Laboratory of Genetic Resources and Evolution, Kunming Institute of Zoology, the Chinese Academy of Sciences, Kunming, China
University of the Chinese Academy of Sciences, Beijing, China

Robert W. Murphy
State Key Laboratory of Genetic Resources and Evolution, Kunming Institute of Zoology, the Chinese Academy of Sciences, Kunming, China
Department of Natural History, Royal Ontario Museum, Toronto, Canada

Xue-Juan Shen
Joint Influenza Research Centre (SUMC/HKU), Shantou University Medical College, Shantou, China

Ya-Ping Zhang
State Key Laboratory of Genetic Resources and Evolution, Kunming Institute of Zoology, the Chinese Academy of Sciences, Kunming, China
Laboratory for Conservation and Utilization of Bio-resources, Yunnan University, Kunming, China

Daejin Hyung
Department of Computer Science, Dong-A University, Busan, Republic of Korea

Chaeyoung Lee and Jin-Hyun Kim
Department of Medical Bioscience, Dong-A University, Busan, Republic of Korea

Dongwoon Yoo, Jai-Heon Lee, Youngsoo Chung and Hong-kyu Choi
Department of Genetic Engineering, Dong-A University, Busan, Republic of Korea

Young-Su Seo
Department of Microbiology, Busan National University, Busan, Republic of Korea

Soon-Chun Jeong
Bio-Evaluation Center, Korea Research Institute of Bioscience and Biotechnology, Cheongwon, Republic of Korea

Ki-Hong Jung
Department of Plant Molecular Systems Biotechnology & Graduate School of Biotechnology, Kyunghee University, Yongin, Republic of Korea

Douglas R. Cook
Department of Plant Pathology, University of California Davis, Davis, California, United States of America

Leila do Nascimento Vieira, Hugo Pacheco de Freitas Fraga, Rubens Onofre Nodari and Miguel Pedro Guerra
Laboratório de Fisiologia do Desenvolvimento e Genética Vegetal, Programa de Pós-graduação em Recursos Genéticos Vegetais, Universidade Federal de Santa Catarina,

Florianó polis, Santa Catarina, Brazil

Helisson Faoro, Rodrigo Luis Alves Cardoso, Emanuel Maltempi de Souza and Fábio de Oliveira Pedrosa
Departamento de Bioquímica e Biologia Molecular, Núcleo de Fixação Biológica de Nitrogênio, Universidade Federal do Paraná, Curitiba, Paraná, Brazil

Marcelo Rogalski
Departamento de Biologia Vegetal, Universidade Federal de Viçosa, Viçosa, Minas Gerais, Brazil

Dany Morisset, Dejan Štebih, Mojca Milavec, Kristina Gruden and Jana Žel
Department of Biotechnology and Systems Biology, National Institute of Biology, Ljubljana, Slovenia

Maia Gurushidze, Goetz Hensel, Stefan Hiekel, Sindy Schedel and Jochen Kumlehn
Leibniz Institute of Plant Genetics and Crop Plant Research (IPK), Plant Reproductive Biology, Gatersleben, Germany

Vladimir Valkov
Leibniz Institute of Plant Genetics and Crop Plant Research (IPK), Plant Reproductive Biology, Gatersleben, Germany
Institute of Genetics and Biophysics, Naples, Italy

Lang Yan, Ying-Hong Gu, Xian-Jun Lai, Yi-Zheng Zhang, Xue-Mei Tan and Haiyan Wang
Key Laboratory of Bio-resources and Eco-environment, Ministry of Education, Sichuan Key Laboratory of Molecular Biology and Biotechnology, Center for Functional Genomics and Bioinformatics, College of Life Sciences, Sichuan University, Chengdu, Sichuan, People's Republic of China

Xiang Tao
Chengdu Institute of Biology, Chinese Academy of Sciences, Chengdu, Sichuan, People's Republic of China

Dongyan Zhao and Ning Jiang
Department of Horticulture, Michigan State University, East Lansing, Michigan, United States of America

Jia Wang
Center for Bioinformatics, College of Life Science and Technology, Huazhong Agricultural University, Wuhan, P.R. China

School of Science, Huazhong Agricultural University, Wuhan, P.R. China

Dijun Chen, Yang Lei, Ji-Wei Chang, Bao-Hai Hao, Feng Xing, Sen Li and Ling-Ling Chen
Center for Bioinformatics, College of Life Science and Technology, Huazhong Agricultural University, Wuhan, P.R. China

Qiang Xu and Xiu-Xin Deng
Key Laboratory of Horticultural Plant Biology of Ministry of Education, Huazhong Agricultural University, Wuhan, P.R. China

Alevtina Ruban
Russian State Agrarian University - Moscow Timiryazev Agricultural Academy, Department of Genetics and Biotechnology, Moscow, Russia
Leibniz-Institute of Plant Genetics and Crop Plant Research (IPK), Chromosome Structure and Function Laboratory, Gatersleben, Germany

Jörg Fuchs, Veit Schubert and Andreas Houben
Leibniz-Institute of Plant Genetics and Crop Plant Research (IPK), Chromosome Structure and Function Laboratory, Gatersleben, Germany

André Marques
Laboratory of Plant Cytogenetics and Evolution, Department of Botany, Universidade Federal de Pernambuco, Recife, Brazil

Alexander Solovie
Russian State Agrarian University - Moscow Timiryazev Agricultural Academy, Department of Genetics and Biotechnology, Moscow, Russia

Olga Raskina
Institute of Evolution, University of Haifa, Laboratory of Plant Molecular Cytogenetics, Haifa, Israel

Ekaterina Badaeva
Engelhardt Institute of Molecular Biology Russian Academy of Sciences, Laboratory of Molecular Karyology, Moscow, Russia

Xiu-Qing Li
Molecular Genetics Laboratory, Potato Research Centre, Agriculture and Agri-Food Canada, Fredericton, New Brunswick, Canada

Donglei Du
Quantitative Methods Research Group, Faculty of Business Administration, University of New Brunswick, Fredericton, New Brunswick, Canada

Eva Herzog and Matthias Frisch
Institute of Agronomy and Plant Breeding II, Justus Liebig University, Giessen, Germany

Karen Christin Falke
Institute for Evolution and Biodiversity, University of Münster, Münster, Germany

Thomas Presterl, Daniela Scheuermann and Milena Ouzunova
KWS Saat AG, Einbeck, Germany

Jonathan Seguin
University of Basel, Department of Environmental Sciences, Institute of Botany, Basel, Switzerland

Fasteris SA, Plan-les-Ouates, Geneva, Switzerland

Rajendran Rajeswaran, Nachelli Malpica-López and Mikhail M. Pooggin
University of Basel, Department of Environmental Sciences, Institute of Botany, Basel, Switzerland

Robert R. Martin
United States Department of Agriculture–Agricultural Research Service, Horticultural Crops Research Laboratory, Corvallis, Oregon, United States of America

Oregon State University, Department of Botany and Plant Pathology, Center for Genome Research and Biocomputing, Corvallis, Oregon, United States of America

Kristin Kasschau and Valerian V. Dolja
Oregon State University, Department of Botany and Plant Pathology, Center for Genome Research and Biocomputing, Corvallis, Oregon, United States of America

Patricia Otten and Laurent Farinelli
Fasteris SA, Plan-les-Ouates, Geneva, Switzerland

Jungeun Lee, Yoonjee Kang, Seung Chul Shin, Hyun Park and Hyoungseok Lee
Division of Life Sciences, Korea Polar Research Institute, Incheon, South Korea

Index